TECHNICAL COMMUNICATION

Process and Product

Sixth Edition

SHARON J. GERSON

DeVry University (Emeritus)

STEVEN M. GERSON

Johnson County Community College

PEARSON

Prentice
Hall

Upper Saddle River, New Jersey
Columbus, Ohio

Library of Congress Cataloging-in-Publication Data

Gerson, Sharon J.
   Technical communication: process and product / Sharon J. Gerson, Steven M. Gerson.
      —6th ed. p. cm.
   Includes bibliographical references and index.
   ISBN 0-13-615475-1
      1.   English language —Technical English. 2.   English language—Rhetoric.
      3.   Technical writing. I. Gerson, Steven M. II. Title.

PE1475.G47 2009
808.0666—dc22

2007040926

Editor in Chief: Vernon R. Anthony
Editor: Gary Bauer
Editorial Assistant: Kathleen Rowland
Production Editor: Kevin Happell
Production Coordination: S4Carlisle Publishing Services
Design Coordinator: Diane Ernsberger
Cover Designer: Candace Rowley
Cover art: Getty
Production Manager: Patricia Tonneman
Director of Marketing: David Gesell
Senior Marketing Manager: Leigh Ann Sims
Marketing Assistant: Alicia Dysert

This book was set in Sabon by S4Carlisle Publishing Services and was printed and
bound by C. J. Krehbiel Company. The cover was printed by Phoenix Color Corp.

Microsoft product screen shot(s) reprinted with permission from Microsoft
Corporation.

Credits and acknowledgments borrowed from other sources and reproduced, with
permission, in this textbook appear on appropriate pages within text.

Pearson Education Ltd.   Pearson Education Australia PTY, Limited
Pearson Education Singapore, Pte. Ltd.   Pearson Education North Asia Ltd.
Pearson Education Canada, Ltd.   Pearson Educación de Mexico, S.A. de C.V.
Pearson Education—Japan   Pearson Education Malaysia, Pte. Ltd.

10 9 8 7 6 5 4

ISBN-13: 978-0-13-615475-4
ISBN-10:   0-13-615475-1

For our daughters, Stacy and Stefani

ABOUT THE AUTHORS

SHARON AND STEVE GERSON are dedicated career professionals who have a combined total of over seventy years teaching experience at the college and university level. They have taught technical writing, business writing, professional writing, and technical communication to thousands of students; attended and presented at dozens of conferences; written numerous articles; and published several textbooks, including *Technical Communication: Process and Product* (sixth edition), *The Red Bridge Reader* (third edition, co-authored by Kin Norman), *Writing That Works: A Teacher's Guide to Technical Writing* (second edition), and *Workplace Communication: Process and Product* (first edition). They are currently preparing a textbook for Prentice Hall entitled *Workplace Communication: Real People, Real Challenges,* to be published in 2009.

In addition to their academic work, Sharon and Steve are involved in business and industry through their business, Steve Gerson Consulting. In this capacity they have worked for companies such as Sprint, AlliedSignal–Honeywell, General Electric, JCPenney, Avon, the Missouri Department of Transportation, H&R Block, Mid-America Regional Council, and Commerce Bank. Their work for these businesses includes writing, editing, and proofreading many different types of technical documents, such as proposals, marketing collateral, reports, and instructions.

Steve also has presented hundreds of hands-on workshops on technical/business writing, business grammar in the workplace, oral presentations in the workplace, and business etiquette. Over 10,000 business and government employees have benefited from these workshops. For the past decade, Steve has worked closely with K–12 teachers. He has presented many well-attended, interactive workshops to give teachers useful tips about technical writing in the classroom.

Both Steve and Sharon have been awarded for teaching excellence and are listed in *Who's Who Among America's Teachers*. Steve is a Society for Technical Communication Fellow. In 2003, Steve was named Kansas Professor of the Year by the Carnegie Foundation for the Advancement of Education.

Their wealth of experience and knowledge is gathered for you in this sixth edition of *Technical Communication: Process and Product*.

PREFACE

Welcome to the sixth edition of *Technical Communication: Process and Product*. In this edition, we have worked to create an even more reader-friendly textbook, coupling technological information with easy-to-follow instructions and real-life examples. In response to changes in the field, technological advances, and comments from a variety of users of the fifth edition, we have made many changes throughout the text. The most important changes include the following:

- Coverage of **audience analysis and contextual writing** integrated throughout the text and supported by numerous writing examples
- **Collaborative writing** now introduced in Chapter 1 and supported throughout the book with collaboration exercises and examples
- Coverage of **contemporary communication channels** such as webinars, wikis, instant messaging, and blogging and how to use them appropriately
- **Persuasive communication** with discussion of innovative argumentation techniques (Chapter 10)
- **Long, formal reports** with an emphasis on the importance and use of research in these reports and discussion of information, analysis, and recommendation (Chapter 16)
- New pedagogical elements to aid learning, including the following:
  - **Problem-Solving Think Pieces** to help students practice critical-thinking skills
  - **FAQs** addressing commonly asked questions about chapter topics
  - **Technology Tips** focusing on the use of Microsoft Word 2007
- Many new **Writing at Work** chapter-opening vignettes featuring real people facing workplace communication challenges and case study assignments
- Many new **before and after** writing examples to show students the importance of revision

Another major new enhancement for the sixth edition is the availability of **Mytechcommlab**, can be packaged with new copies of the textbook. Students will find an interactive textbook incorporating a variety of multimedia technical communication resources such as interactive document examples, Web site and visual rhetoric tutorials, grammar exercises, mechanics, writing diagnostic and instructional tools, and online research tools.

We also changed the word "Writing" in the title of the textbook to "Communication." The title was changed because the communication skills required of today's employees have changed. Today's employers are looking for workers who can write and communicate in a variety of media beyond traditional print documents such as letters, memos, and reports. Employees may be expected to give oral presentations, create online help guides, work with wikis, and participate in teleconferences and webinars. Modern business employees communicate by using Microsoft PowerPoint, voice mail, instant messages, videoconferences, and blogs. Writing remains at the core of communication, and improving writing is still a driving theme in this textbook, but the title change acknowledges the broader context of diverse communication channels on the job. This book embraces the multidimensional aspects of a technical communicator. We invite you to explore the new and enhanced material in this sixth edition.

Sharon Gerson   Steve Gerson

DESIGN AND INSTRUCTIONAL AIDS

To ensure readability and enjoyment of this sixth edition, we have created a colorful new layout and design. In addition, we have added several new pedagogical elements to improve learning, navigation, and access to content. Our end-of-chapter activities will help students apply their knowledge, work collaboratively, and develop critical-thinking skills. The following are examples of the instructional aids included in this edition.

Chapter Design

All chapters in the sixth edition contain the following components:

- Chapter Objectives
- Communication at Work Opening Scenarios
- The Writing Process at Work
- Margin callouts
- Before/After examples
- FAQ Boxes
- Technology Tips
- Spotlights
- Checklists
- Chapter Highlights
- Apply Your Knowledge (end-of-chapter exercises and applications)
  - Case Studies
  - Individual and Team Projects
  - Problem-Solving Think Pieces
  - Web Workshops
  - Quiz Questions

COMMUNICATION AT WORK Opening Scenarios present real people in the workplace and their communication challenges.

MARGIN CALLOUTS clarify and highlight key points in sample documents.

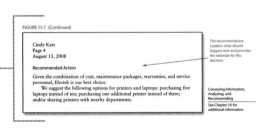

WEB RESOURCES NOTES direct students to the Companion Website for additional samples, activities, and content related to each chapter.

Check Online Resources
www.prenhall.com/gerson
For more information about high-tech, low-tech, and lay audiences, visit our companion Web site.

CROSS-REFERENCE NOTES in the margins help students and faculty find additional content in the textbook on a topic.

Diversity
See Chapter 3 for more discussion of diversity.

DOT-COM UPDATES guide students to articles and Web sites related to content in the textbook.

DOT-COM UPDATES
Go to http://www.shrm.org/ diversity/ (the Society for Human Resource Management) for additional information about diversity in the workplace, including definitions, links, and articles about diversity in the news.

BEFORE/AFTER EXAMPLES visually demonstrate, through comparison and contrast, how revision improves communication.

BEFORE	AFTER
Example—Unfriendly Text	**Example—Reader-Friendly Text**

Example—Unfriendly Text
This year began with an increase, as we sold 4.5 million units in January compared to 3.7 for January 2006. In February we continued to improve with 4.6, compared with 3.6 for the same time in 2007. March was not quite so good, as we sold 4.3 against the March 2007 figure of 3.9. April was about the same with 4.2, compared to 3.8 for April 2007.

Example—Reader-Friendly Text
Comparative Quarterly Sales (in Millions)

	2006	2007	Increase/Decrease
Jan.	3.7	4.5	0.8+
Feb.	3.6	4.6	1.0+
Mar.	3.9	4.3	0.4+
Apr.	3.8	4.2	0.4+

FAQ BOXES address frequently asked questions about technical communication topics.

FAQs: Multiculturalism and Technical Communication

Q: Why is "multiculturalism" important in technical communication? Doesn't everyone in the United States speak English?

A: No, not everyone in the United States speaks English. Look at these facts:
- "50 million people in the U.S., 19 percent of the population, . . . speak a language other than English at home and 22 million . . . have limited English proficiency" (Gardner 2007).
- Between 1990 and 2000, the number of Americans speaking a language other than English at home grew by 15.1 million (a 47 percent increase) and the number with limited English proficiency grew by 7.3 million (a 53 percent increase).

TECHNOLOGY TIPS provide helpful instruction on using Microsoft Word 2007 to communicate. Microsoft product screen shot(s) are reprinted with permission from Microsoft Corp.

TECHNOLOGY TIP
Using Microsoft Word 2007 to Check the Readability Level of Your Text

SPOTLIGHTS highlight real people in the workplace, their on-the-job communication challenges, and their use of multiple communication channels to succeed.

SPOTLIGHT
Conducting research to find solutions
Tom Woltkamp is Senior Manager of Information Solutions for Teva Neuroscience, a global leader in the "development, production and marketing of generic and proprietary branded pharmaceuticals" (**www.tevapharm.com/**). Teva Pharmaceutical Industries is home based in Israel; Teva Neuroscience, a subsidiary of Teva Pharmaceuticals, is based in Kansas City, Missouri, with branches in Florida, Pennsylvania, Texas, North Carolina, California, Minnesota, and Canada.
Approximately 20 employees and consultants report to Tom regarding computer infrastructure (computer networks, help desks, call centers, hardware/software purchases) and application development for programming, software integration, installation, configuration, Web programming, and customized documentation.

CHECKLISTS provide students with an evaluative guide to assist them in revising the many different documents taught in the textbook.

CHECKLIST FOR COLLABORATION

____ 1. Have you chosen a team leader (or has a team leader been assigned)?
____ 2. Do all participants understand the team's goal and their individual responsibilities?
____ 3. Does the team have a schedule, complete with milestones and target due dates?
____ 4. Does the team have compatible hardware and software?
____ 5. in planning the team's project, did you seek consensus?
____ 6. Have all participants been allowed to express themselves?
____ 7. If conflicts occurred, did you table topics for later discussion or additional research?
____ 8. Did you encourage diversity of opinion?
____ 9. Have you avoided confronting people in public, choosing to meet with individuals privately to discuss concerns?
____ 10. If challenges continue, have you reassigned team members?

DEVELOPING COMMUNICATION SKILLS WITH END-OF CHAPTER ACTIVITIES

Each chapter concludes with a variety of **Chapter Highlights, Case Studies, Problem-Solving Think Pieces, Web Workshops,** and **Quiz Questions** that provide students with ample opportunities to apply chapter concepts to real-world situations, work collaboratively, and develop critical-thinking skills. These varied activities appeal to different learning styles, including memorization, application of principles with case studies, research, collaboration, and discovery.

CHAPTER HIGHLIGHTS summarize key points discussed in the chapter.

CHAPTER HIGHLIGHTS
1. Technical communication is written for and about business and industry and focuses on products and services.
2. Technical communication is an important part of your everyday work life. It can consume as much as 31 percent of a typical workweek.
3. Technical communication costs a company both time and money, so employees must write effectively.
4. The top skills employers want include communication skills, honesty, interpersonal skills, a strong work ethic, and teamwork.

CASE STUDIES present real-world scenarios and on-the-job communication challenges.

CASE STUDY
You are the team leader of a work project at Gulfview Architectural and Engineering Services. The team has been involved in this project for a year. During the year, the team has met weekly, every Wednesday at 8:00 A.M. It is now time to assess the team's successes and areas needing improvement.
  Your goal will be to recommend changes as needed before the team begins its second year on this project. You have encountered the following problems.
• One team member, Caroline Jensen, misses meetings regularly. In fact, she has missed at least one meeting a month during the past year.

INDIVIDUAL AND TEAM PROJECTS provide students with an opportunity to apply principles discussed in the chapter, drawing from personal experience, research, or collaboration.

INDIVIDUAL AND TEAM PROJECTS
1. To practice prewriting, take one of the following topics. Then, using the suggested prewriting technique, gather data.
   a. **Reporter's questions.** To gather data for your resume, list answers to the reporter's questions for two recent jobs you have held and for your past and present educational experiences.
   b. **Mind mapping.** Create a mind map for your options for obtaining college financial aid.
   c. **Brainstorming or listing.** List five reasons why you have selected your degree

PROBLEM-SOLVING THINK PIECES help students practice critical-thinking skills.

PROBLEM-SOLVING THINK PIECES
To understand and practice conflict resolution, complete the following assignments.
1. **Attend a meeting.** This could be at your church, synagogue, or mosque; a city council meeting; your school, college, or university's board of trustees meeting; or a meeting at your place of employment. Was the meeting successful? Did it have room for improvement? To help answer these questions, use the following Conflict Resolution in Team Meetings Matrix. Then report your findings to your professor or classmates, either orally or in writing. Write an e-mail message, memo, or report, for example.

WEB WORKSHOPS give students an opportunity to discover topics related to technical communication beyond the classroom.

WEB WORKSHOP
1. How important is technical communication in the workplace? Go online to research this topic. Find five Web sites that discuss the importance of communication in the workplace, and report your discoveries to your teacher and/or class. To do so, write a brief report, memo, or e-mail message. You could also report your information orally.
2. Create a class wiki for collaborative writing. To do so, consider using either of the following sites.
   • http://www.wikispaces.com

QUIZ QUESTIONS allow students to test their knowledge of the chapter's content.

QUIZ QUESTIONS
1. Define *technical communication*.
2. What are five channels of technical communication?
3. List three reasons why technical communication is important in business.
4. What is the percentage of time employees spend writing?
5. What are the top five skills employers want?
6. Define *silo building*.
7. Explain why business depends on teamwork to ensure quality.
8. List five causes for collaborative breakdown, according to Human

CHANGES TO THE ORGANIZATION AND CHAPTERS IN THE SIXTH EDITION

Chapter 1 ("An Introduction to Technical Communication") has been significantly revised. In this chapter, you will find a *Communication at Work* scenario providing a context for technical communication (all chapters in this sixth edition open with a scenario). We then define technical communication and discuss its importance.

New to this chapter, we emphasize collaboration by focusing on dispersed teams, diverse teams, and using groupware to collaborate in virtual teams.

New to this chapter, we introduce the reader to wikis, a technology used for collaborative writing. Emphasis in this chapter is on the importance of teamwork and how to collaborate successfully.

Chapter 2 ("The Communication Process") explains why process—prewriting, writing, and rewriting—will improve your correspondence. We discuss the rationale for the writing process and supply an example of this process in practice. We emphasize the recursive nature of the writing process. The writing process is dynamic, with the three parts (prewriting, writing, and rewriting) often occurring simultaneously. In addition, this chapter explains the benefits and challenges of communication channels, including e-mail, handheld computers, cell phones, Microsoft PowerPoint, and more.

New to this chapter is research from The National Commission on Writing on important skills in technical communication.

Chapter 3 ("The Goals of Technical Communication") provides a detailed discussion of five major objectives in successful technical writing: clarity, conciseness, accuracy, organization, and ethics. Discussion of these goals coupled with numerous illustrations will assist students in understanding the style of writing unique to the field of technical communication.

New to this chapter are many before/after examples to illustrate technical communication objectives and the importance of revision.

Chapter 4 ("Audience Recognition and Involvement") teaches how to communicate to different levels of audience and focuses on the importance of definition, multiculturalism, avoiding biased language, and personalization. We discuss the need to understand diversity—age, gender, culture, and religion—and communicate appropriately.

New to this chapter is a memo written to a high-tech audience, contrasted with a memo written to a low-tech audience. This sample helps students understand how communicating to different audiences requires changes in tone, style, content, and layout.

Chapter 5 ("Research and Documentation") teaches how to find and document information. We discuss how to find research online and use both MLA and APA style sheets.

New to this chapter, we focus on why research is important in technical communication, how to use research in formal reports, and primary and secondary sources.

Chapter 6 ("Routine Correspondence:—Memos, Letters, E-mail, and Instant Messaging") brings together everyday correspondence written on the job, including memos, letters, e-mail, and instant messages. We provide criteria for this everyday correspondence, numerous samples, and checklists.

New to this chapter, we combine our discussion of memos, letters, e-mail, and instant messaging as illustrations of commonly written correspondence in the workplace.

New to this chapter, we compare and contrast memos, letters, and e-mail.

New to this chapter, we provide a sample of an e-mail written for a high-tech audience. We then contrast this sample with a rewritten e-mail designed for multiple levels of audience. Our text and callouts explain in detail the distinctions between the two samples, highlighting the ways in which audience affects communication.

New to this book, we discuss the challenges, benefits, and techniques for writing effective instant messages.

New to this chapter, we discuss the importance of choosing the correct communication channel for routine correspondence.

Chapter 7 ("Employment Communication") helps you find employment, write a resume, send a resume, write a letter of application or e-mail message, and interview effectively. We show you several different types of resumes and explain when each resume type should be used. Also provided in this chapter are samples of resumes, letters of application, e-mail cover messages, and follow-up letters and e-mail. Our checklists will help you to revise these documents, prepare for an interview, and follow up after an interview.

New to this chapter are additional ways to find employment such as job shadowing, attending a job fair, and informational interviews.

New to this chapter is a focus on the importance of adding a summary of qualifications to your resume.

Chapter 8 ("Document Design") explains various ways to create an effective page layout by considering use of highlighting techniques and visual aids. Students will see one document revised several times to illustrate the use of different design elements.

New to this chapter is a discussion of "talking headings" to assist students with the design of longer documents such as reports and proposals.

Chapter 9 ("Graphics") focuses on tables and figures with tips for appropriate usage of visual aids. We provide criteria for creating both tables and numerous different types of figures. Also provided are examples of the many different types of visual aids.

New to this chapter, we show a pictograph designed for a lay audience and contrast the pictograph redesigned as a horizontal bar chart written for a high-tech audience. Our text and callouts help to illustrate the distinctions between the two visual aids.

Chapter 10 ("Communicating to Persuade") is a totally new chapter that focuses on the importance of argument and persuasion in technical communication. We discuss methods for creating effective persuasive documents.

New to this chapter, we have created a memorable approach to organizing persuasive documents, entitled *ARGU* (*a*rouse interest, *r*efute opposing points of view, *g*ive support, *u*rge action).

New to this chapter, we discuss how to avoid logical fallacies.

New to this chapter, we discuss persuasive documents such as sales letters, fliers, and brochures.

Chapter 11 ("Technical Descriptions and Process Analyses") discusses how to write specifications for mechanisms, tools, and pieces of equipment.

New to this chapter, we discuss process analyses which, explain how something works. Included in the chapter is a checklist for process analyses and technical descriptions to assist students with revision.

New to this chapter, we show how technical descriptions can be written to either a high-tech or lay audience.

Chapter 12 ("Instructions") explains how to write instructions that show customers how to use or operate equipment.

New to this chapter, we contrast an instruction written for a high-tech audience with the same instruction written for a low-tech audience.

New to this chapter are professionally written instructions from Crate & Barrel, Microsoft Project 2003, and Hewlett Packard.

Chapter 13 ("Web Sites, Blogging, and Online Help") defines characteristics of the online e-reader and ways in which online communication differs from the printed page. We discuss effective Web site and online help criteria.

New to this chapter is information about blogging, including why to blog, how to blog, and a blogging code of ethics.

Chapter 14 ("Summaries") provides an illustration of a summary and explains why summaries are important in technical communication.

New to this chapter is a completely updated sample summary on "Organizational Implications of Future Development of Technical Communication." We take this summary from rough draft through revision to finished summary. In the revision stage, students can see the use of Microsoft Word's "track changes" feature.

Chapter 15 ("Short, Informal Reports") discusses the similarities and differences among various short reports, including trip reports, progress reports, lab reports, feasibility/recommendation reports, incident reports, investigative reports, and meeting minutes.

New to this chapter, we have included a sample of a short report written for a high-tech audience. We then provide a sample of the same report rewritten and redesigned for a lay audience of multiple readers. Our text and callouts explain fully for students the distinctions between the two documents and the ways in which audience affects content, tone, and layout.

Chapter 16 ("Long, Formal Reports") is an entirely new chapter that discusses the purpose of long, formal research reports in business. We emphasize how using both primary and secondary research can enhance the content and development of a topic.

New to this chapter, we focus on the role of information, analysis, and recommendation in long, formal reports.

New to this chapter, we use a long, formal research report as an illustration, complete with highly effective instructional callouts.

Chapter 17 ("Proposals") focuses on the criteria for writing effective internal and external proposals.

New to this chapter is a proposal created in Microsoft PowerPoint that provides students with an alternative communication channel for written documents.

Chapter 18 ("Oral Communication") contains instruction on the following: conducting everyday oral communication, tips for using the telephone and voice mail, making informal oral presentations, tips for teleconferences and videoconferences, making formal oral presentations, creating Microsoft PowerPoint presentations, using the writing process for oral communication, and using effective visual aids in oral presentations.

New to this chapter is information about webconferencing (webinars).

The Appendix ("Grammar, Punctuation, Mechanics, and Spelling") provides rules and conventions for correct grammar, punctuation, mechanics, and spelling.

YOUR ONE-STOP SOURCE FOR TECHNICAL COMMUNICATION RESOURCES

Mytechcommlab for Technical Communication: Process and Product, Sixth Edition

 This dynamic, comprehensive resource can be packaged at no additional cost with the purchase of a new text. Mytechcommlab comes in two versions: a generic version that requires no instructor involvement and an e-book version that includes instructor gradebook and classroom management tools. Both versions provide a wide array of multimedia tools, all in one place and all designed specifically for technical communicators.

- **Over 80 model documents** provide interactive activities and annotations selected from a variety of professions and purposes (letters, memos, career correspondence, proposals, reports, instructions and procedures, descriptions and definitions, Web sites, and presentations). Mytechcommlab provides a wealth of opportunities for students to understand how real professionals work.

- **Grammar, mechanics, and writing help** assists students who need more practice in basic grammar and usage. Mytechcommlab's grammar diagnostics will generate a study plan linked to the thousands of test items in ExerciseZone, with results tracked by Pearson's exclusive GradeTracker.

- **Research Help: Research Navigator™** helps students research quickly and efficiently. Our program supplies extensive help on the research process and includes four exclusive databases of credible and reliable source material— EBSCO's ContentSelect Academic Journal Database, *The New York Times* Search-by-Subject Archive, the FT.com archives, and a "Best of the Web" Link Library.

- **Pearson Tutor Service: Powered by SMARTTHINKING*** asks you to submit your English papers, essays, or other written work to Pearson Tutor Services for personalized and detailed feedback on how to improve your paper. Highly qualified writing tutors review your writing and return it to you with their feedback for improvement. Please visit http://www.mytechcommlab.com/ for more detailed information. [*Some restrictions apply, and this offer is subject to change without notice.]

- **E-book with Online Reference Sources** allows students access to the textbook within Mytechcommlab. Embedded in the pages are links to online resources, including URLs referenced in the text and the model documents. Also available are a large variety of additional exercises, cases, and supporting material organized by chapter.

To order *Technical Communication: Process and Product*, Sixth Edition, with the generic MyTechCommLab access code, order ISBN: 0-13-501344-5.

To order *Technical Communication: Process and Product*, Sixth Edition, with Mytech commlab with e-book in CourseCompass access code, order ISBN: 0-13-501343-7.

To order the Gersons' Mytechcommlab with e-book in CourseCompass standalone access go to www.prenhall.com and order online.

To order Mytechcommlab with e-book in CourseCompass standalone Access Card through your campus bookstore, ask the bookstore to order ISBN: 0-13-501307-0.

A stand-alone access code can be purchased online at www.prenhall.com

To preview Mytechcommlab, go to http://www.mytechcommlab.com/

COMPANION WEBSITE:
A WEALTH OF NEW ONLINE MATERIALS

We are especially excited about the wealth of new cases, exercises, activities, and documents that have been developed for each chapter and are available at the Companion Website located at www.prenhall.com/gerson. On this site, you also will find dozens of excellent student-written examples of the documents taught in the textbook, complete with suggested assignments.

Online materials for each chapter in the text include the following:

- **Chapter Learning Objectives**— Overview of major chapter concepts
- **Writing Process Exercises**—Prewriting/writing/rewriting assignments
- **Interactive Editing and Revision Exercises**—Interactive documents that allow students to see poorly done and corrected versions of documents with additional assignable document revision exercises
- **Communication Cases**—Real-world situations with links to outside content, and a student response box for students to send answers to the professor
- **Activities and Exercises**—Activities specific to a variety of technical and career fields, allowing students to practice producing communication relevant to their interests
- **Collaboration Exercises**—Assignments designed to provide practice writing and communicating in teams
- **Web Resources**—Links to helpful online resources related to chapter content
- **Document Library**—Additional documents
- **Chapter Quizzes**—Self-grading, multiple-choice quizzes to help students master chapter concepts and prepare for tests

DISTANCE LEARNING SOLUTIONS

Complete online courses, including Mytechcommlab with e-book, are also available in Blackboard, WebCT, eCollege, and CourseCompass course management systems. Contact your local Pearson representative for more information.

INSTRUCTOR'S RESOURCES

Instructor's Manual (ISBN: 0-13-119665-0)

New to this edition is an expanded Instructor's Manual loaded with helpful teaching notes for your classroom. Included in the manual are answers to the chapter quiz questions, a test bank, and instructor notes for assignments and activities located on the Companion Website.

The online Instructor's Manual, Microsoft PowerPoint slides, and additional instructor resources are also available to instructors through this title's catalog page at **www.prenhall.com/gerson.** Instructors can search for a text by author, title, ISBN, or by selecting the appropriate discipline from the pull down menu at the top of the catalog home page. To access supplementary materials online, instructors need to request an instructor access code. Go to **www.prenhall.com,** click the **Instructor Resource Center** link, and then click Register Today for an instructor access code. Within 48 hours of registering you will receive a confirming e-mail including an instructor access code. Once you have received your code, go to the site and log on for full instructions on downloading the materials that you wish to use.

Instructor's Resource CD (ISBN: 0-13-119669-3)

The Instructor's Resource CD includes the following components:

- Test Generator
- PowerPoint Lecture Presentation Package
- Instructor's Manual (in Microsoft Word)

Online Newsletter

In addition to this textbook, the instructor's manual, and the online components of *Technical Communication: Process and Product*, you can have up-to-date information about the field of technical communication in our quarterly newsletters. Contact Prentice Hall at http://www.prenhall.com/gerson to obtain a free subscription to this enlightening newsletter.

ACKNOWLEDGMENTS

We would like to thank the following reviewers and focus group participants for their helpful comments on the fourth and fifth editions of the textbook:

Floyd Brigdon, Trinity Valley Community College

Elizabeth Christensen, Sinclair Community College

Amy Tipton Cortner, Caldwell Community College and Technical Institute

Natalie Daley, Linn-Benton Community College

Myra G. Day, North Carolina State University

Rosemary Day, Central New Mexico Community College

Treven Edwards, SUNY-Pittsburgh

Paul Fattaruso, University of Denver

Cynthia Gillispie-Johnson, North Carolina A&T State University

Arthur Khaw, Kirkwood Community College

Liz Kleinfeld, Red Rocks Community College

Angela W. Lamb, Robeson Community College

Lindsay Lewan, Arapahoe Community College

Caroline Mains, Palo Alto College

Josie Mills, Arapahoe Community College

Sharon Mouss, Oklahoma State University—Okmulgee

Joseph Nocera, Jefferson Community College

Nancy Roberts, Griffin Technical College

Sherry Rosenthal, Community College of Southern Nevada

Joyce C. Staples, Patrick Henry Community College

Brian Still, Texas Tech University

Alan Tessaro, Spartanburg Community College

Chris Thaiss, University of California—Davis

Marc Wilson, Ivy Tech Community College

Esther J. Winter, Central Community College

J. B. Zwilling, ACCCC

We also would like to thank the following people who contributed activities, exercises, and documents to the Companion Website:

- May Beth Van Ness, Terra Community College, for helping to check and revise our existing online material
- Linda Gray, Oral Roberts University, for creating the innovative Interactive Editing and Revision Exercises
- Catharine Schauer, Visiting Professor, Embry-Riddle Aeronautical University, for contributing a variety of interesting activities
- Malanie Rosen Brown, St. Johns River Community College, for contributing interesting activities and case studies
- Connie Cerniglia, Guilford Technical Community College, for contributing a variety of career-related assignments and documents

We would especially like to thank our Editor, Gary Bauer, and our project manager, Wayne Schmidt from ContentConnections, for their efforts, patience, and creativity in helping us bring out this sixth edition.

Sharon J. Gerson   *Steve M. Gerson*

BRIEF CONTENTS

CONTENTS

CHAPTER 1

An Introduction to Technical Communication

COMMUNICATION
at work

In the Gulfview scenario, employees in diverse locations reveal the importance of technical communication.

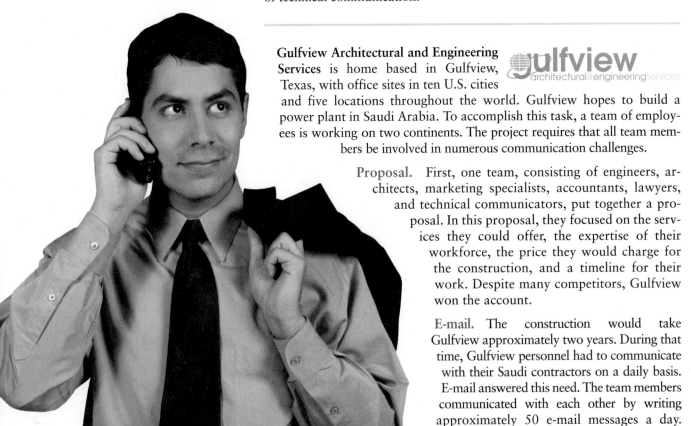

Gulfview Architectural and Engineering Services is home based in Gulfview, Texas, with office sites in ten U.S. cities and five locations throughout the world. Gulfview hopes to build a power plant in Saudi Arabia. To accomplish this task, a team of employees is working on two continents. The project requires that all team members be involved in numerous communication challenges.

Proposal. First, one team, consisting of engineers, architects, marketing specialists, accountants, lawyers, and technical communicators, put together a proposal. In this proposal, they focused on the services they could offer, the expertise of their workforce, the price they would charge for the construction, and a timeline for their work. Despite many competitors, Gulfview won the account.

E-mail. The construction would take Gulfview approximately two years. During that time, Gulfview personnel had to communicate with their Saudi contractors on a daily basis. E-mail answered this need. The team members communicated with each other by writing approximately 50 e-mail messages a day.

Objectives

When you complete this chapter, you will be able to

1. Define and understand technical communication.

2. Use many different channels of oral and written technical communication.

3. Understand the importance of technical communication.

4. Recognize the importance of teamwork in technical communication.

5. Deal with challenges to effective teamwork.

6. Resolve conflicts in collaborative projects.

7. Apply the checklist to team activities.

In these transmittals, the team members focused on construction permits, negotiated costs with vendors, changed construction plans, and asked questions and received answers. They also used these e-mail messages to build rapport with coworkers.

Intranet Web Site and Corporate Blog. To help all parties involved (those in Saudi Arabia as well as Gulfview employees throughout the United States), Gulfview's Information Technology Department built an intranet site and a blog geared specifically toward the power plant project. This firewall-protected site, open to Gulfview employees and external vendors associated with the project, helped all construction personnel submit online forms, get corporate updates, and access answers to frequently asked questions. Many of these FAQs were managed through online help screens with pull-down menus. The blog allowed employees to provide work journals, web logs in which they could comment on construction challenges and get feedback from other employees working with similar issues.

Letters. To secure and revise construction permits, Gulfview personnel had to write formal letters to government officials in Saudi Arabia. In addition, Gulfview employees had to write letters to vendors, asking for quotes.

Reports. Finally, all of the employees involved in the power plant project had to report on their activities. These included

- Progress reports providing updates on the project's status
- Incident reports when job-related accidents and injuries occurred
- Feasibility reports to recommend changes to the project's plan or scope
- Meeting minutes following the many team meetings

Like all companies engaged in job-related projects, Gulfview Architectural and Engineering Services spent much of its time communicating with a diverse audience. The challenges they faced involved teamwork, multicultural and multilingual concerns, a vast array of communication technologies, and a variety of communication channels.

WWW
Check Online Resources

www.prenhall.com/gerson
For more information about samples and activities, visit our companion Web site.

WHAT IS TECHNICAL COMMUNICATION?

Technical communication is written and oral communication for and about business and industry. Technical communication focuses on products and services—how to manufacture them, market them, manage them, deliver them, and use them.

Technical communication is composed primarily in the work environment for supervisors, colleagues, subordinates, vendors, and customers. As either a professional technical communicator, an employee at a company, or a consumer, you can expect to write the following types of correspondence for the following reasons (and many more).

- As a computer information systems (CIS) employee, you work at a 1-800 hotline helpdesk. A call comes from a concerned customer. Your job is to answer that client's questions and follow up with a *one-page e-mail* documenting the problem and your responses.

- You are a technical communicator, working in engineering, biomedical equipment manufacturing, the automotive industry, computer software development, or a variety of other job areas. Your job is to write *user manuals* to explain the steps for building a piece of equipment, performing preventative maintenance, or for shipping and handling procedures.

- As a trust officer in a bank, one of your jobs is to make proposals to potential clients. To do so, you must write a *20- to 30-page proposal* about your bank's services.

- You are a customer. You ordered an automotive part from a national manufacturer. Unfortunately, the part was shipped to you five days later than promised, it arrived broken, and you were charged more than the agreed-upon price. You need to write a *letter of complaint*.

- As the manager of a medical records reporting department, one of your major responsibilities is ensuring that your staff's training is up to date. After all, insurance rules and regulations keep changing. To document your department's compliance, you must write a monthly *progress report* to upper-level management.

- You are a webmaster. Your job is creating a corporate *Web site*, complete with *online help screens*. The Web site gives clients information about your locations, pricing, products and services, mission statement, and job openings. The drop-down help screens provide easy-to-access answers for both customer and employee questions.

- As an entrepreneur, you are opening your own computer-maintenance service (or services for HVAC repair, deck rebuilding, home construction, lawn care, or automotive maintenance). To market your company, you will need to write *fliers, brochures,* or *sales letters*.

- You have just graduated from college (or, you have just been laid off). It's time to get a job. You need to write a *resume* and a *letter of application* to show corporations what assets you will bring to their company.

COMMUNICATION CHANNELS

Technical communication takes many different forms. Not only will you communicate both orally and in writing, but also you will rely on various types of correspondence and technology, dependent upon the audience, purpose, and situation. To communicate successfully in the workplace, you must adapt to many different channels of communication. Table 1.1 gives you examples of different communication channels, both oral and written.

Table 1.2 illustrates how different writers and speakers might use various channels to communicate effectively to both internal and external audiences. Internal audiences consist of the coworkers, subordinates, and supervisors in your workplace; external audiences consist of vendors, customers, and other workplace professionals.

Many communication channels overlap in terms of purpose and audience. If you are requesting information from a vendor, for example, you could write a letter, send an e-mail message, or make a telephone call. However, in other instances, communication channels are more exclusive. You would not want to communicate bad news—such as layoffs, loss of benefits, or corporate closings—to employees by way of mass e-mail messages or televised reports. In these instances, face-to-face meetings would be more appropriate. A key to successful technical communication is choosing the right channel.

TABLE 1.1 Communication Channels

Written Communication Channels	Oral Communication Channels
• E-mail	• Leading meetings
• Memos	• Conducting interviews
• Letters	• Making sales calls
• Reports	• Managing others
• Proposals	• Participating in teleconferences and videoconferences
• Fliers	• Facilitating training sessions
• Brochures	• Participating in collaborative team projects
• Faxes	• Providing customer service
• Internet Web sites	• Making telephone calls
• Intranet Web sites	• Leaving voice-mail messages
• Extranet Web sites	• Making presentations at conferences or to civic organizations
• Instant Messaging	• Participating in interpersonal communication at work
• Blogging	• Conducting performance reviews
• Job information (resumes, letters of application, follow-up letters, interviews)	

TABLE 1.2 Communication Channels—Audience and Purpose

Writers/Speakers	Type of Communication and Communication Channel	Purposes	Internal or External
Human resources/Training	Instructions—either hard copy or online (Internet, intranet, extranet)	Help employees and staff perform tasks	Internal
Marketing personnel	Brochures, sales letters, blogs, and phone calls	Promote new products or services	External
Customers	Inquiry or complaint letters or phone calls	Ask about/complain about products or services	External
Quality assurance	Investigative, incident, or progress reports	Report to regulatory agencies about events	External
Vendors	Newsletters, phone calls, and e-mail	Update clients on new prices, products, or services	External
Management	Oral presentations to department staff	Update employees on mergers, acquisitions, layoffs, raises, or site relocations	Internal
Everyone (management, employees, clients, vendors, governmental agencies)	E-mail, blogging, Web sites, and instant messaging	All conceivable purposes	Internal and External

FIGURE 1.1 Channels "Almost Always" Used in Workplace Communication

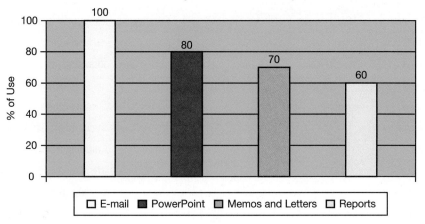

To clarify the use of different technical communication channels, look at Figure 1.1. In a 2004 survey of approximately 120 companies employing over 8 million people, the National Commission on Writing found that employees "almost always" use different forms of writing, including e-mail messages, PowerPoint, memos, letters, and reports ("Writing: A Ticket to Work" 2007, 11).

THE IMPORTANCE OF TECHNICAL COMMUNICATION

The National Commission on Writing concluded that "in today's workplace writing is a 'threshold skill' for hiring and promotion among salaried . . . employees. Survey results indicate that writing is a ticket to professional opportunity, while poorly written job applications are a figurative kiss of death" ("Writing: A Ticket to Work" 2007, 3). Technical communication is a significant factor in your work experience for several reasons.

Business

Technical communication is not a frill or an occasional endeavor. It is a major component of the work environment. Through technical correspondence, employees

- Maintain good customer–client relations (follow-up letters).
- Ensure that work is accomplished on time (directive memos or e-mail).
- Provide documentation that work has been completed (progress reports).
- Generate income (sales letters, brochures, and fliers).
- Keep machinery working (user manuals).
- Ensure that correct equipment is purchased (technical descriptions).
- Participate in teleconferences or videoconferences (oral communication).
- Get a job (resumes).
- Define terminology (online help screens).
- Inform the world about a company's products and services (Internet Web sites and blogs).

Time

In addition to serving valuable purposes in the workplace, technical communication is important because it requires your time. Across all professions, workers spend nearly one-third of their time writing (31 percent). Figure 1.2 shows that 43 percent of the respondents to a survey spend between 11 and 30 percent of their time writing. Another 26 percent of respondents spend between 31 and 50 percent of their time writing (Miller et al. 1996, 10).

FIGURE 1.2 Time Spent Writing

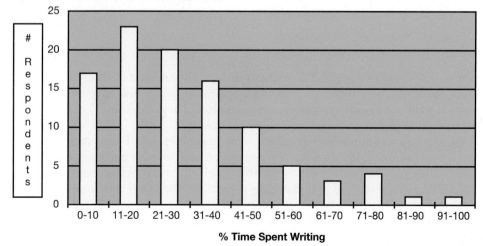

The 31 percent of time spent writing on the job is only a base number—an average across the board. Generally speaking, new hires might spend less time writing on the job. As a supervisor, you will spend more time directing your subordinates through written correspondence.

Money

You have heard it before—time is money. Here are three simple ways of looking at the cost of your technical communication.

- **Cost of correspondence**—A study by Dartnell's Institute of Business Research says that the "average cost of producing and mailing a letter is $19.92" ("Business Identity" 2004). That amount factors in the time it takes a worker to write the letter as well as the cost of the paper, printing, and stamp. If one letter costs almost $20, imagine how much an entire company's correspondence might cost annually, including every employee's e-mail, letters, memos, and reports.
- **Percentage of salary**—Consider how much of your salary is being paid for your communication skills. Let's say you make $35,000 a year. If you are spending 31 percent of your time writing, then your company is paying you approximately $10,850 just to write. That does not include the additional time you spend on oral communication.

    If you are not communicating effectively on the job, then you are asking your bosses to pay you a lot of money for substandard work. Your time spent communicating, both in writing and orally, is part of your salary—and part of your company's expenditures.
- **Cost of training**—Corporations spend money to improve their employees' writing skills. The National Commission on Writing for America's Families, Schools, and Colleges reported, "More than 40 percent of responding firms offer or require training for salaried employees with writing deficiencies. 'We're likely to send out 200–300 people annually for skills upgrade courses like "business writing" or "technical writing," said one respondent.' Based on survey responses, the Commission estimates that remedying deficiencies in writing costs American corporations as much as $3.1 billion annually" ("Writing Skills" 2005).
- **Generating income**—Your communication skills do more than just cost the company money; these talents can earn money for both you and the company. A well-written sales letter, flier, brochure, proposal, or Web site can generate corporate income. Effectively written newsletters to clients and stakeholders can keep customers happy and bring in new clients. Good written communication is not just part of your salary—it helps pay your wages.

Interpersonal Communication

A major component of a successful company is the environment it develops, the tone it expresses, the atmosphere it creates. Successful companies know that effective communication, both written and oral, creates a better workplace. These "soft skills" make customers want to shop with you and employees work for you.

Your technical communication reflects something about you. E-mail messages, letters, memos, or telephone skills are a photograph of you and your company. If you write well, you are telling your audience that you can think logically and communicate your thoughts clearly. When your writing is grammatically correct, or when your telephone tone of voice is calm and knowledgeable, you seem professional to your audience. Technical communication is an extension of your interpersonal communication skills. Coworkers or customers will judge your competence based on what you say and how you say it.

THE IMPORTANCE OF TEAMWORK

Companies have found that teamwork enhances productivity. Teammates help and learn from each other. They provide checks and balances. Through teamwork, employees can develop open lines of communication to ensure that projects are completed successfully.

Collaboration

In business and industry, many user manuals, reports, proposals, PowerPoint presentations, and Web sites are team written. Teams consist of engineers, graphic artists, marketing specialists, and corporate employees in legal, delivery, production, sales, accounting, and management. These collaborative team projects extend beyond the company. A corporate team also will work with subcontractors from other corporations. The collaborative efforts include communicating with companies in other cities and countries through teleconferences, faxes, and e-mail. Modern technical communication requires the participation of "communities of practice": formal and informal networks of people who collaborate on projects based on common goals, interests, initiatives, and activities (Fisher and Bennion 2005, 278).

The National Association of Colleges and Employers lists the "Top Ten Skills Employers Want" (see Figure 1.3). Notice how interpersonal, teamwork, oral communication, and written communication skills take precedence over other abilities.

The Problems with "Silo Building"

Working well with others requires collaboration versus "silo building." The *silo* has become a metaphor for departments and employees that behave as if they have no responsibilities outside their areas. They build bunkers around themselves, failing to collaborate with others. In addition, they act as if no other department's concerns or opinions are valuable.

FIGURE 1.3 Top Ten Qualities/Skills Employers Want

Skill	Rating
1. Communication Skills (Verbal and Written)	4.7
2. Honesty/Integrity	4.7
3. Teamwork Skills (work well with others)	4.6
4. Strong Work Ethic	4.5
5. Analytical Skills	4.4
6. Flexibility/Adaptability	4.4
7. Interpersonal Skills (relate well to others)	4.4
8. Motivation/Initiative	4.4
9. Computer Skills	4.3
10. Detail Oriented	4.3
(5-point scale, where 1 = not at all important and 5 = extremely important)	

(Source: National Association of Colleges and Employers: Job Outlook 2006 Student Version)

Such "stand-alone" departments or people isolate themselves from the company as a whole and become inaccessible to other departments. They "focus narrowly" (Hughes 2003, 9), which creates problems. Poor accessibility and poor communication "can cause duplicate efforts, discourage cooperation, and stifle cross-pollination of ideas" (Hughes 2003, 9).

To be effective, companies need "open lines of communication within and between departments" (Hughes 2003, 9). The successful employee must be able to work collaboratively with others to share ideas. In the workplace, teamwork is essential.

Why Teamwork Is Important

Teamwork benefits employees, corporations, and consumers. By allowing all constituents a voice in project development, teamwork helps to create effective workplaces and ensures product integrity.

Diversity of Opinion. When you look at problems individually, you tend to see issues from limited perspectives—*yours*. In contrast, teams offer many points of view. For instance, if a team has members from accounting, public relations, customer service, engineering, and information technology, then that diverse group can offer diverse opinions. You should always look at a problem from various angles.

Checks and Balances. Diversity of opinion also provides the added benefit of checks and balances. Rarely should one individual or one department determine outcomes. When a team consists of members from different disciplines, those members can say, "Wait a minute. Your idea will negatively impact my department. We had better stop and reconsider."

Broad-Based Understanding. If decisions are made in a silo, by a small group of like-minded individuals, then these conclusions might surprise others in the company. Surprises are rarely good. You always want buy-in from the majority of your stakeholders. An excellent way to achieve this is through team projects. When multiple points of view are shared, a company benefits from broad-based knowledge. Improved communication allows people to see the bigger picture.

Empowerment. Collaboration gives people from varied disciplines an opportunity to provide their input. When groups are involved in the decision-making process, they have a stake in the project. This allows for better morale and productivity.

Team Building. Everyone in a company should have the same goals—corporate success, customer satisfaction, and quality production. Team projects encourage shared visions, a better work environment, a greater sense of collegiality, and improved performance. Employees can say, "We are all in this together, working toward a common goal."

Using a variety of communication channels to achieve collaboration

Rob Studin is the Executive Director of Financial Advisory Services for Lincoln Financial Advisors (LFA). Home based in Philadelphia, LFA's 3,000 advisors and employees provide fee-based financial planning for clients, including "Estate Planning, Investment Planning, Retirement Planning Strategies and Business Owner Planning" (http://www.lfg.com/).

LFA uses four electronic oral communication channels to ensure a consistent, collaborative workforce: teleconferences, videoconferences, webinars, and LFA's Virtual University.

- **Teleconferences:** Rob, who works in LFA's Birmingham office, has six key managers who work in San Francisco, Salt Lake City, Cleveland, Columbus, Rochester, and Baltimore. To communicate with his dispersed team members, "we have a conference call just to touch base. Sometimes we have a formal agenda, and sometimes I just ask, 'What's going on guys?' A casual, weekly teleconference allows us to stay up to date on issues facing us individually or as a group. We collectively understand that six heads are better than one for problem solving."

- **Videoconferences:** You can't communicate effectively with 3,000 people on the telephone. While teleconferencing works well for Rob and smaller groups, when LFA needs to communicate to all of its employees and/or advisors about corporate-wide issues that affect policy, budget, personnel, and strategic planning, face-to-face meetings might be the optimum solution. However, transporting 3,000 people to a central location is neither time efficient nor cost effective. A three-hour meeting might require two days of travel plus hotel, food, and air fares. To save time and money, LFA uses videoconferences.

- **Webinars:** Videoconferences create at least three challenges for LFA employees. First, to participate in a videoconference, the employees must be in a room fitted with LFA's companywide videoconference system. Second, if many people are in the audience and seated at a distance from the TV screen, visibility/readability can be an obstacle. Finally, videoconferences don't allow the audience any hands-on opportunity to practice new skills. Webinars solve these corporate communication challenges. All an individual needs to participate in an online seminar is a computer.

- **LFA's Virtual University:** Most of LFA's webinars are synchronous. All employees are asked to log on at a given time while a webinar host runs the training program. Inevitably, however, an employee can't participate in the webinar when it is initially presented. LFA has solved this problem. Their Virtual University offers asynchronous "training on demand." All training videoconferences and all webinars are recorded and archived. By accessing LFA's password-protected www.LFAplanner.com site, advisors and employees can retrieve training materials at their convenience.

Rob says that he spends approximately 50 percent of his work time communicating via e-mail messages, telephone calls, and teleconferences. For efficiency, cost savings, and consistent communication to a geographically dispersed workforce, LFA has found that multiple, electronic channels help team members achieve their communication goals.

DIVERSE TEAMS . . . DISPERSED TEAMS

Collaborative projects will depend on diverse team members and dispersed team members.

Diverse Teams

Teams will be diverse, consisting of people from different areas of expertise. Your teams will be made up of engineers, graphic artists, accountants, technical communicators, financial advisors, human resource employees, and others. In addition, the team will consist of people who are different ages, genders, cultures, and races.

Diversity

See Chapter 3 for more discussion of diversity.

Dispersed Teams

In a global economy, members of a team project might not be able to work together, face to face. Team members might be located across time and space. They could work in different cities, states, time zones, countries, or different shifts. For example, you might work for your company in New York, while members of your team work for the company at other sites in Chicago, Denver, and Los Angeles. This challenge to collaboration is compounded when you also must team with employees at your company's sites in India, Mexico, France, and Japan. According to a 2005 report, 41 percent of employees at the top international corporations live outside the borders of their company's home country (Nesbitt and Bagley-Woodward 2006, 25).

Using Groupware to Collaborate in Virtual Teams

When employees are dispersed geographically, getting all team members together would be costly in terms of time and money. Companies solve this problem by forming virtual, remote teams that collaborate using electronic communication tools called *groupware*. Groupware consists of software and hardware that helps companies reduce travel costs, allows for telecommuting, and facilitates communication for employees located in different cities and countries.

Groupware includes the following types of hardware and software (Nesbitt and Bagley-Woodward 2006, 28).

- Electronic conferencing tools such as webinars, listservs, chat systems, message and discussion boards, videoconferences, and teleconferences.
- Electronic management tools, such as Digital Dashboards, project management software that schedules, tracks, and charts the steps in a project; Microsoft Outlook's electronic calendaring, which allows you to send a meeting request to dispersed team members, check the availability of meeting attendees, reschedule meetings electronically, forward meeting requests, and cancel a meeting—without ever visiting with your team members face to face.
- Electronic communication tools for writing and sending documents, such as instant messaging, e-mail, blogs, intranets and extranets, and wikis.

Collaborative Writing Tools

Wikis.   What's a wiki? The latest in technology geared toward collaborative writing is a wiki. A wiki "is a website that allows the visitors . . . to easily add, remove, and otherwise edit and change available content, and typically without the need for registration. This ease of interaction and operation makes a wiki an effective tool for mass collaborative authoring" ("Wiki" 2007). In addition, wikis let collaborative writers track "the history of a document as it is revised." Whenever a team member edits text in the "wiki, that new text becomes the current version, while older versions are stored" (Mader 2005). The largest example of a wiki is *Wikipedia*, "the free encyclopedia that anyone can edit. Wikipedia is an encyclopedia collaboratively written by many of its readers" ("What Is Wikipedia?" 2007).

Wikis have entered the workplace. Prentice Hall, the world's largest publishing company, is using wiki technology to create a community-written textbook. *We Are Smarter Than Me* is the first networked book on business. Collaborators used wiki technology to write a "book on how the emergence of community and social networks will change the future rules of business" ("We Are Smarter than Me" 2006).

Who's Using Wikis?

Many companies use wikis for collaborative writing projects. **Yahoo** uses a wiki. Eric Baldeschwieler, director of software development of Yahoo!, says, "Our development team includes hundreds of people in various locations all over the world, so web collaboration is VERY important to us." **Cmed** runs pharmaceutical clinical trials and develops new technology. In this heavily regulated environment, wikis "improved communication and increased the quantity (and through peer review, the quality) of documentation." **Cingular Wireless** "project managers have been encouraged to utilize the site for any issues needing collaborating efforts in lieu of emails." **Disney** uses a wiki "for an engineering team that was rearchitecting the Go.com portal. During this time we found TWiki to be a very effective means of posting and maintaining development specs and notes as well as pointers to resources." **Texas Instrument's** "India design centre" uses Wikis to manage all project-specific information, such as documenting ideas, plans, and status; sharing information with other teams across various work sites; and updating information and content to team members ("Twiki" 2007).

How Can You Use a Wiki?

In your dispersed teams, whether virtual, remote, or mobile, you might use wikis in the following ways to create collaborative documents.

- **Create Web sites.** Wikis help team members easily add pages, insert graphics, create hyperlinks, and add simple navigation.
- **Project development with peer review.** A wiki makes it easy for team members to write, revise, and submit projects, since all three activities can take place in the wiki.
- **Group authoring.** Wikis allow group members to build and edit a document. This creates a sense of community within a group, allows group members to build on each other's work, and provides immediate, asynchronous access to all versions of a document.
- **Track group projects.** Each wiki page lets you track how group members are developing their contributions. The wiki also lets you give feedback and suggest editorial changes.

Figures 1.4, 1.5, and 1.6 are illustrations of a wiki, edited text in a wiki, and a record of edited versions.

FIGURE 1.4 Wiki Page

By clicking on "Edit page," any team member can revise the text.

What the audience sees when the wiki is first opened.

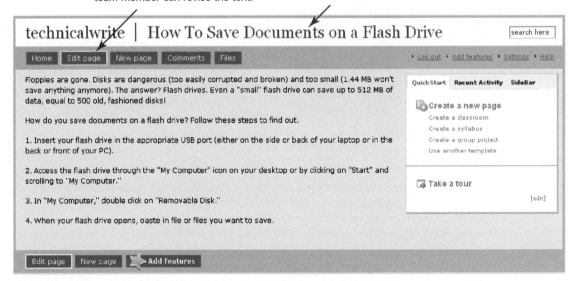

Reprinted with permission of PBWiki.

FIGURE 1.5 Wiki in "Edit" Mode

The "editing" screen allows team members to add, delete, and enhance text (by bolding, italicizing, underlining, adding bullets, etc.,) just as in a Word document.

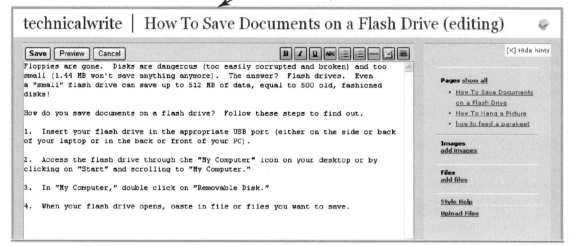

Reprinted with permission of PBWiki.

FIGURE 1.6 Record of Edited Versions

This screen allows team members to compare all versions of the text, thus seeing what has been added, deleted, or enhanced.

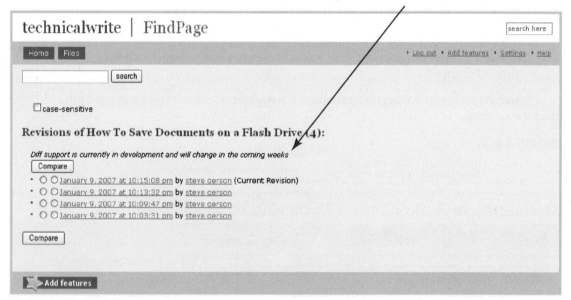

Reprinted with permission of PBWiki.

Google Documents

Another collaborative writing tool you can use easily is Google Documents, useful for document sharing, collaborating on group projects, and publishing to the World Wide Web. Google Documents is free to anyone with a Web browser and an Internet connection. This electronic tool allows for seamless integration of its collaborative features and user-friendly interface.

Using Google Documents, you and group members can edit Word documents, RTF, and HTML files. Teams can work on one document at the same time. Changes made by one writer will be seen by all team members instantly. Google Documents provide you these benefits:

- View a document's revision history.
- Return to earlier versions.
- Edit and view a document.
- Add new team members or delete writers.
- Post documents to a blog or publish a document to a Web page.

Challenges to Effective Teamwork

Any collaborative activity is challenging to manage: Team members do not show up for class or work; one student or employee monopolizes the activity while another individual snoozes; people exert varying amounts of enthusiasm and ability; personalities clash. Some people fight over everything. Occasionally, when a boss participates on a team, employees fear speaking openly. Some team members will not stay on the subject. One team member will not complete her assignments ("Individual's and Teams' Roles and Responsibilities" 2003). Group dynamics are difficult and can lead to performance gaps.

Human Performance Improvement

Human Performance Improvement (HPI) focuses on "root cause analysis" to assess and overcome the barriers inherent in teamwork. To close performance gaps, HPI analyzes the following possible causes for collaborative breakdowns.

1. **Knowledge**—Perhaps employees do not know how to perform a task. They have never acquired the correct skills or do not understand which skills are needed to complete the specific job. Varying skills of team members can impede the group's progress.
2. **Resources**—Think of these possibilities: Tools are broken or missing; the department is out of funds; you do not have enough personnel to do the job; the raw material needed for the job is below par; you ordered one piece of machinery but were shipped something different; you needed 100 items but have only 50 in stock. To complete a project, you often have to solve problems with resources.
3. **Processes**—For teams to succeed in collaborative projects, everyone must know what their responsibilities are. Who reports to whom? How will these reports be handled (orally, in writing)? Who does what job? Are responsibilities shared equally? Structure, of some sort, is needed to avoid chaos, lost time, inefficiency, hurt feelings, and many other challenges to teamwork. To achieve successful collaboration, the team should set and maintain effective procedures.
4. **Information**—A team needs up-to-date and accurate information to function well. If required database information is late or incorrect, then the team will falter. If the information is too high tech for some of the team members, then a lack of understanding may undermine the team effort.
5. **Support**—To succeed in any project, a team needs support. This could be financial, attitudinal, or managerial. When managers from different departments are fighting "turf wars" over ownership of a project, teams cannot succeed. Teams need enough money for staffing, personnel, or equipment.
6. **Wellness**—A final consideration involves the team's health and well-being. People get sick or miss work for health reasons. People have car accidents. If a teammate must miss work for a day or an extended period, this will negatively impact the team's productivity. Stress and absences can lead to arguments, missed deadlines, erratic work schedules, and poor quality.

HPI Intervention Techniques.   After assessing root causes that challenge a team's success, HPI creates intervention options. These might include the following:

- Improved compensation packages
- Employee recognition programs
- Revised performance appraisals
- Improved employee training
- Simulations
- Mentoring or coaching
- Restructured work environments to enhance ergonomics
- Safety implementations
- Strategic planning changes
- Improved communication channels
- Health and wellness options—lectures, on-site fitness consultants, incentives to weight loss, and therapist and social worker interventions

People need help in order to work more effectively with each other. A progressive company recognizes these challenges and steps in to help.

CONFLICT RESOLUTION IN COLLABORATIVE PROJECTS

To ensure that team members work well together and that projects are completed successfully, consider these approaches to conflict resolution.

1. **Choose a team leader**—Sometimes, team leaders are chosen by management; sometimes, team leaders emerge from the group by consensus. However this person gains the position, he or she can serve many valuable purposes. The team leader becomes "point person," the individual whom all can turn to for assistance. He or she can solve problems, seek additional resources, or organize the team effort. For instance, a team leader can give the team direction, interface with management, and/or act as the team's mediator.

2. **Set guidelines**—One reason that conflicts occur is because people do not know what to expect or what is expected of them. On the other hand, if expectations are clear, then several major sources of conflict can be resolved.

    For example, one simple conflict might be related to time. A team member could be unaware of when the meeting will end and schedules another meeting. If that team member then has to leave the first meeting early, disrupting the team's progress, this can cause a conflict.

    To solve this problem, set guidelines. Hold an initial meeting (online or teleconferenced for remote, virtual teams) to define goals and establish guidelines, establish project milestones, or create schedules for synchronous dialogues. Communicate to all team members (before the meeting via e-mail or early in a project) how long the project will last. Also, clarify the team's goals, the chain of command (if one exists), and each team member's responsibilities.

3. **Ensure that all team members have compatible hardware and software**—This is especially important for virtual, remote teams. To communicate successfully, all team members need access to the same e-mail platform. Some members cannot use Yahoo or MSN or Hotmail while others use Outlook. This would cause communication challenges if software is incompatible. The problem is further heightened when video or telephone equipment is different.

4. **Encourage equal discussion and involvement**—A team's success demands that everyone participate. A team leader should encourage involvement and discussion. All team members should be mutually accountable for team results, including planning, writing, editing, proofreading, and packaging the finished project. Be sure that everyone is allowed a chance to give input.

   Conflicts also arise when one person monopolizes the work. If one person speaks excessively, others will feel left out and disregarded. A team leader should ensure equal participation. He or she should call on others for their opinions and ask for additional input from the team.

   In addition, team leaders must limit an overly aggressive team member's participation by saying, "Thanks, John, for your comments. Now, let's see what others have to say." Or, "Wait one second, John. I'll come back to you after we've heard from a few others."

5. **Discourage taking sides**—Discussion is necessary, but conflict will arise if team members take sides. An "us against them" mentality will harm the team effort. You can avoid this pitfall by seeking consensus, tabling issues, creating subcommittees, or asking for help from an outside source (boss or teacher, for example).

6. **Seek consensus**—Not every member of the team needs to agree on a course of action. However, a team cannot go forward without majority approval. To achieve consensus, your job as team leader is to listen to everyone's opinion, seek compromise, and value diversity. Conflict can be resolved by allowing everyone a chance to speak. Once everyone has spoken, then take a vote.

7. **Table topics when necessary**—If an issue is so controversial that it cannot be agreed upon, take a time out. Tell the team, "Let's break for a few minutes. Then we can reconvene with fresh perspectives." Maybe you need to table the topic for the next meeting. Sometimes, conflicts need a cooling-off period.

8. **Create subcommittees**—If a topic cannot be resolved, teammates are at odds, or sides are being taken, then create a subcommittee to resolve the conflict. Let a smaller group tackle the issue and report back to the larger team.

9. **Find the good in the bad**—Occasionally, one team member comes to a meeting with an agenda. This person does not agree with the way things have been handled in the past or the way things are being handled presently. You do not agree, nor do other team members. However, you cannot resolve this issue simply by saying, "That's not how we do things." A disgruntled team member will not accept such a limited viewpoint.

   As team leader, seek compromise. Let the challenging team member speak. Discuss each of the points of dissension. Allow for input from the team. Some of the ideas might have more merit than you originally assumed.

10. **Deal with individuals individually**—From time to time, a team member will cause problems for the group. The teammate might speak out of turn or say inappropriate things. These could include off-color or off-topic comments. A team member might cause problems for the group by habitually showing up late, missing meetings, or monopolizing discussions.

    To handle these conflicts, avoid pointing a finger of blame at this person during the meeting. Do not react aggressively or impatiently. Doing so will lead to the following problems:

    - Your reaction might call more attention to this person. Sometimes people come to meetings late or speak out in a group *just* to get attention. If you react, you might give the individual exactly what he wants.
    - Your reaction might embarrass this person.

- Your reaction might make you look unprofessional.
- Your reaction might deter others from speaking out. You want an open environment, allowing for a free exchange of ideas.

   Speak to any offending team members individually. This could be accomplished at a later date, in your office, or during a coffee break. Speaking to the person later and individually might defuse the conflict.

11. **Stay calm**—Act professionally when dealing with conflict. To resolve conflicts, speak slowly, keep your voice steady and quiet, and stay seated (rising will look too aggressive). You also might want to take notes. This will provide you with a record of the discussion.

12. **Remove, reassign, or replace if necessary**—Finally, if a team member cannot be calmed, cannot agree with the majority, or has too many other conflicts, your best course of action might be to remove, reassign, or replace this individual.

CHECKLIST FOR COLLABORATION

_____ 1. Have you chosen a team leader (or has a team leader been assigned)?

_____ 2. Do all participants understand the team's goal and their individual responsibilities?

_____ 3. Does the team have a schedule, complete with milestones and target due dates?

_____ 4. Does the team have compatible hardware and software?

_____ 5. In planning the team's project, did you seek consensus?

_____ 6. Have all participants been allowed to express themselves?

_____ 7. If conflicts occurred, did you table topics for later discussion or additional research?

_____ 8. Did you encourage diversity of opinion?

_____ 9. Have you avoided confronting people in public, choosing to meet with individuals privately to discuss concerns?

_____ 10. If challenges continue, have you reassigned team members?

CHAPTER HIGHLIGHTS

1. Technical communication is written for and about business and industry and focuses on products and services.
2. Technical communication is an important part of your everyday work life. It can consume as much as 31 percent of a typical workweek.
3. Technical communication costs a company both time and money, so employees must write effectively.
4. The top skills employers want include communication skills, honesty, interpersonal skills, a strong work ethic, and teamwork.
5. Avoid "silo building," isolating yourself on the job.
6. Working in teams allows you to see issues from several points of view.
7. Human Performance Improvement (HPI) solves problems—"gaps"—inherent in teamwork.
8. Problems teams face include varied knowledge levels, differing motives, and insufficient resources.
9. Conflict resolution strategies are essential to a team's success.
10. To resolve conflicts in a team, you should set guidelines, encourage all to participate, and avoid taking sides.

CASE STUDY

You are the team leader of a work project at Gulfview Architectural and Engineering Services. The team has been involved in this project for a year. During the year, the team has met weekly, every Wednesday at 8:00 A.M. It is now time to assess the team's successes and areas needing improvement.

Your goal will be to recommend changes as needed before the team begins its second year on this project. You have encountered the following problems.

- One team member, Caroline Jensen, misses meetings regularly. In fact, she has missed at least one meeting a month during the past year. Occasionally, she missed two or three in a row. You have met with Caroline to discuss the problem. She says she has had child care issues that have forced her to use the company's flextime option, allowing her to come to work later than usual, at 9:00 A.M.
- Another team member, Tasha Stapleton, tends to talk a lot during the meetings. She has good things to say, but she speaks her mind very loudly and interrupts others as they are speaking. She also elaborates on her points in great detail, even when the point has been made.
- A third team member, Sharon Mitchell, almost never provides her input during the meetings. She will e-mail comments later or talk to people during breaks. Her comments are valid and on topic, but not everyone gets to hear what she says.
- A fourth team member, Craig Mabrito, is very impatient during the meetings. This is evident from his verbal and nonverbal communication. He grunts, slouches, drums on the table, and gets up to walk around while others are speaking.
- A fifth employee, Julie Jones, is overly aggressive. She is confrontational, both verbally and physically. Julie points her finger at people when she speaks, raises her voice to drown out others as they speak, and uses sarcasm as a weapon. Julie also crowds people, standing very close to them when speaking.

Assignment

How will you handle these challenges? Try this approach:

- *Analyze* the problem(s). To do so, brainstorm. What gaps might exist causing these problems?
- *Invent* or envision solutions. How would you solve the problems? Consider Human Performance Improvement issues, as discussed in this chapter.
- *Plan* your approach. To do so, establish verifiable measures of success (including time frames and quantifiable actions).

Write an e-mail to your instructor sharing your findings.

INDIVIDUAL AND TEAM PROJECTS
Teamwork—Business and Industry Expectations

Individually or in small groups, visit local banks, hospitals, police or fire stations, city offices, service organizations, manufacturing companies, engineering companies, or architectural firms. Once you and your teammates have visited these sites, have asked your questions (see the following assignments), and have completed your research, share your findings using one of the following methods.

- **Oral**—as a team, give a three- to five-minute briefing to share with your colleagues the results of your research.
- **Oral**—invite employee representatives from other work environments to share with your class their responses to your questions.
- **Written**—write a team memo, letter, or report about your findings.

1. Ask employees at the sites you visit if, how, and how often they are involved in team projects. In your team, assess your findings and report your discoveries.
2. Ask employees at the sites you visit about the challenges they face with conflict resolution. In your team, assess your findings and report your discoveries.
3. Use the Internet and/or your library to research companies that rely on teamwork. Focus on which industries these companies represent and the goals of their team projects. You could also consider the challenges they encounter, their means of resolving conflicts, the numbers of individuals on each team, and whether the teams are cross-functional. Then report these findings to your professor or classmates, either orally or in writing.
4. Visit the Society for Technical Communication (STC) Web site to learn about its membership. See which industries employ technical communicators and determine these writers' job responsibilities. Also, learn which colleges and universities have programs in technical communication and what the programs entail. What else can you learn about technical communication from the STC Web site?
5. Research major publications of technical communication, such as *Intercom, Technical Communication,* and *The Journal of Scientific and Technical Communication.* On what topics do the articles in these journals focus?

PROBLEM-SOLVING THINK PIECES

To understand and practice conflict resolution, complete the following assignments.

1. **Attend a meeting.** This could be at your church, synagogue, or mosque; a city council meeting; your school, college, or university's board of trustees meeting; or a meeting at your place of employment. Was the meeting successful? Did it have room for improvement? To help answer these questions, use the following Conflict Resolution in Team Meetings Matrix. Then report your findings to your professor or classmates, either orally or in writing. Write an e-mail message, memo, or report, for example.

Conflict Resolution In Team Meetings Matrix			
Goals	**Yes**	**No**	**Comments**
1. Were meeting guidelines clear?			
2. Did the meeting facilitator encourage equal discussion and involvement?			
3. Were the meeting's attendees discouraged from taking sides?			
4. Did the meeting facilitator seek consensus?			
5. Were topics tabled if necessary?			
6. Were subcommittees created if necessary?			
7. Did the meeting facilitator find the good in the bad?			
8. Did the meeting facilitator deal with individuals *individually?*			
9. Did the meeting's facilitator stay calm?			

2. Have you been involved in a team project at work or at school? Perhaps you and your classmates grouped to write a proposal, research Web sites, create a Web site, or perform mock job interviews. Maybe you were involved in a team project for another class. Did the team work well together? If so, analyze how and why the team succeeded. If the team did not function effectively, why not? Analyze the gaps between what should have been and what was. To help you with this analysis, use the following Human Performance Index Matrix. Then, report your findings to your instructor or classmates either orally or in writing. Write an e-mail message, memo, or report, for example.

Human Performance Index Matrix			
Potential Gaps	**Yes**	**No**	**Comments**
1. Did teammates have equal and appropriate levels of knowledge to complete the task?			
2. Did teammates have equal and appropriate levels of motivation to complete the task?			
3. Did the team have sufficient resources to complete the task?			
4. Did teammates understand their roles in the process needed to complete the task?			
5. Did the team have sufficient and up-to-date information to complete the task?			
6. Did the team have sufficient support to complete the task?			
7. Did wellness issues affect the team's success?			

WEB WORKSHOP

1. How important is technical communication in the workplace? Go online to research this topic. Find five Web sites that discuss the importance of communication in the workplace, and report your discoveries to your teacher and/or class. To do so, write a brief report, memo, or e-mail message. You could also report your information orally.
2. Create a class wiki for collaborative writing. To do so, consider using either of the following sites.
   - http://www.wikispaces.com
   - http://www.pbwiki.com

QUIZ QUESTIONS

1. Define *technical communication*.
2. What are five channels of technical communication?
3. List three reasons why technical communication is important in business.
4. What is the percentage of time employees spend writing?
5. What are the top five skills employers want?
6. Define *silo building*.
7. Explain why business depends on teamwork to ensure quality.
8. List five causes for collaborative breakdown, according to Human Performance Improvement (HPI).
9. List four HPI intervention options to help a team solve its problems.
10. List three things a team leader can do to ensure successful teamwork.

CHAPTER 2

The Communication Process

In the following scenario, Creative International follows the writing process to create effective communication with its clients.

Connie Jones (President), Mary Michelson (Project Director), and Lori Smith (Director of Sales and Marketing) have made Creative International cutting-edge company. Creative International works

with organizations to define strategic communication goals. A key to their success is following a process "from the beginning to the end of a communication project." They prewrite, write, and rewrite.

Prewriting:

- **Initial Client Contact**—Through telephone calls, e-mail messages, networking, or a preliminary meeting, Creative gathers data to discover the client's needs. In this phase, the Creative team interviews the end users and observes them at work.

- **Clarify Request Meeting**—Meeting face to face with an upper-level decision maker, the Creative team collects information about the end user's needs. Connie, Mary, and Lori don't just say, "Sure, we can do that job." Instead, they ask probing questions, such as "Why do you need that?" "Why do you want that?" "What do you want to communicate to your audience?"

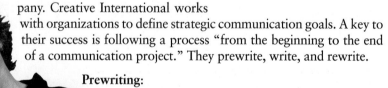

Objectives

When you complete this chapter, you will be able to

1. Understand the writing process including prewriting, writing, and rewriting.
2. Prewrite to examine purposes and goals, to determine audience, to gather data, and to choose the communication channel.
3. Write to organize your information, provide visuals, and format content.
4. Test for usability of your technical communication.
5. Rewrite by adding or deleting information, simplifying terms, moving or reformatting content, changing style of writing, and correcting.
6. Apply the checklist to your technical communication.

- **Proposal Creation**—Following the initial meeting, Creative writes a proposal, complete with schedules, project plans, the project's scope, and a description of the deliverables.

Writing:

- **Design, Development, Production, and Pilot Testing**—Creative creates text, graphics, audio and video training modules for final beta testing. This rough draft verifies that the product works the way everyone expects it to.

Rewriting:

- **Editing**—With input from both coworkers and the end user, Creative revises text by adding details, deleting unnecessary content, and correcting errors.

- **Evaluation and Maintenance**—Through end-user analysis, usability testing, and customer measurement, Creative ensures that the performance needs are met and that training materials are current and valid.

Creative International refers to its "process map" from the beginning to the end of a project. They use process for marketing, for internal communication, and for project planning and management. The writing process that Creative follows is recursive. It includes constant sign-offs and change orders. With input from all parties, during prewriting, writing, and rewriting, Creative provides its customers "communication that provides custom solutions."

Check Online Resources

www.prenhall.com/gerson
For additional samples and activities, visit our companion Web site.

Check out our quarterly newsletters TechCom E-Notes at www.prenhall.com/gerson for dot.com updates, new case studies, insights from business professionals, grammar exercises, and facts about technical communication.

THE WRITING PROCESS: AN OVERVIEW

Technical communication is a major part of your daily work experience. It takes time to construct the correspondence, and your writing has an impact on those around you. A well-written memo, letter, report, or e-mail message gets the job done and makes you look good. Poorly written correspondence wastes time and creates a negative image of you and your company.

However, recognizing the importance of technical communication does not ensure that your correspondence will be well written. How do you effectively write the memo, letter, or report? How do you successfully produce the finished product? To produce successful technical communication, you need to approach writing as a process. The process approach to writing has the following sequence.

Routine Correspondence

See Chapter 6 for more discussion of e-mail, memos, and letters.

Reports

See Chapter 15 for more discussion of reports.

Visual Aids

See Chapter 9 for more discussion of tables and figures.

1. **Prewrite**—Before you can write your document, you must have something to say. Prewriting allows you to spend quality time, prior to writing the correspondence, generating information, considering the needs of the audience, and deciding how best to communicate.

2. **Write**—Once you have gathered your data and determined your objectives, the next step is to state them. You need to draft your document. To do so, you should organize the draft, supply visual aids, and format the content to allow for ease of access.

3. **Rewrite**—The final step, and one that is essential to successful communication, is to rewrite your draft. This step requires that you revise the rough draft. Revision allows you to test for usability and to perfect your memo, letter, report, or any document so you can be a successful communicator.

The writing process is dynamic, with the three parts—prewriting, writing, and rewriting—often occurring simultaneously. You may revisit any of these parts of the process at various times as you draft your document. The writing process is illustrated in Figure 2.1.

FIGURE 2.1 The Writing Process

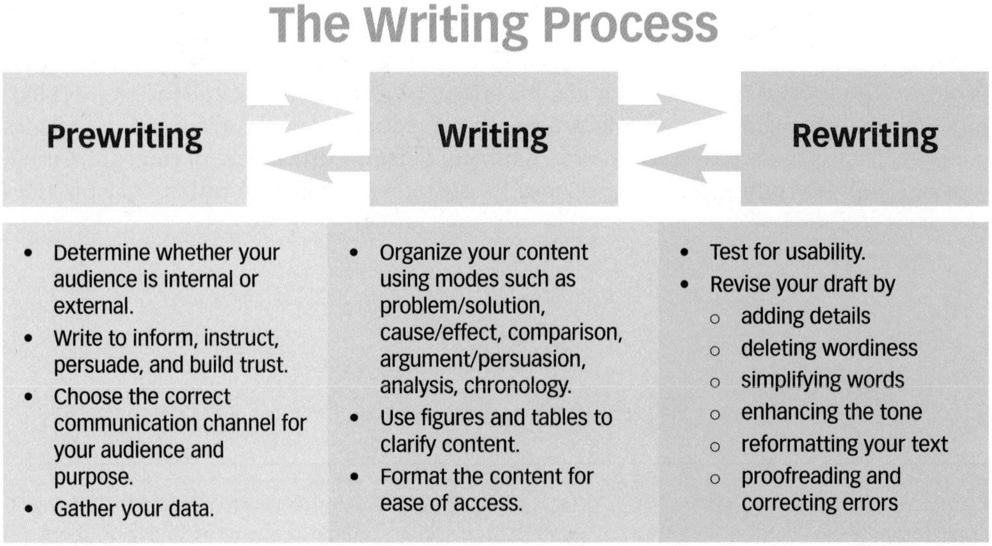

The Writing Process

Prewriting	Writing	Rewriting

• Determine whether your audience is internal or external.
• Write to inform, instruct, persuade, and build trust.
• Choose the correct communication channel for your audience and purpose.
• Gather your data.
	• Organize your content using modes such as problem/solution, cause/effect, comparison, argument/persuasion, analysis, chronology.
	• Use figures and tables to clarify content.
	• Format the content for ease of access.
		• Test for usability.
		• Revise your draft by
		  o adding details
		  o deleting wordiness
		  o simplifying words
		  o enhancing the tone
		  o reformatting your text
		  o proofreading and correcting errors

PREWRITING

Prewriting, the first stage of the process, allows you to plan your communication. If you do not know where you are going in the correspondence, you will never get there, and your audience will not get there with you. Through prewriting, you accomplish the following objectives:

- Examining your purposes
- Determining your goals
- Considering your audience
- Gathering your data
- Determining how the content will be provided

Examine Your Purposes

Before you write the document, you need to know why you are communicating. Are you planning to write because you have chosen to do so of your own accord or because you have been asked to do so by someone else? In other words, is your motivation external or internal?

External Motivation. If someone else has requested the correspondence, then your motivation is external. Your boss, for example, expects you to write a monthly status report, a performance appraisal of your subordinate, or a memo suggesting solutions to a current problem. Perhaps a vendor has requested that you write a letter documenting due dates, or a customer asks that you respond to a letter of complaint. In all of these instances, someone else has asked you to communicate.

Internal Motivation. If you have decided to write on your own accord, then your motivation is internal. For example, you need information to perform your job more effectively, so you write a letter of inquiry. You need to meet with colleagues to plan a job, so you write an e-mail message calling a meeting and setting an agenda. Perhaps you recognize a problem in your work environment, so you create a questionnaire and transmit it via the company intranet. Then, analyzing your findings, you call a meeting to report on them. In all of these instances, you initiate the communication.

Audience

See Chapter 4 for more discussion of audience recognition.

Determine Your Goals

Once you have examined why you are planning to communicate, the next step is to determine your goals in the correspondence or presentation. You might be communicating to

- Persuade an audience to accept your point of view.
- Instruct an audience by directing actions.
- Inform an audience of facts, concerns, or questions you might have.
- Build trust and rapport by managing work relationships.

These goals can overlap, of course. You might want to inform by providing an instruction. You might want to persuade by informing. You might want to build trust by persuading. Still, it is worthwhile looking at each of these goals individually to clarify their distinctions.

Figure 2.2 depicts the interrelationship of these four communication goals.

Communicating to Persuade. If your goal in writing is to change others' opinions or a company's policies, you need to be persuasive. For example, you might want to write a proposal, a brochure, or a flier to sell a product or a service. Maybe you will write your annual progress report to justify a raise or a promotion. As a customer, you might want to write a letter of complaint about poor service. Your goal in each of these cases is to persuade an audience to accept your point of view.

FIGURE 2.2 Communication Goals in Practice

Informing the team about the Web site meeting

Instructing the team about their responsibilities prior to the Web design meeting

Persuading the team by helping them understand the importance of this project

Building trust to ensure that people enjoy working together and that everyone feels empowered

---

DATE: April 15, 2008
TO: Web Design Team
FROM: Doug Yost
SUBJECT: WEB SITE IMPLEMENTATION MEETING

Please attend our first Web site implementation meeting, scheduled for April 20, 11:00 A.M.–1:00 P.M. in Room 204.

To ensure productivity, I am asking that each of you prepare the following prior to our meeting:

1. Josh—inventory our stock product photos. Then determine if we will need to upgrade our graphics for the Web site's online shopping cart. Your job also will be to redesign our corporate logo.
2. Tasha—research our competitors. Find out which components of their Web sites we might need to include in ours. More important, determine new screens we could add to make our site unique.
3. Ychun—contact our site administrator to determine optimum load-up time. This will help Josh and the team decide how many graphics to use.
4. Susan—mock up a storyboard for the proposed Web site. Visit with our staff in sales, accounting, human resources, and information technology to get their ideas.

This is an important meeting, as you all know. Without a Web site, our company has fallen behind the competition. Though our local market share is sound, our national and global sales are at least 56 percent below goal. The quick fix for this is a Web site, which will allow us to reach millions of potential clients at a keystroke. With an outstanding Web site, our company's stock should increase, and that will mean bonuses for all.

I have chosen you four employees for this project not only for your expertise but also because of your proven record of excellence. You have worked well together on past projects. I am confident that again you will excel. Thank you for your talents.

P.S. Lunch is on me. I have chosen a vegetarian pasta and salad to accommodate everyone's nutritional needs.

Communicating to Instruct. Instructions will play a large role in your technical communication activities. As a manager, for example, you often will need to direct action. Your job demands that you tell employees under your supervision what to do. You might need to write an e-mail providing instructions for correctly following procedures. These could include steps for filling out employee forms, researching documents in your company's intranet data bank, using new software, or writing reports according to the company's new standards.

As an employee, you also will provide instructions. As a computer information specialist, maybe you work the 1-800 hotline for customer concerns. When a customer calls about a computer crisis, your job would be to give instructions for correcting the problem. You either will provide a written instruction in a follow-up e-mail or a verbal instruction while on the phone.

Communicating to Inform. Often, you will write letters, reports, and e-mails merely to inform. In an e-mail message, for instance, you may invite your staff to an upcoming meeting. A trip report will inform your supervisor what conference presentations you attended or what your prospective client's needs are. A letter of inquiry will inform a vendor about questions you might have regarding her services. Maybe you will be asked to write a newsletter informing your coworkers about the corporate picnic, personnel birthdays, or new stock options available to employees. In these situations, your goal is not to instruct or persuade. Instead, you will share information objectively.

Communicating to Build Trust. Building rapport (empathy, understanding, connection, and confidence) is a vital component of your communication challenge. As a manager or employee, your job is not merely to "dump data" in your written communication. You also need to realize that you are communicating with coworkers, people with whom you will work every day. To maintain a successful work environment, you want to achieve the correct, positive tone in your writing.

This might require nothing more than saying, "Thanks for the information," or "You've done a great job reporting your findings." A positive tone shows approval for work accomplished and recognition of the audience's time. For more detail on audience recognition and involvement, read Chapter 4. Recognizing the goals for your correspondence makes a difference. Determining your goals allows you to provide the appropriate tone and scope of detail in your communication. In contrast, failure to assess your goals can cause communication breakdowns.

Consider Your Audience

What you say and how you say it is greatly determined by your audience. Are you writing up to management, down to subordinates, or laterally to coworkers? Are you speaking to a high-tech audience (experts in your field), a low-tech audience (people with some knowledge about your field), or a lay audience (customers or people outside your work environment)? Face it—you will not write the same way to your boss as you would to your subordinates. You will not speak the same way to a customer as you would to a team member. You must provide different information to a multicultural audience than you would to individuals with the same language and cultural expectations. You must consider issues of diversity when you communicate.

Diversity in Multicultural Audiences

See Chapter 4 for more discussion of diversity in multicultural audiences.

Gather Your Data

Once you know why you are writing and who is your audience, the next step is deciding what to say. You have to gather data. The page or screen remains blank until you fill it with content. Your communication, therefore, will consider personnel, dates, actions required, locations, costs, methods for implementing suggestions, and so forth. As the writer, it is your obligation to flesh out the detail. After all, until you tell your readers what you want to tell them, they do not know.

There are many ways to gather data. In this chapter, and throughout the textbook, we provide options for gathering information. These planning techniques include the following:

- Answering the reporter's questions
- Mind mapping
- Brainstorming or listing
- Outlining
- Storyboarding
- Creating organization charts
- Flowcharting
- Researching (online or at the library)

Research and Documentation

See Chapter 5 for more discussion of research and documentation.

Each is discussed in greater detail in Table 2.1 (except for research techniques which we discuss in Chapter 5). Table 2.2 lists some good Web sites for online research.

Determine How the Content Will Be Provided—The Communication Channel

After you have determined your audience, your goals, and your content, the last stage in prewriting is to decide which communication channel will best convey your message. Will you write a letter, memo, report, e-mail, instant message, Web site, proposal, instructional procedure, flier, or brochure or will you make an oral presentation?

In Table 2.3 you can review the many channels or methods you may use for communicating your content.

Single Sourcing.   Maybe you will create content that will be used in a variety of communication channels simultaneously. Single sourcing is the act of "producing documents designed to be recombined and reused across projects and various media" (Carter 2003, 317). In a constantly changing marketplace, you will need to communicate your content to many different audiences using a variety of communication channels. For instance, you might need to market your product or service using the Internet, a flier, brochure, newsletters, and sales letter. You might need to write hard-copy user manuals and develop online help screens. To ensure that content is reusable, the best approach would be to write a "single source of text" that will "generate multiple documents for different media" (Albers 2003, 337).

DOT-COM UPDATES

For more information about single sourcing, check out the following links.

- "STC Single Sourcing SIG" http://www.stcsig.org/ss/
- "Single Sourcing: An Introduction" http://www.cherryleaf.com/news_single_sourcing.htm
- "Single Sourcing and Its Applicability to Small Projects" http://www.techscribe.co.uk/techw/single_source.htm

TABLE 2.1 Prewriting Techniques

Answering the Reporter's Questions	**Sample Reporter's Questions**	
By answering *who, what, when, where, why,* and *how,* you create the content of your correspondence.	*Who*	Joe Kingsberry, Sales Rep
	What	Need to know • what our discount is if we buy in quantities • what the guarantees are • if service is provided on-site • if the installers are certified and bonded • if Acme provides 24-hour shipping
	When	Need the information by July 9 to meet our proposal deadline
	Where	Acme Radiators 11245 Armour Blvd. Oklahoma City, Oklahoma 45233 Jkings@acmerad.com
	Why	As requested by my boss, John, to help us provide more information to prospective customers
	How	Either communicate with a letter or an e-mail. I can write an e-mail inquiry to save time, but I must tell Joe to respond in a letter with his signature to verify the information he provides.
Mind Mapping	**Picnic MindMap**	
Envision a wheel. At the center is your topic. Radiating from this center, like spokes of the wheel, are different ideas about the topic. Mind mapping allows you to look at your topic from multiple perspectives and then cluster the similar ideas.	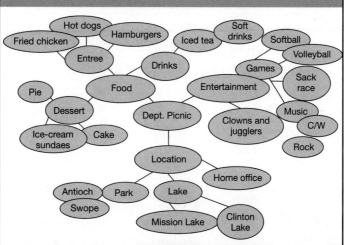	
Brainstorming or Listing	**Improving Employee Morale**	
Performing either individually or with a group, you can randomly suggest ideas (brainstorming) and then make a list of these suggestions. This method, which works for almost all kinds of communication, is especially valuable for team projects.	• Before meetings, ask employees for agenda items (that way, they can feel empowered) • Consider flextime • Review employee benefits packages • Hold yearly awards ceremony for best attendance, highest performance, most cold calls, lowest customer complaints, etc. • Offer employee sharing for unused personal days/sick leave days • Roll over personal days to next calendar year • Include employees in decision-making process • Add more personal days (as a tradeoff for anticipated lower employee raises)	

(Continued)

Outlining

This traditional method of gathering and organizing information allows you to break a topic into major and minor components. This is a wonderful all-purpose planning tool.

Topic Outline

1.0 The Writing Process
   1.1 Prewriting
     • Planning Techniques
   1.2 Writing
     • All-Purpose Organizational Template
     • Organizational Techniques
   1.3 Rewriting
2.0 Criteria for Effective Technical Communication
   2.1 Clarity
   2.2 Conciseness
   2.3 Document Design
   2.4 Audience Recognition
   2.5 Accuracy

Storyboarding

Storyboarding is a visual planning technique that lets you graphically sketch each page or screen of your text. This allows you to see what your document might look like.

Brochure Storyboard

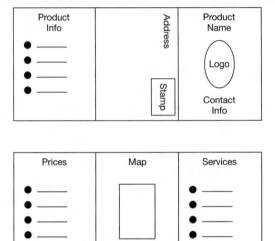

Creating Organization Charts

This graphic allows you to see the overall organization of a document as well as the subdivisions to be discussed.

Organization Chart for Web Site

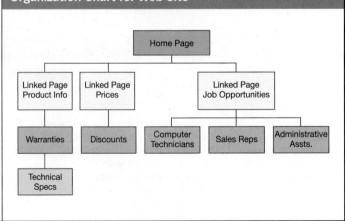

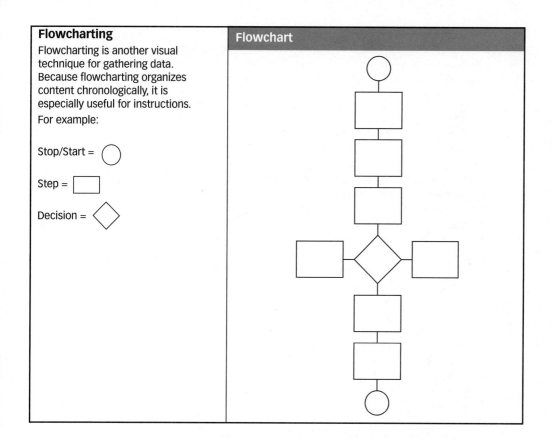

Flowcharting

Flowcharting is another visual technique for gathering data. Because flowcharting organizes content chronologically, it is especially useful for instructions.

For example:

Stop/Start = ◯

Step = ▭

Decision = ◇

Flowchart

TABLE 2.2 Internet Search Engines

Purpose	Sites
Popular online search engines	Yahoo.com, Excite.com, Google.com, AltaVista.com, Lycos.com, alltheweb.com, Ask Jeeves, and HotBot.com
Meta-search engines (multithreaded engines that search several major engines at once)	MetaCrawler.com, Dogpile.com, and Vivisimo.com
Specialty search engines	Findlaw.com focuses on legal resources. Achoo.com lets you access health and medical sites.
Broad academic searches	Librarians' Index to the Internet (http://lii.org) and Infomine.ucr.edu
Business search engines	ZDNet.com, EarthLink.net, Business Week Online (http://www.businessweek.com), and AbusinessResource.com. For information about business news in Great Britain, look at All Search Engines.com (http://www.allsearchengines.co.uk/business_list.htm).
Government search sites	First Gov (http://firstgov.gov/) and Google's Uncle Sam (http://www.google.com/unclesam)
International search sites	Search Engine Colossus, Abyz News Links (international newspapers and magazines), and World Press Review (international perspectives on the United States)
Multipurpose search engines	All Search Engines.com gives you a one-stop search site for exactly what it says: *all search engines* (http://www.allsearchengines.com/).

TABLE 2.3 Options for Providing Content

Communication Channels	Good News/Bad News
E-mail messages and instant messages	*Benefits:* These types of electronic communication are quick and can almost be synchronous. You can have a real-time, electronic chat with one or more readers. Though e-mail messages should be very short (20 or so lines of text), you can attach documents, Web links, graphics, and sound and movie files for review. *Challenges:* E-mail and instant messages tend to be less formal than other types of communication. E-mail might not be private (a company's network administrators can access your electronic communication).
Letters	*Benefits:* Typed on official corporate letterhead stationery, letters are formal correspondence to readers outside your company. *Challenges:* Letters are time consuming because they must be mailed physically. Although you can enclose documents, this might demand costly or bulky envelopes.
Memos	*Benefits:* Memos—internal correspondence to one or several coworkers—allow for greater privacy than e-mail (e-mail can be kept in corporate computer banks and observed by administrators within a company). Even though most memos are limited in length (one or two pages), you can attach or enclose documents. *Challenges:* Memos are both more time consuming than e-mail and less formal than letters.
Reports and proposals	*Benefits:* Reports, internal and external, are usually very formal. They can range in length, from one page to hundreds of pages (proposals and annual corporate reports to stakeholders, for example). Because of their length, reports are appropriate for extremely detailed information. *Challenges:* They can be time consuming to write.
Brochures	*Benefits:* Brochures are appropriate for informal informational and promotional communication to large audiences. *Challenges:* Most brochures are limited to six or so panels, the equivalent of a back and front hard copy. Thus, in-depth coverage of a topic will not occur.
Web sites, intranets, extranets, and blogs	*Benefits:* An Internet Web site or blog can provide informal and public communication to the entire world—anytime, anyplace (with the appropriate technological connections). A company can have a firewall-protected intranet or extranet to allow more private communication for a large, selected audience. Web sites essentially have unlimited size, so you can provide lots of information, and the content can be updated instantaneously by Web designers. A Web site can include links to other sites, animation, graphics, and color. *Challenges:* Audiences need access to the Internet. Blogs could divulge sensitive corporate communication.
Microsoft PowerPoint for oral presentations	*Benefits:* PowerPoint slides enhance written and oral communication, not only making correspondence look more professional but also aiding clarity. A pie chart, bar chart, line graph, or map within a PowerPoint presentation can make complex information more clear. *Challenges:* PowerPoint slides usually convey only key points or a synopsis, rather than very lengthy details.

WRITING

Writing lets you package your data. Once you have gathered your data, determined your objectives, recognized your audience, and chosen the channel of communication, the next step is writing the document. You need to package it (the draft) in such a way that your readers can follow your train of thought readily and can easily access your data. Writing the draft lets you organize your thoughts in some logical, easy-to-follow sequence. Writers usually know where they are going, but readers do not have this same insight. When readers pick up your document, they can read only one line at a time. They know what you are saying at the moment, but they don't know what your goals are. They can only hope that in your writing, you will lead them along logically and not get them lost with unnecessary data or illogical arguments.

Organization

To avoid confusing or misleading your audience, you need to organize your thoughts. As with prewriting, you have many organizational options. In Chapter 3, we discuss organizing according to the following traditional methods of organization.

- Space (spatial organization)
- Chronology
- Importance
- Comparison/contrast
- Problem/solution

These organizational methods are not exclusive. Many of them can be used simultaneously within a memo, letter, report, proposal, or any communication to help your reader follow your train of thought.

Formatting

You also must format your text to allow for ease of access. In addition to organizing your ideas, you need to consider how the text looks on the page. If you give your readers a massive wall of words, they will file your document for future reading and look for the nearest exit. An unbroken page of text is not reader friendly. To invite your readers into the document, to make them want to read the memo, letter, or report, you need to highlight key points and break up monotonous-looking text. You need to ensure that your information is accessible.

Document Design and Formatting

See Chapter 8 for more discussion of document design and formatting.

FAQs

Q: Do writers actually follow a process when they compose correspondence?

A: Most good writers follow a process. It's like plotting your route before a trip. Sure, someone can get in a car without a map, head west (or east or north or south) and find their destination without getting lost, but mapping the route before a trip assures that you won't get lost and waste time.

There's no one way to plot your destination. Prewriting might entail only a quick outline, a few brief notes that list the topics you plan to cover and the order in which you'll cover them. This way, you will know where you're going before you get there.

In addition to creating both brief and sometimes much more detailed outlines, an important part of prewriting is considering your audience. By considering the readers, writers can decide how much detail, definition, or explanation is needed. In fact, thinking about the audience can even help writers determine how many examples or illustrations to include and what details need to be removed from the document.

After prewriting, good writers always perceive of their text as a draft that can be improved. *All* writing can be improved. Improving text requires rewriting. Word processing programs make this essential step in the process easier. Word processing programs let you add, delete, and reformat text. Today, it's impossible to type a document on a word processor without considering the terms highlighted in color (spelling and grammar errors). Thus, editing is an integral part of writing.

REWRITING

Rewriting lets you perfect your writing. After you have prewritten (to gather data, organize your thoughts, and understand your audience) and written your draft, your final step is to rewrite. Revision requires that you look over your draft to determine its usability and correctness.

DOT-COM UPDATES

For more information about usability, check out the following links.

- **Usability First**—a great source for an introduction to usability, plus links and more (http://www.usabilityfirst.com/index.txl)
- **The Usability Professionals' Association**—an international organization focusing on human-centered design. Defines usability, provides case studies, offers more links, etc. (http://www.upassoc.org/)
- **Microsoft Usability**—usability and Web accessibility links, resources, and more case studies (http://www.microsoft.com/usability/default.htm)
- **The Society for Technical Communication's Usability Web Site**—links to usability articles (http://www.stcsig.org/usability/)

Usability Testing

A memo, letter, report, instruction, or Web site is only good if your audience can understand the content and use the information. Usability testing helps you determine the success of your draft. Through usability testing, you decide what works in the draft and what needs to be rewritten. Thus, usability focuses on the following key factors.

- Retrievability—Can the user find specific information quickly and easily?
- Readability—Can the user read and comprehend information quickly and easily?
- Accuracy—Is the information complete and correct?
- User satisfaction—Does the document present information in a way that is easy to learn and remember? (Dorazio 2000)

See the usability checklist on page 35.

Revision Techniques

After testing your document for usability, revise your text by using the following revision techniques.

- *Add* any missing detail for clarity.
- *Delete* dead words and phrases for conciseness.
- *Simplify* unnecessarily complex words and phrases to allow for easier understanding.
- *Move* around information (cut and paste) to ensure that your most important ideas are emphasized.
- *Reformat* (using highlighting techniques) to ensure reader-friendly ease of access.
- *Enhance* the tone and style of the text.
- *Correct* any errors to ensure accurate grammar and content.

We discuss each of these points in greater detail throughout the text.

USABILITY EVALUATION CHECKLIST

Audience Recognition

_____ 1. Are technical terms defined?

_____ 2. Are examples used to explain difficult steps at the reader's level of understanding?

_____ 3. Do the graphics depict correct completion of difficult steps at the reader's level of understanding?

_____ 4. Are the tone and word usage appropriate for the intended audience?

_____ 5. Does a cover page explain for the audience your manual's purpose?

_____ 6. Does the introduction involve the audience and clarify how the reader will benefit?

Development

_____ 1. Are steps precisely developed?

_____ 2. Is all required information provided, including hazards, technical descriptions, warranties, accessories, and required equipment or tools?

_____ 3. Is irrelevant or rarely needed information omitted?

Ease of Use

_____ 1. Can readers easily find what they want because instructions include
- Table of contents
- Glossary
- Hierarchical headings
- Headers and footers
- Index
- Cross-referencing or hypertext links
- Frequently asked questions (FAQs)

Conciseness

_____ 1. Are words, sentences, and paragraphs concise and to the point?

_____ 2. Are the steps self-contained so the reader doesn't have to remember important information from the previous step?

_____ 3. Are the steps cross-referenced to help the reader refer to information provided elsewhere?

Consistency

_____ 1. Is a consistent hierarchy of headings used?

_____ 2. Are graphics presented consistently (same location, same use of figure titles and numbers, similar sizes, etc.)?

_____ 3. Does wording mean the same throughout (technical terms, cautions, warnings, notes, etc.)?

_____ 4. Is the same system of numbering used throughout?

Document Design

_____ 1. Do graphics depict how to perform steps?

_____ 2. Is white space used to make information accessible?

_____ 3. Does color emphasize hazards, key terms, or important parts of a step?

_____ 4. Do numbers or bullets divide steps into manageable chunks?

_____ 5. Are boldface or italics used to emphasize important information?

How Important Is Proofreading?

Do employees in the workplace really care about grammar and mechanics? Is proofreading only important to teachers? Proofreading is absolutely important. Incorrect documentation costs companies money. "A misplaced decimal point resulted in one company paying . . . $120,000 in taxes on a piece of industrial equipment, instead of the $1,200 the firm rightfully owed."

A Chicago-based company purchased an industrial sander for $54,589.62. Unfortunately, when listing the purchase on their year-end taxes, the company reported the purchase price as $5,458,962. This misplaced decimal point equaled a difference of over $5 million. The issue is now in court, of course costing even more money. A single mark of punctuation can be important (Rizzo 2005, A1, A6).

Table 2.4 shows the importance of proofreading. The *National Commission on Writing* highlights what employers and employees consider to be essential skills in technical communication ("Writing: A Powerful Message from State Government" 2005, 19).

Revision is possibly the most important stage in the writing process. If you prewrite effectively (gathering your data, determining your objectives, and recognizing your audience) and write an effective draft, then you are off to a great start. However, if you then fail to rewrite your text, you run the risk of having wasted the time you spent prewriting and writing. Rewriting is the stage in which you make sure that everything is just right. Failure to do so not only can cause confusion for your readers but also can destroy your credibility.

TECHNOLOGY TIPS

Using Microsoft Word 2007 for Rewriting

Word-processing programs help you rewrite your document in many ways.

- **Spell check**—when you misspell a word, often spell check will underline the error in red (as shown in the following example with "grammer" incorrectly spelled). Spell check, unfortunately, will not catch all errors. If you use a word like to instead of too, spell check will not *no* the difference (of course, that should be "know" but spell check did not mark the error). Microsoft Word 2007's **"Review"** tab also provides you access to proofreading help, and allows you to make comments and track changes.

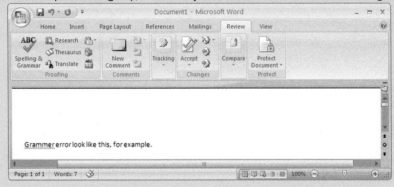

- **Grammar check**—Word processors also can help you catch grammar errors. Grammar check underlines errors in green. When you right-click on the underlined text, the word-processing package will provide an optional correction.

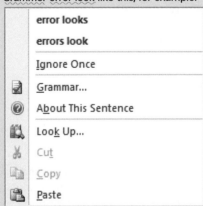

- **Add/Delete**—Word processing makes adding new content and deleting unneeded text very easy. All you need to do is place your cursor where you want to add/delete. Then, to add, you type. To delete, you hit the Backspace key or the Delete key.
- **Move**—The Copy, Cut, and Paste features of word processing allow you to move text with ease.
- **Enhance/Reformat**—In addition to changing the tone of your text, you also can enhance the visual appeal of your document at a keystroke. From the Home tab on your toolbar, you can choose from the Word 2007 Ribbon and include bullets, italics, boldface, font changes, numbered lists, etc.

TABLE 2.4 Essential Skills in Technical Communication

Skills	Extremely Important
Accuracy	87.8%
Spelling, Grammar, Punctuation	71.4%
Clarity	69.4%
Documentation/Support	61.2%
Logical	55.1%
Concise	42.9%
Visual Appeal	12.2%

The process approach to writing—including prewriting, writing, and rewriting (usability testing)—can help you communicate successfully in any work environment or writing situation. In fact, the greatest benefit of process is that it is generic. Process is not designed for any one profession or type of correspondence. No author of a technical communication book can anticipate exactly where you will work, what type of documents you will be required to write, or what your supervisors will expect in your writing. However, we *can* give you a methodology for tackling any communication activity. Writing as a process will help you write any kind of oral or written communication, for any boss, in any work situation.

SPOTLIGHT

An Editor's Use of the Writing Process

Candice Millard, an editor at *Wireless World* and *National Geographic* magazines, and author of *The River of Doubt* sees herself as "the reader's advocate." As an editor, her job is to make sure that readers understand the text she is editing. Candice states, "Because authors invest enormous time and effort in their writing, they often become so attached to their work that they get lost in the details." Ms. Millard's job is to be the detached eye, the objective reader's point of view.

In the writing process, Candice's authors are in charge of the prewriting and writing. Ms. Millard's job focuses on the final stage of the writing process—rewriting. To accomplish this goal, she offers these editorial hints.

- **Start big and get small.** To achieve editorial objectivity, Candice says that she must "step back, get the big picture, and then work toward the details." She asks herself, "What's superfluous, what's confusing, what questions do I have that haven't been answered, and where is clarity needed?"
- **Flesh out the details.** Candice's next job is to "fill in the holes." Sometimes, Candice's biggest challenge is defining her authors' highly technical terminology. When authors depend on jargon, Candice "translates the terms."
- **Slice and dice.** The opposite goal to adding detail for clarity is deleting text. Candice also must "edit out contradictions, weed the irrelevancies, remove the distractions, and excise whatever might ring false."
- **Practice diplomacy.** Candice's authors are all experts in their fields. When these authors write about what they love, every word they use is precious to them; every detail is crucial. Candice, on the other hand, must cut their text to fit space limitations and to meet a reader's limited attention span. Cutting makes "people bristle." Therefore, when editing, she considers how her comments will affect the writer. She "engages in conversation, points out what's good, tempers the bad with the positive, and then weaves in required changes."

As an editor, Candice knows that she isn't just dealing with words; she is working with people.

Prewriting
1. Have you determined the purpose for communicating?
2. Is your goal to inform, instruct, persuade, or build trust?
3. Is your audience high-tech, low-tech, or lay?
4. Have you gathered data through prewriting techniques or re-searching?
5. Have you chosen the correct communication channel?

Writing
6. Have you organized the information (using modes such as spatial, chronology, comparison/contrast, etc.)?
7. Is the content formatted for readability?

Rewriting
8. Have you tested for usability?
9. Have you revised by adding, deleting, simplifying, enhancing tone, and reformatting?
10. Have you proofread for accuracy?

THE WRITING PROCESS AT WORK

Following is a letter produced using the process approach to writing. The document was produced in the workplace by a senior transportation analyst for an international cosmetics firm. He had to write a problem/solution follow-up letter to a sales representative.

Prewriting

A senior transportation analyst received a phone call from an unhappy sales representative. The sales rep had not received a shipment of goods on time, and the shipment was incomplete when it did arrive. While talking to the sales rep, the analyst jotted down notes, as shown in Figure 2.3 (using the listing method of prewriting).

In addition to listing, the transportation analyst used another prewriting technique—reporter's questions. The note tells us *who* the sales rep is (Beth); *what* her Social Security number, phone number, and sales area are; *what* her problem is (late and missing goods); *where* the shipment originated (Denver); *how* much was ordered ($700); and *when* the shipment was due (2 weeks ago). By jotting down this list, the analyst is gathering data. After concluding his discussion with Beth, the analyst contacted his manager to decide

FIGURE 2.3 Listing

```
365-6532

                    Beth Fox
                  449-87-7247
                   Milwaukee

        1. two weeks ago
        2. June $300 short
        3. Troy $700 ordered
        4. Denver
        5. split order
```

FIGURE 2.4 Listing Objectives

Send letter to sales rep.
Send copy to manager.
In letter
 discuss problem encountered.
 show alternative method of shipment
 for better control.
Call manager for further help if needed.

FIGURE 2.5 Rough Draft

Dear Beth—

I appreciate your notifying me of the delay in delivery of your order. Your
orders are coded to ship via Allied Shipping. Allied's stated shipping level for Milwaukee
is next day. ~~However, because of Allied's sorting system,~~ As you have noticed, because of the way Allied
ships packages, it is possible for ~~an~~ a order to become split. a multiple carton To avoid this problem, ~~Alternate~~ an alternative
delivery is possible through ~~an~~ a delivery service we use in your district. To
do this we need an alternative delivery address. ~~Please let me know should you~~
~~need additional~~ If you have any questions or need additional information, please ~~feel free to~~

contact me.

what to do next. This time, the analyst wrote down a list of objectives, as determined by his manager, as shown in Figure 2.4.

The list again answers the reporter's questions: *what* to do (write a letter), *who* gets a copy (manager), *what* to focus on in the letter (we understand your problem; here is an alternative), and *why* to pursue the alternative (better control of shipment).

With data gathered and objectives determined, the analyst was ready to write.

Writing

First, the analyst wrote a rough draft (revising it as he wrote), as shown in Figure 2.5. In this draft, the analyst made subtle changes by adding new detail and deleting unnecessary words. However, he was unsatisfied with this draft, so he tried again (see Figure 2.6).

As is evident from the first two drafts (Figures 2.5 and 2.6), the analyst took the word *rough* seriously. When you draft, do not worry about errors or how the correspondence looks. It is meant to be rough, to free you from worry about making errors. You can correct errors when you revise.

FIGURE 2.6 Second Rough Draft

~~Thank you for your letter regarding the split deliver~~ies.

Thank you for letting me know about the split deliveries of your
                                    last week
campaign 19 orders. Our talk ~~lets me~~ gives me an opportunity not
                          ^

only to expalin the ~~situation problem~~ situation but also to offer help.¶
                                                                        ^
                          Allied   Allied will deliver to your home.
Here's the way ~~the situation~~ works. Allied sorts packages individually
                              ^   ^ However,   packages
rather than as a group. That is, even though we send your ~~orders~~ to

Allied as a unit, ~~they~~ all under your name, it, ~~however,~~ loads its
                  by complete   cartons
trucks not ~~according to~~ order but just as individual ~~boxes~~. ~~Thus,~~
           ^
Because of this,
~~because some~~, occasionally one carton ends up ~~one~~ on one truck while
          carton
the other is shipped separately. You ~~then~~ received such a split order.
         ^

Once the analyst drafted the letter, he typed a clean copy for his manager's approval (Figure 2.7). At this point, the manager added a dateline and added content to the second paragraph. He deleted wordiness in the second and third paragraphs. By deleting the entire fourth paragraph, the manager enhanced the tone of the document (see Figure 2.8).

Rewriting

No writing is ever perfect. Every memo, letter, or report can be improved. Note how the manager improved the analyst's typed draft. When the senior transportation analyst received the revised letter from his manager, he typed and mailed the final version (Figure 2.9).

Once the manager received his copy, he wrote the note you see in the letter's top right corner. When you approach writing as a step-by-step process (prewriting, writing, and rewriting), your results usually are positive—and you will receive positive feedback from your supervisors.

Each company you work for over the course of your career will have its own unique approach to writing memos, letters, and reports. Your employers will want you to do it their way. Company requirements vary. Different jobs and fields of employment require different types of correspondence. However, you will succeed in tackling any writing task if you have a consistent approach to writing. A process approach to writing will allow you to write any correspondence effectively.

FIGURE 2.7 Third Draft

Mrs. Beth Fox
6078 Browntree
Milwaukee, WI 53131

Date has been omitted

Dear Mrs. Fox:

Thanks for letting me know about the split delivery of your Campaign 19 order. Our talk last week gives me an opportunity not only to explain the situation but also to offer help.

Here's the way Allied Shipping works. Allied will deliver to your home; however, Allied sorts packages individually rather than as a group. That is, even though we send your packages as a unit (all under your name), Allied loads its trucks *not by complete order, but just as individual cartons.* Because of this, occasionally, one carton ends up on one truck with another carton shipped separately. You received such a split order. This is an inherent flaw in Allied's system.

Negative tone places blame on a vendor

Because we understand this problem, we have an alternative delivery service for you. Here is our option. Free of charge, you can have your order delivered by our delivery agent, who does not split orders. Our agent, however, will deliver only within a designated area. All we need from you is an alternative address of a friend or relative in the designated delivery area.

I realize that neither of these options is perfect. Still, I wanted to share them with you. Your district manager now can help you decide which option is best for you.

The writer avoids taking responsibility for the problem

Sincerely,

David L. Porter

David L. Porter
Senior Transportation Analyst

FIGURE 2.8 Revised Draft

Mrs. Beth Fox
6078 Browntree
Milwaukee, WI 53131

Dateline?

Dear Mrs. Fox:

Thanks for letting me know about the split delivery of your Campaign 19 order. Our talk last week gives me an opportunity not only to explain the situation but also to offer help.

~~Here's the way Allied Shipping works.~~ Allied will deliver to your home; however, Allied sorts packages individually rather than as a group. ~~That is,~~ even though we send your packages as a unit (all under your name), Allied loads its trucks *not by complete order, ~~but~~* *just as individual cartons.* Because of this, occasionally, one carton ends up on one truck with another carton shipped separately. You received such a split order. This is an inherent flaw in Allied's system.

However, Allied constantly works with us to eliminate these service failures.

whose system has better control of orders

Because we understand this problem, we have an alternative delivery service for you. ~~Here is our option.~~ Free of charge, you can have your order delivered by our delivery agent, ~~who does not split orders~~. Our agent, however, will deliver only within a designated area. ~~All we need from you is an alternative address of a friend or relative in the designated delivery area!~~ *If you would like this service, we will need an alternative delivery address in Milwaukee.* ~~I realize that~~ neither of these options is perfect. Still, I wanted ~~to share them~~ with you. Your district manager ~~now can help~~ you decide which option is ~~best for you.~~

Should you be unable to establish a different delivery address, we will still work with Allied to ensure that you receive home delivery of your complete orders.

Sincerely,

David L. Porter

David L. Porter
Senior Transportation Analyst

FIGURE 2.9 Finished Letter

Carefree Cosmetics
83rd and Preen
Kansas City, MO 64141

September 21, 2005

Mrs. Beth Fox
6078 Browntree
Milwaukee, WI 53131

Great Job, David

Dear Mrs. Fox:

Thanks for letting me know about the split delivery of your Campaign 19 order. Our talk last week gives me an opportunity not only to explain the situation but also to offer help.

Allied will deliver to your home; however, Allied sorts packages individually rather than as a group. Even though we send your packages as a unit (all under your name), Allied loads its trucks not by complete order but by individual cartons. Because of this, occasionally one carton ends up on one truck with another carton shipped separately. You received such a split order. This is an inherent flaw in Allied's system; however, Allied constantly works with us to eliminate these service failures.

Because we understand this problem, we have an alternative delivery service for you. Free of charge, you can have your order delivered by our delivery agent, whose system has better control of orders. Our agent, however, delivers only within a designated area. If you would like this service, we will need an alternate delivery address in Milwaukee.

Should you be unable to establish a different delivery address, we will still work with Allied to ensure that you receive home delivery of your complete orders.

Sincerely,

David L. Porter

David L. Porter
Senior Transportation Analyst

pc: R. H. Handley

CHAPTER HIGHLIGHTS

1. Writing effectively is a challenge for many people. Following the process approach to writing will help you meet this challenge.

2. Prewriting helps you determine your goals, consider your audience, gather your data, examine your purposes, and determine the communication channel.

3. Prewriting techniques will help you get started. Try answering reporter's questions, mind mapping, brainstorming or listing, outlining, storyboarding, creating organization charts, flowcharting, or researching.

4. When you prewrite, you decide whether you are communicating to persuade, instruct, inform, or build trust.

5. To begin writing a rough draft, organize your material, consider the layout and design of the communication, and add visual aids such as tables and figures.

6. You can communicate content through e-mail messages, instant messages, blogging, letters, memos, reports, brochures, proposals, Web sites, and PowerPoint presentations.

7. Perfect your text by testing for usability.

8. Rewrite your document by adding, deleting, simplifying, moving, reformatting, enhancing, and correcting.

9. Proofreading is an essential part of the rewriting step in the writing process. Lack of proofreading causes businesses to lose money.

10. Accuracy is an essential skill in business according to The National Commission on Writing.

APPLY YOUR KNOWLEDGE

CASE STUDIES

1. You are the co-chair of the "Mother's Weekend" at your sorority, fraternity, or other school organization. Using mindmapping and listing, brainstorm the activities, menus, locations, decorations, dates, and fees for this weekend's festivities. Brainstorm the pros and cons of hosting the weekend at your sorority or fraternity house or at a hotel or restaurant.

Assignment

Write an outline showing the decisions you've made regarding the topics above. Then, write a short memo or e-mail to your organization's executive board sharing your findings.

2. You work for the Oneg, Oregon, City Planning Department. Your boss, Carol Haley, has received complaints recently from citizens concerned about a wastewater facility being built in their neighborhood. The homeowners are worried about odors, chemical runoff in nearby Tomahawk Creek, decreases in home values, and a generally diminished quality of life in the neighborhood. The wastewater facility will be built. Despite the citizens' concerns, City Planning has decided that the city needs and will profit from the plant. Nonetheless, you must respond to these complaints, acting upon the citizens' issues.

For odor abatement, the wastewater management company plans to control fumes and particulate matter through the use of cross flow and wet scrubbers, thermal oxidizers, absorption materials, and bio-filters. Many of the concerns regarding runoff and home values can be solved through improved land management and ecological restoration. By planting more reeds, bushes, and trees in the green space between the homes and the proposed plant, runoff can be absorbed more efficiently and green barriers will improve home values. Finally, you have learned that the wastewater company wants to be a good neighbor. To do so, it plans to become actively involved in the community by building more parks, playgrounds, hike/bike trails, and by stocking the nearby pond.

Assignment

In small teams or as individuals, write an e-mail to the boss, Carol Haley, detailing the problems and suggesting solutions. Be sure to consider page layout and space limitation presented by technology.

3. Electronic City is a retailer of DVDs, televisions, CDs, computer systems, cameras, telephones, fax machines, printers, and more. Electronic City needs to create a Web site to market its products and services. The content for this Web site should include the following:

Prices	Store hours	Warranties	Service agreements
Job opportunities	Installation fees	Extended holiday hours	Discounts
Technical support	Product information	Special holiday sales	Delivery fees

Assignment

Review the list of Web site topics above. Using an organizational chart, decide how to group these topics. Which will be major links on the Web site's navigation bar? Which will be topics of discussion within each of the major links?

Once you have organized the links, sketch the Web site by creating a storyboard.

4. You are the special events planner in the Marketing Department at Thrill-a-Minute Entertainment Theme Park. You and your project team need to plan the grand opening of the theme park's newest sensation ride—The Horror—a wooden roller coaster that boasts a 10 g drop. What activities should your team plan to market and introduce this special event?

Assignment

Using at least three of the planning techniques discussed in this chapter, gather ideas for a day-long event to introduce The Horror. Report your findings as follows:

- In the brief report to your Marketing Department boss, explain why you are writing, give options for the event and clarify which techniques you used to gather ideas, and sum up by recommending what you think are the best marketing approaches.
- Write an e-mail to your teacher providing options for the event and explaining which techniques you used to gather data.
- Give an oral presentation in class providing options for the event and explaining which techniques you used to gather data.

INDIVIDUAL AND TEAM PROJECTS

1. To practice prewriting, take one of the following topics. Then, using the suggested prewriting technique, gather data.
   a. **Reporter's questions.** To gather data for your resume, list answers to the reporter's questions for two recent jobs you have held and for your past and present educational experiences.
   b. **Mind mapping.** Create a mind map for your options for obtaining college financial aid.
   c. **Brainstorming or listing.** List five reasons why you have selected your degree program or why you have chosen the school you are attending.
   d. **Outlining.** Outline your reasons for liking or disliking a current or previous job.
   e. **Storyboarding.** If you have a personal Web site, use storyboarding to graphically depict the various screens. If you do not have such a site, use storyboarding to graphically depict what your site's screens would include.

f. **Creating organizational charts.** What is the hierarchy of leadership or management at your job or college organization (fraternity, sorority, club, or team)? To graphically depict who is in charge of what and who reports to whom, create an organizational chart.

g. **Flowcharting.** Create a flowchart of the steps you followed to register for classes, buy a car, or seek employment.

h. **Researching.** Go online or find a hard copy of the *Occupational Outlook Handbook.* Then, research a career field that interests you. Reading the *Occupational Outlook Handbook,* find out the nature of the work, working conditions, employment opportunities, educational requirements, and pay scale.

2. Using the techniques illustrated in this chapter, edit, correct, and rewrite the following flawed memo.

DATE: April 3, 2008
TO: William Huddleston
FROM: Julie Schopper
SUBJECT: TRAINING CLASSES

Bill, our recent training budget has increased beyond our projections. We need to solve this problem. My project team has come up with several suggestions, you need to review these and then get back to us with your input. Here is what we have come up with.

We could reduce the number of training classes, fire several trainers, but increase the number of participants allowed per class. Thus we would keep the same amount of income from participants but save a significant amount of money due to the reduction of trainer salaries and benefits. The downside might be less effective training, once the trainer to participant ratio is increased. As another option, we could outsource our training. This way we could fire all our trainers which would mean that we would save money on benefits and salaries, as well as offer the same number of training sessions, which would keep our trainer to participant ratio low.

What do you think. We need your feedback before we can do anything so even if your busy, get on this right away. Please write me as soon as you can.

PROBLEM-SOLVING THINK PIECE

In an interview, a company benefits manager said that she spent over 50 percent of her workday on communication issues. These included the following:

- Consulting with staff, answering their questions about retirement, health insurance, and payroll deductions
- Meeting weekly with human resources (HR) colleagues
- Collaborating with project team members
- Preparing and writing quarterly reports to HR supervisors
- Teleconferencing with third-party insurance vendors regarding new services and/or costs
- E-mailing supervisors and staff, in response to questions
- Calling and responding to telephone calls
- Faxing information as requested
- Writing letters to vendors and staff to document services

Though she had to use various methods of both written and oral communication, the communication channels each have benefits and drawbacks. E-mailing, for example, has pluses and minuses (convenience over depth of discussion, perhaps). Think about each of the communication options above. Using the table below, list the benefits of each particular type of communication versus the drawbacks.

Communication Channels	Benefits	Drawbacks	Possible Solutions
One-on-one discussions			
Group meetings			
Collaborative projects			
Written reports			
Teleconferences			
E-mail			
Phone calls			
Faxes			
Letters			

WEB WORKSHOP

1. Proofreading is a key component of successful technical communication. Access the following Web sites and read what these sites suggest as editing/proofreading hints. Compare the content to your approaches to proofreading and editing. Write an e-mail message or memo summarizing your findings.
   - Literacy Education Online, http://leo.stcloudstate.edu/acadwrite/genproofed.html
   - Purdue University's Online Writing Lab, http://owl.english.purdue.edu/handouts/general/gl_edit.html
   - University of North Carolina, http://www.unc.edu/depts/wcweb/handouts/proofread.html
2. The Society for Technical Communication provides a link to professional articles about usability testing: http://www.stcsig.org/usability/. Read any of the articles found in this Web site and report on your findings in an e-mail message or memo.

QUIZ QUESTIONS

1. What are the three main parts of the writing process?
2. What are four ways you can provide technical communication content?
3. What can you achieve by prewriting?
4. What is the difference between external and internal motivation?
5. Why should you consider your audience before you begin writing?
6. What are four different prewriting techniques?
7. Why do you consider format when you write a business document?
8. What are four rewriting techniques?
9. What happens when you fail to revise accurately?
10. How can the writing process help ensure that you become a successful writer?
11. What are four goals of technical communication?
12. What are three search engines?
13. How do reporters' questions differ from mind mapping?
14. What is usability testing?
15. How can software help you to rewrite your documents?

CHAPTER 3

The Goals of Technical Communication

COMMUNICATION *at work*

In the CompToday scenario, employees address the needs of an audience by writing text that is clear, concise, accurate, and ethical.

CompToday is a computer hardware company located in Boston, Massachusetts. Like all publicly traded companies, CompToday must meet the Sarbanes-Oxley Act, passed by Congress in response to accounting scandals following the Enron, Tyco, and WorldCom illegalities. As of June 14, 2004, this act makes all corporate executives responsible for their companies' accounting practices. Penalties for failing to abide by this act are severe, including heavy fines as well as the potential for prison sentences.

CompToday, like many other companies, realizes that maintaining ethics and legalities in business is not just a matter of being a good citizen in the community. Maintaining ethics in the work environment demands a great deal of written communication. A key component of the Sarbanes-Oxley Act, combating ethical problems in the workplace, is the need for extensive documentation, including the following:

- Written policies provided for all corporate employees, clarifying in low-tech terms the intent of the act as well as its standards
- Clear instructional procedures (hard-copy and online) guiding employees in the steps they must follow to abide by Sarbanes-Oxley

When you complete this chapter, you will be able to

1. Achieve clarity by providing details and answering reporter's questions.
2. Delete wordiness in sentences and paragraphs to achieve conciseness.
3. Simplify word usage.
4. Learn proofreading techniques to ensure accuracy.
5. Organize your thoughts to help readers better understand documents.
6. Communicate ethically orally, in writing, and through graphics.
7. Evaluate your technical communication for clarity, conciseness, accuracy, organization, and ethics using a checklist.

- Concise online help screens providing answers to FAQs (Frequently Asked Questions) relevant to Sarbanes-Oxley
- Accurate monthly reports documenting key financial activities, such as accounts payables, cash disbursement, purchases, and earnings
- Annual reports submitted to and signed by outside auditors

To achieve this daunting goal, CompToday's chief executive officer William Huddleston has reassigned a number of technical communicators from his Corporate Communication Department. They had been working on other projects, such as the company's Web site, instructional manuals, and online help screens. Now, they are tasked with new responsibilities—creating the documentation required by CompToday to meet the Sarbanes-Oxley requirements.

This hurts the company's bottom-line profit margin. After all, new writers must be hired to complete unfinished writing projects while the reassigned technical communicators work toward compliance with the act. Maintaining an ethical workplace can be costly. However, practicing good ethics and abiding by the law benefits everyone—employees, stockholders, clients, and vendors.

Source: Adapted from Harkness 2004, 16–18.

Check Online Resources

www.prenhall.com/gerson
For more information about clarity, conciseness, accuracy, organization, and ethics, visit our companion Web site.

Check out our quarterly newsletters TechCom E-Notes at www.prenhall.com/gerson for dot.com updates, new case studies, insights from business professionals, grammar exercises, and facts about technical communication.

ACHIEVING CLARITY IN TECHNICAL COMMUNICATION

The ultimate goal of good technical communication is clarity. If you write a memo, letter, report, user manual, or Web site that is unclear to your readers, problems can occur. Unclear technical communication can lead to missed deadlines, damaged equipment, inaccurate procedures, incorrectly filled orders, or danger to the end user. To avoid these problems and many others caused by unclear communication, write to achieve clarity.

Provide Specific Detail

One way to achieve clarity is by supplying specific, quantified information. If you write using vague, abstract adjectives or adverbs, such as *some* or *recently*, your readers will interpret these words in different ways. The adverb *recently* will mean 30 minutes ago to one reader, yesterday to another, and last week to a third reader. This adverb, therefore, is not clear. The same applies to an adjective like *some*. You write, "I need some information about the budget." Your readers can only guess what you mean by *some*. Do you want the desired budget increase for 2008 or the budget expenditures for 2007?

Look at the following example of vague writing caused by imprecise, unclear adjectives. (Vague words are underlined.)

BEFORE

Our <u>latest</u> attempt at molding preform protectors has led to <u>some</u> positive results. We spent <u>several</u> hours in Dept. 15 trying different machine settings and techniques. <u>Several</u> good parts were molded using two different sheet thicknesses. Here's a summary of the findings.

First, we tried the <u>thick</u> sheet material. At 240°F, this thickness worked well. Next, we tried the <u>thinner</u> sheet material. The <u>thinner</u> material is less forgiving, but after a <u>few</u> adjustments we were making good parts. Still, the <u>thin</u> material caused the most handling problems.

The engineer who wrote this report realized that it was unclear. To solve the problem, she rewrote the report, quantifying the vague adjectives.

AFTER

During the week of 10/4/08, we spent approximately 12 hours in Dept. 15 trying different machine settings, techniques, and thicknesses to mold preform mold protectors. Here is a report on our findings.

<u>0.030" Thick Sheet</u>
At 240°F, this thickness worked well.
<u>0.015" Thick Sheet</u>

This material is less forgiving, but after decreasing the heat to 200°F, we could produce good parts. Still, material at 0.015" causes handling problems.

Your goal as a technical communicator is to express yourself clearly. To do so, state your exact meaning through specific, quantified word usage (measurements, dates, monetary amounts, and so forth).

Answer the Reporter's Questions

A second way to write clearly is to answer the reporter's questions—who, what, when, where, why, and how. The best way we can emphasize the importance of answering these reporter's questions is by sharing with you the following memo, written by a highly placed executive, to a newly hired employee.

BEFORE

Date: March 5, 2008
To: Staff
From: Earl Eddings, Manager
Subject: Research

Please prepare to plan a presentation on research. Make sure the information is very detailed. Thanks.

That's the entire memo. The questions are, "What doesn't the newly hired employee know?" "What additional information would that employee need to do the job?" "What needs clarifying?"

To achieve successful communication, the writer needs to answer reporter's questions. *What* is the subject of the presentation and the research? *Who* is the audience? The word "Staff" is too encompassing. Will all of the staff be involved in this project? *Why* is the presentation being made? That is, what is the rationale or motivation for this presentation? *When* will the presentation be made, and *how* much detail is "very detailed"? *Where* will the presentation take place? The audience for this memo has a right to ask for clarity on one more point: *what* exactly is the reader supposed to do? Will the reader of this memo make a presentation, plan a presentation, or prepare to plan a presentation?

In contrast, the "After" memo achieves clarity by answering reporter's questions.

AFTER

Date: March 5, 2008   What
To: Melissa Hider
From: Earl Eddings   Who
Subject: Research for Homeland Security Presentation

Please make a presentation on homeland security for the Weston City Council. This meeting is planned for March 18, 2008, in Conference Room C, from 8:00 a.m.-5:00 p.m.   When   Where

We have a budget of $6,000,000. Thus, to use these funds effectively, our city must be up to date on the following concerns:

1. Bomb-detection options
2. Citizen preparedness   Why
How
3. Defense personnel training

Use our new PowerPoint software to make your presentation. With your help, I know Weston and the KC metro area will benefit.

Melissa knows *where* the meetings will be held (Conference Room C); *who* she will be speaking to (Weston City Council); *what* she will accomplish (making a presentation about homeland security); *when* the meeting will occur (March 18, 2008); *why* she's involved (to determine how to use funds effectively and make a presentation about her findings); and *how* the presentation will be most effective (the use of PowerPoint to create an effective presentation).

Use Easily Understandable Words

Another key to clarity is using words that your readers can understand easily. Avoid obscure words and be careful when you use acronyms, abbreviations, and jargon.

Avoiding Obscure Words. A good rule of thumb is to *write to express, not to impress; write to communicate, not to confuse.* If your reader must use a dictionary, you are not writing clearly.

Read the following unclear example.

---

example

Words like "nonduplicatable" and phrases like "execute without abnormal termination" are hard to understand. For clarity, write "do not duplicate secure messages" and "JCL system testing will ensure that applications continue to work."

> The following rules are to be used when determining whether or not to duplicate messages.
>
> - Do not duplicate nonduplicatable messages.
> - A message is considered nonduplicatable if it has already been duplicated.
>
> Your job duties will be to ensure that distributed application modifications will execute without abnormal termination through the creation of production JCL system testing.

Following is a list of difficult, out-of-date terms and the modern alternatives.

Obscure Words	Alternative Words
aforementioned	already discussed
initial	first
in lieu of	instead of
accede	agree
as per your request	as you requested
issuance	send
this is to advise you	I'd like you to know
subsequent	later
inasmuch as	because
ascertain	find out
pursuant to	after
forward	mail
cognizant	know
endeavor	try
remittance	pay
disclose	show
attached herewith	attached
pertain to	about
supersede	replace
obtain	get

Defining Acronyms, Abbreviations, and Jargon. In addition to obscure, unclear words, a similar obstacle to readers is created by acronyms, abbreviations, and jargon.

We have all become familiar with common acronyms such as *scuba* (<u>s</u>elf-<u>c</u>ontained <u>u</u>nderwater <u>b</u>reathing <u>a</u>pparatus), *radar* (<u>ra</u>dio <u>d</u>etecting <u>a</u>nd <u>r</u>anging), *NASA* (<u>N</u>ational <u>A</u>eronautics and <u>S</u>pace <u>A</u>dministration), *FICA* (<u>F</u>ederal <u>I</u>nsurance <u>C</u>ontributions <u>A</u>ct), and *MADD* (<u>M</u>others <u>A</u>gainst <u>D</u>runk <u>D</u>riving)—single words created from the first letters of multiple words. We are comfortable with abbreviations like *FBI* (<u>F</u>ederal <u>B</u>ureau of <u>I</u>nvestigation), *JFK* (<u>J</u>ohn <u>F</u>. <u>K</u>ennedy), *NFL* (<u>N</u>ational <u>F</u>ootball <u>L</u>eague), *IBM* (<u>I</u>nternational <u>B</u>usiness <u>M</u>achines), and *LA* (<u>L</u>os <u>A</u>ngeles). Some jargon (in-house language) has become so common that we reject it as a cliché. Baseball jargon is a good example. It is hard to tolerate sportscasters who speak baseball jargon, describing line drives as "frozen ropes" and fast balls as "heaters."

More often than not, acronyms, abbreviations, and jargon cause problems, not because they are too common but because no one understands them. Your technical communication loses clarity if you depend on them. You have to decide when to use acronyms, abbreviations, and jargon and how to use them effectively. One simple rule is to define your terms. You can do so either parenthetically or in a glossary. Rather than just writing *CIA*, write *CIA* (*Cash in Advance*). Such parenthetical definitions, which are only used once per correspondence, don't take a lot of time and won't offend your readers. Instead, the result will be clarity.

If you use many potentially confusing acronyms or abbreviations, or if you need to use a great deal of technical jargon, then parenthetical definitions might be too cumbersome. In this case, supply a separate glossary.

A *glossary* is an alphabetized list of terms, followed by their definitions, as in Figure 3.1.

FIGURE 3.1   Glossary

CPA	Certified Public Accountant
FICA	Federal Insurance Contributions Act (Social Security taxes)
Franchise	Official establishment of a corporation's existence
Gross pay	Pay before deductions
Line of credit	Amount of money that can be borrowed
Net pay	Pay after all deductions
Profit and loss statement	Report showing all incomes and expenses for a specified time

Use Verbs in the Active Voice Versus the Passive Voice.

BEFORE

It has been decided that Joan Smith will head our Metrology Department.

The preceding sentence is written in the *passive voice* (the primary focus of the sentence, *Joan Smith*, is acted on rather than initiating the action). Passive voice causes two problems.

1. Passive constructions are often unclear. In the preceding sentence, *who* decided that Joan Smith will head the department? To solve this problem and to achieve clarity, replace the vague indefinite pronoun *it* with a precise noun: "Kin Norman decided that Joan Smith will head our Metrology Department."
2. Passive constructions are often wordy. Passive sentences always require helping verbs (such as *has been*). In the "after," revised sentence, the helping verb has been disappears.

AFTER

Kin Norman decided that Joan Smith will head our Metrology Department.

The revision ("Kin Norman decided that Joan Smith will head our Metrology Department") is written in the *active voice*. When you use the active voice, your subject *(Kin Norman)* initiates the action.

Another common problem with passive voice construction concerns prepositions. Look at the following "before" example.

BEFORE

Passive voice created when an inanimate object ("overtime") takes precedence over people ("workers")

→ Overtime is favored by hourly workers.

Again, this sentence, written in the passive voice, uses a helping verb *(is)*, has the doer *(hourly workers)* acted on rather than initiating the action, and also includes a preposition *(by)*. The sentence is wordy. Revised, the sentence reads as follows:

AFTER

Active voice created when people take precedence over inanimate objects

→ Hourly workers favor overtime.

By omitting the helping verb *is* and deleting the preposition *by*, the sentence is less wordy and more precise. Occasionally, you can use the passive voice when an individual is less important than an inanimate object.

EXAMPLE

The individual, *salesperson*, is not as important as the inanimate object, *software*.

→ The tax preparation software can be learned easily even when the user is new to our CPA firm.

You might also use passive voice when the individual is unknown.

EXAMPLE

The individual, *user*, is unnamed.

→ The tax preparation software can be learned easily even when the user is new to our CPA firm.

The importance of clarity, conciseness, and accuracy

How important is clarity, conciseness, and accuracy in technical communication? For answers to this question, look at the varied communication channels used by Joel Blobaum, Community Relations Manager at The Missouri Department of Transportation (MoDOT). MoDOT, which plans, builds, inspects, and maintains over 32,000 miles of roadway, is the sixth largest state highway system in the nation.

As Community Relations Manager, Joel conducts public meetings and writes newsletters, news releases, and daily work zone reports. His audience includes approximately 1,200 current and retired MoDOT employees and thousands of external readers, consisting of citizens, public works employees, and local and state elected officials.

- Elected officials base legislative decisions and funding allocations on Joel's documentation.
- Business consultants and vendors depend on MoDOT news for job and bidding opportunities.
- Community members engage in meetings and surveys to provide feedback for MoDOT construction plans.

In his hard-copy, electronic, and online communication, Joel must be clear, concise, and accurate. Why? Look at the following clear and concise documentation from Joel's daily "Road Zones Today" updates, provided online in MoDOT's Web site:

I-70: Shoulder-repair crews will work in the left lane of westbound I-70 between I-470 and Lee's Summit Road from 9 A.M. to 3 P.M. Friday. Bridge-maintenance crews will close the right lane of westbound I-70 between Grain Valley and west of Adams Dairy Parkway from 9 A.M. to 3 P.M. Monday.

Route 291: Pavement-repair crews will work on southbound Route 291 between Kentucky Road and Route 78 from 9 A.M. to 2 P.M. weekdays through March 16, and on northbound Route 291 between Route 78 and Route 24 from 9 A.M. to 2 P.M. weekdays March 12–30.

I-35 at Route 69: Bridge-maintenance crews will close the right lane of northbound I-35 between Route 69 and north of I-435 from 9 A.M. to 3 P.M. Friday.

In each of the examples above, Joel writes short paragraphs, provides specific details, and answers the reporter's questions (who, what, when, where, why, and how). His audience can access this information quickly and use the news to plot their driving routes around the city. Thus, Joel's content must be accurate. If his information is incorrect, and people get snarled in traffic, bad things can occur—accidents, delays, and poor public relations.

DOT-COM UPDATES

For more information about active versus passive voice, check out the following link.

- http://www.plainlanguage. gov/whatisPL/ govma dates/memo.cfm. President Clinton's "Presidential Memorandum on Plain Language" mandates plain language in government documents, focusing specifically on clarity and conciseness.

SIMPLIFYING WORDS, SENTENCES, AND PARAGRAPHS FOR CONCISENESS

A second major goal in technical communication is conciseness, providing detail in fewer words. Conciseness is important for at least three reasons.

Conciseness Saves Time

Technical communication in the workplace is time consuming. American workers spend on average 31 percent of their time writing. Therefore, conciseness in writing can help save some of this time.

Conciseness Aids Clarity

Concise writing can aid comprehension. If you dump an enormous number of words on your readers, they might give up before finishing your correspondence or skip and skim so much that they miss a key concept. Wordy writing will lead your readers to think, "Oh no! I'll never be able to finish that. Maybe I can skim through it. I'll probably get enough information that way." Conciseness, on the other hand, makes your writing more appealing to your readers. They'll think, "Oh, that's not too bad; I can read it easily." If they can read your correspondence easily, they will read it with greater interest and involvement. This, of course, will aid their comprehension and increase readability—the audience's ability to easily read the text.

Technology Demands Conciseness

Technology is impacting the size of your technical communication. The size of the screen makes the difference, and screen sizes are shrinking. Thus, when you write, you need to consider the way in which technology limits your space. Today, more and more, effective technical communication must be concise enough to *fit in a box*.

Notice how the size of the "box" containing the following communication affects the way you package your content (see Figure 3.2 and Table 3.1).

FIGURE 3.2   The Shrinking Size of Technical Communication

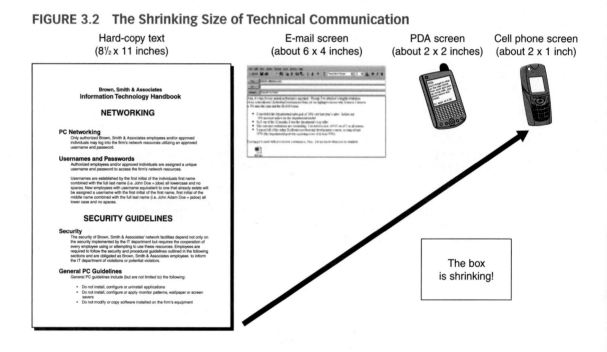

TABLE 3.1 Size Characteristics of Communication Channels

Communication Choice	Overall Size of the Screen	Lines per Page or Screen	Characters per Line
Hard-copy paper	8½ × 11 inches	55	70–80
E-mail screen	4 × 6 inches	20–22	60–70
PDA screen	2 × 2 inches	10–15	30–35
Cell phone screen	2 × 1 inch	4–6	10–15

Resumes.   A one-page resume is standard. One hard-copy page measures $8\frac{1}{2} \times 11$ inches. That is a box. In fact, this box (one page of text) allows for only about 55 lines of text, and each line of text allows for only about 70 to 80 characters. (A "character" is every letter, punctuation mark, or space.)

E-mail Messages.   In contrast, much of today's written communication in the workplace is accomplished through e-mail messages. In Chapter 1, we quote statistics stating that close to 100 percent of business employees report using e-mail every day ("Writing: A Ticket to Work," 2007).

Because e-mail screens tend to be smaller (around only 4×6 inches) than the typical computer screen, e-mails should be brief and concise. Yes, you can scroll an e-mail message endlessly, but no one wants to do that. In fact, the reason that readers like a one-page resume is that one page allows for what is known as the "W-Y-S-I-W-Y-G" factor ("What You See Is What You Get"). Readers like to see what they will be getting in the correspondence. In contrast, if you make your readers scroll endlessly in e-mail, they do not see what they get. This causes problems. Therefore, a good e-mail message should fit in the box, letting the reader see the entire content at one glance. It should be limited to about 20 lines of text.

Multimedia Messaging Equipment.   The technological impact is even more dramatic when you consider the screen size for handheld multimedia messaging equipment such as PDAs, cell phones, and combination products. The Verizon enV™ screen display allows for 11 lines of text. The Palm Treo Smartphone™ and Blackberry have screen sizes that are about 2×2 inches. In "Serving the Electronic Reader," Linda E. Moore says, "More and more e-readers are accessing documents using cell phones, PDAs, and other wireless devices" (2003, 17). Technical communicators in the next decade will "have to figure out how to create content that works on a four-line cell phone screen" (Perlin 2001, 4–8).

Online Help Screens.   Another example of technical communication that must fit in a box is online help screens. If you work in information technology or computer sciences, you might need to create or access online help screens. If you do so, then your text will be limited by the size of these screens.

Microsoft PowerPoint.   Finally, you also must fit your technical communication within a box (or boxes) when you use Microsoft PowerPoint software.

Technology requires that you write concisely. If you are filling out an online form, you will be limited by the size of the form's fields. If you are writing correspondence that will be viewed as a PowerPoint presentation, an online help screen, or e-mail sent to someone's PDA or cell phone, you need to limit the size of your text. Therefore, as you write, consider the impact of technology and write concisely.

Online Help

See Chapter 13 for more discussion of online help.

PowerPoint

See Chapter 18 for more discussion of PowerPoint.

Limit Paragraph Length

Let's look at some poor writing—writing that is wordy, time consuming to read, and not easily comprehensible.

BEFORE

Please prepare to supply a readout of your findings and recommendations to the officer of the Southwest Group at the completion of your study period. As we discussed, the undertaking of this project implies no currently known incidences of impropriety in the Southwest Group, nor is it designed specifically to find any. Rather, it is to assure ourselves of sufficient caution, control, and impartiality when dealing with an area laden with such potential vulnerability. I am confident that we will be better served as a company as a result of this effort.

This "before" paragraph is not easy to understand. The difficulty is caused by the lengthy paragraph, words, and sentences. An excessively long paragraph is ineffective. In a long paragraph, you force your reader to wade through many words and digest large amounts of information. This hinders comprehension. In contrast, short, manageable paragraphs invite reading and help your reader understand your content.

A paragraph in a technical document should consist of no more than four to six typed lines, or no more than 50 words. Sometimes you can accomplish these goals by cutting your paragraphs in half; find a logical place to stop a paragraph and then start a new one. Even the previous poorly written example can be improved in this way.

AFTER

Please prepare to supply a readout of your findings and recommendations to the officer of the Southwest Group at the completion of your study period. As we discussed, the undertaking of this project implies no currently known incidences of impropriety in the Southwest Group, nor is it designed specifically to find any.

Rather, it is to assure ourselves of sufficient caution, control, and impartiality when dealing with an area laden with such potential vulnerability. I am confident that we will be better served as a company as a result of this effort.

The writing is still difficult to understand, but at least it is a bit more manageable. You can read the first paragraph, stop, and consider its implications. Then, once you have grasped its intent, you can read the next paragraph and try to tackle its content. The paragraph break gives you some room to breathe.

Limit Word and Sentence Length

In addition to the length of the example paragraph, the writing is flawed because the paragraph is filled with excessively long words and sentences. This writer has created an impenetrable wall of haze—the writing is foggy. In fact, we can determine how foggy this prose is by assessing it according to Robert Gunning's fog index.

Using the Fog Index.   The following is Gunning's mathematical way of determining how foggy your writing is.

1. Count the number of words in successive sentences. Once you reach approximately 100 words, divide these words by the number of sentences. This will give you an average number of words per sentence.

2. Now count the number of long words within the sentences you have just reviewed. Long words are those with three or more syllables. You cannot count (a) proper names, like Leonardo DaVinci, Christopher Columbus, or Alexander DeToqueville; (b) long words that are created by combining shorter words, such as *chairperson* or *firefighter*; or (c) three-syllable verbs created by *–ed* or *–es* endings, such as *united* or *arranges*. Discounting these exceptions, count the remaining multisyllabic words. (A good example of a multisyllabic word is the word *mul-ti-syl-lab-ic*.)

3. Finally, to determine the fog index, add the number of words per sentence and the number of long words. Then multiply your total by 0.4.

Given this system, let's see how the original difficult paragraph (page 57) scores. The paragraph is composed of 92 words in four sentences. Thus, the average number of words per sentence is 23. The paragraph contains 16 multisyllabic words (*recommendations, officer, completion, period, undertaking, currently, incidences, impropriety, specifically, sufficient, impartiality, area, potential, vulnerability, confident*, and *company*).

23	(words per sentence)
+16	(multisyllabic words)
39	(total)
39	(total)
× 0.4	(fog factor)
15.6	(fog index)

What does a fog index of 15.6 mean? Look at Table 3.2. It shows that the paragraph is written at a level midway between college junior and senior, definitely above the danger line.

A fog index of 15.6 enters the danger zone for two reasons.

- Approximately 28 percent of Americans graduate from college ("College Degree" 2005). Thus, if you are writing at a college level, you could be alienating approximately 72 percent of your audience.

- College graduates do not read well. According to the National Assessment of Adult Literacy, "The average American college graduate's literacy in English declined significantly over the past decade." Only 31 percent of college graduates could read lengthy, complex English texts and draw complicated inferences (Dillon 2005).

Given these facts, many businesses ask their employees to write at a sixth- to eighth-grade level. To accomplish this, you would have to strive for an average of approximately 15 words per sentence and no more than 5 multisyllabic words per 100 words.

15 (words per sentence) + 5 (multisyllabic words) = 20 × 0.4 (fog factor)
= 8.0 (fog index level)

You cannot always avoid multisyllabic words. Scientists would find it impossible to write if they could never use words like *electromagnetism*, *nitroglycerine*, *telemetry*, or *trinitrotolulene*. The purpose of a fog index is to make you aware that long words and sentences create reading problems. Therefore, although you cannot always avoid long words, you should be careful when using them. Similarly, you cannot always avoid lengthy sentences. However, try not to rely on sentences over 15 words long. Vary your sentence lengths, relying mostly on sentences less than 15 words long.

TABLE 3.2   Fog Index and Reading Level

	Fog Index	By Grade	By Magazine
	17	College graduate	No popular magazine scores this
	16	College senior	high.
	15	College junior	
	14	College sophomore	
Danger Line	13	College freshman	
	12	High school senior	*Atlantic Monthly*
	11	High school junior	*Time* and *Newsweek*
	10	High school sophomore	*Reader's Digest*
	9	High school freshman	*Good Housekeeping*
	8	Eighth grade	*Ladies' Home Journal*
	7	Seventh grade	Modern romances
	6	Sixth grade	Comics

Using Microsoft Word 2007 to Check the Readability Level of Your Text

Microsoft Word 2007 allows you to check the "readability" level of your writing. Doing so will let you know the following:

- How many words you've used
- How many sentences you've written
- How many words per sentence
- If you've used passive voice constructions
- The grade level of your writing

1. Click on the **Microsoft Office Button**, and then click **Word Options**.

2. Click on **Proofing**.

3. When the following screen pops up, select **Check grammar with spelling** and **Show readability statistics**. Then click **OK**.

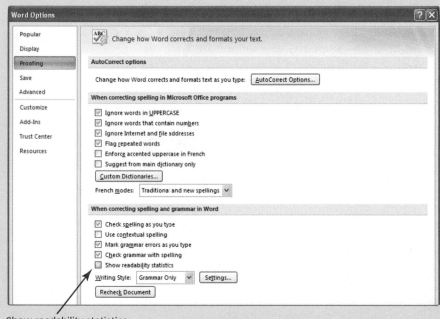

Show readability statistics

Once you have enabled this readability feature, open a file and check the spelling. When Word has finished checking the spelling and grammar, you will see a display similar to the one below.

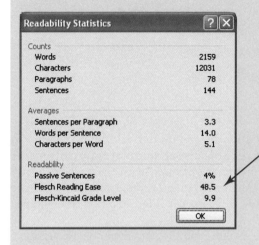

In this example, the text consisted of 2159 words, 78 paragraphs, and 144 sentences. This averaged out to 14 words per sentence, equaling about 10th grade level writing. Four percent of the sentences were written in passive voice.

Use the Meat Cleaver Method of Revision.   One way to limit the number of words per sentence is to cut the sentence in half or thirds. The following sentence, which contains 44 words, is too long.

BEFORE

To maintain proper stock balances of respirators and canister elements and to ensure the identification of physical limitations that may negate an individual's previous fit-test, a GBC-16 Respirator Request and Issue Record will need to be submitted for each respirator requested for use.

AFTER

Please submit a GBC-16 Respirator Request and Issue Record for each requested respirator. We then can maintain proper respirator and canister element stock balances. We also can identify physical limitations that may negate an individual's previous fit-test.

Using the meat cleaver approach makes this sentence more concise and easier to understand. The "before" sentence, now rewritten as three sentences, is easier to read.

Avoid Shun Words.   In the preceding examples, the original sentence contained 44 words; the revised version is composed of three sentences totaling 38 words. Where did the missing six words go?

One way to write more concisely is to shun words ending in *–tion* or *–sion*—words ending in a "shun" sound. For example, the original sentence reads "to ensure the identifica*tion* of physical limita*tions*." To revise this, you could write, "identify physical limitations." That's three words versus seven words in the original version. Deleting four words in this case reduces wordiness by over 50 percent. "Shun" words are almost always unnecessarily wordy.

Let's try another example. Instead of writing, "I want you to take into considera*tion* the following," you could write, "Consider the following." That is three words versus nine, a 66.6 percent savings in word count.

Look at the following "shun" words and their concise versions:

Shun Words	Concise Versions
came to the conclu*sion*	concluded (or decided)
with the excep*tion* of	except for
make revi*sions*	revise
investiga*tion* of the	investigate
consider implementa*tion*	implement
utiliza*tion* of	use

Avoid Camouflaged Words. Camouflaged words are similar to "shun" words. In both instances, a key word is buried in the middle of surrounding words (usually helper verbs or unneeded prepositions). For example, in the phrase "with the exception of," the key word *except* is camouflaged behind the unneeded *with*, *the*, *-tion*, and *of*. Once we prune away these unneeded words, the key word *except* is left, making the sentence less wordy.

Camouflaged words are common. Here are some examples and their concise versions.

Camouflaged Words	Concise Versions
make an *amend*ment to	amend
make an *adjust*ment of	adjust
have a *meet*ing	meet
*thank*ing *you* in advance	thank you
for the purpose of *discuss*ing	discuss
arrive at an *agree*ment	agree
at a *later* moment	later

Avoid the Expletive Pattern. Another way to write more concisely is to avoid the following expletives.

- *there* is, are, was, were, will be
- *it* is, was

Both these expletives (*there* and *it*) lead to wordy sentences. For example, consider the following sentence.

---

example | There are three people who will work for Acme.

This sentence can be revised to read, "Three people will work for Acme." The original sentence contains nine words; the revision has six words. Deleting the expletive *there* omits three words. Deleting three words in one sentence might not seem like an achievement; however, if you can delete three words from every sentence, the benefits add up.

The expletive *it* creates similar wordiness, as in the following sentence.

---

example | It has been decided that ten engineers will be hired.

By deleting the expletive *it*, the sentence reads, "Ten engineers will be hired." The original sentence contained ten words; the revision has only five words.

Omit Redundancies.   Redundancies are words that say the same thing. Conciseness is achieved by saying something once rather than twice. For example, in each of the following instances, the boldface words are redundant.

during **the year of** 2008
>   (Obviously 2008 is a year; the words *the year of* are redundant.)

in **the month of** December
>   (As in the preceding example, *the month of* is redundant; what else is December?)

needless to say
>   (If it's needless to say, why say it?)

The computer will cost **the sum of** $1,000.
>   (One thousand dollars *is* a sum.)

the results **so far achieved** prove
>   (A result, by definition, is something that has been achieved.)

our **regular** monthly status reports require
>   (Monthly status reports must occur every month; regularity is a prerequisite.)

We collaborated **together** on the project.
>   (One can't collaborate alone!)

the **other** alternative is to
>   (Every alternative presumes that some option exists.)

This is a **new** innovation.
>   (As opposed to an old innovation?)

the consensus **of opinion** is to
>   (The word *consensus* implies opinion.)

Avoid Wordy Phrases.   Sentences may be wordy not because you have been redundant or because you have used "shun" words, camouflaged words, or expletives. Sometimes sentences are wordy simply because you've used wordy phrases.

Here are examples of wordy phrases and their concise revisions.

Wordy Phrases	Concise Revisions
in order to purchase	to buy
at a rapid rate	fast (or state the exact speed)
it is evident that	evidently
with regard to	about
in the first place	first
a great number of times	often (or state the number of times)
despite the fact that	although
is of the opinion that	thinks
due to the fact that	because
am in receipt of	received
enclosed please find	enclosed is
as soon as possible	by 11:30 A.M.
in accordance with	according to
in the near future	soon
at this present writing	now
in the likely event that	if
rendered completely inoperative	broken

ACHIEVING ACCURACY
IN TECHNICAL COMMUNICATION

Clarity and conciseness are primary objectives of effective technical communication. However, if your writing is clear and concise but incorrect—grammatically or contextually—then you have misled your audience and destroyed your credibility. To be effective, your technical communication must be *accurate*.

Accuracy in technical communication requires that you *proofread* your text. The examples of inaccurate technical communication below are caused by poor proofreading (we have underlined the errors to highlight them).

First City Federal Savings and Loan
1223 Main
Oak Park, Montana

October 12, 2008

Mr. and Mrs. David Harper
2447 N. Purdom
Oak Park, Montana

Dear Mr. and Mrs. <u>Purdom</u>:

Note that the savings and loan incorrectly typed the customer's street rather than the last name.

National Bank
1800 Commerce Street
Houston, TX

September 9, 2008

Adler's Dog and <u>Oat</u> Shop
8893 Southside
Bellaire, TX

Dear <u>Sr.</u>:
In response to your request, your account with us has been <u>close</u> out. We <u>are</u> <u>submitted</u> a check in the amount of $468.72 (your existing balance). If you have any questions, please <u>fill</u> free to <u>conact</u> us.

In addition to all the other errors, it should be "Dog and <u>C</u>at Shop," of course. The errors make the writer look incompetent.

To ensure accurate writing, use the following proofreading tips.

1. **Let someone else read it**—We miss errors in our own writing for two reasons. First, we make the error because we don't know any better. Second, we read what we think we wrote, not what we actually wrote. Another reader might help you catch errors.

2. **Use the gestation approach**—Let your correspondence sit for a while. Then, when you read it, you'll be more objective.

3. **Read backwards**—You can't do this for content. You should read backwards only to slow yourself down and to focus on one word at a time to catch typographical errors.

4. **Read one line at a time or print it out**—Use a ruler or scroll down your PC screen to isolate one line of text. Again, this slows you down for proofing. You can also proofread more effectively by printing the document and then reading it for errors, line by line.

5. **Read long words syllable by syllable**—How is the word *responsiblity* misspelled? You can catch this error if you read it one syllable at a time (re-spon-si-bl-i-ty).

6. **Use technology**—Computer spell checks are useful for catching most errors. They might miss proper names, homonyms (*their*, *they're*, or *there*) or incorrectly used words, such as *device* to mean *devise*.

7. **Check figures, scientific and technical equations, and abbreviations**—If you mean $400,000, don't write $40,000. Double-check any number or calculations. If you mean to say *HCl* (hydrochloric acid), don't write *HC* (a hydrocarbon).

8. **Read it out loud**—Sometimes we can hear errors that we cannot see. For example, we know that *a outline* is incorrect. It just sounds wrong. *An outline* sounds better and is correct.

9. **Try scattershot proofing**—Let your eyes roam around the page at random. Sometimes errors look wrong at a glance. If you wander around the page randomly reading, you often can isolate an error just by stumbling on it.

10. **Use a dictionary**—If you are uncertain, look it up.

If you commit errors in your technical communication, your readers will think one of two things about you and your company: (1) They will conclude that you are unprofessional, or (2) they will think that you are lazy. In either situation, you lose. Errors create a negative impression at best; at worst, a typographical error relaying false figures, calculations, amounts, equations, or scientific or medical data can be disastrous.

ORGANIZING TECHNICAL COMMUNICATION

If you are clear, concise, and accurate, but no one can follow your train of thought because your text rambles, you still haven't communicated effectively. Successful technical communication also must be well organized. No one method of organization always works. Following are five patterns of organization that you can use to help clarify content.

Spatial

If you are writing a technical specification to describe the parts of a machine or a plot of ground, you might want to organize your text spatially. You would describe what you see as it appears in space—left to right, top to bottom, inside to outside, or clockwise. These spatial sequences help your readers visualize what you see and, therefore, better understand the physical qualities of the subject matter. They can envision the layout of the land you describe or the placement of each component within the machine.

For example, let's say you are a contractor describing how you will refinish a basement. Your text reads as follows:

Spatial and Chronological Organization

See Chapter 11 for more discussion of spatial and chronological organization in technical descriptions and process analyses.

> At the basement's north wall, I will build a window seat 7′ long by 2′ wide by 2′ high. To the right of this seat, on the east wall, I will build a desk 4′ high by 5′ long by 3′ wide. On the south wall, to the left of the door, I will build an entertainment unit the height of the wall including four, 4′ high by 4′ wide by 2′ deep shelving compartments. The west wall will contain no built-ins. You can use this space to display pictures and to place furniture.

example

Note how this text is written clockwise, uses points of the compass to orient the reader, and includes the transitional phrases "to the right" and "to the left" to help the reader visualize what you will build. That's spatial organization.

Chronological

Whereas you would use spatial organization to describe a place, you would use chronology to document time, steps in a process analysis, or the steps in an instruction. For example, an emergency medical technician (EMT) reporting services provided during an emergency call would document those activities chronologically.

example

> At 1:15 P.M., we arrived at the site and assessed the patient's condition, taking vitals (pulse, respiration, etc.). At 1:17 P.M., after stabilizing the patient, we contacted the hospital and relayed the vitals. By 1:20 P.M., the patient was on an IV drip and en route to the hospital. Our vehicle arrived at the hospital at 1:35 P.M. and hospital staff took over the patient's care.

Chronology also would be used to document steps in an instruction. No times would be provided as in the EMT report. In contrast, the numbered steps would denote the chronological sequence a reader must follow.

Importance

Your page of text is like real estate. Certain areas of the page are more important than others—location, location, location. If you bury key data on the bottom of a page, your reader might not see the information. In contrast, content placed approximately one-third from the top of the page and two-thirds from the bottom (eye level) gets more attention. The same applies to a bulleted list of points. Readers will focus their attention on the first several points more than on the last few. Knowing this, you can decide which ideas you want to emphasize and then place that information on the page accordingly. Organize your ideas by importance. Place the more important ideas above the less important ones.

The following agenda is incorrectly organized.

Agenda

- Miscellaneous ideas
- Questions from the audience
- Refreshments
- Location, date, and time
- Subject matter
- Guest speakers

Which of these points is most important? Certainly the first two items are not important. A better list would be organized by importance, as follows:

Agenda

- Subject matter
- Guest speakers
- Location, date, and time
- Refreshments
- Questions from the audience
- Miscellaneous ideas

TABLE 3.3 Housing Costs

Item	Features	Cost
The Broadmoor	4 bedrooms, 3½ baths	
	2-car garage	
	fully equipped kitchen	$300,000
The Aspen	4 bedrooms, 3½ baths	
	finished basement	
	3-car garage	
	fully equipped kitchen	$340,000
The Regency	4 bedrooms, 3½ baths	
	patio deck	
	finished basement with ½ bath	
	3-car garage	
	fully equipped kitchen	$380,000

Comparison/Contrast

Many times in business you will need to document options and ways in which you surpass a competitor. These require that you organize your text by comparison/contrast. You compare similarities and contrast differences. For example, if you are writing a sales brochure, you might want to present your potential client alternatives regarding services, personnel, timetables, and fee structures. Table 3.3 shows how comparison/contrast provides the client options for cost and features. Each housing option is comparable in that the developer provides a four-bedroom, three-bath home with fully equipped kitchen. However, the homes contrast regarding garage size, deck availability, and basement. These options then affect the cost. The table, organized according to comparison/contrast, helps the reader understand these distinctions.

Problem/Solution

Proposals and sales letters are often problem/solution oriented. When you write a proposal, for instance, you are proposing a solution to an existing problem. If your proposal focuses on new facilities, your reader's current building must be flawed. If your proposal focuses on new procedures, your reader's current approach to doing business must need improvement. Similarly, if your sales letter promotes a new product, your customers will purchase it only if their current product is inferior.

To clarify the value of your product or service, therefore, you should emphasize the readers' need (their problem) and show how your product is the solution. Note how the following summary from a proposal is organized according to problem/solution.

Your city's 20-year-old wastewater treatment plant does not meet EPA requirements for toxic waste removal or ozone depletion regulations. This endangers your community and lessens property values in its neighborhoods.

Anderson and Sons Engineering Company has a national reputation for upgrading wastewater treatment plants. Our staff of qualified engineers will work in partnership with your city's planning commission to modernize your facilities and protect your community's values.

example

The summary's first paragraph identifies the problem. The second paragraph promotes the solution. The problem/solution organization clarifies the writer's intent.

RECOGNIZING THE IMPORTANCE OF ETHICAL COMMUNICATION

Here is the scenario. You are a technical communicator responsible for producing a maintenance manual. Your boss tells you to include the following sentence.

example | NOTE: Our product has been tested for defects and safety by trained technicians.

When read literally, this sentence is true. The product has been tested, and the technicians are trained. However, you know that the product has been tested for only 24 hours by technicians trained on-site without knowledge of international regulations. So where's the problem? As a good employee, you are required to write what your boss told you, right? Even though the statement is not completely true, legally you can include it in your manual, correct?

The answer to both questions is no! Actually, you have an ethical responsibility to write the truth. Your customers expect it, and it is in the best interests of your company. Equally important is that including the sentence in your manual is illegal. Although the sentence is essentially true, it implies something that is false. Readers will assume that the product has been *thoroughly* tested by technicians who have been *correctly* trained. Thus, the sentence deceives the readers. Such comments are "actionable under law" if they lead to false impressions (Wilson 1987, WE-68). If you fail to properly disclose information, including dangers, warnings, cautions, or notes like the sentence in question, then your company is legally liable.

Knowing this, however, does not make writing easy. Ethical dilemmas exist in corporations. The question is, what should you do when confronted with such problems?

One way to solve this dilemma is by checking your actions against these three concerns: legal, practical, and ethical. For example, if you plan to write operating instructions for a mechanism, will your text be

1. *legal,* focusing on liability, negligence, and consumer protection laws?
2. *practical,* because dishonest technical communication backfires and can cause the company to lose sales or to suffer legal expenses?
3. *ethical,* written to promote customer welfare and avoid deceiving the end user? (Bremer et al. 1987, 76–77)

These are not necessarily three separate issues. Each interacts with the other. Our laws are based on ethics and practical applications.

Legalities

If you're uncertain, that's what lawyers are for. When asked to write text that profits the company but deceives the customer, for instance, you might question where your loyalties lie. After all, the boss pays the bills, but your customers might also be your next-door neighbors. Such conflicts exist and challenge all employees. What do you do? You should trust your instincts and trust the laws. Laws are written to protect the customer, the company, and you—the employee. If you believe you are being asked to do something illegal that will harm your community, seek legal counsel.

Practicalities

Even though it might appear to be in the best interests of the company to hide potentially damaging information from customers, such is not the case. First, as a technical communicator your goal is candor. That means you must be truthful, stating the facts. It also means you must not lie, keeping silent about facts that are potentially dangerous (Girill 1987, 178–79). Second, practically speaking, the best business approach is *good* business. The

Hazard Alert Messages

See Chapter 12 for more discussion of hazard alert messages.

goal of a company is not just making a profit, but making money the right way—"good ethics is good business" (Guy 1990). What good is it to earn money from a customer who will never buy from you again or who will sue for reparation? That is not practical.

Ethicalities

Here's the ultimate dilemma. Defining ethical standards has been challenging for professional organizations. As Shirley A. Anderson-Hancock, manager for the Society for Technical Communication (STC) Ethical Guidelines Committee, states, "Everyone who participates in discussions of ethical issues may offer a different perspective, and many viewpoints may be legitimate" (1995, 6). Due to the difficulty of clearly defining what is and what is not ethical, the STC struggled for three years before publishing its guidelines in 1995 and updated them in 1998 (see Figure 3.3). These ethical principles, as the STC Ethical Guidelines Committee admits, had to be "sufficiently broad to apply to different working environments for the next several years" (Anderson-Hancock 1995, 6).

One way to clarify these necessarily broad standards is by looking at the STC Code for Communicators, which reads as follows ("Code for Communicators" 2004).

Code for Communicators

As a technical communicator, I am the bridge between those who create ideas and those who use them. Because I recognize that the quality of my services directly affects how well ideas are understood, I am committed to excellence in performance and the highest standards of ethical behavior.

I value the worth of the ideas I am transmitting and the cost of developing and communicating those ideas. I also value the time and effort spent by those who read or see or hear my communication.

I therefore recognize my responsibility to communicate technical information truthfully, clearly, and economically.

My commitment to professional excellence and ethical behavior means that I will
- Use language and visuals with precision.
- Prefer simple, direct expression of ideas.
- Satisfy the audience's need for information, not my own need for self-expression.
- Hold myself responsible for how well my audience understands my message.
- Respect the work of colleagues, knowing that a communication problem may have more than one solution.
- Strive continually to improve my professional competence.
- Promote a climate that encourages the exercise of professional judgment and that attracts talented individuals to careers in technical communication.

FIGURE 3.3   Code for Communicators

Ethical Principles for Technical Communicators

As technical communicators, we observe the following ethical principles in our professional activities.

Legality

We observe the laws and regulations governing our profession. We meet the terms of contracts we undertake. We ensure that all terms are consistent with laws and regulations locally and globally, as applicable, and with STC ethical principles.

Honesty

We seek to promote the public good in our activities. To the best of our ability, we provide truthful and accurate communications. We also dedicate ourselves to conciseness, clarity, coherence, and creativity, striving to meet the needs of those who use our products and services. We alert our clients and employers when we believe that material is ambiguous. Before using another person's work, we obtain permission. We attribute authorship of material and ideas only to those who have made an original and substantive contribution. We do not perform work outside our job scope during hours compensated by clients or employers, except with their permission; nor do we use their facilities, equipment, or supplies without their approval. When we advertise our services, we do so truthfully.

Confidentiality

We respect the confidentiality of our clients, employers, and professional organizations. We disclose business-sensitive information only with their consent or when legally required to do so. We obtain releases from clients and employers before including any business-sensitive materials in our portfolios or commercial demonstrations or before using such materials for another client or employer.

Quality

We endeavor to produce excellence in our communication products. We negotiate realistic agreements with clients and employers on schedules, budgets, and deliverables during project planning. Then we strive to fulfill our obligations in a timely, responsible manner.

Fairness

We respect cultural variety and other aspects of diversity in our clients, employers, development teams, and audiences. We serve the business interests of our clients and employers as long as they are consistent with the public good. Whenever possible, we avoid conflicts of interest in fulfilling our professional responsibilities and activities. If we discern a conflict of interest, we disclose it to those concerned and obtain their approval before proceeding.

Professionalism

We evaluate communication products and services constructively and tactfully, and seek definitive assessments of our own professional performance. We advance technical communication through our integrity and excellence in performing each task we undertake. Additionally, we assist other persons in our profession through mentoring, networking, and instruction. We also pursue professional self-improvement, especially through courses and conferences.

Approved by the STC Board of Directors
September 1998.

(STC "Ethical Principles for Technical Communicators")

Use Language and Visuals with Precision.   In a survey we conducted comparing technical communicators and teachers of technical communication, we discovered an amazing finding: Professional technical communicators rate grammar and mechanics higher than teachers do (Gerson and Gerson 1995). On a five-point scale (5 equaling "very important"), writers rated grammar and mechanics 4.67, whereas teachers rated grammar and mechanics only 3.54. That equals a difference of 1.13, which represents a 22.6 percent divergence of opinion.

Given these numbers, would we be precise in writing, "Teachers do not take grammar and mechanics as seriously as writers do"? The numbers accurately depict a difference of opinion, and the 22.6 percent divergence is substantial. However, these figures do not assert that teachers ignore grammar. To say so is imprecise and would constitute an ethical failure to present data accurately. Even though writers are expected to highlight their client's values and downplay their client's shortcomings, technical communicators ethically cannot skew numbers to accomplish these goals (Bowman and Walzer 1987). Information must be presented accurately, and writers must use language precisely.

Precision also is required when you use visuals to convey information. Look at Figures 3.4 and 3.5. Both figures show that XYZ's sales have risen. As with language, the writer is ethically responsible for presenting visual information precisely.

Another major consideration regarding the precise use of language and visuals involves the role of intellectual property laws as they relate to the Internet. If something is on the Internet, including graphics or text, we can just take it—right? We can download any graphic or just appropriate any text if it's online, right?

BEFORE

FIGURE 3.4   Imprecise Depiction of Sales Growth

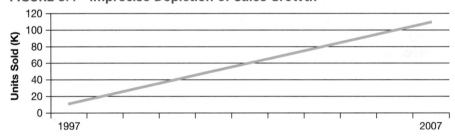

AFTER

FIGURE 3.5   Precise Depiction of Sales Growth

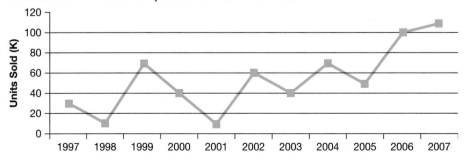

Of course, these assumptions are false. "Every element of a Web page—text, graphics, and HTML code—is protected by U.S. and international copyright laws, whether or not a formal copyright application has been submitted" (LeVie 2000, 20–21). The U.S. Copyright Office's "Circular 66, Copyright Registration for Online Works" states the following:

> Copyright protects original authorship fixed in tangible form. For works transmitted online, the copyrightable authorship may consist of text, artwork, music, audiovisual material (including any sounds), sound recordings, etc. Copyright does not protect ideas, procedures, systems, or methods of operation. Under U.S. law, copyright protection subsists from the time the work is fixed in any tangible medium of expression from which it can be perceived, reproduced, or otherwise communicated, either directly or with the aid of a machine or device.

DOT-COM UPDATES

For more information about online copyright laws, check out the following links.

- http://www.bitlaw.com/index.html. BitLaw provides information about Internet patent, copyright, trademark, and legal issues.
- http://www.copyright.gov/. The United States Copyright Office

If you and your company "borrow" from an existing Internet site, thus infringing upon that site's copyright, you can be assessed actual or statutory damages. The Copyright Act allows the owner of Web material to do the following—stop infringement and obtain damages and attorney fees. In addition to financial damages, your company, if violating intellectual property laws, could lose customers, damage its reputation, and lose future capital investments.

To protect your property rights, you should

- Assume that any information on the Internet is covered under copyright protection laws unless proven otherwise.
- Obtain permission for use from the original creator of graphics or text.
- Cite the source of your information.
- Create your own graphics and text.
- Copyright any information you create.
- Place a copyright notice at the bottom of your Web site (Johnson 1999, 17).

Prefer Simple, Direct Expression of Ideas. While writing an instructional manual, you might be asked to use legal language to help your company avoid legal problems (Bowman and Walzer 1987). Here is a typically obtuse warranty that legally protects your company.

example

Acme's liability for damages from any cause whatsoever, including fundamental breach, arising out of this Statement of Limited Warranty, or for any other claim related to this product, shall be limited to the greater of $10,000 or the amount paid for this product at the time of the original purchase, and shall not apply to claims for personal injury or damages to personal property caused by Acme's negligence, and in no event shall Acme be liable for any damages caused by your failure to perform your responsibilities under this Statement of Limited Warranty, or for loss of profits, lost savings, or other consequential damages, or for any third-party claims.

Can your audience easily grasp this warranty's lengthy sentence structure and difficult-to-understand words? No, the readers will become frustrated and will fail to recognize what is covered under such a warranty. Although you are following your boss's directions, you are not fulfilling one of your ethical requirements to use simple and direct language.

Technical communication can be legally binding *and* easy to understand. The first requirement does not negate the second. As a successful technical writer who values "the time and effort spent by those who read or see or hear [your] communication," you want to aid communication by using simple words and direct expression of ideas whenever possible ("Code for Communicators" 2004).

Satisfy the Audience's Need for Information, Not My Own Need for Self-Expression. The importance of simple language also encompasses the third point in STC's Code for Communicators: satisfying "the audience's need for information, not my own need for self-expression." The previously mentioned warranty might be considered poetic in its sentence structure and sophisticated in its word usage. It does not communicate what the audience needs, however. Such elaborate and convoluted writing will satisfy only one's need for self-expression: That is not the goal of effective technical communication.

Electronic mail (e-mail) presents another opportunity for unethical behavior. Because e-mail is so easy to use, many company employees use it too frequently. They write e-mail for business purposes, but they also use e-mail when writing to family and friends. Employees know not to abuse their company's metered mail by using corporate envelopes and stamps to send in their gas bills or write thank-you notes to Aunt Rose. These same employees, however, will abuse the company's e-mail system (on company time) by writing e-mail messages to relatives coast to coast. They are not satisfying their business colleagues' need for information; instead, they are using company-owned e-mail systems for their own self-expression (Hartman and Nantz 1995, 60). That is unethical.

Hold Myself Responsible for How Well My Audience Understands My Message. As a technical writer, you must place the audience first. When you write a precise proposal or instruction, using simple words and syntax, you can take credit for helping the reader understand your text. Conversely, you must also accept responsibility if the reader fails to understand.

Ethical standards for successful communication require the writer to always remember the readers—the real people who read manuals to put together their children's toys, who read corporate annual reports to understand how their stocks are doing, and who read proposals to determine whether to purchase a service. An ethical writer remembers that these real people will be frustrated by complex instructions, confused by inaccessible stock reports, and misled by inaccurate proposals. It is unethical to forget that your writing can frustrate people and even endanger them. It is unethical to use your writing to mislead or to confuse. In contrast, as an ethical writer, you need to treat these readers like neighbors, people you care about, and then write accordingly. Take the time to check your facts, present your information precisely, and communicate clearly so that your readers—your friends, coworkers, and clients—are safe and satisfied. You are responsible for your message (Barker et al. 1995).

Respect the Work of Colleagues. Each of the ethical considerations already discussed relates to your clients, the readers of your technical communication. This fifth point differs in two ways: First, it relates to coworkers or professional colleagues; second, it highlights the importance of online ethics, the ethical dilemmas presented by electronic communications on the Internet. We can discuss these two points simultaneously.

Much of today's technical communication takes place online, electronically. Electronic communication creates three unique reasons for an increased focus on ethics. As the technical communicator, you must ethically consider confidentiality, courtesy, and copyright when working with the Internet or any private electronic network.

- **Confidentiality.** The 1974 Privacy Act allows "individuals to control information about themselves and to prevent its use without consent" (Turner 1995, 59). The Electronic Communication Privacy Act of 1986, which applies all federal wiretap laws to electronic communication, states that e-mail messages can be disclosed only "with the consent of the senders or recipients" (Turner 1995, 60). However, both of these laws can be abused easily on the Internet. First, neither law specifically defines "consent." Data such as your credit records can be accessed without your knowledge by anyone with the right hardware and software. Your confidentiality can be breached easily.

Electronic Communication Channels

See Chapter 13 for more discussion of electronic communication channels.

Second, the 1986 Electronic Communication Privacy Act fails to define "the sender." In the workplace, a company owns the e-mail system, just as it owns other more tangible items such as desks, computers, and file cabinets. Companies have been held liable for electronic messages sent by employees. Thus, many corporations consider the contents of one's e-mail and one's e-mailbox company property, not the property of the employee. With ownership comes the right to inspect an employee's messages. Although you might write an e-mail message assuming that your thoughts are confidential, this message can be monitored without your knowledge (Hartman and Nantz, 61).

These instances might be legal, but are they ethical? Even though a company can eavesdrop on your e-mail or access your life's history through databases, that doesn't mean it should. As an employee or corporate manager, you should respect another's right to confidentiality. Ethically, you should avoid the temptation to read someone else's e-mail or to access data about an individual without consent.

E-mail

See Chapter 6 for more discussion of e-mail.

DOT-COM UPDATES

For more information about ethical codes of conduct, check out the following links.

- http://onlineethics.org/codes/. Ethical codes from societies for engineers and scientists
- http://www.iit.edu/departments/csep/. The Center for the Study of Ethics in the Professions—a collection of many different codes of conduct for industries such as agriculture, business, communications, finance, government, and management

- **Courtesy.** As already noted, e-mail is not as private as you might believe. Whatever you write in your e-mail correspondence can be read by others. Given this reality, you should be very careful about what you say in e-mail. Specifically, you don't want to offend coworkers by "flaming" (writing discourteous messages). Before you criticize a coworker's ideas or ability, and before you castigate your employer, remember that common courtesy, respect for others, is ethical (Adams et al. 1995, 328).

- **Copyright.** A final ethical consideration relates to copyright laws. Every English teacher you have ever had has told you to avoid *plagiarism* (stealing another writer's words and ideas). Plagiarism, unfortunately, is an even greater problem on the Internet. The Internet is perhaps the world's largest library without walls; it is almost a universal commons where anyone can set up a soap box and speak (or print) his opinion.

This incredible ability to disseminate massive amounts of information presents problems. If you were an unethical technical communicator, you could access data from the Internet and easily print it as your own. After all, it is hard to trace the identity of a writer on the Net, and Internet information can be downloaded by anyone (Adams et al. 1995, 328).

An ethical communicator will not fall prey to such temptation. If you have not written it, give the other author credit. Words are like any other possession. Taking words and ideas without attributing your source through a footnote or parenthetical citation is wrong. You should respect copyright laws.

Strive Continually to Improve My Professional Competence. Promote a Climate that Encourages the Exercise of Professional Judgment. We've combined these last two tenets from the STC's Code for Communicators because together they sum up all of the preceding ethical considerations. Professional competence, for example, includes one's ability to avoid plagiarizing. Competence should also include professional courtesy, respect for another's right to confidentiality, a sense of responsibility for one's work, and the ability to write clearly and precisely. Ultimately, your competence is dependent upon professional judgment. Do you know the difference between right and wrong? Recognizing the distinction is what ethics is all about.

As a writer, you will always be confronted by a multitude of options, such as loyalty to your company, responsible citizenship, need for a salary, accountability to your client and your coworkers, and personal integrity. You must weigh the issues—ethically, legally, and practically—and then write according to your conscience.

Strategies for Making Ethical Decisions

When confronted with ethical challenges, try following these writing strategies (Guy 1990, 165).

a. *Define the problem.* Is the dilemma legal, practical, ethical, or a combination of all three?

b. *Determine your audience.* Who will be affected by the problem? Clients, coworkers, management? What is their involvement, what are their individual needs, and what is your responsibility—either to the company or to the community?

c. *Maximize values; minimize problems.* Ethical dilemmas always involve options. Your challenge is to select the option that promotes the greatest worth for all stakeholders involved. You won't be able to avoid all problems. The best you can hope for is to minimize those problems for both your company and your readers while you maximize the benefits for the same stakeholders.

d. *Consider the big picture.* Don't just focus on short-term benefits when making your ethical decisions. Don't just consider how much money the company will make now, how easy the text will be to write now. Focus on long-term consequences as well. Will what you write please your readers so that they will be clients for years to come? Will what you write have a long-term positive impact on the economy or the environment?

e. *Write your text.* Implement the decision by writing your memo, letter, proposal, manual, or report. When you write your text, remember to

- Use precise language and visuals.
- Use simple words and sentences.
- Satisfy the audience's need for information, not your own need for self-expression.
- Take responsibility for your content, remembering that real people will follow your instructions or make decisions based on your text.
- Respect your colleagues' confidentiality, be courteous, and abide by copyright laws.
- Promote professionalism and good judgment.

CHECKLIST FOR CLARITY, CONCISENESS, ACCURACY, AND ETHICS IN TECHNICAL COMMUNICATION

Clarity

_____ 1. Have you answered the reporter's questions (who, what, when, where, why, and how)?

_____ 2. Have you provided specific information, avoiding vague word usage?

_____ 3. Have you defined acronyms, abbreviations, and jargon?

_____ 4. Have you avoided passive voice, striving for active voice?

Conciseness

_____ 5. Have you limited the number of syllables per word?

_____ 6. Have you limited the number of words per sentence?

_____ 7. Have you limited the number of lines per paragraph?

Accuracy

_____ 8. Have you proofread your technical communication?

Organization

_____ 9. Have you used modes, such as spatial, importance, chronology, comparison/contrast, and problem/solution to organize your technical communication?

Ethics

_____ 10. Have you considered the ethical implications of your technical communication?

CHAPTER HIGHLIGHTS

1. If your technical communication is unclear, your reader may misunderstand you and then do a job wrong, damage equipment, or contact you for further explanations.
2. Use details to ensure reader understanding. Whenever possible, specify and quantify your information.
3. Answering *who, what, when, where, why*, and *how* (the reporter's questions) helps you determine which details to include.
4. For some audiences, you should avoid acronyms, abbreviations, and jargon.
5. Avoid words that are not commonly used (legalisms, outdated terms, etc.).
6. Write to express, not impress—to communicate, not to confuse.
7. Avoid passive voice constructions, which tend to lengthen sentences and confuse readers.
8. Writing concisely helps save time for you and your readers.
9. Shorter paragraphs are easier to read, so they hold your reader's attention.
10. When possible, use short, simple words (always considering your reader's level of technical knowledge).
11. Apply readability formulas to determine your text's degree of difficulty.
12. Proofreading is essential to effective technical communication.
13. Well-organized documents are easy to follow.
14. Different organizational patterns—spatial, chronological, importance, comparison/contrast, and problem/solution—can help you explain material.
15. Consider whether or not your technical communication is legal, practical, and ethical.

APPLY YOUR KNOWLEDGE

CASE STUDIES

1. CompToday computer hardware company must abide by the Sarbanes-Oxley Act, passed by Congress in response to accounting scandals. This act specifically mandates the following related to documentation standards.

   - **Section 103: Auditing, Quality Control, and Independence Standards and Rules.** Companies must "prepare, and maintain for a period of not less than 7 years, audit work papers, and other information related to any audit report, in sufficient detail to support the conclusions reached in such report."

   - **Section 401(a): Disclosures in Periodic Reports; Disclosures Required.** "Each annual and quarterly financial report . . . must be presented so as not to contain an untrue statement or omit to state a material fact necessary in order to make the pro forma financial information not misleading."

   Beverly Warden, technical documentation specialist at Comp Today, is responsible for managing the Sarbanes-Oxley reports. She is being confronted by the following ethical issues.

   a. To help Beverly prepare the first annual report, her chief financial officer (CFO) has given her six months of audits (January through June). These prove that the company is meeting its accounting responsibilities. However, Beverly's report covers the entire year, including July through December. Section 103 states that the report must provide "sufficient detail to support the conclusions reached in [the] report."

Are the company's first six months of audits sufficient? If Beverly writes a report stating that her company is in compliance, is she abiding by her Society for Technical Communication Ethical Principles, which state that a technical writer's work is "consistent with laws and regulations"?

What are her ethical, practical, and legal responsibilities?

- Share your findings in an oral presentation.
- Write a letter, memo, report, or e-mail stating your opinion regarding this issue.

b. Beverly's CFO also has told her that during the year, the company fired an outside accounting firm and hired a new one to audit the company books. The first firm expressed concerns about several bookkeeping practices. The newly hired firm, providing a second opinion after reviewing the books, concluded that all bookkeeping practices were acceptable. The CFO sees no reason to mention the first firm.

Section 401(a) states that reports must contain no untrue statements or omit to state a material fact. The STC Ethical Principles also say that her writing must be truthful and accurate, to the best of her ability. Beverly can report factually that the new accounting firm finds no bookkeeping errors. Should she also report the first accounting firm's assessment? Is that a material fact? If she omits any mention of the first accounting firm, as her boss suggests, is she meeting both her STC technical writer's responsibilities and the needs of Sarbane-Oxley?

What are her ethical, practical, and legal responsibilities?

- Share your findings in an oral presentation.
- Write a letter, memo, report, or e-mail stating your opinion regarding this issue.

2. The Blue Valley Wastewater Treatment Plant processes water running to and from Frog Creek, a water reservoir that passes through the North Upton community. This water is usually characterized by low alkalinity (generally, <30 mg/l), low hardness (generally, <40 mg/l), and minimal water discoloration. Inorganic fertilizer nutrients (phosphorus and nitrogen) are also generally low, with limited algae growth.

Despite the normal low readings, algae-related tastes and odors occur occasionally. Although threshold odors range from three to six, they have risen to ten in summer months. Alkalinity rises to <50 mg/l, hardness to <60 mg/l, and discoloration intensifies. Taste and odor problems can be controlled by powdered activated carbon (PAC); nutrient-related algae growth can be controlled by filtrated ammonia. Both options are costly.

The odors and tastes are disturbing North Upton residents. The odors are especially bothersome to outdoor enthusiasts who use the trails bordering Frog Creek for biking and hiking. Residents also worry about the impact of increased alkalinity on fish and turtles, many of which are dying, further creating an odor nuisance. Frog Creek is treasured for its wildlife and recreational opportunities.

The Blue Valley Wastewater Treatment Plant is under no legal obligation to solve these problems. The alkalinity, hardness, color, and nutrient readings are all within regulated legal ranges. However, community residents are insistent that their voices be heard and that restorative steps be taken to improve the environment around their homes.

This is an ethical and practical dilemma. In response to this dilemma, divide into small groups to write one of the following documents.

- You are Blue Valley Wastewater Treatment Plant's director of public relations. Write a letter to the city commission stating your plant's point of view. (We discuss letters in Chapter 6.)
- You are North Upton community's resident representative. Write a letter to the city commission stating your community's point of view.
- You are an employee for the Blue Valley Wastewater Treatment Plant. Write a memo to the plant director suggesting ways in which the plant could solve this environmental and community relations problem. (We discuss memos in Chapter 6.)
- You are the Blue Valley Wastewater Treatment Plant's director. Write a memo to the engineering supervisor (your subordinate) stating how to solve this environmental and community relations problem.

These documents could be organized by the problem/solution or comparison/contrast methods. Whichever method you choose, consider the strategies for making ethical decisions, discussed in this chapter.

INDIVIDUAL AND TEAM PROJECTS

Achieving Clarity

The following sentences are unclear. They will be interpreted differently by different readers. Revise these sentences—replace the vague, impressionistic words with more specific information.

1. We need this information as soon as possible.
2. The machinery will replace a flawed piece of equipment in our department.
3. Failure to purchase this will have a negative impact.
4. Weather problems in the area resulted in damage to the computer systems.
5. The most recent occurrences were caused by insufficient personnel.

Avoiding Obscure Words

Obscure words make the following sentences difficult to understand. Improve the sentences by revising the difficult words and making them more easily understood.

1. As Very Large Scale Integration (VLSI) continues to develop, a proliferation of specialized circuit simulators will be utilized.
2. As you requested at the commencement of the year, I am forwarding my regular quarterly missive.
3. Though Randolph was cognizant of his responsibility to advise you of any employment aberrations, he failed to abide by this mandate.
4. Herewith is an explanation of our rationale for preferring services, pursuant to your request.
5. Please be advised that the sale constitutes a successful closure.

Using the Active Voice Versus the Passive Voice

Use of the passive voice often leads to vague, wordy sentences. Revise the following sentences by writing them in the active voice.

1. Implementation of this procedure is to be carried out by the Accounting Department.
2. Benefits derived by attending the conference were twofold.
3. The information was demonstrated and explained in great detail by the training supervisor.
4. Discussions were held with representatives from Allied, who supplied analytical equipment for automatic upgrades.
5. Also attended was the symposium on polymerization.

Limiting Paragraph Length to Achieve Conciseness

You can achieve clarity and conciseness if you limit the length of your paragraphs. An excessively long paragraph (beyond six typed lines) requires too much work for your reader.

Revise the following paragraph to make it more reader friendly.

> As you know, we use electronics to process freight and documentation. We are in the process of having terminals placed in the export departments of some of our major customers around the country so they may keep track of all their shipments within our system. I would like to propose a similar tracking mechanism for your company. We could handle all of your export traffic from your locations around the country and monitor these exports with a terminal located in your home office. This could have many advantages for you. You could generate an export invoice in your export department, which could be transmitted via the computer to our office. You could trace your shipments more readily. This would allow you to determine rating fees more accurately. Finally, your accounting department would benefit. All in all, your export operations would achieve greater efficiency.

Reducing Word Length to Achieve Conciseness

Multisyllabic words can create long sentences. To limit sentence length, limit word length. Find shorter words to replace the following words.

1. advise
2. anticipate
3. ascertain
4. cooperate
5. determine
6. endeavor
7. inconvenience
8. indicate
9. initially
10. presently

Reducing Sentence Length to Achieve Conciseness

Each of the following sentences is too long. Revise them using the techniques suggested in this chapter: Use the meat cleaver approach; avoid shun words, camouflaged words, and expletives; omit redundancies; and delete wordy phrases.

1. In regard to the progress reports, they should be absolutely complete by the fifteenth of each month.
2. I wonder if you would be so kind as to answer a few questions about your proposal.
3. I am in receipt of your memo requesting an increase in pay and am of the opinion that it is not merited at this time due to the fact that you have worked here for only one month.
4. On two different occasions, I have made an investigation of your residence, and I believe that your sump pump might result in damage to your neighbor's adjacent property. I have come to the conclusion that you must take action to rectify this potential dilemma, or your neighbor might seek to sue you in a court of law.
5. If there are any questions that you might have, please feel free to contact me by phone.

Using Organization

1. *Spatial:*
   a. Using spatial organization, write a paragraph describing your classroom, office, work environment, dorm room, apartment, or any room in your house.
   b. Using spatial organization, write an advertisement describing the interior of a car, the exterior of a mechanism or tool, a piece of clothing, or a motorcycle.
2. *Chronological:* Organizing your text chronologically, write a report documenting your drive to school or work, your activities accomplished in class or at work, your discoveries at a conference or vacation, or your activities at a sporting event.
3. *Importance:* A fashion merchandising retailer asked her buyers to purchase a new line of clothing. In her memo, she provided them the following list to help them accomplish their task. Reorganize the list by importance, and justify your decisions.

Date: January 15, 2008
To: Buyers
From: Sharon Baker
Subject: Clothing Purchases

It is time again for our spring purchases. This year, let's consider a new line of clothing. When you go to the clothing market, focus on the following:

- Colors
- Materials
- Our customers' buying habits
- Price versus markup potential
- Quantity discounts
- Wholesaler delivery schedules

Good luck. Your purchases at the market are what make our annual sales successful.

4. *Comparison/contrast:* Visit two auto dealerships, two clothing stores, two restaurants, two music shops, two prospective employers, two colleges, and so on. Based on your discoveries, write a report using comparison/contrast to make a value judgment. Which of the two cars would you buy, which of the two restaurants would you frequent, and at which of the two music shops would you purchase CDs?

5. *Problem/solution case study:* You work for Acme Electronics as an electrical engineer. Carol Haley, your boss, informs you that your department's electronic scales are measuring tolerances inaccurately. You are asked to study the problem and determine solutions. In your study, you find that one scale (ID #1893) is measuring within 90 percent of tolerance; another scale (ID #1887) is measuring within 75 percent of tolerance; a third scale (ID #1890) is measuring within 60 percent of tolerance; a final scale (ID #1885) is measuring within 80 percent of tolerance. Standards suggest that 80 percent is acceptable. To solve this problem, the company could purchase new scales ($2,000 per scale); reduce the vibration on the scales by mounting them to the floor ($1,500 per scale); or reduce the vibration around the scales by enclosing the scales in plexiglass boxes ($1,000 per scale).

Write a memo to your boss detailing your findings (the problems) and suggesting the solutions.

PROBLEM-SOLVING THINK PIECES
Considering Ethics

1. The Society for Technical Communication constantly is trying to redefine its Ethical Principles. As a class, how would you define the word *ethics*? Brainstorm new definitions as they apply to technical communication and come to a class consensus. Then in small groups, based on your class's definition of *ethics*, do the following:
   - List five or more examples of ethical responsibilities you believe technical communicators should have when writing memos, letters, reports, proposals, Web sites, e-mail messages, instant messages, or instructions (other than those already discussed in this chapter).

- List five or more examples of failures to abide by ethical responsibilities you have seen either in writing or in other types of media (recordings, movies, television programs, newspapers, magazines, news reports, etc.).
- Technical communication is factual, as are newspaper, magazine, radio, and television news reports; recordings, movies, and television programs are art forms. Do the same ethical considerations apply for all types of media? If there are differences, explain your answer.

2. Bring to class examples of ethically flawed communication. These could include poorly written warranties (which are too difficult to understand) or dangers, warnings, and cautions (which do not clearly identify the potential for harm). You might find misleading annual reports or unethical advertisements. Poor examples could even include graphics that are visually misleading. Then in small groups, rewrite these types of communication or redraw the graphics to make them ethical.

3. Every day, the Internet is presenting ethical dilemmas to lawmakers and to the populace. Congress is currently debating laws to curb unethical practices online. These unethical practices include hacking, pornography, solicitation, infringements on confidentiality, hate mail, inappropriate advertisements, spam, and unauthorized viewing of e-mail.

   To update your knowledge of ethics problems online, research any of the topics listed. Then, present your findings as follows:
- Write an individual or group report on your findings.
- Give an individual or group oral presentation on your findings.
- Write a summary of the article(s) you've researched.

WEB WORKSHOP

1. To learn more about plain language and how it relates to the importance of conciseness, visit http://www.plainlanguage.gov/. This site provides a definition of plain language, governmental mandates, and before and after comparisons. Click on any of the links in this site and report your findings either orally or in an e-mail message.

2. The Internet plays a major role in technical communication. You might create Internet, extranet, or intranet sites, communicate through online blogs, or consider downloading information and graphics from the Internet. What is legal, and what isn't? To learn more about online copyright laws, visit BitLaw (http://www.bitlaw.com/index.html) or the U.S. Copyright Office (http://www.copyright.gov/). Research issues related to the Internet and report your findings either orally or in writing.

3. Communicating ethically is both the legal and moral responsibility of a technical communicator. To learn more about ethical codes of conduct, visit either of the following sites.
- http://onlineethics.org/codes/ (Provides ethical codes from societies for engineers and scientists)
- http://www.iit.edu/departments/csep/ (The Center for the Study of Ethics in the Professions)

   Research a code of conduct from your major field of interest and report your findings either orally or in writing.

QUIZ QUESTIONS

1. Why is clarity the ultimate goal of technical communication?
2. How can you achieve clarity?
3. How will answering the reporter's questions help you achieve clarity?
4. Why should you avoid using acronyms, abbreviations, and jargon in your technical communication?
5. In what two places in a business document can you define a term?

6. Why should you strive for conciseness in your technical communication?
7. Why is a long paragraph ineffective in your business documents?
8. What is a fog index?
9. Why should you use shorter sentences?
10. What are three causes of wordy sentences?
11. What are four proofreading tips?
12. When will you organize a document spatially?
13. Why would you organize chronologically?
14. What is the effect of organization by importance?
15. When is comparison/contrast used effectively in a business document?
16. How can you use problem/solution in a document?
17. What do you consider when you focus on the legal aspect of a document?
18. What do you consider when you focus on the practical aspect of a document?
19. What do you consider when you focus on the ethical aspect of a document?
20. What are three things you consider when you follow the guide for ethical standards?

CHAPTER 4

Audience Recognition and Involvement

COMMUNICATION
at work

In this scenario, Home and Business Mortgage works to meet the company's commitment to diversity through its technical communication.

Home and Business Mortgage (HBM), a Phoenix, Arizona, company, prides itself on being a good neighbor, an asset to the community. Its goal is to ensure that the company meets all governmental personnel regulations, including those required by the Equal Employment Opportunity Commission (EEOC), the Family Medical Leave Act (FMLA), and the Americans with Disabilities Act (ADA). HBM is committed to achieving diversity in its workplace.

In fact, the company is so focused on its appreciation of and commitment to diversity that HBM's technical writers have created the following:

- Bilingual brochures, newsletters, and annual reports
- Brochures geared toward different stakeholders—senior citizens (age 55 and up), African-American, Hispanic, and Native American homebuyers, women, and young singles (18–25)
- A Web site with photographs depicting Phoenix's diverse population

Objectives

When you complete this chapter, you will be able to

1. Achieve audience recognition.

2. Distinguish among high-tech, low-tech, lay, and multiple audiences.

3. Define terms for different audience levels.

4. Avoid biased language.

5. Recognize diversity including multiculturalism and cross-culturalism.

6. Follow the guidelines for effective multicultural communication.

7. Achieve audience involvement.

8. Evaluate your audience using the audience checklist.

HBM's Web site is unique in another important way. It has sought to abide by W3C WAI mandates. The World Wide Web Consortium (W3C) hosts the Web Accessibility Initiative (WAI). This initiative states that "the power of the Web is its universality. Access by everyone regardless of disability is an essential aspect." HBM's webmasters, in agreement with the WAI, realize that graphic-laden Web sites, as well as those with audio components, are not accessible to people with disabilities such as hearing and visual impairment.

To meet this WAI platform, HBM's webmasters have achieved the following at the company's Web site.

- Avoided online audio

- Limited online video and used the Alt attribute to describe each visual's function

- Increased the font of Web text to 20-point versus the standard 12- to 14-point font

- Avoided all designer fonts, such as cursive, which are hard to read online

- Avoided red and green for color emphasis because these colors are hard to read for audiences with color blindness

- Avoided frames that not only fail to load on all computers but also create readability challenges online

HBM is honored to serve all of its customers, regardless of race, age, or religion. One way to achieve the company's commitment to diversity is through its technical communication.

When you write a memo, letter, or report, someone reads it. That individual or group of readers is your audience. To compose effective technical communication, you should achieve audience recognition and audience involvement.

Check out our quarterly newsletters TechCom E-Notes at www.prenhall.com/gerson for dot.com updates, new case studies, insights from business professionals, grammar exercises, and facts about technical communication.

AUDIENCE RECOGNITION

In the business world, you will never write or speak in a vacuum. When you write your memo, e-mail, letter, report, or brochure, someone will read it. When you give an oral presentation, convene a meeting, communicate with customers in a salesroom, or make a speech at a conference, consider the following questions.

- Who is your audience?
- What does this reader or listener know?
- What does this reader or listener not know?
- What must you write or say to ensure that your audience understands your point?
- How do you communicate to more than one person (multiple audiences)?
- What is this person's position in relation to your job title?
- What diversity issues (gender, sexual orientation, cultural, multicultural) must you consider?

If you do not know the answers to these questions, your communication might contain jargon or acronyms the reader will not understand. The tone of the memo may be inappropriate for management (too dictatorial) or for your subordinates (too relaxed). Your verbal communication might not consider your audience's unique culture and language. To communicate successfully, you must recognize your audience's level of understanding. You also should factor in your audience's unique traits, which could impact how successfully you communicate.

See Table 4.1 for a summary of audience variables to consider when communicating.

Knowledge of Subject Matter

Check Online Resources

www.prenhall.com/gerson
For more information about high-tech, low-tech, and lay audiences, visit our companion Web site.

What does your audience know about the subject matter? Does the person work closely with you on the project? That would make the audience a high-tech peer. Does the audience have general knowledge of the subject matter, but an area of expertise that is elsewhere? That would make the audience a low-tech peer. Is the audience totally uninvolved in the subject matter? That would make the audience a lay audience. Finally, could your audience be a combination of these types? Then, you would be communicating with an audience with multiple levels of expertise.

High-Tech Audiences. High-tech readers work in your field of expertise. They might work directly with you in your department, or they might work in a similar

TABLE 4.1 Audience Variables

Knowledge of Subject Matter
- High tech
- Low tech
- Lay
- Multiple

Issues of Diversity
- Age
- Gender
- Race
- Religion
- Sexual orientation
- Language
- Culture of origin—multicultural or cross-cultural

capacity for another company. Wherever they work, they are your colleagues because they share your educational background, work experience, or level of understanding.

If you are a computer programmer, for example, another computer programmer who is working on the same system is your high-tech peer. If you are an environmental engineer working with hazardous wastes, other environmental engineers focusing on the same concerns are your high-tech peers.

Once you recognize that your reader is high tech, what does this tell you? High-tech readers have the following characteristics.

- They are experts in the field you are writing about. If you write an e-mail message to a department colleague about a project you two are working on, your associate is a high-tech peer. If you write a letter to a vendor requesting specifications for a system she markets, that reader is a high-tech expert. If you write a journal article geared toward your colleagues, they are high tech.

- Because their work experience or education is comparable to yours, high-tech readers share your level of understanding. Therefore, they will understand high-tech jargon, acronyms, and abbreviations. You do not have to explain to an electronic technician, for example, what MHz means. Defining *megahertz* for this high-tech reader would be unnecessary. We recently read a procedure written by a manager to her engineers explaining how to write departmental technical reports. In the memo, the boss specified that the engineers should use "metric SI units in their reports." Because the boss recognized that her audience was high tech, she did not have to define SI as "System International."

- High-tech readers require minimal detail regarding standard procedures or scientific, mathematical, or technical theories. Two physicists, when discussing the inflation theory of the universe's creation, would not labor over the importance of quarks and leptons (subatomic particles). Although this jargon is unintelligible to most of us, to high-tech peers no explanations are required.

- High-tech colleagues read to discover new technical knowledge or for updates regarding the status of a project.

- High-tech readers need little background information regarding a project's history or objectives unless the specific subject matter of the correspondence is new to them. If, for example, you are writing a status report to your first-line supervisor, who has been involved in a project since its inception, then you will not need to flesh out the history of the project. On the other hand, if you have a new supervisor or if you are updating a colleague new to your department, even though these readers are high tech, you will need to provide background data.

If you work in an environment in which you write only to high-tech peers, you are a rare and lucky individual. Writing to high-tech readers is rather easy because you can use acronyms, abbreviations, jargon, complex graphics, and so on. You usually do not have to define terms or provide background information. Writing within a technical environment, however, also requires that you write to low-tech readers.

Low-Tech Audience.   Low-tech readers include your coworkers in other departments. Low-tech readers also might include your bosses, subordinates, or colleagues who work for other companies. For instance, if you are a biomedical equipment technician, then the accountant, personnel director, or graphic artists in your company are low-tech peers. These individuals have worked around your company's equipment and, therefore, are familiar with your technology. However, they do not understand the intricacies of this technology.

Your bosses are often low tech because they no longer work closely with the equipment. Although they might have been technicians at one time, as they moved higher into management they moved further away from technology. Your subordinates might be low tech because their levels of education or work experience are less than yours.

Although your colleagues at other companies might have your level of education or work experience, they could be low tech if they are not familiar with your company's procedures or in-house jargon, acronyms, and abbreviations.

When you write memos, letters, and reports to low-tech readers, remember that they share the following characteristics.

- Low-tech readers are familiar with the technology, but their job responsibilities are peripheral to the subject matter. They either work in another department, manage you, work under your supervision, or work outside your company.

- Because low-tech readers are familiar with your subject matter, they understand some abbreviations, jargon, and technical concepts. To ensure that readers understand your content, therefore, define your terms. An abbreviation such as VLSI cannot stand alone. Define it parenthetically: VLSI (very large scale integration). Similarly, technical jargon such as *silicon foundry* needs a follow-up explanation. You should write instead, "A silicon foundry, as the name implies, is a factory or manufacturer that casts into silicon an integrated chip based on a customer's specifications."

  In addition, technical concepts must be defined for low-tech readers. For example, whereas high-tech readers understand the function of pressure transducers, a low-tech reader needs further information, such as "Pressure transducers: Solid-state components sense proximal pressure in the patient tubing circuit. The transducers convert this pressure value into a proportional voltage for the control system."

You can also use a Glossary to define terms.

- Since the low-tech reader is not in your normal writing "loop"—that is, not someone to whom you write often regarding your field of expertise—you need to provide more background information. When you submit a status report to upper-level management, for example, you cannot just begin with work accomplished. You need to explain why you are working on the project (objectives, history), who is involved (other personnel), when the project began and its scheduled end date, and how you are accomplishing your goals. Low-tech readers understand the basic concepts of your work, but they have not been involved in it daily. Fill them in on past history.

Lay Audience.   Customers and clients who neither work for your company nor have any knowledge about your field of expertise are your lay audience members. If you work in network communications for a telephone company, for example, and you write a letter to a client regarding a problem with the company's phone line, your audience is a lay reader. If your field of expertise is biomedical equipment and you write a procedures manual for the patient end user, you are writing to a lay audience.

If you are an automotive technician writing a service report for a customer, that customer is a lay reader. Although you understand your technology, your reader, who uses the phone or the medical equipment or the car, is not an expert in the field. These readers are using your equipment or require your services, but the technology topic is not within their daily realm of experience.

This makes writing for a lay audience difficult. It is easy to write to a high-tech reader who thoroughly understands the technology you are discussing. However, writing to a low-tech coworker or a lay audience totally outside your field of expertise is demanding.

To write successfully to a lay audience, remember that these readers share the following characteristics.

- Lay readers are unfamiliar with your subject matter. Because they do not understand your technology, write simply. Your tone should not insult your lay reader by sounding remedial or patronizing. To avoid this problem, explain your topic clearly. You achieve clarity through precise word usage, depth of detail, and simple graphics.
- Because your lay readers do not understand your technology or work environment, avoid using in-house jargon, abbreviations, or acronyms. Avoid high-tech terms, or define them thoroughly.
- Lay readers will need background information. If you leap into a discussion about a procedure without explaining to your lay readers why they should perform each step, they will not understand the causes or rationale. High-tech and possibly even low-tech readers might not need such explanations, but the lay reader needs clarity.

For example, look at the maintenance procedure, which is part of a user's manual provided for the purchaser of an audio recorder (Figure 4.1).

Next, look at the procedure for cable preparation from an installation manual geared toward high-tech readers (Figure 4.2).

A lay reader would have no idea what is going on in Figure 4.2. Why are we performing this act? What's the purpose? What will be achieved? Perhaps these questions are irrelevant for the technician, who knows the background or rationale, but a lay audience will be confused without an introductory overview. In addition, a lay audience will not know the terms *gasket, cable dielectric,* or *center conductor*. These terms are high-tech jargon, which a lay reader will not understand. What dimensions are we supposed to achieve? The high-tech reader, who is familiar with this operation, knows what is meant by "proper length," but not a lay audience. Finally, when is a clamp seated "properly in gasket"? Again, a lay audience needs these points clarified.

Composing effective technical communication requires that you recognize the differences among high-tech, low-tech, and lay audiences. If you incorrectly assume that all readers are experts in your field, you will create problems for yourself as well as for your readers. If you write using high-tech terms to low-tech or lay audiences, your readers will be confused and anxious. You will waste time on the phone clarifying the points that you did not make clear in the technical document.

Multiple Audiences.   Correspondence is not always sent to just one type of audience. Your correspondence may have an audience with multiple levels of expertise, or you may write to all of the aforementioned audiences simultaneously. For example, when writing a report, most people assume that the first-line supervisor will be the only reader. This might not be the case, however. The first-line supervisor could send a copy of your report to the manager, who could then submit the same report to the executive officer. Similarly, your first-line supervisor might send the report to your colleagues or to your subordinates. Your report might be sent out to other lateral departments. Figure 4.3 shows the possibilities.

Writing correspondence for multiple readers with different levels of understanding and reasons for reading creates a challenge. When you add the necessity of using an appropriate tone for all of these varied readers, the writing challenge becomes even greater.

FIGURE 4.1 Cleaning the Head Section

MAINTENANCE

Cleaning the Head Section

The heads, capstan, and pinch rollers get dirty easily. If this head section becomes dirty, the high-frequency sound will not be reproduced and the stereo balance will be impaired. This hurts your system's sound quality.

To avoid these problems, clean your system's head section regularly by following these simple steps:

1. Push the STOP/EJECT button to open the cassette door.
2. Dip a cleansing swab into the cleaning fluid.
3. Wipe the heads, capstan, and pinch rollers with the swab.
4. Allow 30 seconds to dry.

Note:

- Do not hold screwdrivers, metal objects, or magnets close to the heads.
- When demagnetizing the heads, be sure the unit's POWER switch is in the OFF position.

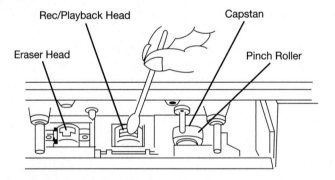

The introductory paragraph clarifies for the lay reader why she should clean the head section, instead of incorrectly assuming that the reader will understand without being told. This procedure also uses a simple graphic to clearly depict what action is required.

FIGURE 4.2 Cable Preparation

1. Place nut and gasket over cable and cut jacket to dimension shown.
2. Comb out braid and fold out. Cut cable dielectric to dimension shown. Tin center conductor.
3. Pull braid wires forward and taper toward center conductor. Place clamp over braid and push back against cable jacket.
4. Fold back braid wires as shown, trim to proper length (D), and form over clamp as shown. Solder contact to center conductor.
5. Insert cable and parts into connector body. Make sure sharp edge of clamp seats properly in gasket. Tighten nut.

This example provides no background about why to perform the action, no clarity about how to perform the steps, and no graphics to help the reader visualize the procedure. This procedure is written for a high-tech audience.

FIGURE 4.3 Examples of Possible Audiences for Correspondence

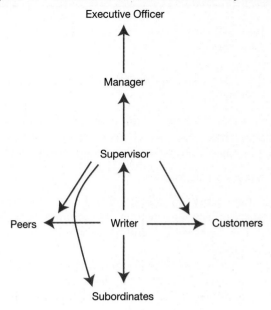

How do you meet such a challenge? The first key to success is recognizing that multiple audiences exist and that they share the following characteristics.

- Your intended audience may not be your only readers. Others might receive copies of your correspondence.
- Some of the multiple readers will be unfamiliar with the subject matter. Proper background data (objectives, overviews) will clarify the history of the report for these readers. In a letter, memo, report, or e-mail message, this background information cannot be too elaborate. Often, all you do is provide a reference line suggesting where the readers can find out more about the subject matter if they wish (e.g., "Reference: Operations Procedure 321 dated 9/21/08."). In longer reports, background data will appear in the summary or abstract, as well as in the report's introduction.

  Summaries, abstracts, introductions, and references are especially valuable when you consider one more fact regarding technical writing: reports, memos, and letters are kept on file. When you write the correspondence initially, you can assume that your audience has knowledge of the subject matter. Months (or years) later when the report is retrieved from the files, however, will your readers still be familiar with the topic? Will you have the same readers? Certain people who did not work for the company at the time of the original writing will read your correspondence. These future readers need background information.

Routine Correspondence and Reports

See Chapters 6 and 15 for additional information.

- Multiple readers have diverse understandings of your technology. Some will be high tech; others will be low tech. This requires that you define jargon, abbreviations, and acronyms. As mentioned earlier in this chapter, you can define your terms either parenthetically within your text or in a glossary, depending on the length of the correspondence. Short e-mail messages, memos, or letters allow for only parenthetical definitions. Long reports, Web sites, and manuals allow for glossaries and extended definitions.

Long Reports, Web Sites, and Manuals

See Chapters 12, 13 and 16 for additional information.

FIGURE 4.4 Effective Memo for Multiple Audiences

High-tech terms such as *operating procedure*, *engineering notice*, and *product quality requirements* are followed by parenthetical abbreviations—(OP), (EN), and (PQR). To cover all readers in the multiple audience, the writer correctly used both the written-out terms and the abbreviations.

The tone of the memo is appropriate for readers with different levels of responsibility. Although it contains no direct commands, which might have been offensive to management, the memo is still assertive. The matter-of-fact tone simultaneously suggests to management that these actions will be carried out and informs subordinates to do so.

DATE:	July 7, 2008
TO:	Distribution
FROM:	Rochelle Kroft
SUBJECT:	REVISION OF OPERATING PROCEDURE (OP) 354 DATED 5/31/08

The reissue of this procedure was the result of extensive changes requested by Engineering, Manufacturing, and Quality Assurance. These procedural changes will be implemented immediately according to Engineering Notice (EN) 185.

Some important changes are as follows:

1. *Substitutions.* An asterisk (*) can no longer be used to identify substitutable items in requirement lists. Engineering will authorize substitutions in work orders (WOs). Substitutions also will be specified by item numbers rather than by generic description names.
2. *Product Quality Requirements (PQRs).* These will be included in WOs either by stating the requirements or by referring to previous PQRs.
3. *Oral Instructions.* When oral instructions for process adjustments are given, the engineer must be present in the department.

All managers will review OP 354 and EN 185. Next, the managers will review the changes with their supervisors to make sure that each supervisor is aware of his or her responsibilities. These reviews will occur immediately.

Distribution:

Rob Harken	Manuel Ramos
Julie Burrton	Jeannie Kort
Hal Lang	Jan Hunt
Sharon Myers	Earl Eddings

- Correspondence geared toward multiple readers must have a matter-of-fact, businesslike tone. You should not be too authoritative because upper-level management might read the memo, letter, or report. You should not be too ingratiating because lower-level subordinates might also read the correspondence.

The memo in Figure 4.4 is written to a multiple audience. "To: Distribution" is a common way to direct correspondence to numerous readers. The multiple audience, then, affects the memo's content.

Figure 4.5 is a memo to low-tech security personnel, informing them of procedures to follow if they discover suspicious substances on site.

Figure 4.6 is a memo to high-tech HVAC employees, informing them of the capabilities of a new HVAC security system.

FIGURE 4.5 Memo to Low-Tech Audience

DATE: June 15, 2008
TO: Security Personnel
FROM: Bill Schneider, Facilities Manager
SUBJECT: Security of Heating, Ventilating, and Air Conditioning

When you are patrolling the facilities and grounds, be on the alert for suspicious substances, such as fine powders, residues, fog, mist, oily liquids or unusual odors. Not every liquid or odor is necessarily suspicious. Just focus on those with unexplainable origins. Also watch for discomfort in our employees, such as two or more people experiencing difficulty breathing. These signs could be evidence of dangerous chemicals in our HVAC system.

If you discover problems, follow these procedures:

- DO NOT touch the substance.
- Evacuate all employees from the affected area.
- Contact our HVAC personnel and ask them to shut down all systems to avoid spreading contamination.
- Secure the area.
- Report the situation to management, including the location, number of affected employees, and a description of the substance.

As first responders, your alertness and professionalism are essential to ensure the safety of our 5,000 employees.

This memo to low-tech readers deals with a very important subject (security in the face of potential terrorism). The writer presents only what the audience needs to know in order to complete their job responsibilities. The writer conveys this content simply, using short sentences, a list, and easy-to-understand words. Finally, the tone is conversational and friendly.

FIGURE 4.6 Memo to High-Tech Audience

DATE: June 15, 2008
TO: HVAC Personnel
FROM: Bill Schneider, Facilities Manager
SUBJECT: Security of HVAC

Due to increased concerns about the possibility of terrorist threats, our company is installing an interconnected wireless HVAC system and wireless security system. These two systems will be interconnected through a common wireless technology, using the same frequency, modulation and protocols.

Wireless Technology
The HVAC system uses wireless control systems, including RF, IR, and protocols such as 802.1 Ix, Bluetooth, and $IRDA_5$. We will install one wireless thermostat with a transceiver for controlling the HVAC system.

Improved HVAC Capabilities
This new security system will include the following:

- Wireless CO sensors and wireless smoke or fire detectors.
- Wireless control keypads for communicating with and controlling the security system.
- IR sensors to detect the presence of employees in rooms.
- Room controls to regulate the temperature of unoccupied rooms.
- GUIs to provide programmable thermostats.
- An air quality sensor configured to monitor toxicity levels and presence of foreign substances.

Next week, you will receive training for installation and monitoring of this new security system.

This memo, written to high-tech readers, uses terminology without explanation or definition. Its tone is businesslike and directive.

DEFINING TERMS FOR DIFFERENT AUDIENCE LEVELS

Communicating with audiences at different levels, including high-tech, low-tech, lay, and multiple readers, is challenging because each industry has its own specialized vocabulary. Usually your high-tech readers work most closely with you and share your expertise; colleagues understand your high-tech jargon. Therefore, you can use acronyms and abbreviations for these high-tech readers without providing definitions. However, it is your responsibility as a technical communicator to define terms that might be unfamiliar to audiences at other levels. You can accomplish this goal by using any or all of the following:

- Words or phrases that define your terms, presented either parenthetically or in a glossary
- Sentences that define your terms, either in a glossary or following the unfamiliar terms
- Extended definitions of one or more paragraphs

Low-tech readers at your company have different areas of expertise. They are unfamiliar with your high-tech acronyms and abbreviations, but they are familiar with your work environment. Therefore, parenthetical or brief definitions in a glossary should suffice. You could define ATM parenthetically (as asynchronous transfer mode). Then, your low-tech reader will not misconstrue the abbreviation as automatic teller machine. You could also use a glossary.

example		
HTTPS	Hypertext Transfer Protocol, Secure	
TDD	Telecommunication device for the deaf	
TTY	Teletypewriter	

Lay readers are further removed from your immediate work world. Customers, city council members, end users of your equipment, and vendors have no knowledge of your high-tech jargon. Merely defining your terms will not provide these readers with the necessary information. You could provide either a follow-up sentence or an extended definition. If you provide a sentence definition, include the following:

Term + Type + Distinguishing characteristics

The **term** consists of the words, abbreviations, or acronyms you are using. The **type** explains the class into which your term belongs. For example, a "car" is a "thing," but the word *thing* applies to too many types of "things," including bananas, hammers, and computers. You need to classify your term more precisely. What type of thing is a car? "Vehicle" can be a more precise classification, but a motorcycle, a dirigible balloon, and a submarine are also types of vehicles. That's when the **distinguishing characteristics** become important. How does a car differ from other vehicles? Perhaps the distinguishing characteristics of a car would include "land-driven, four-wheeled vehicles." The definition would read as follows:

example	
A *car* is a *vehicle* that has four wheels and is *driven on land*.	
Term + Type + Distinguishing characteristics	

Using a sentence to define HTTP, you would write

<table>
<tr>
<td>*Hypertext Transfer Protocol* is a computer access code that provides secure communications on the Internet, an intranet, or an extranet.</td>
<td>**example**</td>
</tr>
</table>

When you need to provide an extended definition of a paragraph or more, in addition to providing the term, type, and distinguishing characteristics, also consider including examples, procedures, and descriptions. Look at the following definition of a voltmeter.

<table>
<tr>
<td>The voltmeter is an instrument used to measure voltage. The voltmeter usually consists of a magnet, a moving coil, a resistor, and control springs. Types of voltmeters include the microvoltmeter, millivoltmeter, and kilovoltmeter, which measure voltages with a span of 1 billion to 1. By connecting between the points of a circuit, voltmeters measure potential difference.</td>
<td>**example**</td>
</tr>
</table>

Here are other ways to define your terms.

- Place the definition in the front matter or in an appendix of a document (such as a manual).
- Use endnotes in a report with other researched material.
- Use pop-up screens for online help incorporated into your Internet, intranet, or extranet site. The reader can access the definitions by clicking on hypertext links. (See Figure 4.7.)

FIGURE 4.7 Pop-up Definition

Use a gutter margin

1. On the **File** menu, click **Page Setup,** and then click the **Margins** tab.

2. In the **Gutter** box, enter a value for the gutter margin.

3. Under **Gutter position,** click **Left** or **Top.**

Tip
To set gutter margins for part of a document, select the text, and then change the gutter margins as usual. In the **Apply to** box, click **Selected text.** Word automatically inserts section breaks before and after

section break
A mark you insert to show the end of a section. A section break stores the section formatting elements, such as the margins, page orientation, headers and footers, and sequence of page numbers. A section break appears as a double dotted line that contains the words "Section Break."

Hypertext link to pop-up

Pop-up box, defining the term ("Section break"), the type ("a mark"), and the unique aspects of the term ("stores the section formatting elements . . ."), etc.

TABLE 4.2 How to Communicate to Different Audience Levels

Audience Level	How to Communicate	Sample
High Tech	Use jargon, acronym, or abbreviation alone.	The wastewater is being treated for DBPs.
Low Tech	Use jargon, acronym, or abbreviation with a parenthetical definition.	The wastewater is being treated for DBPs (disinfection by-products).
Lay	Use jargon, acronym, or abbreviation with a parenthetical definition *and* a brief explanation or extended definition.	The wastewater is being treated for DBPs (disinfection by-products), such as acid, methane, chlorine, and ammonia.

Table 4.2 shows techniques for communicating with audiences who have different levels of technical knowledge. These various techniques for defining terms are neither foolproof nor mandatory. You are always the final judge of how much information to provide for your intended readers. Remember, however, that if your readers fail to understand your content, then you have failed to communicate. No one will complain if you define your terms, but readers who are uncertain about your meaning either will call you for assistance, wasting your time, or will perform tasks incorrectly, wasting their time. It is better to provide too much information than not enough.

BIASED LANGUAGE—ISSUES OF DIVERSITY

Your audience will not be composed of people just like you. In contrast, your readers or listeners will more than likely be diverse. Diversity includes "gender, race/ethnicity, religion, age, sexual orientation, class, physical and mental characteristics, language, family issues, [and] departmental diversity" (Grimes and Richard 2003, 8).

Think of it this way: You work for a city government, and your audience is the citizenry. Who comprises your city's populace? Or, you work in a hospital. Who will visit your facility? Or, you work in a retail establishment. Who will shop there? Or, you work in a company, any company. Who are your coworkers? They are people of many different interests, levels of knowledge, and backgrounds—and they are all valuable to your business.

Why should you be concerned about a diverse audience?

1. **Diversity is protected by the law**—Prejudicial behavior and discrimination on the job will not be tolerated. "Non-compliance with Equal Opportunity or Affirmative Action legislation can result in fines and/or loss of contracts" (McInnes 2003).

2. **Respecting diversity is the right thing to do**—People should be treated equally, regardless of their age, gender, sexual orientation, culture, or religion.

3. **Diversity is good for business**—"An environment where all employees feel included and valued yields greater commitment and motivation" ("What Is The 'Business Case' For Diversity?" 2003). Clients and customers prefer shopping in an environment devoid of prejudice. In addition, "buying power . . . is represented by people from all walks of life. . . . To ensure that . . . products and services are designed to appeal to this diverse customer base, 'smart' companies are hiring people from those walks of life—for their specialized insights and knowledge" (McInnes 2003).

4. **A diverse workforce keeps companies competitive**—Talent does not come in one color, nationality, or belief system. Instead, talent is "represented by people from a vast array of backgrounds and life experiences. Competitive companies cannot allow discriminatory preferences and practices to impede them from attracting the best available talent" (McInnes 2003).

DOT-COM UPDATES

Go to http://www.shrm.org/diversity/ (the Society for Human Resource Management) for additional information about diversity in the workplace, including definitions, links, and articles about diversity in the news.

Diversity management is such an important concern that "American businesses, educational institutions, and governmental agencies have spent untold millions on multicultural awareness and diversity training. The intended outcome: organizations that value and celebrate differences as well as similarities, thereby creating a more harmonious and productive work/study environment" (Jordan 1999, 1).

This is further verified in "the results of the 1998 *Society for Human Resource Management Survey of Diversity Initiatives*, [which stated that] 84 percent of human resource professionals at Fortune 500 companies say their top-level executives think diversity management is important. At organizations outside of the Fortune 500, 67 percent of human resource professionals said diversity management is important to their organizations' high-level executives" ("What Is The 'Business Case' For Diversity?" 2003).

WWW
Check Online Resources

www.prenhall.com/gerson
For more information about issues of diversity, visit our companion Web site.

MULTICULTURALISM
Multicultural Communication

Another feature of diversity in technical communication is today's global economy. Your company will market its products or services worldwide. International business requires multicultural communication, the sharing of written and oral information between businesspeople from many different countries (Nethery 2003). Who is doing business internationally? Almost everyone!

DOT-COM UPDATES

To learn more about these companies' global communication concerns, visit their Web sites.

- http://www2.coca-cola.com/ourcompany/around world.html
- http://www.microsoft.com/Office/previous/xp/multilingual/
- http://www2.bv.com/about_us/office_locations.aspx
- http://www.hallmark.com

FAQs: Multiculturalism and Technical Communication

Q: Why is "multiculturalism" important in technical communication? Doesn't everyone in the United States speak English?

A: No, not everyone in the United States speaks English. Look at these facts:

- "50 million people in the U.S., 19 percent of the population, . . . speak a language other than English at home and 22 million . . . have limited English proficiency" (Gardner 2007).
- Between 1990 and 2000, the number of Americans speaking a language other than English at home grew by 15.1 million (a 47 percent increase) and the number with limited English proficiency grew by 7.3 million (a 53 percent increase).

Q: How diverse is the United States?

A: Your audience will be diverse—and the "majority" in the United States is changing.

- "America's population has become increasingly diverse as Hispanics, African Americans, Asians and people from many other segments rapidly contribute to the cultural richness of our population" (Melgoza 2007).
- Hispanics in 2006 number 44.2 million and are expected to surge to 51.4 million by 2011.
- Asian/Pacific Islanders will surpass 13 million in 2006.
- African Americans will number nearly 36 million in 2006.
- Asians, Blacks, and Hispanics accounted for more than 70 percent of population growth, and Hispanics accounted for more than 50 percent since 2000.
- By 2011, the population in the three largest ethnic groups will be more than 107 million and Hispanics will represent nearly half of that population.
- On July 1, 2006, America had 113.5 million households. Of this total, nearly 13 million were Hispanic, nearly 11.5 million African American, and nearly 4 million Asian (Melgoza, 2007).

The Global Economy.   Table 4.3 shows how production and communication at various major corporations are affected by the global economy.

The Challenges of Multicultural Communication.   The Internet and e-mail affect global communication and global commerce constantly. With these technologies, companies can market their products internationally and communicate with multicultural clients and coworkers at a keystroke. An international market is great for companies because a global economy increases sales opportunities. However, international commerce also creates written and oral communication challenges. Because companies that work internationally must communicate with all of their employees and clients, communication must be multilingual.

For example, Medtronic, a leading medical technology company, does business in 120 countries. Many of those countries mandate that product documentation be written in the local language. To meet these countries' demands, Medtronic translates its manuals into 11 languages: French, Italian, German, Spanish, Swedish, Dutch, Danish, Greek, Portuguese, Japanese, and Chinese (Walmer 1999, 230).

Multilingual reports create unique communication challenges, as John K. Courtis and Salleh Hassan note. Will each language version be identical "in terms of reading ease" and "in terms of content"? "Does the official first language version signal any

TABLE 4.3 Global Communication

Company	Impact
Coca Cola	Coca Cola produces 300 different brands in 200 countries. Seventy percent of their income is generated from outside the United States ("Around the World" 2003).
Microsoft	Microsoft realizes that "increasing numbers of companies need to have their computing systems support multiple languages. [Thus,] the Microsoft Windows XP Professional operating system [addresses] a multilingual environment by providing different versions and options . . . for all . . . users regardless of their location or language." To accommodate this multicultural need, Windows XP Professional is written in 25 different languages ("Multilingual Features in Windows XP Professional" 2001).
General Motors	General Motors, which employs about 355,000 people around the world, realizes that "integrity transcends borders, language, and culture." Thus, it seeks to "establish common goals, monitor progress and share best practices across [its] global operations" (General Motors 2003).
Black & Veatch	Black & Veatch, a global engineering company, has locations in Argentina, Australia, Botswana, Brazil, China, Czech Republic, Egypt, Germany, Hong Kong, India, Indonesia, Korea, Kuwait, Malaysia, Mexico, Philippines, Poland, Saudi Arabia, Singapore, South Africa, Swaziland, Taiwan, Thailand, Turkey, United Arab Emirates, United Kingdom, United States, Vietnam, Zambia, and Zimbabwe.
Hallmark Cards	Margaret Keating, vice president-operations for Hallmark Cards Inc., oversees Hallmark's North American manufacturing and distribution, graphics, global procurement, and global operations. She says, "Communication is critical with a group [of 6,000 employees] geographically dispersed." "She spends a lot of time making communication clear, concise and compelling. That it has to be translated into different languages makes efficient communication even more important." Technical communication at Hallmark "has to make sense to the audience, and the audiences are very different" (Cardarella 2003).

advantages to investors over second or third language versions? Are different language versions prepared with the same attention to tone, style, and emphasis?" "Does the external audit firm read each language version and test for accuracy, comparability and completeness?" (2002, 395).

Multicultural Team Projects. What about international, multilingual project work teams? If, for example, your U.S. company is planning to build a power plant in China, you will work with Chinese engineers, financial planners, and regulatory officials. To do so effectively, you will need to understand that country's

- Verbal and nonverbal communication norms
- Management styles
- Decision-making procedures
- Sense of time and place
- Local values, beliefs, and attitudes

Our natural instinct is to evaluate people and situations according to our sense of values, our cultural perspectives. That is called ethnocentrism—a belief that one's own culture represents the norm. Such is not the case. The world's citizenry does not share the same perspectives, beliefs, values, political systems, social orders, languages, or habits. Successful communication takes into consideration language differences, nonverbal communication differences, and cultural differences.

Due to the multicultural makeup of your audience, you must ensure that your writing, speaking, and nonverbal communication skills accommodate language barriers and cultural customs. The classic example of one company's failure to recognize the importance of translation concerns a car that was named Nova. In English, *nova* is defined as a star that spectacularly flares up. In contrast, *no va* in Spanish is translated as "no go," a poor advertisement for an automobile.

Communicating Globally—In Your Neighborhood

Cross-Cultural Communication. Multiculturalism will affect you not only when you communicate globally, but also when you are confronted with communication challenges in your own city and state. "Marlene Fine . . . argues that 'the challenge posed by the increasing cultural diversity of the U.S. workforce is perhaps the most pressing challenge of our times'" (Grimes and Richard 2003, 9). Another term for this challenge is *cross-cultural communication,* writing and speaking between businesspeople of two or more different cultures within the same country (Nethery 2003).

How big a challenge is this? "About 19 million people in the United States are not proficient in English" (Sanchez 2003, A1). In addition, look at the statistics regarding America's melting pot shown in Figure 4.8.

These numbers are estimated to change. A U.S. Department of Labor study of the civilian labor force by age, sex, race, and Hispanic origin projects that in 2014, the White population will fall in percentage to 67 percent, the Black population will fall to 12 percent, and the Asian population will stay the same at 5 percent. The Hispanic population will grow to 16 percent. See Figure 4.9.

One challenge presented by the increasingly multicultural nature of our society and workplace is language. Language barriers are an especially challenging situation for hospitals, police and fire personnel, and governmental agencies where failure to communicate effectively can have dangerous repercussions.

One hospital reported that it has "13 staff members supplying Spanish, Arabic and Somali translations. . . . In 2001, nearly 21,000 Spanish interpretations were performed at [the hospital]. The numbers for 2002 . . . exceeded 29,000 interpretations." However, this hospital's successful use of translators to help with doctor–patient communication is rare. "Only about 14 percent of U.S. hospitals provide training for volunteer translators." Most hospitals depend on the patient's relatives. "In one case studied, an 11-year-old sibling translated. The child made 58 mistakes" (Sanchez 2003, A4). Imagine how such errors can negatively impact healthcare and medical records.

The communication challenges are not only evident for employees in healthcare and other community infrastructures. Industries as diverse as banking, hospitality (restaurants and hotels), construction, agriculture, meat production and packing, and insurance also face difficulties when communicating with clients and employees for whom English is a second language.

FIGURE 4.8 Occupational Employment in Private Industry by Race/Ethnic Group, United States, 2005

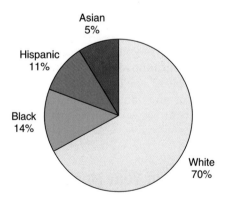

The U.S. Equal Employment Opportunity Commission. May 9, 2005. http://www.eeoc.gov/stats/jobpat/2003/national.html.

FIGURE 4.9 Occupational Employment in Private Industry by Race/Ethnic Group, United States, 2014

Asian 5%

Hispanic 16%

Black 12%

White 67%

U.S. Department of Labor. December 7, 2005. http://www.bls.gov/news.release/ecopro.t07.htm.

SPOTLIGHT

Why do some corporations provide foreign language training for their employees?

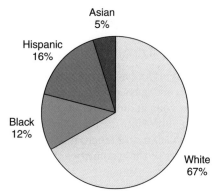

Phil Wegman, Program Director of Skills Enhancement for the Center for Business and Technology, sighed deeply and said, "I receive two to three calls *every day* from companies, desperate for Spanish language training. They need to teach their supervisors how to communicate more effectively with customers as well as with employees for whom English is a second language."

Phil's client base covers literally hundreds of fields, including dental and medical, criminal justice, public safety, transit, government, and education, as well as industries such as hospitality management (hotels and restaurants), construction, casinos, manufacturing, warehousing, banking, retail, childcare, and accounting.

Police, nurses, physicians, and paramedics, for example, have asked Phil to provide them occupational language training that is "work specific and real life." Then, when an emergency situation occurs, the safety personnel can ask, in Spanish, French, German, or Chinese, "Where does it hurt?"

McDonalds, Hardees, and Burger King supervisors need to know enough Spanish to be able to say to their staff, "Here's how you ring up this sale," "Here's how you make this meal," or "Here's how you clean this piece of equipment." Of equal importance, these companies have asked Phil to teach their supervisors how to answer customer questions or to take customer orders. Many of these fast-food restaurants have on staff a worker who is fluent in a second language. However, the pace is so hectic in fast food that the restaurants can't always pull that one employee off the job to handle the situation. Thus, *all* supervisors need enough language skills to provide direct commands, make simple statements, or answer common questions in a foreign language.

Sometimes, the needs are immediate . . . and life threatening. Phil says, "When a construction or manufacturing company is rolling out a new piece of equipment, where the learning curve is enormous, language can be a huge barrier. Maybe a company has either employees for whom English is a second language, or that company does business in a foreign country. Fluency becomes mandatory to ensure productivity, quality, and personal safety." If the new machinery is operated incorrectly, due to a failure to understand the English instructions, the end products might not meet standards and, of greater concern, injuries can occur.

There is one more key component to language knowledge, Phil states. "When a company teaches its employees a different language, the company honors that culture. The company tells its client base, 'We're trying to learn who you are and what makes you unique.'" The company is learning cross-cultural information to minimize barriers; the company is saying to its staff and customers, "We respect who you are and we want to work with you. That's just good business."

GUIDELINES FOR EFFECTIVE MULTICULTURAL COMMUNICATION

To achieve effective multicultural communication, follow these guidelines.

Define Acronyms and Abbreviations

Acronyms and abbreviations cause most readers a problem. Although you and your immediate colleagues might understand such high-tech usage, many readers will not, especially when your audience is not native to the United States. For example, corporate employees often abbreviate the job title "system manager" as *sysmgr*. However, in German, the title is *system leiter;* in French, it's *le responsable*. The abbreviation *sysmgr.* would make no sense in either of these countries (Swenson 1987, WE-193).

Avoid Jargon and Idioms

The same dilemma applies to *jargon* and *idioms,* words and phrases that are common expressions in English but that could be meaningless outside our borders. Every day in the United States, we use "on the other hand" as a transitional phrase and "in the black" or "in the red" to denote financial status. What will these idioms mean in a global market?

Similarly, the computer industry says, "The system crashed so we rebooted." A literal translation of this jargon into Chinese, German, or French might confuse the readers (Swenson 1987, WE-194).

Distinguish Between Nouns and Verbs

Many words in English act as both nouns and verbs. This is especially true with computer terms, such as *file, scroll, paste, code,* and *help.* If your text will be translated, make sure that your reader can tell whether you are using the word as a noun or a verb (Rains 1994).

Watch for Cultural Biases and Expectations

Your text will include words and graphics. As a technical communicator, you need to realize that many colors and images that connote one thing in the United States will have different meanings elsewhere. For example, the idioms "in the red" and "in the black" will not necessarily communicate your intent when they are translated. Even worse, the colors black and red have different meanings in different cultures. Red in the United States connotes danger; therefore, "in the red" suggests a financial problem. In China, however, the word *red* has a positive connotation, which would skew your intended meaning. The word *black* often implies death and danger, yet "in the black" suggests financial stability. Such contradictions could confuse readers in various countries.

Animals represent another multicultural challenge. In the United States, we say you're a "turkey" if you make a mistake, but success will make you "soar like an eagle." The same meanings don't translate in other cultures. Take the friendly piggy bank, for example. It represents a perfect image for savings accounts in the United States, but pork is a negative symbol in the Mideast. If you are "cowed" by your competition in the United States, you lose. Cows, in contrast, represent a positive and sacred image in India (Horton 1993, 686–93).

Be Careful When Using Slash Marks

Does the slash mark mean "and," "or" or both "and/or"? The word *and* means "both," but the word *or* means "one or the other," not necessarily both. If your text will be translated

to another language, will the translator know what you meant by using a slash mark? To avoid this problem, determine what you want to say and then say it (Rains 1994).

Avoid Humor and Puns

Humor is not universal. In the United States, people talk about regional humor. If a joke is good in the South but not in the North, how could that same joke be effective overseas? Microsoft's software package Excel is promoted by a logo that looks like an X superimposed over an L. This visual pun works in the United States because we pronounce the letters X and L just as we would the names of the software package. If your readers are not familiar with English, however, they might miss this clever sound-alike image (Horton 1993, 686).

Realize That Translations May Take More or Less Space

Paper Size.   If your writing will be conveyed not on paper but on disk or on the Internet, you must consider software's line-length and screen-length restrictions. For example, a page of hard-copy text in the United States will consist of approximately 55 lines that average 80 characters per line. How could this present a problem? The standard sheet of paper in the United States measures $8\frac{1}{2} \times 11$ inches. In contrast, the norm in Europe for standard-sized paper is A4—210×297 millimeters, or 8.27×11.69 inches.

Why is the size of paper important? This has nothing to do with language barriers, right? Here is the problem: If you format your text and graphics for an $8\frac{1}{2} \times 11$-inch piece of paper in Atlanta, travel to London, download the files on a computer there, and hit Print, you might find that what you get is not what you hoped for. The line breaks will not be the same. You will not be able to three-hole punch and bind the text. The margins will be off, and so will the spacing on your flowchart or table (Scott 2000, 20–21).

Web Sites.   An even greater problem occurs when you are writing for the Internet. On a Web site, you will provide a navigation bar and several frames. Why is this a problem? A page of English text translates to the same length in any language, right? The answer is no. The word count of a document written in English will expand more than 30 percent when translated into some European languages. For example, the four-letter word *user* has 12 characters in German (Hussey and Homnack 1990, RT-46). In Table 4.4, notice how English words become longer when translated into other languages (Horton 1993, 691).

The Swiss government is trying to curb what it defines as "the encroachment of English." To do so, the government's French Linguistics Service is asking its citizens to avoid using the word *spam,* instead opting for *courier de masse non sollicite,* meaning

TABLE 4.4 Translations Increase Word Length

English word	Translations into Other Languages
Print	*Impression* (French)
File	*Archivo* (Spanish)
View	*Visualizzare* (Italian)
Help	*Assistance* (French)
E-mail	*Courriel* (Swiss)

TABLE 4.5 Different Ways of Understanding and Writing the U.S. Date 05/03/08

Country	Date
United States	May 3, 2008
United Kingdom	March 5, 2008
France	5 mars 2008
Germany	5. Marz 2008
Sweden	08–05–03
Italy	5.3.08

"unsolicited bulk mail" ("Swiss Fight Encroachment of English" 2002, A16). Whereas the above examples add length to a document, "Chinese character writing captures the subject matter more compactly than English. Consequently, the length of the Chinese translation is usually about 60% of that needed by the English version" (Courtis and Hassan 2002, 397). You must consider length to accommodate translations.

Avoid Figurative Language

Many of us use sports images to figuratively illustrate our points. We "tackle" a chore; in business, a "good defense is the best offense"; we "huddle" to make decisions; if a sale isn't made, you might have "booted" the job; if a sale is made, you "hit a home run." Each of these sports images might mean something to native speakers, but they may not communicate worldwide. Instead, say what you mean, using precise words (Weiss 1998, 14).

Be Careful with Numbers, Measurements, Dates, and Times

Numbers and Measurements.   If your text uses measurements, you are probably using standard American inches, feet, and yards. However, most of the world measures in metrics. Thus, if you write 18 high × 20 wide × 30 deep, what are the measurements? There is a huge difference between $18 \times 20 \times 30$ inches and $18 \times 20 \times 30$ millimeters.

Dates.   In the United States, we tend to abbreviate dates as MM/DD/YY: 05/03/08. In the United Kingdom, however, this could be perceived as March 5, 2008, instead of May 3, 2008. See Table 4.5 for additional examples.

Time. Time is another challenge. Table 4.6 shows how different countries write times. In addition to different ways of writing time, you must also remember that even within the United States, 1:00 P.M. does not mean the same thing to everyone. Is that central time, Pacific time, mountain time, or eastern time?

where saving face is a cultural norm, potentially offending a superior would be a problem. Finally, "pronto" is cowboy slang. To communicate internationally, you must consider each country's cultural norms—and be careful to avoid offense or confusion. In contrast, the following example corrects these communication problems.

AFTER

To...	Jose Guerrero, Mexico Office; Yong Kim, Hong Kong Office; Hans Rittmaster, Berlin Office
Cc...	
Bcc...	
Subject:	AGENDA FOR TELECONFERENCE
Attachments:	

| Normal | A | Arial | 10 | A | B | I | U | ≡ ≡ ≡ | ⋮≡ ⋮≡ ⸬ ⸬ | ◄¶ ¶► |

We need to complete our team project. Doing so will allow our respective companies to meet our client's deadline. A teleconference is our best way to communicate, given everyone's diverse locations. I have made all the technical arrangements. The teleconference is scheduled for March 7, 2007, at 12:00 noon, Pacific standard time. This is the only time that is at least somewhat convenient for all of us. During the teleconference, we will discuss the following:

- Restructured design—rather than build the part at 8" x 10" x 23" as planned (all dimensions in inches), we should consider a smaller design.

- Shipping method—a new carrier or vendor might save us money and time. Flyrite Overnight has increased its shipping fees by 25 percent. What do you think? I am open to your creative ideas.

Brainstorm with your coworkers before our teleconference so we can use our time effectively. My boss, Sue Cottrell, needs our suggestions by March 8, 2007, 4:00 P.M. central standard time. With your help, I know our company will make the correct decisions.

Leonard Liss
1212 W. 148th
New York, NY
1-800-555-1212

SEXIST LANGUAGE

Check Online Resources

www.prenhall.com/gerson
For more information about sexist language, visit our companion Web Site.

Many of your readers will be women. This does not constitute a separate audience category. Women readers will be high tech or low tech, management or subordinate. Thus, you don't need to evaluate a woman's level of understanding or position in the chain of command any differently than you do for readers in general.

Recognize, however, that women constitute over half the workforce. As such, when you write, you should avoid *sexist language*, which is offensive to all readers.

Let's focus specifically on ways in which sexism is expressed and techniques for avoiding this problem. Sexism creates problems through omission, unequal treatment, and stereotyping, as well as through word choice.

Ignoring Women or Treating Them as Secondary

When your writing ignores women or refers to them as secondary, you are expressing sexist sentiments. The following are examples of biased comments and their nonsexist alternatives.

Biased	Unbiased
Radium was discovered by a woman, Marie Curie.	Radium was discovered by Marie Curie.
When setting up his experiment, the researcher must always check for errors.	When setting up experiments, the researcher must always check for errors.

| As we acquired scientific knowledge, men began to examine long-held ideas more critically. | As we acquired scientific knowledge, people began to examine long-held ideas more critically. |

Modifiers that describe women in physical terms not applied to men treat women unequally.

Biased	**Unbiased**
The poor women could no longer go on; the exhausted men . . .	The exhausted men and women could no longer go on.
Mrs. Acton, a statuesque blonde, is Joe Granger's assistant.	Jan Acton is Joe Granger's assistant.

Stereotyping

If your writing implies that only men do one kind of job and only women do another kind of job, you are stereotyping. For example, if all management jobs are held by men and all subordinate positions are held by women, this is sexist stereotyping.

Biased	**Unbiased**
Current tax regulations allow a head of household to deduct for the support of his children.	Current tax regulations allow a head of household to deduct for child support.
The manager is responsible for the productivity of his department; the foreman is responsible for the work of his linemen.	Management is responsible for departmental productivity. Supervisors are responsible for their personnel.
The secretary brought her boss his coffee.	The secretary brought the boss's coffee.
The teacher must be sure her lesson plans are filed.	The teacher must file all lesson plans.

Sexist language disappears when you use pronouns and nouns that treat all people equally.

Pronouns

Pronouns such as *he, him,* or *his* are masculine. Sometimes you read disclaimers by manufacturers stating that although these masculine pronouns are used, they are not intended to be sexist. They are only used for convenience. This is an unacceptable statement. When *he, him,* and *his* are used, a masculine image is created, whether such companies want to admit it.

To avoid this sexist image, avoid masculine pronouns. Instead, use the plural, generic *they* or *their.* You also can use *he or she* and *his or her. (S)he* is not a good compromise, because it's awkward. Sometimes you can solve the problem by omitting all pronouns.

Biased	**Unbiased**
Sometimes the doctor calls on his patients in their homes.	Sometimes the doctor calls on patients in their homes.
The typical child does his homework after school.	Most children do their homework after school.
A good lawyer will make sure that his clients are aware of their rights.	A good lawyer will make sure that clients are aware of their rights.

Gender-Tagged Nouns

Use nouns that are nonsexist. To achieve this, avoid nouns that exclude women and denote that only men are involved.

Gender-Tagged, Biased Nouns	Unbiased
mankind	people
manpower	workers, personnel
the common man	the average citizen
wise men	leaders
businessmen	businesspeople
policemen	police officers
firemen	firefighters
foreman	supervisor
chairman	chairperson, chair
stewardess/steward	flight attendant
waitress/waiter	server

Consider the following examples of sexist and nonsexist writing in a letter advertising landscaping services. In the first letter the sexist writing is boldfaced, whereas in the second example the sexist terms are replaced with nonsexist boldfaced words.

BEFORE

Sexist Writing

DiBono and Sons Landscaping
1349 Elm
Oakland Hills, IA 20981

April 2, 2008

Owner
TurboCharge, Inc.
2691 Sommers Rd.
Iron Horse, IA 20992

Dear **Sir:**

Interested in beautifying your company property, with no care or worries? We have been in business for over 25 years helping **men** just like you. Here's what we can offer:

1. Shrub Care—your **workmen** will no longer need to water shrubs. We'll take care of that with a sprinkler system geared toward your business's unique needs.
2. Seasonal System Checks—don't worry about your **foreman** having to ask **his** workers to turn on the sprinkler system in the spring or turn it off in the winter. Our **repairmen** take care of that for you as part of our contract.
3. New Annual Plantings—we plant rose bushes and daffodils that are so pretty, **your secretary will want to leave her desk and pick a bunch for her** office. Of course, you might want to do that for **her** yourself (☺).
4. Trees—we don't just plant annuals. We can add shade to your entire property. Just think how beautiful your parking lot will look with elegant elms, oaks, and maples. Then, when you have that company picnic, your **employees and their wives** will be surrounded by nature.

Call today for prices. Let our skilled **craftsmen** work for you!

Sincerely,

Richard DiBono

AFTER

Nonsexist Revision

DiBono and Sons Landscaping
1349 Elm
Oakland Hills, IA 20981

April 2, 2008

Owner
TurboCharge, Inc.
2691 Sommers Rd.
Iron Horse, IA 20992

Subject: Sales Information

Interested in beautifying your company property, with no care or worries? We have been in business for over 25 years helping **business owners** just like you. Here's what we can offer:

- Shrub Care—your **employees** will no longer need to water shrubs. We'll take care of that with a sprinkler system geared toward your business's unique needs.
- Seasonal System Checks—don't worry about your **supervisor** having to **ask his or her workers** to turn on the sprinkler system in the spring or turn it off in the winter. Our **staff** takes care of that for you as part of our contract.
- New Annual Plantings—we plant rose bushes and daffodils that are so pretty, **all of your employees** will want vase-filled flowers on **their** desks.
- Trees—we don't just plant annuals. We can add shade to your entire property. Just think how beautiful your parking lot will look with elegant elms, oaks, and maples. Then, when you have that company picnic, your **employees and their families and friends** will be surrounded by nature.

Call today for prices. Let our skilled **experts** work for you!

Sincerely,

Richard DiBono

AUDIENCE INVOLVEMENT

In addition to audience recognition, effective technical communication demands audience involvement. You not only need to know whom you are writing to in your technical correspondence (audience recognition), but also you need to involve your readers—lure them into your writing and keep them interested. Achieving audience involvement requires that you strive for personalized tone and reader benefit.

Personalized Tone

Companies do not write to companies; people write to people. Modern technical communication is often person to person. Remember that when you write your memo, letter, report, or procedure, another person will read it, so you want to achieve a personalized tone to involve your reader. This personalization can be accomplished in any of the following ways.

- Pronoun usage
- Names
- Contractions

Pronouns.   The best way to personalize correspondence is through pronoun usage. If you omit pronouns in your technical communication, the text will read as if it has been computer generated, absent of human contact. When you use pronouns in your technical communication, you humanize the text. You reveal that the memo, letter, report, or procedure is written by people, for people.

The generally accepted hierarchy of pronoun usage is as follows:

Pronoun	Focus
You	
Your	The reader
We	
Us	The team
Our	
I	
Me	The ego
My	

The first group of pronouns, *you* and *your*, is the most preferred. When you use *you* or *your,* you are speaking directly to your readers on a one-to-one basis. The readers, whoever they are, read the word *you* or *your* and see themselves in the pronoun, envisioning that they are being spoken to, focused on, and singled out. In other words, *you* or *your* appeals to the reader's sense of self-worth. We all like to be thought of as special and worthy of the writer's attention. By focusing on the pronouns *you* and *your* in technical communication, you make your readers feel special and involved in the writing transaction.

The second group of pronouns (*we, us,* and *our*) uses team words to connote camaraderie and group involvement. These pronouns are especially valuable when writing to multiple audiences or when writing to subordinates. *We, us,* and *our* imply to the readers that "we're all in this together." Such a team concept helps motivate by making the readers feel an integral part of the whole.

The third group (*I, me,* and *my*) denotes the writer's involvement. If overused, however, these pronouns can connote egocentricity ("All I care about is *me, me, me.*"). Because of this potential danger, emphasize *you* and *your* and downplay *I, me,* and *my.* Strive for a two-to-one ratio. For every *I, me,* or *my,* double your use of *you* or *your.* We are not saying that you should avoid using first-person pronouns. You cannot write without involving yourself through *I* or *my.* We are just saying that an overuse of these first-person pronouns creates an egocentric image, whereas an abundance of *you* orientation achieves audience involvement.

Let's look at an automobile manufacturer's user manual, which omits pronouns and, thus, reads as if it's computer generated.

BEFORE

Claims Procedure
To obtain service under the Emissions Performance Warranty, take the vehicle to the company dealer as soon as possible after it fails an I/M test along with documentation showing that the vehicle failed an EPA-approved emissions test.

Without any pronouns, the version has no personality. It's dull and stiff, and it reads as if no human beings are involved.

Compare this flat, dry, dehumanized example with the more personalized version.

AFTER

> **Claims Procedure**
>
> How do you get service under the Emissions Performance Warranty? To get service under this warranty, take your car to the dealer as soon as possible after it has failed an EPA-approved test. Be sure to bring along the document that shows your car failed the test.

When pronouns are added in this version, the writing involves the reader on a personal level.

Many writers believe that the first version is more professional. However, professionalism does not require that you write without personality or deny the existence of your reader. The pronoun-based writing in the second version is more friendly. Friendliness and humanism are positive attributes in technical communication.

The second version also ensures reader involvement. As a professional, you want your readers (customers, clients, colleagues) to be involved, for without these readers no transaction occurs. You end up writing in a vacuum and thus denying your audience, which is not the goal of technical communication.

This desire for reader involvement through pronoun usage is evident in other examples of technical communication. Here are a few lines from a legal contract, which again shows that professional writers are striving to involve readers through pronoun usage.

example

> Throughout this policy, *you* and *your* refer to the "named insured" shown in the Declaration. *We, us,* and *our* refer to the company providing the insurance.

Using pronouns takes a conscious effort. Technical communicators want to erase the old-fashioned style of writing that denies the reader's existence or depersonalizes the reader through stiff euphemisms such as the "named insured." Instead, writers want to involve the readers and humanize the text. Pronouns soften the technical edge of such communication.

The following is an example of dry, depersonalized correspondence and a more friendly revision.

example

Depersonalized	Friendly
Dear Sir:	Dear Mr. Riojas:
With regard to lost policy 123, enclosed is a lost policy form. Complete this form and return it ASAP. Upon receipt, the company will issue a replacement policy.	Your lost policy 123 can be replaced easily. I've enclosed a form to help us replace it for you. All you need to do is fill it out for us. As soon as we get it, we'll send you your replacement policy.

Names

Another way to personalize and achieve audience involvement is through use of names—incorporating the reader's name in your technical communication. By doing so, you create a friendly reading environment in which you speak directly to your audience.

Note in the examples that follow, that the reader's first name is used in the first example and the reader's last name in the second example. When do you use first names versus last names? The decision is based on your closeness to or familiarity with your reader. If you know your reader well, have worked together for a while, and know the

---

DATE: June 22, 2008
TO: Steve McMann
FROM: Maureen Pierce
SUBJECT: CABLE PURCHASE ORDERS

Steve, attached are requisitions for cable replacement for California. These are held pending our analysis and your approval:

Requisition No.	Amount	Location
1045	$3,126	Waconia
1825	$4,900	Chaska
2561	$3,829	Chanhassen

When you sign the attached requisition forms, Steve, we'll get right on the purchases.

Pfeiffer Consulting

February 13, 2008

Jennifer Shanmugan
Home Health Care
1238 Roe
Phoenix, AZ 34143

Subject: Proposed Business Writing Seminar

Thank you, Ms. Shanmugan, for inviting us to help your colleagues with their business writing. We're looking forward to working with you.

Here's a reminder of our agreed arrangements:
Date: February 22, 2008
Time: 8:00 A.M.–5:00 P.M.
Site: Cedar Inn

If we can provide any other information, please let us know.

person will not be offended, then use the first name. However, if you do not know the reader well, have not worked together, or worry that the reader might be offended, then use the surname instead. In either instance, calling a reader by name will involve that reader and personalize your writing.

Contractions

One way to achieve a conversational tone in technical communication is to use contractions occasionally (*we're, let's, here's, it's, can't,* etc.). Contractions now are accepted in composition classes as well as in technical communication. We all use them every day, and they make technical communication more approachable and more friendly.

The following examples prove our point. The first is from a user manual; the second is from a memo written by a CEO and distributed to an entire company.

In addition to the contractions, note that the text achieves audience involvement through pronouns.

CONNECTING THE DG-40 TO YOUR COMPUTER

Now that your printer is working, it's time to hook it up to your computer. It's best to turn off both the computer and the printer when you hook them up.

Remember that each computer communicates differently with a printer. If your computer communicates through a parallel interface, all you'll need is a cable. If your computer requires another kind of interface, then you'll need an interface board.

If you don't know what a parallel interface is, your computer manual or dealer will tell you.

This letter epitomizes motivation through personalization. The text involves the readers by making them feel a part of the team (through pronouns) and softens the technical edge of the writing (through conversational contractions).

Dear Friends,

I can't tell you how proud I am of the extraordinary effort each of you has put forth toward reaching our goal. As you've probably heard by now, we attained 99.89 percent of our March 31, 2008, ship schedule.

We did not quite reach our gold-ring goal of 100 percent ship performance, but we are so close that we're going to celebrate as if we did. We are committed to continuous improvement. I'd like to invite you to a special celebration this Tuesday.

Let's take a moment to congratulate ourselves before we get back to work. We've got more units to ship and more gold rings to reach. When we work together as a team, we can make significant achievements.

Reader Benefit

A final way to achieve audience involvement is to motivate your readers by giving them what they want or need. We're not saying that you should make false promises. Instead, you should show your audience how they will benefit from your technical communication. You can do this one of two ways: explain the benefit or use positive words and verbs.

Explain the Benefit.   Until you tell your readers how they will benefit, they may not know. In your letter, memo, report, or manual, state the benefit clearly. You can do this anywhere, but you're probably wise to place the statement of benefit either early (first paragraph or abstract/summary) or late (last paragraph or conclusion). Placing the benefit early in the writing will interest readers and help ensure that they continue to read, remaining alert throughout the rest of the document. Placing the benefit at the end could provide a motivational close, leaving readers with a positive impression when they finish.

Routine Correspondence, Reports, and Manuals
See Chapters 6, 12, and 15 for additional information.

For example, when writing a procedure, you want your readers to know that by performing the steps appropriately, they will avoid mechanism breakdowns, maintain tolerance requirements, achieve a proper fit, or ensure successful equipment operation. By following the procedures, they will reap a benefit.

The following examples of procedures show how reader benefit is conveyed.

Instructions for Poured Foundations

A poured foundation will provide a level surface for mounting both the pump and motor. Carefully aligned equipment will provide you a longer and more easily maintained operation.

`example`

Instructions for Drive Belt Installation

When installing your Amox Drive Belt, you need to achieve a belt tension of 3/8" deflection for a 9-lb. to 13-lb. load applied at the center span. Check tension frequently during the first 30 hours of operation. Correct belt tension is important for maximum belt life in a large drive system.

`example`

Memos can also develop reader benefit to involve audiences. Figure 4.10 achieves audience involvement through reader benefit in many ways.

Use Positive Words and Verbs.   In the preceding examples geared toward reader benefit, the motivation or value is revealed through positive words or verbs. A sure way to involve your audience is to use positive words throughout your correspondence or to inject verbs (power writing) into your text. Positive words give your writing a warm glow; verbs give it punch. Positive words and verbs can sell the reader on the benefits of your subject matter.

Positive Words				
advantage	effective	happy	profitable	successful
asset	efficient	please	satisfied	thank you
benefit	enjoyable	pleased	succeed	value
confident	favorable	pleasure	success	worthwhile

To clarify how important such words are, let's look at some examples of negative writing, followed by positive revisions.

Negative	Positive
• We cannot process your request. You failed to follow the printed instructions.	• So that we may process your request rapidly, please fill in line 6 on the printed form.
• The error is your fault. You keep your books incorrectly and cannot complain about our deliveries. If you would cooperate with us, we could solve your problem.	• To ensure prompt deliveries, let's get together to review our bookkeeping practices. Would next Tuesday be convenient?
• We have received your letter complaining about our services.	• Thank you for writing to us about our services.
• Your bill is now three weeks overdue. Failure to pay immediately will result in lower credit ratings.	• If you're as busy as we are, you've probably misplaced our recent bill (mailed three weeks ago). Please send it in soon to maintain your high credit ratings.
• The invoice you sent was useless by the time it arrived. You wasted my time and money.	• We're proud of our ability to maintain schedules. But we need your help. When you return invoices by the fifteenth, we save time and you save money.

FIGURE 4.10 Memo Achieving Audience Involvement

The benefit is evident in the positive words—*improved, strengthen, reduce, timely, accurate, happier, pride,* and *contributions.* These words motivate readers by making them perceive a benefit from their actions.

MEMORANDUM

DATE: December 2, 2007
TO: Fred Mittleman
FROM: Renata Shuklaper
SUBJECT: IMPROVED BILLING CONTROL PROCEDURES

Three times in the last quarter, one of the billing cycles excluded franchise taxes. This was a result of the CRT input being entered incorrectly.

To strengthen internal controls and reduce the chance of future errors, the following steps will be implemented, effective January 1, 2008.

1. Input   The control screen entry form will be completed by the accounting clerk.
2. Entry   The CRT entry clerk will enter the control screen input and use local print to produce a hard copy of the actual input.
3. Verification   The junior accountant will compare the hard copy to the control screen and initial if correct.
4. Notification   The junior accountant will tell data processing to start the billing process.

Reader benefit is emphasized in the last paragraph. The pronouns *we* and *our* in the last sentence involve the readers in the activity. A team concept is implied.

The new procedure will provide more timely and accurate billing to our customers. Customer service will be happier with accounting, and we will have more pride in our contributions.

Making something positive out of something negative is a challenge, but the rewards for doing so are great. If you attack your readers with negatives, you lose. When you involve your readers through positive words, you motivate them to work with you and for you.

Verbs also motivate your readers. The following verbs will motivate your readers to action.

accomplish	develop	insure	raise
achieve	direct	lead	recommend
advise	educate	maintain	reduce
analyze	ensure	manage	reestablish
assess	establish	monitor	serve
assist	guide	negotiate	supervise
build	help	operate	support
conduct	implement	organize	target
construct	improve	plan	train
control	increase	prepare	use
coordinate	initiate	produce	
create	install	promote	

The memo in Figure 4.11 uses power verbs to motivate the reader.

FIGURE 4.11 Memo Using Power Verbs

DATE: July 22, 2008
TO: Martha Collins
FROM: Bob Kaplan
SUBJECT: ASSIGNMENT OF CONTRACT ADMINISTRATION
               PROJECT

Martha, thank you for accepting the Contract Administration Project. To accomplish our goal of competitive bidding, I need your help. Please follow these guidelines with your team:

1. Analyze other operating companies to see how they ensure neutrality when administering contracts.
2. Coordinate our consultants to guarantee that they achieve impartiality.
3. Develop standards for all operating procedures across all departments.
4. Prepare a report for our advisory team. They'll need it by the end of the month.

Your successful completion of this project, Martha, will help us improve all future contract administrations. I'm confident that you'll do a great job, and I know our company and employees will benefit.

This memo achieves audience involvement by speaking directly to the reader ("Martha"), by using pronouns, by including contractions to soften the tone, by using verbs that give the memo punch, and by employing positive words to motivate the reader to action.

To communicate effectively with readers, ask yourself the following questions. Then circle the appropriate responses or fill in the needed information.

AUDIENCE CHECKLIST

_____ 1. What is my audience's level of understanding regarding the subject matter?
- High tech
- Low tech
- Lay
- Multiple

_____ 2. Given my audience's level of understanding, have I written accordingly?
- Have I defined my acronyms, abbreviations, and jargon?
- Have I supplied enough background data?
- Have I used the appropriate types of graphics?

_____ 3. Who else might read my correspondence?
- How many people?
- What are their levels of understanding?
  - High tech
  - Low tech
  - Lay
- Will my audience be multicultural?

_____ 4. What is my role in relation to my audience?
- Do I work for the reader?
- Does the reader work for me?
- Is the reader a peer?
- Is the reader a client?

_____ 5. What response do I want from my audience? Do I want my audience to act, respond, confirm, consider, decide, or file the information for future reference?

_____ 6. Will my audience act according to my wishes? What is my audience's attitude toward the subject (and me)?
- Negative
- Positive
- Noncommittal
- Uninformed

_____ 7. Is my audience in a position of authority to act according to my wishes (to make the final decision)? If not, who will make the decision?

_____ 8. What are my audience's personality traits?
- Slow to act
- Eager
- Receptive
- Questioning

- Organized
- Disorganized
- Reticent

_____ 9. Have I motivated my audience to act?
- Have I involved my reader in the correspondence by using pronouns and his or her name?
- Have I shown my reader the benefit of the proposal by using positive words and verbs?
- Have I considered any questions or objections my audience might have?

_____ 10. Have I considered my audience's preferences regarding style?
- Will my reader accept contractions?
- Is use of a first name appropriate?
- Should I use my reader's last name?

_____ 11. Have I avoided sexist language?
- Have I used _their_ or _his or her_ to avoid sexist singular pronouns such as _his_?
- Have I used generic words such as _police officer_ versus _policeman_?
- Have I avoided excluding women, writing sentences such as, "All present voted to accept the proposal," versus the sexist sentence, "All the men voted to accept the proposal"?
- Have I avoided patronizing, writing sentences such as, "Mr. Smith and Mrs. Brown wrote the proposal," versus the sexist sentence, "Mr. Smith and Judy wrote the proposal"?

_____ 12. Have I considered diversity?
- Have I avoided any language that could offend various age groups, people of different sexual orientations, people with disabilities, or people of different cultures and religions?
- Have I considered that people from different countries and people for whom English is a second language will be involved in the communication? This might mean that I should do the following:
  - Clarify time and measurements.
  - Define abbreviations and acronyms.
  - Avoid figurative language and idiomatic phrases unique to one culture.
  - Avoid humor and puns.
  - Consider each country's cultural norms.

CHAPTER HIGHLIGHTS

1. High-tech readers understand acronyms, abbreviations, and jargon, but not all of your readers will be high tech.
2. Low-tech readers need glossaries or parenthetical definitions of technical terms.
3. Lay readers usually will not understand technical terms. Consider providing these readers with extended definitions.
4. In addition to considering high-tech, low-tech, and lay audience levels, you also should consider issues of diversity, such as gender, race, religion, age, and sexual orientation.
5. Multiple audiences have different levels of technical knowledge. This fact affects the amount of technical content and terms you should include in your document.
6. The definition of a technical term usually includes its type and distinguishing characteristics, but can also include examples, descriptions, and procedures.
7. For multicultural audiences, define terms, avoid jargon and idioms, and consider the context of the words you use. Avoid cultural biases, be careful with humor, and allow space for translation.
8. More than 50 percent of your audience will be female, so avoid sexist language.
9. To get your audience involved in the text, personalize it by using pronouns, the reader's name, and contractions.
10. Show your reader how he or she will benefit from your message or proposal.

APPLY YOUR KNOWLEDGE

CASE STUDY
Home and Business Mortgage

Home and Business Mortgage (HBM) has suffered several lawsuits recently. A former employee sued the company, contending that it practiced "ageism" by promoting a younger employee over him. In an unrelated case, another employee contended that she was denied a raise due to her ethnicity. To combat these concerns, HBM has instituted new human resources practices. The goal is to ensure that the company meets all governmental personnel regulations, including those required by the Equal Employment Opportunity Commission (EEOC), the Family Medical Leave Act (FMLA), and the Americans with Disabilities Act (ADA). HBM is committed to achieving diversity in its workplace.

HBM now must hire a new office manager for one of its branch operations. It has three outstanding candidates. Following are their credentials.

- **Carlos Gutierrez**—Carlos is a 27-year-old recent recipient of an MBA (masters in business administration) from an acclaimed business school. At this university, he learned many modern business applications, including capital budgeting, human resource management, organizational behavior, diversity management, accounting and marketing management, and team management strategies. He has been out of graduate school for only two years, but his work during that time has been outstanding. As an employee at one of HBM's branches, he has already impressed his bosses by increasing the branch's market share by 28 percent through innovative marketing strategies he learned in college. In addition, his colleagues enjoy working with him and praise his team-building skills. Carlos has never managed a staff, but he is filled with promise.

- **Cheryl Huff**—Cheryl is a 37-year-old employee of a rival mortgage company, Farm-Ranch Equity. She has a BGS (bachelor of general studies) from a local university, which she acquired while working full time in the mortgage/real estate business. At Farm-Ranch, where she has worked for over 15 years, she has moved up the ladder, acquiring many new levels of responsibility. After working as an executive assistant for 5 years, she then became a mortgage payoff clerk for 2 years, a loan processor for 2 years, a mortgage sales manager for 4 years, and, most recently, a residential mortgage closing coordinator for the last 2 years. In the last 2 positions, she managed a staff of 5 employees.

  According to her references, she has been dependable on the job. Her references also suggest that she is a forceful taskmaster. Though she has accomplished much on the job, her subordinates have chaffed at her demanding expectations. Her references suggest, perhaps, that she could profit from some improved people skills.

- **Rose Massin**—Rose is 47 years old and has been out of the workforce for 12 years. During that absence, she raised a family of 3 children. Now, the youngest child is in school and Rose wants to reenter the job market. She has a bachelor's degree in business, and was the former office manager of this HBM branch. Thus, she has management experience, as well as mortgage experience.

  While she was branch manager, she did an outstanding job. This included increasing business, working well with employees and clients, and maintaining excellent relationships with lending banks and realtors. She was a highly respected employee and was involved in civic activities and community volunteerism. In fact, she is immediate past-president of the local Rotary club. Though she has been out of the workforce for years, she has kept active in the city and has maintained excellent business contacts. Still, she's a bit rusty on modern business practices.

Assignment

Whom will you hire from these three candidates? Based on the information provided, make your hiring decision. This is a judgment call. Be sure to substantiate your decision with as much proof as possible. Then share this finding as follows:

- **Oral Presentations**—give a 3–5 minute presentation to share with your colleagues the results of your decision.
- **Written**—write an e-mail, memo, or report about your findings.

Exercises

Avoiding Sexist Language

Revise the following sentences to avoid sexist language.

1. All the software development specialists and their wives attended the conference.
2. The foremen met to discuss techniques for handling union grievances.
3. Every technician must keep accurate records for his monthly activity reports.
4. The president of the corporation, a woman, met with her sales staff.
5. Throughout the history of mankind, each scientist has tried to make his mark with a discovery of significant intellectual worth.
6. The chairman of the meeting handled the debate well.
7. All manmade components were checked for microscopic cracks.
8. The supervisors have always brought the girls flowers on Mother's Day.
9. Mr. Smith (the CEO) introduced his staff: Mr. Jones (the vice president), Mr. Brown (the chief engineer), and Judy (the executive secretary).
10. The policemen and firemen rushed to the scene of the accident.

Achieving Audience Involvement

The following sentences are either dull, dry, and impersonal or egocentric.

1. Revise them to achieve audience involvement through personalization, adding pronouns, names, or contractions. Strive for a two-to-one ratio, using *you* twice for every *I*.
   a. The company will require further information before processing this request.
   b. It has been decided that a new procedure must be implemented to avoid further mechanical failures.
   c. The department supervisor wants to extend a heartfelt thanks for the fine efforts expended.
   d. I think you have done a great job. I want you to know that you have surpassed this month's quota by 12 percent. I believe I can speak for the entire department by saying thank you.
   e. If the computer overloads, simultaneously press Reset and Control. Wait for the screen command. If it reads "Data Recovered," continue operations. If it reads "I/O Error," call the computer resource center.
2. Revise the following sentences to achieve reader benefit, using positive or power verbs.
   a. We cannot lay your cable until you sign the attached waiver.
   b. John, don't purchase the wrong program. If we continue to keep inefficient records, our customers will continue to complain.
   c. You have not paid your bill yet. Failure to do so might result in termination of services.
   d. If you incorrectly quote and paraphrase, you will receive an *F* on the assignment.
   e. Your team has lost 6 of their last 12 games.
3. Rewrite the following flawed correspondence. Its tone is too negative and commanding, regardless of the audience. If the audience is composed of subordinates, the tone is too dictatorial. Rather than encouraging team building, the correspondence borders on officiousness. If the audience is composed of coworkers or upper-level management, then the tone creates an even greater offense. Soften the tone to achieve better audience involvement and motivation.

DATE: October 15, 2008
TO: Distribution
FROM: Darryl Kennedy
SUBJECT: FOURTH QUARTER GOALS

Due to a severe lack of discipline, the company failed to meet third quarter goals. To avoid repeating this disaster for the fourth quarter, this is what I think you all must do—ASAP.

1. Demand that the sales department increase cold calls by 15 percent.
2. Require weekly progress reports by all sales staff.
3. Penalize employees when reports are not provided on time.
4. Tell managers to keep on top of their staff, prodding them to meet these goals.

Remember, when one link is weak in the chain, the entire company suffers. DON'T BE THE WEAK LINK!

Defining Terms for Different Audience Levels

1. Find examples of definitions provided in computer word-processing programs, e-mail packages, Internet dictionaries, and online help screens. Determine whether these examples provide the *term,* its *type,* and its *distinguishing characteristics.* Are the definitions effective? If so, explain why. If not, rewrite the definitions for clarity.

2. Find examples of definitions in manuals such as your car's user manual, a manual that accompanied your computer, or manuals packaged with your radio or alarm clock, coffeemaker, VCR, or lawn mower. Note where these definitions are located. Are they placed parenthetically within the text or in a glossary? Next, determine whether the definitions provide the *term,* its *type,* and its *distinguishing characteristics*. Are the definitions effective? If so, explain why. If not, rewrite the definitions for clarity.

3. Find examples of definitions in your textbooks. Where are these definitions placed—parenthetically within the text or in a glossary? Next, determine whether the definitions provide the *term,* its *type,* and its *distinguishing characteristics*. Are the definitions effective? If so, explain why. If not, rewrite the definitions for clarity.

4. Select 5 to 10 terms, abbreviations, or acronyms from your area of expertise or interest (network communications, accounting, electronic engineering, automotive technology, biomedical records, data processing, etc.). First, define each term using words and phrases. Next, define each term in a sentence, conveying the *term,* its *type,* and its *distinguishing characteristics*. Finally, define each term using a longer paragraph. In these longer definitions, include examples, descriptions, or processes.

Recognizing Issues of Diversity

1. Rewrite the following sentences for multicultural, cross-cultural audiences.
   a. Let's meet at 8:30 P.M.
   b. The best size for this new component is $16 \times 23 \times 41$.
   c. To keep us out of the red, we need to round up employees who can put their pedal to the metal and get us out of this hole.
   d. We need to produce flyers/brochures to increase business.
   e. The meeting is planned for 07/09/08.

2. Visit local banks, hospitals, police or fire stations, or city officials. Ask employees at these sites what diversity challenges they have encountered and how they have tried to solve these problems. These challenges could be language barriers presented by multiculturalism or cross-culturalism. Perhaps the issue is diversity hiring to avoid ageism or sexism. Maybe you will discover problems in the workplace created by diversity hiring, such as incompatible salaries or meeting Equal Opportunity regulations. Report on your findings either orally or in writing.

3. Rewrite the following flawed correspondence. Be sure to consider your multicultural or cross-cultural audience's needs.

TO: Andre Castro, Barcelona; Sunyun Wang, Singapore; Nachman Sumani, Tel Aviv
FROM: Ron Schaefer, New York
SUBJECT: Brainstorming

I need to pick your brains, fellows. We've got a big one coming up, a killer deal with a major European player. Before I can make the pitch, however, let's brainstorm solutions. The client needs a proposal by 12/11/08, so I need your input before that date. Give me your ideas about the following:

1. What should we charge for our product, if the client buys in bulk?

2. What's our turnaround time for production? I know that your people tend to work slowly, so can you hurry the team up if I promise delivery in six weeks?

E-mail me your feedback by tomorrow, 1:00 P.M. my time, at the latest. Trust me, guys. If we boot this one, everyone's bonuses will be in a sling.

Team Projects

1. In small groups of individuals from like majors, list 10 high-tech terms (jargon, acronyms, or abbreviations) unique to your degree programs. Then, envisioning a lay audience, parenthetically define and briefly explain these terms.

   To test the success of your communication abilities, orally share these high-tech terms with other students who have different majors. First, state the high-tech term to see if they understand it. If they don't, provide the parenthetical definition. How much does this help? Do they understand now? If not, add the third step—the brief explanation.

   How much information do the readers need before they understand your high-tech terms?

2. Individually, find examples of writing geared to high-tech, low-tech, and lay audiences. To do so, read professional journals, find procedures and instructions, look at marketing brochures, read trade magazines, or ask colleagues and coworkers for memos, letters, or reports.

   Once you have found these examples, bring them to class. In small groups, discuss your findings to determine whether they are written for high-tech, low-tech, or lay readers. Use the Audience Evaluation Form that follows to record your decisions. Share these with all class members.

Audience Evaluation Form

Example Number	1	2	3	4	5
Type (CIRCLE ONE)	HIGH LOW LAY	HIGH LOW LAY	HIGH LOW LAY	HIGH LOW LAY	HIGH LOW LAY

Criteria
Language
• Abbreviations
• Acronyms
• Jargon (defined?)
 	
Content
• General
• Specific
• Background
 	
Tone
• Formal (high)
• Less formal (low)
• Least formal (lay)
 	
Format
• Highlighting
• Graphics
  — Type
  — Complexity
  — Color
  — Labeling					

3. Once you've made your decisions regarding team project 2, select one of the high-tech examples and rewrite it for a low-tech or lay audience. To do so, work with your group to define your high-tech terms, improve the tone through pronouns and positive words, enhance the page layout through highlighting techniques, and add appropriate graphics.

4. Rewrite the following flawed correspondence. Be sure to achieve effective audience understanding and involvement. To do so, avoid sexist language and define high-tech terms. Search online to find any abbreviations or acronyms that need defining. Remember: Though the immediate readers might understand all of the high-tech terminology, other readers might not. These additional audiences could include lawyers, employees from other departments, or low-tech clients.

West Central Auditors
"Your Technical Engineering and Energy Resource Experts"
1890 River
Pocato, Idaho 89022

March 12, 2008

Marks-McGraw, Inc.
2145 Oceanview
Clackamas, Oregon

Gentlemen:

We will visit your plant next week for the TEA. To ensure that our visit goes smoothly, we plan this procedure:

- Your plant foreman will have his engineer assemble his CAD drawings for the power plant at Brush Prairie.
- Our men will review these drawings against TE specs, such as specific cost estimates for energy usage and energy data.
- Our ER group will then help your men plan and prepare your facilities for technology changes mandated by the DOE's office of the EERE.
- We also want to review your engineer's plans for any building envelopes. Make sure he brings all relevant correspondence he has had with his clients.

This audit should take approximately seven man-hours. It's a man-sized job. If we have to go off site for lunch or breaks, the job will take even longer. Therefore, please ask your secretary to provide drinks and food for our six representatives. Tell her that we have no dietary concerns, so anything she provides will be greatly appreciated.

Sincerely,

Jim Wynn, Team Manager

5. In small groups, read the following two examples and the flowchart to determine whether they are written to a high-tech or low-tech audience. To justify your answer, consider the use of acronyms, abbreviations, or jargon; the presentation of background data; the length of sentences and paragraphs; and the way in which the text has been formatted (how it looks on the page).

example

This documentation summarizes the items to be tested and the methodology utilized in the testing routine. The test plan tested four new 8536 Switcher features: OIM channel commands, OIM disconnect commands, OIM dump commands, and OIM log-off messages.

The patch files associated with the base must disable the features; consequently, only testing to ensure that this is operational will be addressed.

Software development personnel, after testing the features, will deliver copies of this report and product acceptance releases to program management, who are then responsible for writing functional specifications and submitting them to administration for manufacturing and marketing.

Purpose. This document identifies the time to be tested and the methods used for testing. The tests center on the 8536 Switcher (a network consisting of two nodes) and CN (connect node) bases. The 8536 Switcher and CN bases perform the same function as our currently marketed G9 and G12 systems.

OIM (Operator Interface Module) Features Tested:

- OIM channel commands
- OIM disconnect commands
- OIM dump commands
- OIM log-off messages

Testing Focus. To work appropriately, these features must be disabled by our patch files if an error occurs. Therefore, our testing focus will be the disabling capabilities.

Responsibilities. After testing the features, our software development group will send copies of this report and product acceptance releases to program management. They then will write the functional specifications and send them to administration. Administration is responsible for manufacturing and marketing. The following flowchart graphically depicts this sequence.

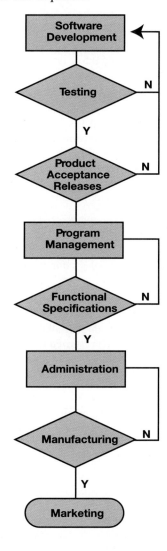

PROBLEM-SOLVING THINK PIECES

1. Dove Hill, GA, experienced severe thunderstorms on March 15, 2008. GAI (Goodwin & Associates Insurance Co.) insured much of the territory affected by this storm. Over 1,200 houses insured by GAI had water damage. As written, the homeowner's policy provided full-replacement coverage for water damaged. In the homeowner's policies, water damage was limited to situations such as the following:

   - A hailstorm smashes a window, permitting hail and rain to access a home.
   - A heavy rain soaks through a roof, allowing water to drip through a ceiling.
   - A broken water pipe spews water into a home.

   In contrast, homes which suffered water damage due to flooding were not covered, unless the homeowners also had taken out a separate flood damage policy from the National Flood Insurance Program. For insurance purposes, "flood" is defined as "the rising of a body of water onto normally dry land." For example, flood damage can include a river overflowing its banks or a heavy rain seeping into a basement. Of the 1,200 homes suffering water damage, only 680 homes qualified for full replacement coverage. The remaining homes suffered flood damage and were thus not covered.

Assignment

Write a letter from GAI to a homeowner denying coverage for their water damage claim. You are writing as a high-tech insurance professional to a lay reader. You want to maintain a good business relationship, despite the bad news. Build rapport while specifying in lay terms the denial.

2. Financial Trust & Annuity, a bank with over 100 sites throughout the Southwest, is growing at approximately 15 percent per year. To accommodate this growth, the bank needs to hire more tellers. Good business demands that FTA hire diversity: people with disabilities, employees from many races and cultures, men and women, and a variety of age ranges. Doing so will help ensure that the bank's tellers represent people from all walks of life and that the bank will design its services to appeal to a diverse customer base. In addition, another goal is to hire a percentage of employees who can speak Spanish, Arabic, and Chinese. This will help FTA communicate with a key constituency of its emerging customer base.

Assignment

You are the Human Resources Manager. To meet FTA's hiring needs, write a job advertisement that includes a statement addressing the bank's commitment to diversity.

3. Funky Duds, a cutting-edge clothing chain that caters to young adults, has encountered some negative publicity recently. Newspapers and television reports have commented that Funky D's seems to hire only "young, good-looking, blonde, blue-eyed" sales help. Funky D's prides itself on being a provider of chic clothing for all segments of society, regardless of race, creed, or color. However, the bad publicity has started to hurt sales and the chain's image. Something must be done to ensure that the clothing chain's hiring practices mirror society.

Assignment

Write a letter to all of Funky Ds' sales managers. In this letter, explain the problem, tell why hiring diversity is important (based on this chapter's content and any other information you might gather through research or personal experience), and provide guidelines for future sales help employment.

4. Sandy Susman works as a corporate trainer for WaxOn/WaxOff Computer Company. Sandy needs to invite employees to attend a work session focusing on benefits package changes. In an e-mail to all employees, including new hires as well as long-term workers and management, Sandy writes the following: "Here are the key changes guys. First you better pay attention to Holiday RFT, Sick/Hrs, Salary Continuations, STD Payments, AD&D, and Supplemental Life (GTL). These will affect your pay ASAP." Several employees contacted Sandy's boss, stating that Sandy's e-mail was not effective and even somewhat offensive.

Assignment

As Sandy's boss, write an e-mail to Sandy explaining what was wrong with the memo's tone and content. In this e-mail, rewrite the problematic sentence to show Sandy how its tone and content could have been improved.

5. To support yourself while going to college, you work weekends at a local drug store. Last Saturday, you experienced several challenges while working at the cash register. During the day, you checked out three hearing impaired customers who needed information about items in the store. You also worked with two senior citizens who needed help with the store's new "do-it-yourself" photography developing technology. Finally, a non-native speaker needed help getting a prescription filled but you had difficulty communicating with this customer.

Assignment

Write an e-mail to the drug store's branch manager. In this e-mail, explain the problems encountered, tell how you handled these situations, and ask for advice for future encounters.

WEB WORKSHOP

1. You are the director of human resources for your company. Your job is to write a nondiscrimination policy. To do so, access an online search engine to find other companies' nondiscrimination policies. Compare and contrast what you find. Then, based on your research, write your company's policy.

2. You are head of international relations at your corporation. Your company is preparing to go global. To ensure that your company is sensitive to multicultural concerns, research the cultural traits and business practices at five countries of your choice. To do so, access an online search engine and type in "multicultural business practices in _____" (specify the country's name). Report your findings either orally or in a memo, report, or e-mail message.

QUIZ QUESTIONS

1. What are the four main types of audiences?
2. What are the differences among the types of audiences?
3. What are three places for definitions in technical documents?
4. What are the main parts of a definition?
5. Why should you consider a multicultural audience when writing a technical document?
6. What guidelines should you follow to write to a multicultural audience?
7. What is sexist language?
8. What are three ways to avoid sexist writing?
9. How can you achieve audience involvement?
10. What are two things you can do to express reader benefit?

CHAPTER 5

······································

Research and Documentation

COMMUNICATION *at work*

The Oak Springs, Iowa, scenario illustrates the use of primary and secondary research before writing construction plans for a new road.

The city of Oak Springs, Iowa, needs to improve a 10-mile stretch of road that runs east and west through the town. The winding, two-lane road was built in 1976. Since then, the city has grown with several businesses and new housing along this road, and the road no longer is sufficient for the increased traffic load. The road must be expanded to four lanes and straightened. To do so, new sewer lines, sidewalks, easements, esplanades, and lighting must be added. However, this construction will impact current homeowners (whose land, through eminent domain, can be expropriated without their consent). In addition, part of the envisioned road construction will impact a wildlife refuge for waterfowl.

Before construction plans can be made, city engineers must conduct research in order to determine needs and considerations of the public. First, the engineers will perform primary research by interviewing city residents, as well as employees from the city's parks and recreation, police, fire, and transportation departments.

Objectives

When you complete this chapter, you will be able to

1. Understand why to conduct research in your technical communication.

2. Use both *primary* and *secondary* research in your technical communication.

3. Locate information in the library and online.

4. Read and analyze sources of information for your report.

5. Form an idea for a research report.

6. Write a research report.

7. Document your sources of information.

8. Evaluate your researched material using a checklist.

Then, the engineers will conduct secondary research. They will read statutes regarding eminent domain, real estate, environmental considerations, state wildlife refuges, zoning, planning, and land use.

Only after conducting primary and secondary research will the engineers be able to produce a construction plan to the Oak Springs city management for consideration and approval.

Check Online Resources

www.prenhall.com/gerson
For more information about research, visit our companion Web site.

WHY CONDUCT RESEARCH?

Research skills are important in your school or work environment. You may want to perform research to better understand a technical term or concept; locate a magazine, journal, or newspaper article for your supervisor; or find data on a subject to prepare an oral or written report. Technology is changing so rapidly that you must know how to do research to stay up to date.

You can research information using online catalogs; online indexes and databases; CD-ROM indexes and databases; reference books in print, online, and CD-ROM format; and by using Internet search engines and directories. Reference sources vary and are numerous, so this chapter discusses only research techniques.

Research, a major component of long, formal business reports, helps you develop your report's content. Often, your own comments, drawn from personal experience, will lack sufficient detail, development, and authority to be sufficiently persuasive. You need research to do the following:

- Create content
- Support commentary and content with details
- Prove points
- Emphasize the importance of an idea
- Enhance the reliability of an opinion
- Show the importance of a subject to the larger business community
- Address the audience's need for documentation and substantiation

USING RESEARCH IN REPORTS

Long, Formal Reports

See Chapter 16 for more discussion of long, formal reports.

You can use researched material to support and develop content in your formal reports. Use quotes and paraphrases to develop your content. Technical communicators often ask how much of a formal report should be *their* writing, as opposed to researched information. A general rule is to lead into and out of every quotation or paraphrase with your own writing. In other words,

- Make a statement (your sentence).
- Support this generalization with a quotation or paraphrase (referenced material from another source).
- Provide a follow-up explanation of the referenced material's significance (your sentences).

If you follow this approach, you can successfully blend researched material with other content in the report. Doing so allows you to minimize excessive reliance on research.

Research Including Primary and Secondary Sources

Research material generally breaks down into two categories: primary and secondary research. *Primary research* is performed or generated by you. You do not rely on books or periodicals for this type of research. Instead you create original research by preparing a survey or a questionnaire targeting a group of respondents, by networking to discover information from other individuals, by visiting job sites, or by performing lab experiments. You may perform this research to determine for your company the direction a new marketing campaign should take, the importance of diversity in the workplace, the economic impact of relocating the company to a new office site, the usefulness of a new product, or the status of a project. You may also need to interview people for their input about a particular topic. For example, your company might be considering a new approach to take for increased security on employee computers. You could ask employees for a record of their logs which would highlight the problems they have encountered with their computer security. With primary research, you will be generating the information based on data or information from a variety of sources that might include observations, tests of equipment, interviews, networking, surveys, and questionnaires.

When you conduct *secondary research*, you rely on already printed and published information taken from sources including books, periodicals, newspapers, encyclopedias, reports, proposals, or other business documents. You might also rely on information taken from a Web site or a blog. All of this secondary research requires parenthetical source citations (discussed later in this chapter).

SPOTLIGHT

Conducting research to find solutions

Tom Woltkamp is Senior Manager of Information Solutions for Teva Neuroscience, a global leader in the "development, production and marketing of generic and proprietary branded pharmaceuticals" **(www.tevapharm.com/)**. Teva Pharmaceutical Industries is home based in Israel; Teva Neuroscience, a subsidiary of Teva Pharmaceuticals, is based in Kansas City, Missouri, with branches in Florida, Pennsylvania, Texas, North Carolina, California, Minnesota, and Canada.

Approximately 20 employees and consultants report to Tom regarding computer infrastructure (computer networks, help desks, call centers, hardware/software purchases) and application development for programming, software integration, installation, configuration, Web programming, and customized documentation.

On an ongoing basis, Teva employees in finance, human resources, sales, and the company's call centers need new software applications. They call on Tom's group to find solutions to their needs. The research and documentation process for Teva's software development is as follows:

- **Project Charter**—these brief reports "drive the research." In a page or two, the Teva departments with whom Tom works "write down what they want." They clarify their needs, as Tom says, by stating "here's my problem," "here's what I think I want to do about it," and "here's what I think the benefits will be."

That's when Tom begins his research into finding solutions for their software needs. Tom and his team conduct both primary and secondary research:

- **Questionnaires**—Tom and his group create FAQ checklists where they "play 20 questions to find what our end users want." They meet with their clients and discuss the project parameters.
- **Internet Searches**—after the initial questionnaire, which helps zero in on the client's needs, Tom and his team search the Internet for "knowledge and solutions." They want to find out more about the topic of research and see if any products already exist that can meet their needs or software they can "tweak" for customization.
- **Consultation**—another avenue of research involves meeting with consultants—experts in the field—for their take on the topic.
- **Interviews**—Tom works with "key partners," including banks, vendors, professional organizations, and patients using the company's pharmaceuticals. By interviewing these constituents, Tom can learn more about how other entities might have solved similar problems or how the customers using a drug can be better served.
- **Online and Hand-Copy Journals**—an excellent secondary source of research for Tom is professional journals, such as *PC Magazine, Call Center Magazine*, and *CRM Magazine* (which focuses on customer relations management).

Tom and his team spend approximately 20 to 30 percent of their work time on research. Why? Research allows Tom to make sure that the software solutions he and his team design and develop are correct. That is, research helps Tom ensure that his products are on target to meet the client's needs, will be time and cost efficient, and meet national, international, and industry standards. Research helps Tom make certain that he can achieve customer satisfaction and provide value to his organization.

CRITERIA FOR WRITING RESEARCH REPORTS

As with all types of technical communication, writing using researched material requires that you

- Recognize your audience.
- Use an effective style, appropriate to your reader and purpose.
- Use effective formatting techniques for reader-friendly ease of access.

Audience

Audience

See Chapter 4 for more discussion of audience.

When writing a research report, you first must recognize the level of your audience. Are your readers high tech, low tech, or lay? If you are writing to your boss, consider his or her level of technical expertise. Your boss probably is at least a low-tech reader, but you must determine this based on your own situation. Doing so will help you determine the amount of technical definition necessary for effective communication.

You must decide next whether your reader will understand the purpose of your research report. This will help you determine the amount of detail needed and the tone to take (persuasive or informative). For instance, if your reader has requested the information, you will not need to provide massive amounts of background data explaining the purpose of your research. Your reader has probably helped you determine the scope and purpose. On the other hand, if your research report is unsolicited, your first several paragraphs must clarify your rationale. You will need to explain why you are writing and what you hope to achieve.

If your report is solicited, you probably know that your reader will use your research for a briefing, an article to be written for publication, a technical update, and so forth. Thus, your presentation will be informative. If your research report is unsolicited, however, your goal is to persuade your reader to accept your hypotheses substantiated through your research.

Effective Style

Research reports should be more formal than many other types of technical communication. In a research report, you are compiling information, organizing it, and presenting your findings to your audience using documentation. Because the rules of documentation are structured rigidly (to avoid plagiarism and create uniformity), a research report is also carefully structured.

The tone need not be stuffy, but you should maintain an objective distance and let the results of your research support your contentions. Again, considering your audience and purpose will help you decide what style is appropriate. For example, if your boss has requested your opinion, then you are correct in providing it subjectively. However, if your audience has asked for the facts—and nothing but the facts—then your writing style should be more objective.

Formatting

Highlighting Techniques

See Chapter 8 for more discussion of document design.

Reading a research report is not always an easy task. As the writer, you must ensure that your readers encounter no difficulties. One way to achieve this is by using effective formatting. Reader-friendly ease of access is accomplished when you use highlighting techniques such as bullets, numbers, headings, subheadings, and graphics (tables and figures).

In addition, effective formatting includes the following:

- Overall organization (including an introduction, discussion, and conclusion)
- Internal organization (various organizational patterns, such as problem/solution, comparison/contrast, analysis, and cause/effect)

- Parenthetical source citations
- Works cited—documentation of sources

Each of these areas is discussed in detail throughout this chapter.

THE WRITING PROCESS AT WORK

the writing process

Prewriting	Writing	Rewriting

• Decide whether you are writing to inform, instruct, persuade, or build rapport.
• Determine whether your audience is high tech, low tech, lay, or multiple.
• Gather information from primary and secondary sources of research.
	• Use an outline to organize your content.
	• Organize your content using modes such as problem/solution, cause/effect, comparison, argument/persuasion, analysis, chronology.
	• Document your sources correctly.
		• Revise your draft by
		  • adding details
		  • deleting wordiness
		  • simplifying words
		  • enhancing the tone
		  • reformatting your text
		  • proofreading and correcting errors
		• Avoid plagiarism by the accuracy of your quotes, paraphrases, summaries, and works cited (references).

Communication Process

See Chapter 2 for discussion of the communication process.

Writing a research report, as with other types of technical correspondence, is easiest when you follow a process. Rather than just wandering into a library and hoping that the correct book or periodical will leap off a shelf and into your hands or expecting your online search to reveal useful information immediately, approach your research systematically. Prewrite (to gather your data and determine your objectives), write a draft, and rewrite to ensure that you meet your goals successfully. The writing process is dynamic, with the steps frequently overlapping.

Prewriting Research Techniques

1. **Select a general topic (or the topic you have been asked to study)**—Your topic may be a technical term, phrase, innovation, or dilemma. If you're in emergency medical technology, for example, you might want to focus on the current problems with emergency medical service. If you're in telecommunications, you might write about satellite communication. If you work in electronics, select a topic such as robotics. If your field is computer science, you could focus on artificial intelligence.

2. **Spot-check sources of information**—Check a library or online sources to find material that relates to your subject. A quick review of your library's online periodical databases, such as ProQuest, Infotrac, or other equivalent sources, will help you locate periodical articles. Your library may also have a print edition of

the *Reader's Guide to Periodical Literature* and other similar specialized periodical indexes. Most libraries now have online catalogs to help you easily search for books on your topic. A keyword search of Internet metasearch engines will give you an idea of how much information might be readily available through the Internet.

3. **Establish a focus**—After you have chosen a topic for which you can find available source material, decide what you want to learn about your topic. A focus statement can guide you. In other words, if you are interested in emergency medical technology, you might write a focus statement such as the following:

example

I want to research current problems with emergency medical services, including variances in training required, delays in timely response to emergency calls, and limited number of vehicles available.

For telecommunications, you could write:

example

I want to research satellite communication maintenance, reliability, and technical innovations.

If you are an electronics technician, you might write:

example

I want to discover the uses, impact on employment, and expenses of robotics applications.

Finally, if you are a computer technician, you might write:

example

I want to discover the pros and cons regarding artificial intelligence.

With focus statements such as these, you can begin researching your topic, concentrating on articles pertinent to your topic.

4. **Research your topic**—You may feel overwhelmed by the prospect of research. But there are many sources that, once you know how to use them, will make the act of research less overwhelming.

Books All books owned by a library are listed in catalogs, usually online. Books can be searched in online catalogs in a variety of ways: by title, by author, by subject, by keyword, or by using some combination of these. No matter how you search for a book, the resulting record will look something like the following example.

Artificial intelligence : robotics and machine evolution / David Jefferis.

Database:	DeVry Institute of Technology
Main Author:	Jefferis, David.
Title:	Artificial intelligence : *robotics* and machine evolution / David Jefferis.
Primary Material:	Book
Subject(s):	<u>*Robotics*—Juvenile literature.</u>
	<u>Artificial intelligence—Juvenile literature.</u>
	<u>*Robotics.*</u>
	<u>Robots.</u>
	<u>Artificial intelligence.</u>
Publisher:	New York : Crabtree Pub. Co., 1999.
Description:	32 p. : col. ill. ; 29 cm.
Series:	Megatech
Notes:	Includes index.
	An introduction to the past, present, and future of artificial intelligence and *robotics,* discussing early science fiction predictions, the dawn of AI, and today's use of robots in factories and space exploration.
Database:	DeVry Institute of Technology
Location:	Dallas Main Stacks
Call Number:	<u>TJ211.2.J44 1999</u>
Number of Items:	1
Status:	Not Changed

Periodicals Use online, CD-ROM, or print periodical indexes to find articles on your topic. Online indexes can be searched in a variety of ways: by title, by author, by subject, by keyword, or by using a combination of these. No matter how you search, the resulting record will look something like the following example.

Help ?

DH Pro Eric Carter
Bicycling; Emmaus; Mar 2001; Andrew Juskaitis;

Volume:	42
Issue:	2
Start Page:	14
ISSN:	00062073
Subject Terms:	Athletes
	Bicycling
	Organizations
	Bicycle racing

Personal Names: Carter, Eric
Abstract:
*Juskaitis discusses bicycling with downhill national champion Eric Carter. Only after riding with Carter did he realize he's also at the forefront of changing the face of gravity-fed mountain **bike racing**. Carter has joined with promoter Rick Sutton to push cycling's race organizations to implement four-to-six-rider racing on highly specialized courses.*

Full Text: . . . (omitted here for copyright issues)

The preceding example from an online periodical database tells us that an article on the subjects **Athletes, Bicycling, Organizations,** and **Bicycle racing** can be found in the March 2001 issue of *Bicycling* magazine, volume 42, issue 2, beginning on page 14. The article is titled "DH Pro Eric Carter" and was written by Andrew Juskaitis. The article contains information about a person named Eric Carter. An abstract, or summary of the article, is given, followed by the full text of the article itself.

Indexes to General, Popular Periodicals Most libraries provide access to at least one of a number of indexes covering popular, nontechnical literature and newspaper articles from a variety of subject fields. There are online, CD-ROM, and print counterparts for most of these. The online and CD-ROM indexes provide the full text of many of the articles.

examples

- *Reader's Guide to Periodical Literature*
- *Periodicals Research I or II*
- *Ebscohost*
- *SIRS Researcher* (emphasizes social issues)
- *Newsbank* (emphasizes newspaper articles)

Indexes to Specialized, Scholarly, or Technical Periodicals Many libraries provide access to one or more specialized indexes covering literature from a variety of disciplines. There are online, CD-ROM, and print counterparts for most of these. The online and CD-ROM indexes provide the full text of many of the articles.

- **Applied Science & Technology Index.** Covers engineering, aeronautics and space sciences, atmospheric sciences, chemistry, computer technology and applications, construction industry, energy resources and research, fire prevention, food and the food industry, geology, machinery, mathematics, metallurgy, mineralogy, oceanography, petroleum and gas, physics, plastics, the textile industry and fabrics, transportation, and other industrial and mechanical arts.
- **Business Periodicals Index.** Covers major U.S. publications in marketing, banking and finance, personnel, communications, computer technology, and so on.
- **ABI/Inform.** Covers business and management.
- **General Science Index.** Covers the pure sciences, such as biology and chemistry.
- **Social Sciences Index.** Covers psychology, sociology, political science, economics, and other social sciences topics.
- **ERIC** (*Education Resources Information Center*). Provides bibliography and abstracts about educational research and resources. Available free through the Internet.
- **MEDLINE.** Covers medical journals and allied health publications.
- **PsycINFO.** Covers psychology and behavioral sciences.
- **NEXIS/LEXIS.** Includes the full text of newspaper articles, reports, transcripts, law journals, and legal reporters and other reference sources in addition to general periodical articles.
- **SIRS** (*Social Issues Resources*). Provides bibliography and full-text articles about social issues.

The Internet Perhaps one of the largest sources of research available today is the Internet. Millions of documents from countless sources are found on the Internet. You can find material on the Internet published by government agencies, organizations, schools, businesses, or individuals (see Table 5.1). The list of options grows daily. For example, nearly all newspapers and news organizations have online Web sites.

Finding Information Online

To find information online, use directories, search engines, or metasearch engines.

Directories Directories such as Yahoo, AltaVista, HotBot, and Excite let you search for information from a long list of predetermined categories, including the following:

Arts	Government	Politics and Law
Business	Health and Medicine	Recreation
Computers	Hobbies	Science
Education	Money and Investing	Sports
Entertainment	News	Society and Culture

To access any of these areas, click on the appropriate category and then "drill down," clicking on each subcategory until you get to a useful site.

Search Engines Search engines such as Google, Northern Light, or Ask Jeeves let you search millions of Web pages by keywords. Type a word, phrase, or name in the appropriate blank space and press the Enter key. The search engine will search through

TABLE 5.1   A Sample of Internet Research Sources

Search Engines, Directories, Metasearch Engines	Online References	Online Libraries	Online Newspapers	Online Magazines	Online Government Sites
Yahoo	*Webster's Dictionary*	Library of Congress	*New York Times*	*National Geographic*	United Nations
Excite	*Roget's Thesaurus*	New York Public Library	CNN	*HotWired*	The White House
Lycos	*Britannica Online Encyclopedia*	Cleveland Public Library	*USA Today*	*Atlantic*	The IRS
Go.com	*Encyclopedia Smithsonian*	Gutenberg Project	*Washington Post*	*The New Republic*	U.S. Postal Service
Alta Vista			*Kansas City Star*	*U.S. News Online*	FirstGov
MetaSearch	*The Internet Almanac*	Most city and university libraries	Most city newspapers	*Time Magazine*	Most states' supreme courts, legislatures, executive offices, and local governments
MetaCrawler				*Ebony Online*	
Google	*The Old Farmer's Almanac*			*Slate*	
Britannica.com					
Northern Light					
Ask Jeeves					
Dogpile					

documents on the Internet for "hits," documents that match your criteria. One of two things will happen: Either the search engine will report "no findings," or it will report that it has found thousands of sites that might contain information on your topic.

In the first instance, "no findings," you'll need to rethink your search strategy. You may need to check your spelling of the keywords or find synonyms. For example, if you want to research information about online writing, you could try typing "writing online," "online writing," "electronic writing," "writing electronically," and other similar terms. In the second instance, finding too many hits, you'll need to narrow your search. For example, if you are researching illegalities in baseball, you cannot type in "baseball." That's too broad. Instead, try "Shoeless Joe Jackson," "Pete Rose," "crime in baseball," "baseball scandals," and so on.

Metasearch Engines   A metasearch engine, such as Dogpile, lets you search for a keyword or phrase in a group of search engines at once, saving you the time of searching separately through each search engine.

Researching the Internet presents at least two problems other than finding information. First, is the information you have found trustworthy? Paperbound newspapers, journals, magazines, and books go through a lengthy publication process involving editing and review by authorities. Not all that's published on the Internet is so professional. Be wary. What you read online needs to be filtered through common sense. Second, remember that although a book, magazine, newspaper, or journal can exist unchanged in print form for years, Web sites change constantly. A site you find today online will not necessarily be the same tomorrow. That's the nature of electronic communication.

In addition to these sources, you can consult the following for help: U.S. government publications, databases, and your reference librarian.

5. **Read your researched material and take notes**—Once you have researched and located a source (whether it is a book, magazine, journal, or newspaper article), study the material. For a book, use the index and the table of contents to locate your topic. When you have found it, refer to the pages indicated and skim, reading selectively.

For shorter documents, such as magazine or journal articles or online materials, you can read closely. Reading and rereading the source material is an essential step in understanding your researched information. After you have studied the document thoroughly, go through it page by page and briefly summarize the content.

For a short magazine article, you can make marginal notes on a photocopy. For instance, read a paragraph and then briefly summarize its main point(s) in several words. Such notations are valuable because they are easy to make and provide a clear and concise overview of the article's focus.

You can also take notes on 3×5 inch cards. If you do this, be sure to write only one fact, quotation, or paraphrase per card, along with the author's name. This will let you organize the information later according to whatever organizational sequence (chronological narrative, analysis by importance, comparison/contrast, problem/solution, cause/effect, etc.) you prefer. Prepare a bibliography of your sources on 3×5 inch cards as well, one source per card.

The following are examples of a **bibliography card** and a **note card** with quotation, respectively.

Stephens, Guy M. "To Market, to Market." *Satellite Communications.* November 2004: 15–16.

Future need for C-band

"'We believe there is going to be an important need for C-band capacity into the '90s, into the next century, because it's an ideal way to serve many users, particularly video users,' Koehler said."

Stephens, "To Market," p. 15.

example

On a *summary* note card, you condense original material by presenting the basic idea in your own words. You can include quotations if you place them in quotation marks, but do not alter the organizational pattern of the original. A summary is shorter than the original.

Future need for C-band
C-band is going to be increasingly important in the '90s because of the increasing number of video owners.

Stephens, "To Market," p. 15.

example

On a *paraphrase* note card, you restate the original material in your own words without condensing. The paraphrase is essentially the complete version rewritten. A paraphrase is the same length as the original.

Future need for C-band

The author asserts the continued relevance of C-band even in the 1990s and beyond. C-band is the top performer, especially for people who use videos.

Stephens, "To Market," p. 15.

example

---

6. **Isolate the main points**—After you complete the analysis of the document (whether you use note cards and/or marginal comments), isolate the main points discussed in the books or periodicals. You will find that, of the major points in an article or book, sometimes only three or four ideas will be relevant for your topic. Choose the ones discussed repeatedly or those that most effectively develop the ideas you want to pursue.

7. **Write a statement of purpose**—Once you have chosen two to four main ideas from your research, write a purpose statement that expresses the direction of your research. For example, one student wrote the following statement of purpose after performing research on superconductivity.

example

> The purpose of this report is to reveal the future for this exciting product, which depends on further progress of technology, the development of easily accessible and cost-efficient materials, and industry's need for the final product.

8. **Create an outline**—After you have written a purpose statement, formulate an outline. An outline will help you organize your paragraphs and ensure that you stay on track as you develop your ideas through quotes and paraphrases. Figures 5.1 and 5.2 are examples of topic and sentence outlines.

FIGURE 5.1 Topic Outline

```
   I.  Sensors used to help robots move
       A.  Light
       B.  Sound
       C.  Touch
  II.  Touch sensor technology (microswitches)
       A.  For gripping
       B.  For maintaining contact with the floor
 III.  Optical sensors (LED/phototransistors)—Like bowling alley
       foul-line sensors
       A.  Less bulky/connected to computer interface
       B.  Not just for gripping, but for locating objects by following this
           sequence:
           1.  Scan gripper to locate object
           2.  Move gripper arm left and right to center object
           3.  Move gripper forward to grasp
           4.  Close gripper
       C.  Problem—What force to use for gripping?
  IV.  Force sensors
       A.  Spring and microswitch
       B.  Optical encoder discs—Microprocessors determine speed of discs
           to determine force necessary
       C.  Integrated circuits with strain gauge and pressure-sensitive paint
       D.  Pressure sensors built with conductive foam
   V.  Conclusion
```

FIGURE 5.2 Sentence Outline

I. Because robots must move, they need sensors. These sensors could include light sensors, sound sensors, and touch sensors.

II. Touch sensors can have the following technology:
   A. Microswitches can be used for gripping.
   B. Microswitches can also be used for maintaining contact with the floor. This would keep the robot from falling down stairs, for example.

III. Optical sensors might be better than microswitches.
   A. LED/phototransistors are less bulky than switches.
   B. When connected to a computer interface, optical sensors also can help a robot locate an object as well as grip it.
   C. Here is the sequence followed when using optical sensors:
      1. The robot's grippers scan the object to locate it.
      2. The robot moves its gripper arms left and right to center the object.
      3. The robot moves forward to grip the object.
      4. The robot closes the gripper.
   D. The only problem faced is what force should be used when gripping.

IV. Force sensors can solve this problem.
   A. A combination spring and microswitch can be used to determine the amount of force required.
   B. Optical encoder discs can be used also. A microprocessor determines the speed of the disc to determine the required force.
   C. Integrated circuits with strain gauges and pressure-sensitive paint can be used to determine force.
   D. Another pressure sensor can be built from conductive foam.

V. To conclude, all these methods of tactile sensing comprise a field of inquiry important to robotics.

Writing

You now are ready to write your research report.

1. **Review your research**—Prior to writing your report, look back over your research sources to make sure that you're satisfied with what you've found. Do you have enough information to develop your points thoroughly? Is the information you've found what you want? If not, it's time to do more research. If you're content with what you've discovered in your research, you can start drafting your text.

2. **Organize your report effectively**—When you are ready to write, provide an introductory paragraph, discussion (body) paragraphs, a conclusion or recommendation, and your works cited page.

Introduction   Begin with something to arouse your reader's interest. This could include a series of questions, an anecdote, a quote, or data pertinent to your topic. Then use this to lead into your statement of purpose.

Discussion   The number of discussion paragraphs will depend on the number of divisions and the amount of detail necessary to develop your ideas. Use quotes and paraphrases to develop your content.

Conclusion/Recommendations   In a final paragraph, summarize your findings, draw a conclusion about the significance of these discoveries, and recommend future action.

Citing Sources

On a final page, provide an alphabetized list of your research sources illustrated on page 143, MLA works cited and APA references pages.

3. **Document your sources correctly**—Your readers need to know where you found your information and from which sources you are quoting or paraphrasing. Therefore, you must document this information. To document your research correctly, you must (a) provide parenthetical source citations and (b) supply a works cited page (Modern Language Association) or a references page (American Psychological Association).

Sample MLA Works Cited Page

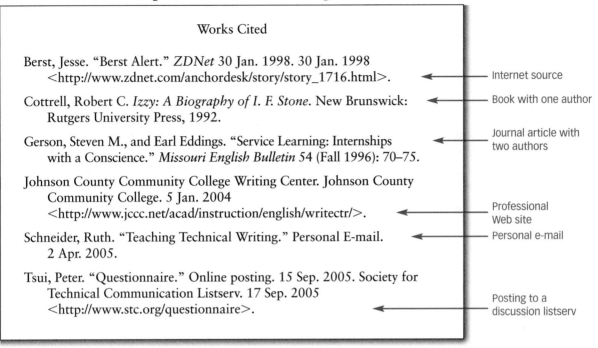

Works Cited

Berst, Jesse. "Berst Alert." *ZDNet* 30 Jan. 1998. 30 Jan. 1998
    <http://www.zdnet.com/anchordesk/story/story_1716.html>.
    — Internet source

Cottrell, Robert C. *Izzy: A Biography of I. F. Stone.* New Brunswick:
    Rutgers University Press, 1992.
    — Book with one author

Gerson, Steven M., and Earl Eddings. "Service Learning: Internships
    with a Conscience." *Missouri English Bulletin* 54 (Fall 1996): 70–75.
    — Journal article with two authors

Johnson County Community College Writing Center. Johnson County
    Community College. 5 Jan. 2004
    <http://www.jccc.net/acad/instruction/english/writectr/>.
    — Professional Web site

Schneider, Ruth. "Teaching Technical Writing." Personal E-mail.
    2 Apr. 2005.
    — Personal e-mail

Tsui, Peter. "Questionnaire." Online posting. 15 Sep. 2005. Society for
    Technical Communication Listserv. 17 Sep. 2005
    <http://www.stc.org/questionnaire>.
    — Posting to a discussion listserv

Sample APA References Page

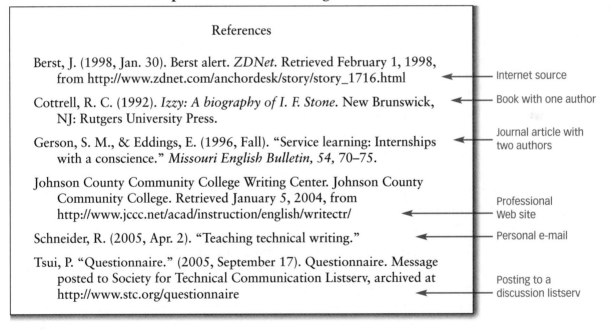

References

Berst, J. (1998, Jan. 30). Berst alert. *ZDNet.* Retrieved February 1, 1998,
    from http://www.zdnet.com/anchordesk/story/story_1716.html
    — Internet source

Cottrell, R. C. (1992). *Izzy: A biography of I. F. Stone.* New Brunswick,
    NJ: Rutgers University Press.
    — Book with one author

Gerson, S. M., & Eddings, E. (1996, Fall). "Service learning: Internships
    with a conscience." *Missouri English Bulletin, 54,* 70–75.
    — Journal article with two authors

Johnson County Community College Writing Center. Johnson County
    Community College. Retrieved January 5, 2004, from
    http://www.jccc.net/acad/instruction/english/writectr/
    — Professional Web site

Schneider, R. (2005, Apr. 2). "Teaching technical writing."
    — Personal e-mail

Tsui, P. "Questionnaire." (2005, September 17). Questionnaire. Message
    posted to Society for Technical Communication Listserv, archived at
    http://www.stc.org/questionnaire
    — Posting to a discussion listserv

Parenthetical Source Citations   The Modern Language Association (MLA) and the American Psychological Association (APA) use a simplified form for source citations. Before 1984, footnotes and endnotes were used in research reports. In certain instances, this form of documentation is still correct. If your boss or instructor requests footnotes or endnotes, you should still use these forms. However, the most modern approach to source citations requires only that you cite the source of your information parenthetically after the quotation or paraphrase.

MLA Format

One author After the quotation or paraphrase, parenthetically cite the author's last name and the page number of the information.

> "Viewing the molecular activity required state-of-the-art electron microscopes" (Heinlein 193).

Note that the period follows the parenthesis, not the quotation. Note also that no comma separates the name from the page number and that no lowercase *p* precedes the number.

Two authors After the quotation or paraphrase, parenthetically cite the authors' last names and the page number of the information.

> "Though *Gulliver's Travels* preceded *Moll Flanders*, few scholars consider Swift's work to be the first novel" (Crider and Berry 292).

Three or more authors Writing a series of names can be cumbersome. To avoid this, if you have a source of information written by three or more authors, parenthetically cite one author's name, followed by *et al.* (Latin for "and others") and the page number.

> "Baseball isn't just a sport; it represents man's ability to meld action with objective—the fusion of physicality and spirituality" (Norwood et al. 93).

Anonymous works If your source has no author, parenthetically cite the shortened title and page number.

> "Robots are more accurate and less prone to errors caused by long hours of operation than humans" ("Useful Robots" 81).

APA Format

One author If you do not state the author's name or the year of the publication in the lead-in to the quotation, include the author's name, year of publication, and page number in parenthesis, after the quotation.

> "Izzy's stay in Palestine was hardly uneventful" (Cottrell, 2003, p. 118).

(Page numbers are included for quoted material. The writer determines whether page numbers are included for source citations of summaries and paraphrases.)

Two authors When you cite a source with two authors, always use both last names with an ampersand (&).

> "Line charts reveal relationships between sets of figures" (Gerson & Gerson, 2005, p. 158).

Three or more authors When your citation has more than two authors but fewer than six, use all the last names in the first parenthetical source citation. For subsequent citations, list the first author's name last followed by *et al.*, the year of publication, and for a quotation, the page number.

> "Two-party politics might no longer be the country's norm next century" (Conners et al., 2002, p. 2).

Anonymous works When no author's name is listed, include in the source citation the title or part of a long title and the year. Book titles are underlined or italicized, and periodical titles are placed in quotation marks.

> Two-party politics might be a thing of the past (*Winning Future Elections,* 2006).
>
> Many memos and letters can be organized in three paragraphs ("Using Templates," 2007).

Works Cited   Parenthetical source citations are an abbreviated form of documentation. In parentheses, you tell your readers only the names of your authors and the page numbers on which the information can be found. Such documentation alone would be insufficient. Your readers would not know the names of the books, the names of the periodicals, or the dates, volumes, or publishing companies. This more thorough information is found on the works cited page or references page, a listing of research sources alphabetized either by author's name or title (if anonymous). This is the last page of your research report.

Your entries should follow MLA or APA standards.

MLA Works Cited

A book with one author

Cottrell, Robert C. *Smoke Jumpers of the Civilian Public Service in World War II*. London: McFarland and Co., Inc., 2006.

A book with two or three authors

Heath, Chip, and Dan Heath. *Made to Stick: Why Some Ideas Survive and Others Die*. New York: Random House, 2007.

A book with four or more authors

Nadell, Judith, et al. *The Macmillan Writer*. Boston: Allyn and Bacon, 1997.

A book with a corporate authorship

Corporate Credit Union Network. *A Review of the Credit Union Financial System: History, Structure, and Status and Financial Trends*. Kansas City: U.S. Central, 2007.

A translated book

Phelps, Robert, ed. *The Collected Stories of Colette*. Trans. Matthew Ward. New York: Farrar, Straus Giroux, 1983.

An entry in a collection or anthology

Hamilton, Kendra. "What's in a Name?" *America Now: Short Readings from Recent Periodicals*. Ed. Robert Atwan. New York: Bedford/St. Martin's, 2005. 12–20.

A signed article in a journal

Davis, Rachel. "Getting—and Keeping Good Clients." *Intercom* (April 2007): 8–12.

A signed article in a magazine

Rawe, Julie. "A Question of Honor." *Time* 28 May 2007: 59–60.

A signed article in a newspaper

Gertzen, Jason. "University to Go Wireless." *The Kansas City Star* 29 Mar. 2007: C3.

An unsigned article

"Diogenes Index." *Time* 23 Sep. 1996: 22.

Encyclopedias and almanacs

"Internet." *The World Book Encyclopedia*. 2000 ed. Chicago: World Book.

Computer software

Drivers and Utilities. Computer software. Dell, Inc., 2002–2004.

Internet source

"Top Ten Qualities/Skills Employers Want." *Job Outlook 2006 Student Version*. National Association of Colleges and Employers, 2005: 5. http://career.clemson.edu/ pdf_docs/NACE_JO6.pdf.

E-mail

Schneider, Ruth. "Teaching Technical Communication." Personal E-mail. 2 Apr. 2006.

CD-ROM

McWard, James. "Graphics On-line." TW/Inform. CD-ROM. New York: EduQuest, 2006.

Personal Web site

Moore, Ta'Isha. Home page. 29 Dec. 2007 http://www.jccc.net/home/depts/1504.

Professional Web site

Johnson County Community College Writing Center. Johnson County Community College. 5 Jan. 2007 http://www.jccc.net/acad/instruction/english/writectr/.

Posting to a discussion listserv

Tsui, Peter. "Questionnaire." Online posting. 15 Sep. 2005. Society for Technical Communication Listserv. 17 Sep. 2005 http://www.stc.org/questionnaire.

APA References
A book with one author

Cottrell, R. C. (2006). *Smoke jumpers of the civilian public service in World War II.* London: McFarland and Co., Inc.

A book with two authors

Heath, C. & Heath, D. (2007). *Made to stick: Why some ideas survive and others die.* New York: Random House.

A book with three or more authors

Nadell, J., McNeniman, L., & Langan, J. (1997). *The Macmillan writer.* Boston: Allyn & Bacon.

A book with a corporate authorship

Corporate Credit Union Network. (2007). *A review of the credit union financial system: History, structure, and status and financial trends.* Kansas City, MO: U.S. Central.

A translated book

Phelps, R. (Ed.). (1983). *The collected stories of Colette* (M. Ward, Trans.). New York: Farrar, Straus & Giroux.

An entry in a collection or anthology

Hamilton, K. (2005). What's in a Name? In R. Atwan (Ed.), *America now: Short readings from recent periodicals* (pp. 12–20). New York: Bedford/St. Martin's.

A signed article in a journal

Davis, R. (2007, April). Getting—and keeping good clients. *Intercom*, 8–12.

A signed article in a magazine

Rawe, J. (2007, May 28). A question of honor. *Time*, 59–60.

A signed article in a newspaper

Gertzen, J. (2007, March 29). University to go wireless. *The Kansas City Star*, p. C3.

An unsigned article

Diogenes index. (1996, September 23). *Time*, 22.

Encyclopedias and almanacs

Internet. (2000). *The world book encyclopedia.* Chicago: World Book.

Computer software

Drivers and Utilities [Computer software]. (2002–2004). Dell, Inc.

Internet source

Top ten qualities/skills employers want. (2006). *Job Outlook 2006 Student Version.* Retrieved May 31, 2007, from National Association of Colleges and Employers. http://career.clemson.edu/pdf_docs/NACE_JO6.pdf.

CD-ROM

McWard, J. (2006). Graphics on-line [CD-ROM]. TW/Inform. New York: EduQuest.

Personal Web site

Moore, T. Home page. Retrieved December 29, 2007, from http://www.jccc.net/home/depts/1504

Professional Web site

Johnson County Community College Writing Center. Johnson County Community College. Retrieved January 5, 2007, from http://www.jccc.net/acad/instruction/english/writectr/

Posting to a discussion listserv

Tsui, P. "Questionnaire." (2005, September 17). Questionnaire. Message posted to Society for Technical Communication Listserv, archived at http://www.stc.org/questionnaire

Alternative Style Manuals Although MLA and APA are popular style manuals, others are favored in certain disciplines. Refer to these if you are interested or required to do so.

- *U.S. Government Printing Office Style Manual,* 29th edition. Washington, DC: Government Printing Office, 2000.
- *The Chicago Manual of Style*, 15th edition. Chicago: University of Chicago Press, 2003.
- Turabian, Kate L. *A Manual for Writers of Term Papers, Theses, and Dissertations,* 7th edition. Chicago: University of Chicago Press, 2007.

4. **Develop your ideas**—You have learned how to organize your report (through an introduction, discussion, and conclusion/recommendation) and how to document your sources of research (through parenthetical source citations and a works cited or references page). Writing your research report also requires that you use your research effectively to develop your ideas. Successful use of research demands that you correctly quote, paraphrase, or summarize.

Rewriting

As with all types of writing, drafting the text of your research report is only the second stage of the writing process. To ensure that your report is effective, revise your draft as follows:

1. **Add new detail for clarity and persuasiveness**—Too often, students and employees assume that they have developed their content thoroughly when, in fact, their assertions are general and vague. This is especially evident in research reports. You might provide a quotation to prove a point, but is this documentation sufficient? Have you truly developed your assertions? If an idea within your report seems thinly presented, either add another quotation, paraphrase, or summary for additional support or explain the significance of the researched information.

2. **Delete dead words and phrases and researched information that does not support your ideas effectively**—Good writing in a work environment is economical writing. Thus, as always, your goal is to communicate clearly and concisely. Delete words that serve no purpose, maintaining a low fog index. In addition, review your draft

Conciseness

See Chapter 3 for more discussion of clarity and conciseness.

Using Microsoft Word 2007 for Documentation

Microsoft Word 2007 provides students and business employees many new tools related to research and documentation. When you click on the **References** tab, you will find ways to create tables of contents, insert footnotes, insert citations, and create either bibliographies or works cited notations.

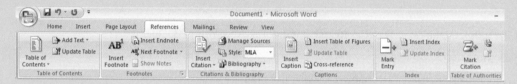

For example, to create parenthetical and bibliographic citations, using MLA, follow these steps:

1. Click on the **Insert Citation** down arrow and **Add New Source.**

The following **Create Source** screen will pop up (the screen fields will be blank; we have added the appropriate information—author's name, title, publishing company, city, and date).

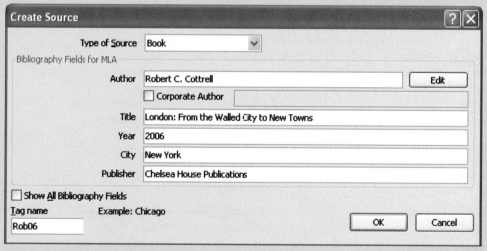

2. Click **OK** and the parenthetical source citation will be inserted.

3. To automatically insert the bibliographical information, click on the **Bibliography** down arrow and then on **Bibliography** (for MLA or **Works Cited** for APA, for example).

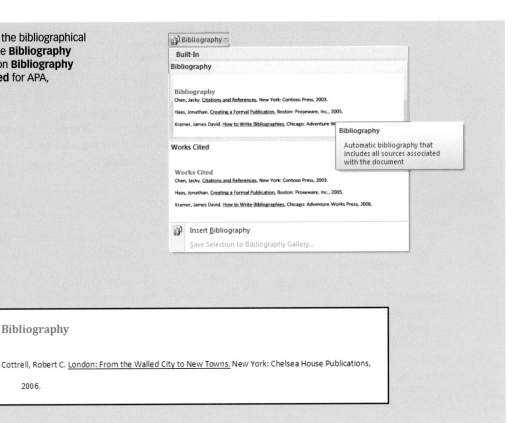

The following will appear:

> Bibliography
>
> Cottrell, Robert C. <u>London: From the Walled City to New Towns.</u> New York: Chelsea House Publications, 2006.

for clarity of focus. The goal of a research report is not to use whatever researched information you've found wherever it seems valid. Instead, you want to use quotations, paraphrases, and summaries only when they help develop your statement of purpose. If your research does not support your thesis, it is counterproductive and should be eliminated. In the rewriting stage, delete any documented research that is tangential or irrelevant.

3. **Simplify your words for easy understanding**—The goal of technical writing is to communicate, not to confuse. Write to be understood. Don't say *grain-consuming animal units* if you mean *chickens*. Don't call the July 2000 stock market crash a *fourth-quarter equity retreat*.

4. **Move information within your report to ensure effective organization**—How have you organized your report? Did you use a problem/solution format? Did you use comparison/contrast or cause/effect? Is your report organized as a chronological narrative or by importance? Whichever method you've used, you want to be consistent. To ensure consistency, rewrite by moving any information that is misplaced.

   Organization

   See Chapter 3 for additional information on methods of organization.

5. **Reformat your text for reader-friendly ease of access**—Look at any technical journal. You will notice that the writers have guided their readers through the text by using headings and subheadings. You will also notice that many journals use graphics (pie charts, bar graphs, line drawings, flowcharts, etc.) to clarify the writer's assertions. You should do the same. To help your readers follow your train of thought, reformat any blocks of wall-to-wall words. Add headings, subheadings, itemized lists, white space, and graphics.

6. **Correct any errors**—This represents your greatest challenge in writing a research report. You not only must be concerned with grammar and mechanics, as you are when writing a memo, letter, or report, but also with accurate quoting, paraphrasing, summarizing, parenthetical source citations, and works cited.

When revising, pay special attention to these concerns. If you quote, paraphrase, or summarize incorrectly, you run the risk of plagiarizing. If you fail to provide correct parenthetical source citations or works cited, you will make it impossible for your readers to find these same sources of information in their research or to check the accuracy of your data. Research demands accuracy and reliability.

RESEARCH CHECKLIST

_____ 1. Have you considered your purpose and audience in choosing a research topic?

_____ 2. Have you limited the focus of your research topic?

_____ 3. Have you used primary and secondary research sources?

_____ 4. Have you researched using books, periodicals, and the Internet?

_____ 5. Do you have enough material to sufficiently develop your topic?

_____ 6. Did you use note cards to gather your information?

_____ 7. Have you used an outline to organize your content?

_____ 8. Have you used both direct quotations and paraphrases?

_____ 9. Have you quoted and paraphrased correctly to avoid plagiarism?

_____ 10. Have you documented your sources according to either the MLA or APA formats to avoid plagiarism?

CHAPTER HIGHLIGHTS

1. You can research a topic either in a library or online at your computer.
2. You need to consider the audience's level of technical knowledge when you write a research paper.
3. Primary and secondary research can be conducted to assist you in developing content for reports.
4. Narrowing a topic can help you find sources of information.
5. A focus statement lets you determine the direction of your document.
6. Use discrimination and consider the source when you research on the Internet.
7. Careful source citations help you avoid plagiarism.
8. The Modern Language Association (MLA) and the American Psychological Association (APA) are two widely used style manuals for citing sources.
9. On a summary note card, condense the original material by using your own words.
10. On a paraphrase note card, restate the original material in your own words without condensing.

APPLY YOUR KNOWLEDGE

CASE STUDIES

1. The city of Oak Springs, IA, needs to improve a 10-mile stretch of road that runs east and west through the town. Before construction plans can be made, city engineers must conduct research in order to determine needs and considerations of the public. First, the engineers will perform primary research by interviewing city residents, as well as employees from the city's parks and recreation, police, fire, and transportation departments. Then, the engineers will conduct secondary research. They will read statutes regarding eminent domain, real estate, environmental considerations, state wildlife refuges, zoning, planning, and land use.

Assignment

- To conduct primary research, create a survey questionnaire. Ask residents what amenities they would want on an improved roadway through their city, their stance on increased taxes or bond issues, and concerns they might have about changes to the existing road.
- To conduct secondary research, go online to find out about Iowa's laws or stance regarding eminent domain, real estate, environmental considerations, state wildlife refuges, zoning, planning, and land use.

2. You plan to create a new business, which you want to be designated as Minority and Women Owned (MWBD). Before you can incorporate your company, you must research the following:
   - How you can become certified as a MWBD.
   - The standards you must uphold.
   - The percentage of women and/or minorities that you must employ.
   - The benefits of operating such a business.

Assignment

Research these topics (and any more that interest you) either online for secondary research or interview a MWBD business owner for primary research. Then, write an e-mail message or memo to your professor about your findings.

INDIVIDUAL AND TEAM PROJECTS

1. Correctly format and alphabetize a works cited page that contains the following entries:
   - An anonymously written magazine article
   - A magazine article signed by two authors
   - A journal article signed by one author
   - A book with three or more authors
   - A book with an editor
   - A signed newspaper editorial
   - An online document
   - A CD-ROM document
   - An e-mail message
   - A professional Web site

2. Summarize in one sentence any paragraph from this textbook. Provide a parenthetical source citation and works cited information.

3. Read a one- or two-page article from a magazine, journal, or online source. Then practice note taking. Writing in the margins or between paragraphs, briefly note the key point(s) made in each paragraph. (These notes can be limited to one or two words.)

4. Using a one- or two-page article from a magazine or journal, practice note taking on 3×5 inch cards. To do so, first write the correct works cited information on one card. Then, on separate cards, take notes about approximately four key ideas discussed in the article. Write only one note per card; give the card a title for future reference; provide either quotations or paraphrases; and then write the author's name, the article's title, and the page number on the bottom of the card.

5. Select a technical topic from your major field or your job and write a research report. You might want to consider a controversy in your area of interest (such as the greenhouse effect, hazardous waste management, or computer viruses) or the impact of a technical innovation (such as micromachines or the Internet).

PROBLEM-SOLVING THINK PIECES

1. Many communities have recycling projects that allow residents to recycle paper products, cans, and plastic. Not all businesses recycle, however. Research the benefits of recycling, determine how a business or businesses could implement a corporate recycling plan, and write a report recommending action based on your research.

2. In today's global economy, understanding and accommodating multiculturalism and cross-culturalism in business is important. Research the following:
   - The unique challenges that cultural diversity presents to businesses
   - How companies have responded to these challenges

   Write a report recommending why and how a business or businesses can help employees develop cultural awareness.

3. Many companies track the time their employees spend either surfing the Web or sending and receiving personal e-mail while at work. Research the following:
   - Software that companies use to track employee electronic communication usage
   - Corporate guidelines for employee use of company-owned electronic communication hardware and software
   - The legal and ethical ramifications of an employee's private use of corporate-owned e-mail and Internet access
   - The legal and ethical ramifications of an employer eavesdropping on an employee's Web usage

   Write a research report on your findings and provide a corporate guideline for both employee and employer electronic communication responsibilities.

4. Corporate training is big business. Many companies hire outside consultants or staff company training departments to teach employees new skills. These could include training workshops on diet and exercise to improve work efficiency, techniques for avoiding e-mail viruses or screening e-mail spam, resume writing for transitional employees, time management, leadership skills, improved oral presentations, improved customer service skills, basic word processing, or business accounting for non-accountants.

   What training class does your company need? Research possible topics and training approaches. How have other companies offered this training? What benefits do employees derive from this training? How does the company benefit? What are the costs for this training (personnel, time, equipment, and so forth)? Then, write a proposal or an instructional training module based on your research.

5. Entrepreneurialism is one of the fastest growing sectors in business. Many people are opening their own businesses. What does it take to open your own business? Before you can write an effective business plan and seek financing from a bank, you must research the project.

Choose a new business venture, selling a product or service of your choice. What would it cost to open this business? What would be your best location, or should your business be online? What certifications or licensing is needed? How many personnel would you need? What equipment is necessary? Who would be your clientele?

Based on research, write a proposal that is appropriate for presentation to a bank. In this proposal, present your business plan for a new entrepreneurial opportunity.

WEB WORKSHOP

1. FirstGov.com allows you to research a wide variety of topics, such as education and jobs, benefits and grants, consumer protection, environment and energy, science and technology, and public safety and health. This Web site also provides information on breaking news. Access FirstGov.com and research a topic relevant to your career goals. Write a memo or report to your instructor summarizing your findings.

2. Go to an online news magazine, such as *Slate*, *Time*, or *U.S. News Online*. Type a topic of interest in the magazine's search engine. Research this topic and make a brief oral presentation to your class about the information you have gathered.

3. Access a search engine such as Google, Ask Jeeves, or Dogpile. Type in a topic relating to your major field. For example, for computer information technology, medical records and health information, or accounting, look for job openings in your field to learn about salary ranges, benefits, application requirements, and so forth. Report your findings either in an oral presentation to the class or in a memo to your instructor.

4. Using an online newspaper, such as *The New York Times*, *CNN*, or *USA Today*, research business and technology news. Find out the major news stories of the day which relate to your career path. Report your findings in an oral presentation to the class or in a memo to your instructor.

QUIZ QUESTIONS

1. In what places can you locate researched information?
2. Explain why you need to consider audience when you write a researched report.
3. What is the difference between primary and secondary research?
4. What is the appropriate writing style of a research report?
5. What is the difference between a quote and a paraphrase?
6. How does a focus statement direct you when you write a research report?
7. Explain plagiarism. How can you avoid plagiarizing?
8. Why must you document your sources?
9. What is the difference between a summary note card and a paraphrase note card?
10. How does APA differ from MLA when you cite a book with one author?

CHAPTER 6

Routine Correspondence—
Memos, Letters, E-Mail,
and Instant Messaging

COMMUNICATION *at work*

In this scenario, a biotechnology company frequently corresponds through letters, memos, e-mail, and instant messages.

CompuMed, a wholesale provider of biotechnology equipment, is home based in Reno, NV. **CompuMed's** CEO, Jim Goodwin, plans to capitalize on emerging nanotechnology to manufacture and sell the following:

- Extremely lightweight and portable heart monitors and ventilators
- Pacemakers and hearing aids, 1/10 the size of current products on the market
- Microscopic biorobotics which can be injected in the body to manage, monitor, and/or destroy blood clots, metastatic activities, arterial blockages, alveoli damage due to carcinogens or pollutants, and scar tissues creating muscular or skeletal immobility

CompuMed is a growing company with over 5,000 employees located in two dozen cities and three states. To manage this business, supervisors and employees write on average over 5 letters, 10 memos, 50 e-mail messages, and numerous instant messages a day.

The letters are written to many different audiences and serve various purposes. **CompuMed** must write letters for

Objectives

When you complete this chapter, you will be able to

1. Understand the differences among memos, letters, and e-mail messages.
2. Follow an all-purpose template to write memos, letters, and e-mail.
3. Use memo samples as guidelines for memo components, organization, writing style, and tone.
4. Evaluate your memos, letters, and e-mail messages with checklists.
5. Correctly use the eight essential letter components: the writer's address, the date, an inside address for the recipient, a salutation, the body of the letter, a complimentary close, and the writer's signed and typed names.
6. Write different types of letters, including the following:
   - Inquiry
   - Cover (Transmittal)
   - Complaint
   - Adjustment
   - Bad News
7. Understand the components of successful e-mail messages.
8. Use e-mail samples as guidelines for effective e-mail components, organization, writing style, and tone.
9. Recognize techniques for successfully using instant messages in the workplace.
10. Follow the writing process—prewriting, writing, and rewriting—to create memos, letters, and e-mail.

---

employee files, to customers, job applicants, outside auditors, governmental agencies involved in biotechnology regulation, insurance companies, and more. They write

- **Letters of inquiry** to retailers seeking product information (technical specifications, pricing, warranties, guarantees, credentials of service staff, and so forth)
- **Cover letters** prefacing **CompuMed's** proposals
- **Complaint letters** written to parts manufacturers if and when faulty equipment and materials are received in shipping, and *adjustment letters* to compensate retailers when problems occur
- **Order letters** to computer and biotechnology retailers

CompuMed's managers and employees also write memos to accomplish a variety of goals:

- Document work accomplished
- Call meetings and establish meeting agendas
- Request equipment from purchasing
- Preface internal proposals

To accomplish the majority of their routine correspondence, **CompuMed's** employees write many e-mail messages each day. These messages serve different purposes. Some e-mail are conversational. Other e-mail messages, however, must be professional in their style, organization, and content. This is especially true for e-mail messages sent to clients, vendors, and customers outside the company. These e-mail messages focus on timelines, deadlines, prices for service, meeting arrangements, cost breakdowns, procedural steps, and a host of other topics.

Finally, when **CompuMed** employees are working at distant locations, on the road, in hotels, or at the airport, they will use instant messages to ask each other quick questions or to casually check up on the status of a project.

Routinely, **CompuMed** employees spend a great deal of their time writing memos, letters, e-mail messages, and instant messages.

Check out our quarterly newsletters TechCom E-Notes at www.prenhall.com/gerson for dot.com updates, new case studies, insights from business professionals, grammar exercises, and facts about technical communication.

THE IMPORTANCE OF MEMOS, LETTERS, AND E-MAIL

On a day-to-day basis, employees routinely write memos, letters, and e-mail messages. The National Commission on Writing, in their *Writing: A Ticket to Work . . . Or a Ticket Out, A Survey of Business Leaders*, published in 2007, states that e-mail is "ubiquitous in the American economy," that "more than half of all responding companies also report the following forms of communication as required 'frequently' or 'almost always': technical reports (59%), formal reports (62%), and memos and correspondence (70%)" (11). Figure 6.1 shows the significance of e-mail, memos, and letters in the workplace.

This survey of "120 major American corporations employing nearly 8 million people" (3) clearly tells us that you routinely can expect to write many e-mail messages, letters, and memos on the job.

WHICH COMMUNICATION CHANNEL SHOULD YOU USE?

Memos, letters, and e-mail messages are three common types of communication channels. Other communication channels include reports, Web sites, blogs, PowerPoint presentations, oral communication, and instant messages. When should you write an e-mail message instead of a memo? When should you write a memo instead of a letter? Is an instant message appropriate to the situation? You will make these decisions based on your audience (internal or external), the complexity of your topic, the speed with which your message can be delivered, and security concerns.

For example, e-mail is a convenient communication channel. It is easy to write a short e-mail message, which can be sent almost instantaneously to your audience at the click of a button. However, e-mail might not be the best communication channel to use.

Communication Channels

See Chapter 1 for additional information.

If you are discussing a highly sensitive topic such as a pending merger, corporate takeover, or layoffs, an e-mail message would be less secure than a letter sent in a sealed envelope.

You might need to communicate with employees working in a manufacturing warehouse. Not all of these employees will necessarily have an office or access to a computer. If you sent an e-mail message, how would they access this correspondence? A memo posted in the break room would be a better choice of communication channel.

FIGURE 6.1 Percentage of Employees Who Consider E-mail, Memos, Letters, and Reports "Extremely Important"

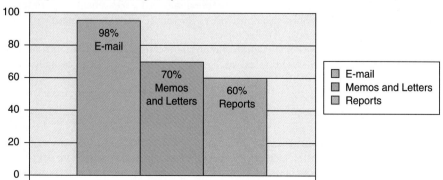

THE DIFFERENCES AMONG MEMOS, LETTERS, AND E-MAIL

To clarify the distinctions among memos, letters, and e-mail, review Table 6.1.

TABLE 6.1   Memos versus Letters versus E-mail

Characteristics	Memos	Letters	E-mail
Destination	Internal: correspondence written to colleagues within a company.	External: correspondence written outside the business.	Internal *and* external: correspondence written to friends and acquaintances, coworkers within a company, and clients and vendors.
Format	Identification lines include "Date," "To," "From," and "Subject." The message follows.	Includes letterhead address, date, reader's address, salutation, text, complimentary close, and signatures.	Identification lines: To and Subject. The Date and From are computer generated. Options include cc (complimentary copy), Ref (reference), and Distribution (other recipients of the e-mail message).
Audience	Generally high tech or low tech, mostly business colleagues.	Generally low tech and lay readers, such as vendors, clients, stakeholders, and stockholders.	Multiple readers due to the internal and external nature of e-mail.
Topic	Generally topics related to internal corporate decisions; abbreviations and acronyms often allowed.	Generally topics related to vendor, client, stakeholder, and stockholder interests; abbreviations and acronyms usually defined.	A wide range of diverse topics determined by the audience.
Complexity and Length of Communication	Memos usually are limited to a page of text. If you need to write longer correspondence and develop a topic in more detail, you might consider using a different communication channel, such as a short report.	Letters usually are limited to a page of text, though you might write a two- or three-page report using a letter format. If you need to develop a topic in greater detail than can be conveyed in one to three pages, you might want to use a different communication channel, such as a longer, formal report.	An effective e-mail message usually is limited to one viewable screen (requiring no scrolling) or two screens. E-mail, generally, is not the best communication channel to use for complex information or long correspondence. If your topic demands more depth than can be conveyed in a screen or two, you might want to write a report instead.
Tone	Informal due to peer audience.	More formal due to audience of vendors, clients, stakeholders, and stockholders.	A wide range of tones due to diverse audiences. Usually informal when written to friends, informal to coworkers, more formal to management.
Attachments or Enclosures	Hard-copy attachments can be stapled to the memo. Complimentary copies (cc) can be sent to other readers.	Additional information can be enclosed within the envelope. Complimentary copies (cc) can be sent to other readers.	Computer word processing files, HTML files and Web links, PDF files, RTF files, or downloadable graphics can be attached to e-mail. Complimentary copies can be sent to other readers. Size of these files is an issue, because large documents can crash a reader's system. A good rule is to limit files to 750 kilobytes (K).
Delivery Time	Determined by a company's in-house mail procedure.	Determined by the destination (within the city, state, or country). Letters could be delivered within 3 days but may take more than a week.	Often instantaneous, usually within minutes. Delays can be caused by system malfunctions or excessively large attachments.

TABLE 6.1 (Continued)

Characteristics	Memos	Letters	E-mail
Security	If a company's mail delivery system is reliable, the memo will be placed in the reader's mailbox. Then, what the reader sees on the hard-copy page will be exactly what the writer wrote. Security depends on the ethics of coworkers and whether the memo was sent in an envelope.	The U.S. Postal Service is very reliable. Once the reader opens the envelope, he or she sees exactly what the writer wrote. Privacy laws protect the letter's content.	E-mail systems are not secure. E-mail can be tampered with, read by others, and sent to many people. E-mail stays within a company's computer backup system and is the property of the company. Therefore, e-mail is not private.

FAQs: Memos vs. E-Mail

Q: Why write a memo? Haven't memos been replaced by e-mail?

A: E-mail is rapidly overtaking memos in the workplace, but employees still write memos for the following reasons.

1. Not all employees work in offices or have access to computers. Many employees who work in warehouses or in the field cannot easily access an e-mail account. They must depend on hard-copy documentation like memos.

2. Not all companies have e-mail. This may be hard to believe in the twenty-first century, but still it's a fact. These companies depend on hard-copy documentation like memos.

3. Many unions demand that hard-copy memos be posted on walls, in break rooms, in offices, and elsewhere, to ensure that all employees have access to important information. Sometimes, unions even demand that employees initial the posted memos, thus acknowledging that the memos have been read.

4. Some information cannot be transmitted electronically via e-mail. A bank we've worked with, for example, sends hard-copy cancelled checks as attachments to memos. They cannot send the actual cancelled check via e-mail.

5. E-mail messages are easy to disregard. We get so many e-mail messages (many of them spam) that we tend to quickly delete them. Memos, in contrast, make more of an official statement. People might take hard-copy memos more seriously than e-mail messages.

MEMOS

Reasons for Writing Memos

Memos are an important means by which employees communicate with each other. Memos, hard-copy correspondence written within your company, are important for several reasons.

First, you will write memos to a wide range of readers. This includes your supervisors, coworkers, subordinates, and multiple combinations of these audiences. Memos usually are copied (cc: complimentary copies) to many readers, so a memo sent to your boss could be read by an entire department, the boss's boss, and colleagues in other departments.

Because of their frequency and widespread audiences, memos could represent a major component of your interpersonal communication skills within your work environment.

Check Online Resources

www.prenhall.com/gerson

For more information about memos, visit our companion Web site.

Furthermore, memos are very flexible and can be written for many different purposes:

- **Documentation**—expenses, incidents, accidents, problems encountered, projected costs, study findings, hirings, firings, reallocations of staff or equipment
- **Confirmation**—a meeting agenda, date, time, and location; decisions to purchase or sell; topics for discussion at upcoming teleconferences; conclusions arrived at; fees, costs, or expenditures
- **Procedures**—how to set up accounts, research on the company intranet, operate new machinery, use new software, apply online for job opportunities through the company intranet, create a new company Web site, or solve a problem
- **Recommendations**—reasons to purchase new equipment, fire or hire personnel, contract with new providers, merge with other companies, revise current practices, or renew contracts
- **Feasibility**—studying the possibility of changes in the workplace (practices, procedures, locations, staffing, equipment, or missions/visions)
- **Status**—daily, weekly, monthly, quarterly, biannually, yearly statements about where you, the department, or the company is regarding many topics (sales, staffing, travel, practices, procedures, or finances)
- **Directive (delegation of responsibilities)**—informing subordinates of their designated tasks
- **Inquiry**—asking questions about upcoming processes or procedures
- **Cover**—prefacing an internal proposal, long report, or other attachments

Criteria for Writing Memos

Memos contain the following key components.

- Memo identification lines—Date, To, From, and Subject
- Introduction
- Discussion
- Conclusion
- Audience recognition
- Appropriate memo style and tone

Figure 6.2 shows an ideal, all-purpose organizational template that works well for memos, letters, and e-mail.

FIGURE 6.2 All-Purpose Template for Memos, Letters, and E-mail

Introduction: A lead-in or overview stating *why* you are writing and *what* you are writing about.

Discussion: Detailed development, made accessible through highlighting techniques, explaining *exactly what*   •   •   •

Conclusion: State *what* is next, *when* this will occur, and *why* the date is important.

Subject Line.   The subject line summarizes the memo's content. One-word subject lines do not communicate effectively, as in the following flawed subject line. The "Before" sample has a *topic* (a what) but is missing a *focus* (a what about the what).

BEFORE	AFTER
Subject: Sales	Subject: Report on Quarterly Sales

Introduction.   Once you have communicated your intent in the subject line, get to the point in the introductory sentence(s). Write one or two clear introductory sentences which tell your readers *what* topic you are writing about and *why* you are writing. The following example invites the reader to a meeting, thereby communicating *what* the writer's intentions are. It also tells the reader that the meeting is one of a series of meetings, thus communicating *why* the meeting is being called.

example

In the third of our series of sales quota meetings this quarter, I'd like to review our productivity.

Discussion.   The discussion section allows you to develop your content specifically. Readers might not read every line of your memo (tending instead to skip and skim). Thus, traditional blocks of data (paragraphing) are not necessarily effective. The longer the paragraph, the more likely your audience is to avoid reading. Make your text more reader friendly by itemizing, using white space, boldfacing, creating headings, or inserting graphics.

BEFORE	AFTER
Example—Unfriendly Text	**Example—Reader-Friendly Text**
This year began with an increase, as we sold 4.5 million units in January compared to 3.7 for January 2006. In February we continued to improve with 4.6, compared with 3.6 for the same time in 2007. March was not quite so good, as we sold 4.3 against the March 2007 figure of 3.9. April was about the same with 4.2, compared to 3.8 for April 2007.	*Comparative Quarterly Sales (in Millions)*

Comparative Quarterly Sales (in Millions)

	2006	2007	Increase/Decrease
Jan.	3.7	4.5	0.8+
Feb.	3.6	4.6	1.0+
Mar.	3.9	4.3	0.4+
Apr.	3.8	4.2	0.4+

Conclusion.   Conclude your memo with "thanks" and/or directive action. A pleasant conclusion could motivate your readers, as in the following example. A directive close tells your readers exactly what you want them to do next or what your plans are (and provides dated action).

If our quarterly sales continue to improve at the current rate, we will double our sales expectations by 2008. Congratulations! Next Wednesday (12/22/07), please provide next quarter's sales projections and a summary of your sales team's accomplishments.

Audience Recognition.   Because letters go outside your company, your audience is usually a low-tech or lay reader, demanding that you define your terms specifically. In memos your in-house audience is easier to address (usually high tech or low tech). You often can use more acronyms and internal abbreviations in memos than you can in letters.

You will write the message to "Distribution" (listing a group of readers) or send the memo to one reader but "Cc" (send a "carbon copy" or "complimentary copy") to other readers. Thus, you might be writing simultaneously to your immediate supervisor (high tech), to his or her boss (low tech), to your colleagues (high tech), and to a CEO (low tech). To accommodate multiple audiences, use parenthetical definitions, such as Cash in Advance (CIA) or Continuing Property Records (CPR).

Audience Recognition and Involvement

See Chapter 4 for additional information.

Style and Tone. Because memos are usually only one page long, use simple words, short sentences, specific detail, and highlighting techniques. In addition, strive for an informal, friendly tone. Memos are part of your interpersonal communication abilities, so a friendly tone will help build rapport with colleagues.

Style, Development, Tone, and Highlighting Techniques

See Chapters 3 and 8 for additional information.

In memos, audience determines tone. For example, you cannot write directive correspondence to supervisors mandating action on their part. It might seem obvious that you can write directives to subordinates, but you should not use a dictatorial tone. Though the subordinates are under your authority, they must still be treated with respect. You will determine the tone of your memo by deciding if you are writing vertically (up to management or down to subordinates) or laterally (to coworkers), as shown in Figure 6.3.

BEFORE	AFTER
Example 1—Unfriendly, Demanding Style	**Example 2—Friendly, Personal Style**
We will have a meeting next Tuesday, Jan. 11, 2008. Exert every effort to attend this meeting. Plan to make intelligent comments regarding the new quarter projections.	Let's meet next Tuesday (Jan. 11, 2008). Even if you're late, I'd appreciate your attending. By doing so you can have an opportunity to make an impact on the new quarter projections. I'm looking forward to hearing your comments.

FIGURE 6.3 Vertical and Lateral Communication within a Company

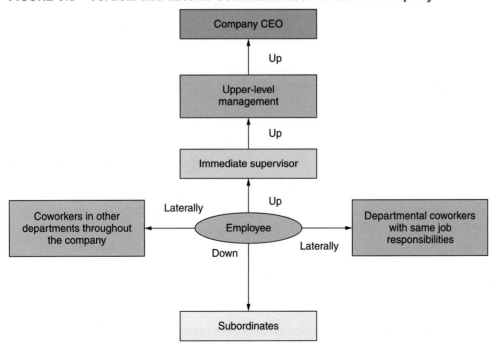

Sample Memos

See Figures 6.4, 6.5, and 6.6 for sample memos.

FIGURE 6.4   Comparison/Contrast Feasibility Memo

"Distribution" indicates that the memo is being sent to a number of employees. Notice that at the bottom of the memo, a distribution list is provided.

The introduction states the purpose of this memo, providing dates, personnel, options, and intended action.

The memo's discussion analyzes the criteria used to decide which radio to purchase. Tables are used to ensure reader-friendly ease of access.

The conclusion summarizes the importance of the action. It also ends in a personalized and positive tone to ensure reader involvement and to build rapport.

MEMORANDUM

DATE: December 12, 2008
TO: Distribution
FROM: Luann Brunson
SUBJECT: Replacement of Maintenance Radios

On December 5, the manufacturing department supervisor informed the purchasing department that our company's maintenance radios were malfunctioning. Purchasing was asked to evaluate three radio options (the RPAD, XPO 1690, and MX16 radios). Based on my findings, I have issued a purchase order for 12 RPAD radios.

The following points summarize my findings.

1. *Performance*
During a one-week test period, I found that the RPAD outperformed our current XPO's reception. The RPAD could send and receive within a range of 5 miles with minimal interference. The XPO's range was limited to 2 miles, and transmissions from distant parts of our building broke up due to electrical interference.

2. *Specifications*
Both the RPAD and the MX16 were easier to carry, because of their reduced weight and size, than our current XPO 1690s.

	RPAD	XPO 1690	MX16
Weight	1 lb.	2 lbs.	1 lb.
Size	5" x 2"	8" x 4"	6" x 1"

3. *Cost of Equipment*
The RPAD is our most cost-effective option because of quantity cost breaks and maintenance guarantees.

	RPAD	XPO 1690	MX16
Cost per unit	$70.00	$215.00	$100.00
Cost per doz.	$750.00	$2,580.00	$1,100.00
Guarantees	1 year	6 months	1 year

Purchase of the RPAD will give us improved performance and comfort. In addition, we can buy 12 RPAD radios for approximately the cost of 4 EXPOs. If I can provide you with additional information, please call. I'd be happy to meet with you at your convenience.

Distribution: M. Ellis   M. Rhinehart   T. Schroeder
                P. Michelson   R. Travers   R. Xidis

FIGURE 6.5   Cover Memo Prefacing Attachments

Memo   CompuMed

DATE: November 11, 2008
TO: CompuMed Management
FROM: Bill Baker, Human Resources Director
SUBJECT: Information about Proposed Changes to Employee Benefits
             Package

As of January 1, 2009, CompuMed will change insurance carriers. This will
affect all 5,000 employees' benefits packages. I have attached a proposal,
including the following:

1. Reasons for changing from our current carrier   page 2
2. Criteria for our selection of a new insurance company pages 3–4
3. Monthly cost for each employee   pages 5–6
4. Overall cost to CompuMed   page 7
5. Benefits derived from the new healthcare plan   page 8

Please review the proposal, survey your employees' responses to our suggestions,
and provide your feedback. We need your input by December 1, 2008. This will
give the Human Resources Department time to consider your suggestions and
work with insurance companies to meet employee needs.

Enclosure: Proposal

> Introducing your memo by stating important information that will affect many people will capture your reader's attention.

> Concluding your memo by providing dated action and the reasons for this request will clarify your needs for the audience.

FIGURE 6.6   Good News Memo Promoting an Employee

Date: March 23, 2008
To: Jan Hunt, Second Shift Line Manager
From: Hailey Osmond, Manager Operations
Subject: Recommendation for Early Promotion

Congratulations! We are proud to offer you early promotion, Jan. You
have earned a grade raise to E30 for the following reasons:

- *Productivity.* Your line personnel produced 2,000 units per month
  throughout this quarter.
- *Efficiency.* You maintained a 95 percent manufacturing efficiency
  rating.
- *Supervisory skills.* You received only four grievances; your annual
  performance appraisals showed that your subordinates appreciated
  your motivational management techniques.
- *Customer satisfaction.* Numerous customers have written us
  letters commending you for your outstanding customer
  service.

Because of your excellent work, you will receive your pay increase
the first of next month. You deserve it. Good work, Jan.

> In the introduction, explain *why* you're writing and *what* you're writing about. Because your goal is to convey good news, begin with positive word usage (such as "Congratulations," "proud," and "earned").

> Provide the details that explain *exactly what* has justified the promotion. This memo achieves specificity by quantifying: "2,000 units," "95 percent," and "only four grievances."

> In your conclusion, state *what* you plan next and *when* this action will occur. Consider ending with a final positive comment, such as "Good work." Personalization is always appropriate in good news. You can accomplish this by using pronouns, such as "you" and "your" and by referring to the reader by name ("Jan").

The memo checklist will give you the opportunity for self-assessment and peer evaluation of your writing. Input from peers can be an important way for you to gauge the response to your memo, determine if content should be added or deleted, and check for correctness.

MEMO CHECKLIST

_____ 1. Does the memo contain identification lines (Date, To, From, and Subject)?

_____ 2. Does the subject line contain a topic and a focus?

_____ 3. Does the introduction clearly state
- Why this memo has been written?
- What topic the memo is discussing?

_____ 4. Does the body explain exactly what you want to say?

_____ 5. Does the conclusion
- Tell when you plan a follow-up or when you want a response?
- Explain why this dated action is important?

_____ 6. Are highlighting techniques used effectively for document design?

_____ 7. Is the memo concise?

_____ 8. Is the memo clear,
- Achieving specificity of detail?
- Answering reporter's questions?

_____ 9. Does the memo recognize audience,
- Defining acronyms or abbreviations where necessary for various levels of readers (high tech, low tech, and lay)?

_____ 10. Did you avoid grammatical errors? Errors will hurt your professionalism. See Appendix for grammar rules and exercises.

TECHNOLOGY TIPS

Using Memos and Letter Templates in Microsoft Word 2007

1. Click on the <u>Office</u> Button located on the top left of your toolbar and scroll to <u>New</u>.

The following window will pop up.

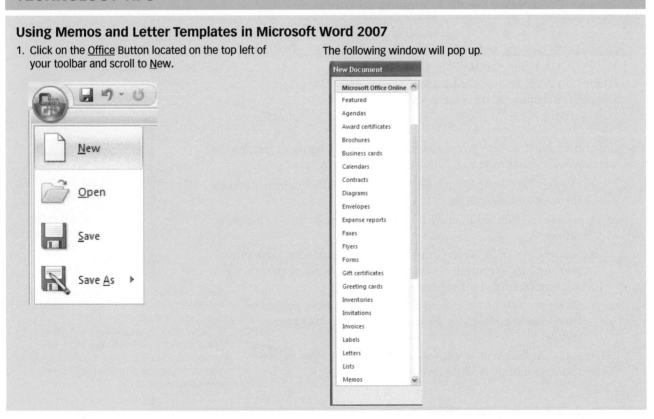

2. Click on the type of document you want to write, such as Letters or Memos.
When you choose the communication channel, either of the following windows will pop up.

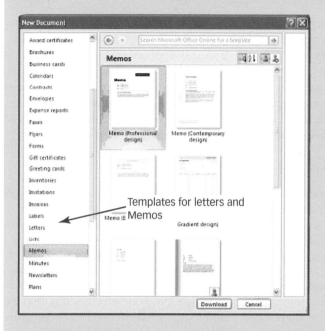

Templates for letters and Memos

You can choose from one of 10 memo templates or one of literally hundreds of letter templates. Each of these templates gives you an already-designed letter format, complete with spacing, font selection, and layout. In addition, these templates provide fields in which you merely need to type the appropriate information (address, company name, date, salutation, complimentary close, your name and title, etc.).

These templates and wizards are both good and bad. They remind you which components can be included in a memo or letter, they make it easy for you to include these components, and they let you choose ready-made formats.

The templates also can create some problems. First, they are somewhat limiting in that they dictate what information you should include and where you should put this information. The content and placement of this information might contradict your teacher's or boss's requirements. Second, the templates are prescriptive, limiting your choice of font sizes and types. Our advice is to use these templates and wizards with caution.

FAQs: Letters vs. E-Mail

Q: Why write a letter? Haven't letters been replaced by e-mail?

A: Though e-mail is quick, it might not be the best communication channel, for the following reasons.

1. E-mail might be too quick. In the workplace, you will write about topics that require a lot of thought. Because e-mail messages can be written and sent quickly, people too often write hurriedly and neglect to consider the impact of the message.

2. E-mail messages tend to be casual, conversational, and informal. Not all correspondence, however, lends itself to this level of informality. Formal correspondence related to contracts, for example, requires the more formal communication channel of a letter. The same applies to audience. You might want to write a casual e-mail to a coworker, but if you were writing to the president of a company, the mayor of a city, or a foreign dignitary, a letter would be a better, more formal choice of communication channel.

3. E-mail messages tend to be short. For content requiring more detail, a longer letter would be a better choice.

4. We get so many e-mail messages a day that they are easy to disregard—even easy to delete. Letters carry more significance. If you want to ensure that your correspondence is read and perceived as important, you might want to write a letter instead of an e-mail.

5. Letters allow for a "greater paper trail" than e-mail. Most employees' e-mail inboxes fill up quickly. To clean these inboxes up, people tend to delete messages that they don't consider important. In contrast, hard-copy letters are wonderful documentation.

LETTERS

Reasons for Writing Letters

Letters are external correspondence that you send from your company to a colleague working at another company, a vendor, a customer, a prospective employee, and stakeholders and stockholders. Letters leave your work site (as opposed to memos, which stay within the company).

Sales Letters and Persuasive Writing

See Chapter 10 for additional information.

Because letters are sent to readers in other locations, your letters not only reflect your communication abilities but also are a reflection of your company. This section provides letter components, formats, criteria, and examples to help you write the following kinds of letters: inquiry, cover (transmittal), complaint, adjustment, and bad news.

Essential Components of Letters

Font Selection and Readability

See Chapter 8 for additional information.

Your letter should be typed or printed on 8 ½ × 11 inch paper. Leave 1 to 1½ inch margins at the top and on both sides. Choose an appropriately businesslike font (size and style), such as Times New Roman or Arial (12 point). Though "designer fonts," such as Comic Sans and Shelley Volante, are interesting, they tend to be harder to read and less professional.

Your letter should contain the essential components shown in Figure 6.7.

Writer's Address. This section contains either your personal address or your company's address. If the heading consists of your address, then you will include your street address, the city, state, and zip code. The state may be abbreviated with the appropriate two-letter abbreviation.

If the heading consists of your company's address, you will include the company's name, street address, and city, state, and zip code.

Check Online Resources

www.prenhall.com/gerson

For more information about letters, visit our companion Web site.

Date. Document the month, day, and year when you write your letter. You can write your date in one of two ways: May 31, 2008 or 31 May 2008. Place the date one or two spaces below the writer's address.

Reader's Address. Place the reader's address two lines below the date.

- Reader's name (If you do not know the name of this person, begin the reader's address with a job title or the name of the department.)
- Reader's title (optional)
- Company name
- Street address
- City, state, and zip code

Salutation. The traditional salutation, placed two spaces beneath the reader's address, is *Dear* and your reader's last name, followed by a colon (Dear Mr. Smith:).

You can also address your reader by his or her first name if you are on a first-name basis with this person (Dear John:). If you are writing to a woman and are unfamiliar with her marital status, address the letter *Dear Ms. Jones*. However, if you know the woman's marital status, you can address the letter accordingly (Dear Miss Jones *or* Dear Mrs. Jones:).

Letter Body. Begin the body of the letter two spaces below the salutation. The body includes your introductory paragraph, discussion paragraph(s), and concluding paragraph. The body should be single spaced with double spacing between paragraphs. Whether you indent the beginning of paragraphs or leave them flush with the left margin is determined by the letter format you employ.

Complimentary Close. Place the complimentary close, followed by a comma, two spaces below the concluding paragraph. Typical complimentary closes include "Sincerely," "Yours truly," and "Sincerely yours."

FIGURE 6.7 Essential Letter Components

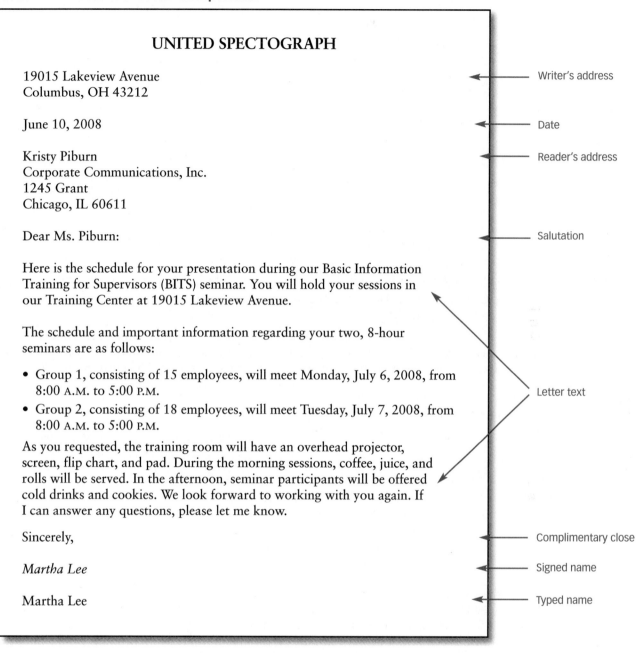

UNITED SPECTOGRAPH

19015 Lakeview Avenue
Columbus, OH 43212 — Writer's address

June 10, 2008 — Date

Kristy Piburn
Corporate Communications, Inc.
1245 Grant
Chicago, IL 60611 — Reader's address

Dear Ms. Piburn: — Salutation

Here is the schedule for your presentation during our Basic Information Training for Supervisors (BITS) seminar. You will hold your sessions in our Training Center at 19015 Lakeview Avenue.

The schedule and important information regarding your two, 8-hour seminars are as follows:

- Group 1, consisting of 15 employees, will meet Monday, July 6, 2008, from 8:00 A.M. to 5:00 P.M.
- Group 2, consisting of 18 employees, will meet Tuesday, July 7, 2008, from 8:00 A.M. to 5:00 P.M.

As you requested, the training room will have an overhead projector, screen, flip chart, and pad. During the morning sessions, coffee, juice, and rolls will be served. In the afternoon, seminar participants will be offered cold drinks and cookies. We look forward to working with you again. If I can answer any questions, please let me know. — Letter text

Sincerely, — Complimentary close

Martha Lee — Signed name

Martha Lee — Typed name

Signed Name. Sign your name legibly beneath the complimentary close.

Typed Name. Type your name four spaces below the complimentary close. You can type your title one space beneath your typed name. You also can include your title on the same line as your typed name, with a comma after your name.

Optional Components of Letters

In addition to the letter essentials, you can include the following optional components.

Subject Line. Place a subject line two spaces below the reader's address and two spaces above the salutation.

Dr. Ron Schaefer
Linguistics Department
Southern Illinois University
Edwardsville, IL 66205

Subject: Linguistics Conference Registration Payment

Dear Dr. Schaefer:

You also could use a subject line instead of a salutation.

Linguistics Department
Southern Illinois University
Edwardsville, IL 66205

Subject: Linguistics Conference Registration Payment

A subject line not only helps readers understand the letter's intent but also (if you are uncertain of your reader's name) avoids such awkward salutations as "To Whom It May Concern," "Dear Sirs," and "Ladies and Gentlemen." In the simplified format, both the salutation and the complimentary close are omitted, and a subject line is included.

New-Page Notations.   If your letter is longer than one page, cite your name, the page number, and the date on all pages after page 1. Place this notation either flush with the left margin at the top of subsequent pages or across the top of subsequent pages. (You must have at least two lines of text on the next page to justify another page.)

Left margin, subsequent page notation   Across top of subsequent pages

Mabel Tinjaca Page 2 May 31, 2008	Mabel Tinjaca 2 May 31, 2008

Writer's and Typist's Initials.   If the letter was typed by someone other than the writer, include both the writer's and the typist's initials two spaces below the typed signature. The writer's initials are capitalized, the typist's initials are typed in lowercase, and the two sets of initials are separated by a colon. If the typist and the writer are the same person, this notation is not necessary.

Sincerely,

W. T. Winnery

WTW: mm

Enclosure Notation.   If your letter prefaces enclosed information, such as an invoice or report, mention this enclosure in the letter and then type an enclosure notation two spaces below the typed signature (or two spaces below the writer and typist initials). The enclosure notation can be abbreviated "Enc."; written out as "Enclosure"; show the number of enclosures, such as "Enclosures (2)"; or specify what has been enclosed—"Enclosure: January Invoice."

Copy Notation.   If you have sent a copy of your letter to other readers, show this in a copy notation. A complimentary copy is designated by a lowercase "cc." List the other readers' names following the copy notation. Type the copy notation two spaces below the typed signature or two spaces below either the writer's and typist's initials or the enclosure notation.

Sincerely,

Brian Altman
Enclosure: August Status Report
cc: Marcia Rittmaster and Larry Rochelle

Formatting Letters

Three common types of letter formats include **full block** (Figure 6.8), **full block with subject line** (Figure 6.9), and **simplified** (Figure 6.10). Two popular and professional formats used in business are full block and full block with subject line. With both formats, you type all information at the left margin without indenting paragraphs, the date, the complimentary close, or signature. The full block with subject line differs only with the inclusion of a subject line.

Another option is the simplified format. This type of letter layout is similar to the full block format in that all text is typed margin left. The two significant omissions include no salutation ("Dear _____:") and no complimentary close ("Sincerely,"). Omitting a salutation is useful in the following instances:

- You do not know your reader's name (NOTE: Avoid the trite salutation, "To Whom It May Concern:")
- You are writing to someone with a nongender-specific name (Jesse, Terry, Stacy, Chris, etc.) and you do not know whether to use "Mr.," "Mrs.," or "Ms."

SPOTLIGHT

Why are letters important in a governmental organization?

Dr. Georgia Nesselrode is Director of Government Training for Mid-America Regional Council's Government Training Institute (GTI). The GTI provides training workshops for local government officials (elected and appointed) and their employees. Dr. Nesselrode writes approximately 30 letters each month.

In fact, as Georgia says, she writes letters when other companies might write e-mail messages. Why not write e-mail for routine communication? "Letters are an integral part of our communication for several reasons:

- Local government is more traditional than many corporations.
- Letters provide a better paper trail of documentation.
- Letters are more formal and have greater impact than e-mail.
- Many of my letters preface large, hard-copy attachments.
- The letters convey information that does not always require immediate action (a benefit of e-mail). Instead, my letters often communicate content for future consideration."

Dr. Nesselrode writes cover letters prefacing contracts, letters of recommendation for colleagues, inquiries, responses to inquiries, sales letters promoting her programs, updates, letters of confirmation to potential clients and vendors, and on occasion, 100 percent "yes adjustment letters" to seminar participants who were dissatisfied with a training workshop or facilitator.

Her audiences include her local government training liaisons and training workshop instructors. She writes to elected and appointed officials such as mayors, police chiefs, fire captains, city managers, and municipal employees who work in their city's health, accounting, tax, water, road, and parks and recreation departments.

Because she writes to governmental agencies, officials, and employees, Dr. Nesselrode ensures the correspondence is concise, informative, and presents a professional image. In fact, her letters must be perfect. She represents an entire metropolitan area's governmental training, and that's a reflection on city management.

FIGURE 6.8 Full Block Format

1 to 1 1/2 inch margins on all sides of the letter

State Health Department

1890 Clark Road
Jefferson City, MO 67220

2–4 spaces above and below the date

June 6, 2008

Dale McGraw, Manager
Elmwood Mobile Home Park
Elmwood, MO 64003

2–4 spaces above and below the salutation

Dear Mr. McGraw:

Single space within the paragraphs.

Double space between the paragraphs.

On April 19, 2008, Ryan Duran and I, environmental specialists from the Health Department, conducted an inspection of the Elmwood Mobile Home Park Wastewater Treatment Facility. The purpose was to assess compliance with the following: the state's Clean Water Law, Clean Water Commission regulations, and your facility's plan for pollution control. The inspection also would allow the state to promote proper operation of Wastewater Facilities and to provide technical assistance where needed to the Elmwood Mobile Homes management.

Though the Elmwood Mobile Home pollution control plan had expired in 2007, a consent judgment was issued by the state's Attorney Generals Office. The county court stipulated a timeline for correction by connection to an available sewer system. Your mobile home park's wastewater system has continually discharged to the Little Osage River. A copy of the abatement order, which requires that monthly discharge monitoring reports (DMRs) be submitted by the 28th of the month following the reporting periods, is attached. All DMRs for the previous twelve months have been received, and reported pollution parameters are not within limits. Due to the plant's performance, the stream was placed on the 2000, 303 (d) stream for impairment by the Elmwood Mobile Home.

As part of the inspection, a review of the facility's DMR was conducted. Twenty-four-hour composite samples were collected using a composite sampler. Attached are the results of the 24-hour composite samples collected on April 20, 2008. Every one of the problems documented is an infraction that must be addressed.

Within 30 days of receipt of this letter, please submit to the Health Department written documentation describing steps taken to correct each of the concerns identified in the attachments. Also include engineering reports, and submit a timeframe to eliminate the problems. Thank you for your cooperation.

Sincerely,

2 spaces before "Sincerely"

4 spaces between "Sincerely" and the typed signature

Harvey Haddix
Environmental Manager

Attachment

FIGURE 6.9 Full Block Format with Subject Line

State Health Department
1890 Clark Road   Jefferson City, MO 67220

June 6, 2008

Dale McGraw, Manager
Elmwood Mobile Home Park
Elmwood, MO 64003

Subject: Pollution Control Inspection

Dear Mr. McGraw:

On April 19, 2008, Ryan Duran and I, environmental specialists from the Health Department, conducted an inspection of the Elmwood Mobile Home Park Wastewater Treatment Facility. The purpose was to assess compliance with the following: the state's Clean Water Law, Clean Water Commission regulations, and your facility's plan for pollution control. The inspection also would allow the state to promote proper operation of Wastewater Facilities and to provide technical assistance where needed to the Elmwood Mobile Homes management.

Though the Elmwood Mobile Home pollution control plan had expired in 2007, a consent judgment was issued by the state's Attorney Generals Office. The county court stipulated a timeline for correction by connection to an available sewer system. Your mobile home park's wastewater system has continually discharged to the Little Osage River. A copy of the abatement order, which requires that monthly discharge monitoring reports (DMRs) be submitted by the 28th of the month following the reporting periods, is attached. All DMRs for the previous twelve months have been received, and reported pollution parameters are not within limits. Due to the plant's performance, the stream was placed on the 2000, 303 (d) stream for impairment by the Elmwood Mobile Home.

As part of the inspection, a review of the facility's DMR was conducted. Twenty-four-hour composite samples were collected using a composite sampler. Attached are the results of the 24-hour composite samples collected on April 20, 2008. Every one of the problems documented is an infraction that must be addressed.

Within 30 days of receipt of this letter, please submit to the Health Department written documentation describing steps taken to correct each of the concerns identified in the attachments. Also include engineering reports, and submit a timeframe to eliminate the problems. Thank you for your cooperation.

Sincerely,

Harvey Haddix
Environmental Manager

Attachment

FIGURE 6.10   Simplified Format Omitting "Dear . . ." and "Sincerely"

State Health Department
1890 Clark Road
Jefferson City, MO 67220

June 6, 2008

Dale McGraw, Manager
Elmwood Mobile Home Park
Elmwood, MO 64003

Subject: Pollution Control Inspection

On April 19, 2008, Ryan Duran and I, environmental specialists from the Health Department, conducted an inspection of the Elmwood Mobile Home Park Wastewater Treatment Facility. The purpose was to assess compliance with the following: the state's Clean Water Law, Clean Water Commission regulations, and your facility's plan for pollution control. The inspection also would allow the state to promote proper operation of Wastewater Facilities and to provide technical assistance where needed to the Elmwood Mobile Homes management.

Though the Elmwood Mobile Home pollution control plan had expired in 2007, a consent judgment was issued by the state's Attorney Generals Office. The county court stipulated a timeline for correction by connection to an available sewer system. Your mobile home park's wastewater system has continually discharged to the Little Osage River. A copy of the abatement order, which requires that monthly discharge monitoring reports (DMRs) be submitted by the 28th of the month following the reporting periods, is attached. All DMRs for the previous twelve months have been received, and reported pollution parameters are not within limits. Due to the plant's performance, the stream was placed on the 2000, 303 (d) stream for impairment by the Elmwood Mobile Home.

As part of the inspection, a review of the facility's DMR was conducted. Twenty-four-hour composite samples were collected using a composite sampler. Attached are the results of the 24-hour composite samples collected on April 20, 2008. Every one of the problems documented is an infraction that must be addressed.

Within 30 days of receipt of this letter, please submit to the Health Department written documentation describing steps taken to correct each of the concerns identified in the attachments. Also include engineering reports, and submit a timeframe to eliminate the problems. Thank you for your cooperation.

Harvey Haddix
Environmental Manager

Attachment

The Administrative Management Society (AMS) suggests that if you omit the salutation, you also should omit the complimentary close. Some people feel that omitting the salutation and the complimentary close will make the letter cold and unfriendly. However, the AMS says that if your letter is warm and friendly, these omissions will not be missed. More importantly, if your letter's content is negative, beginning with "Dear" and ending with "Sincerely" will not improve the letter's tone or your reader's attitude toward your comments.

The simplified format includes a subject line to aid the letter's clarity.

CRITERIA FOR DIFFERENT TYPES OF LETTERS

Though you might write different types of letters, including letters of inquiry, cover, complaint, response, adjustment, or sales, consider using the all-purpose memo, letter, and e-mail template from Figure 6.2 (on page 159) to format your correspondence.

Letter of Inquiry

If you want information about degree requirements, equipment costs, performance records, turnaround time, employee credentials, or any other matter of interest to you or your company, then write a letter requesting that data. Letters of inquiry require that you be specific. For example, if you write, "Please send me any information you have on your computer systems," you are in trouble. You will either receive any information the reader chooses to give you or none at all. Look at the following flawed letter of inquiry from a biochemical waste disposal company.

BEFORE

> Dear Mr. Jernigan:
>
> Please send us information about the following filter pools:
> 1. East Lime Pool
> 2. West Sulphate Pool
> 3. East Aggregate Pool
>
> Thank you.

The reader replied as follows:

> Dear Mr. Scholl:
>
> I would be happy to provide you with any information you would like. However, you need to tell me what information you require about the pools.
>
> I look forward to your response.

The first writer, recognizing the error, rewrote the letter as follows:

AFTER

> Dear Mr. Jernigan:
>
> My company, Jackson County Hazardous Waste Disposal, Inc., needs to purchase new waste receptacles. One of our clients used your products in the past and recommended you. Please send us information about the following:
>
> 1. Lime Pool—costs, warranties, time of installation, and dimensions
> 2. Sulphate Pool—costs, material, and levels of acidity
> 3. Aggregate Pool—costs, flammability, maintenance, and discoloration
>
> We plan to install our pools by March 12. We would appreciate your response by February 20. Thank you.

Providing specific details makes your letter of inquiry effective. You will save your readers time by quantifying your request.

To compose your letter of inquiry, include the following:

Introduction. Clarify your intent in the introduction. Until you tell your readers why you are writing, they do not know. It is your responsibility to clarify your intent and explain your rationale for writing. Also tell your reader immediately what you are writing about (the subject matter of your inquiry). You can state your intent and subject matter in one to three sentences.

Discussion. Specify your needs in the discussion. To ensure that you get the response you want, ask precise questions or list specific topics of inquiry. You must quantify. For example, rather than vaguely asking about machinery specifications, ask more precisely about "specifications for the 12R403B Copier." Rather than asking, "Will the roofing material cover a large surface?" you need to quantify—"Will the roofing material cover 150 × 180 feet?"

Conclusion. Conclude precisely. First, explain when you need a response. Do not write, "Please respond as soon as possible." Provide dated action and tell the reader exactly when you need your answers. Second, to sell your readers on the importance of this date, explain why you need answers by the date given.

Figure 6.11 will help you understand the requirements for effective letters of inquiry.

Cover (Transmittal) Letters

In business, you are often required to send information to a client, vendor, or colleague. You might send multipage copies of reports, invoices, drawings, maps, letters, memos, specifications, instructions, questionnaires, or proposals.

A cover letter accomplishes two goals. First, it lets you tell readers up front what they are receiving. Second, it helps you focus your readers' attention on key points within the enclosures. Thus, the cover letter is a reader-friendly gesture geared toward assisting your audience. To compose your cover letter, include the following:

Introduction. In the introductory paragraph, tell your reader why you are writing and what you are writing about. What if the reader has asked you to send the documentation? Do you still need to explain why you are writing? The answer is yes. Although the reader requested the information, time has passed, other correspondence has been written, and your reader might have forgotten the initial request.

Discussion. In the body of the letter, you can accomplish two things. You tell your reader either exactly what you have enclosed or exactly what of value is within the enclosures. In both instances, provide an itemized list or easily accessible, short paragraphs.

FIGURE 6.11 Letter of Inquiry Using the Simplified Format

CompuMed

8713 Hillview   Reno, NV 32901   1-800-551-9000   Fax: 1-816-555-0000

September 12, 2008

Sales Manager
OfficeToGo
7622 Raintree
St. Louis, MO 66772

Subject: Request for Product Pricing and Shipping Schedules

My medical technology company has worked well with OfficeToGo for the past five years. However, in August I received a letter informing me that OTG had been purchased by a larger corporation. I need to determine if OTG remains competitive with other major office equipment suppliers in the Reno area.

Please provide the following information:

1. What discounts will be offered for bulk purchases?
2. Which freight company will OTG now be using?
3. Who will pay to insure the items ordered?
4. What is the turnaround time from order placement to delivery?
5. Will OTG be able to deliver to all my satellite sites?
6. Will OTG technicians set up the equipment delivered, including desks, file cabinets, bookshelves, and chairs?
7. Will OTG be able to personalize office stationery onsite, or will it have to be outsourced?

Please respond to these questions by September 30 so I can prepare my quarterly orders in a timely manner. I continue to expand my company and want assurances that you can fill my growing office supply needs. You can contact me at the phone number provided above or by e-mail (jgood@CompuMed.com). Thank you for your help.

Jim Goodwin
Owner and CEO

In the introduction, explain why you are writing and introduce yourself to establish the letter's context.

In the discussion, specify your needs. To ensure accuracy of response, ask precise questions.

In the conclusion, state when you need a response and explain why this date is important. Providing contact information will help the reader respond.

Conclusion. Your conclusion should tell your readers what you want to happen next, when you want this to happen, and why the date is important.

See Figure 6.12 on page 176 for an example of a cover letter from a healthcare provider.

Complaint Letters

You are a purchasing director at an electronics firm. Although you ordinarily receive excellent products and support from a local manufacturing firm, two of your recent orders have been filled incorrectly and included defective merchandise. You don't want to have to look for a new supplier. You should express your complaint as pleasantly as possible.

To compose your complaint letter, include the following:

Introduction. In the introduction, politely state the problem. Although you might be angry over the service you have received, you want to suppress that anger. Blatantly negative comments do not lead to communication; they lead to combat. Because

FIGURE 6.12   Cover Letter in Block Format

AMERICAN HEALTHCARE
1401 Laurel Drive
Denton, TX 76201
November 11, 2008

Jan Pascal
Director of Outpatient Care
St. Michael's Hospital
Westlake Village, CA 91362

Dear Ms. Pascal:

Thank you for your recent request for information about our specialized outpatient care equipment. American Healthcare's stair lifts, bath lifts, and vertical wheelchair lifts can help your patients. To show how we can serve you, we have enclosed a brochure including the following information:

	Page
• Maintenance, warranty, and guarantee information	1–3
• Technical specifications for our products, including sizes, weight limitations, colors, and installation instructions	4–6
• Visuals and price lists for our products	7–8
• An American Healthcare order form	9
• Our 24-hour hotline for immediate service	10

Early next month, I will call to make an appointment at your convenience. Then we can discuss any questions you might have. Thank you for considering American Healthcare, a company that has provided exceptional outpatient care for over 30 years.

Sincerely,

Toby Sommers

Enclosure

A positive tone in the introduction builds rapport and informs the reader why this letter is being written: in response to a request.

An itemized body clarifies what is in the enclosure. Adding page numbers in the list helps readers find the information in the enclosed material.

The conclusion provides dated action, noting when a follow-up call will be made.

---

angry readers will not necessarily go out of their way to help you, your best approach is diplomacy.

To strengthen your assertions, in the introduction, include supporting details, such as the following: serial numbers, dates of purchase, invoice numbers, check numbers, names of salespeople involved in the purchase, and/or receipts. When possible, include copies documenting your claims.

Discussion. In the discussion paragraph(s), explain in detail the problems experienced. This could include dates, contact names, information about shipping, breakage information, an itemized listing of defects, and poor service.

Be specific. Generalized information will not sway your readers to accept your point of view. In a complaint letter, you suffer the burden of proof. Help your audience understand the extent of the problem. After documenting your claims, state what you want done and why.

Conclusion. End your letter positively. Remember, you want to ensure cooperation with the vendor or customer. You also want to be courteous, reflecting your company's professionalism. Your goal should be to achieve a continued rapport with your reader. In this concluding paragraph, include your contact information and the times you can best be reached.

See Figure 6.13 for a sample complaint letter to an automotive supplies company.

FIGURE 6.13 Complaint Letter in Block Format

1234 18th Street
Galveston, TX 77001
May 10, 2008

Mr. Holbert Lang
Customer Service Manager
Gulfstream Auto
1101 21st Street
Galveston, TX 77001

Dear Mr. Lang:

On February 12, 2008, I purchased two shock absorbers in your automotive department. Enclosed are copies of the receipt and the warranty for that purchase. One of those shocks has since proved defective.

The introduction includes the date of purchase (to substantiate the claim) and the problems encountered.

I attempted to exchange the defective shock at your store on May 2, 2008. The mechanic on duty, Vernon Blanton, informed me that the warranty was invalid because your service staff did not install the part. I believe that your company should honor the warranty agreement and replace the part for the following reasons:

1. The warranty states that the shock is covered for 48 months and 48,000 miles.
2. The warranty does not state that installation by someone other than the dealership will result in warranty invalidation.
3. The defective shock absorber is causing potentially expensive damage to the tire and suspension system.

In the body, explain what happened, state what you want done, and justify your demand. This letter develops its claim with warranty information.

I can be reached between 1 p.m. and 6 p.m. on weekdays at 763-9280 or at 763-9821 anytime on weekends. I look forward to hearing from you. Thank you for helping me with this misunderstanding.

Conclude your letter by providing contact information and an upbeat, pleasant tone.

Sincerely,

Carlos De La Torre

Enclosures (2)

Creating a Positive Tone. Audiences respond favorably to positive words. If you use negative words, you could offend your reader. In contrast, positive words will help you control your readers' reactions, build goodwill, and persuade your audience to accept your point of view.

Choose your words carefully. Even when an audience expects bad news, they still need a polite and positive response.

The positive words in Table 6.2 and positive verbs in Table 6.3 will help you create a pleasant tone and build audience rapport.

Table 6.4 gives you a *before* and *after* view of negative sentences rewritten using a positive tone.

Adjustment Letters

Responses to letters of complaint, also called adjustment letters, can take three different forms. Table 6.5 shows the differences among these three types of adjustment letters.

Check Online Resources

www.prenhall.com/gerson

For more information about sample complaint letters, visit our companion Web site.

Tone and Technical Communication

See Chapter 4 for additional information.

TABLE 6.2 Positive Words

advantage	efficient	meaningful
asset	enjoyable	please
benefit	favorable	positive
certain	good	profit
confident	grateful	quality
constructive	happy	successful
contribution	helpful	thank you
effective	improvement	value

TABLE 6.3 Positive Verbs

accomplish	improve
achieve	increase
assist	initiate
assure	insure (ensure)
build	maintain
coordinate	organize
create	plan
develop	produce
encourage	promote
establish	satisfy
help	train
implement	value

TABLE 6.4 Negative versus Positive Sentences

Before	After
1. The error is your fault. You scheduled incorrectly and cannot complain about our deliveries. If you would cooperate with us, we would work with you to solve this problem.	1. To improve deliveries, let's work together on our companies' scheduling practices.
2. I regret to inform you that we will not replace the motor in your dryer unless we have proof of purchase.	2. When you provide us proof of purchase, we will be happy to replace the motor in your dryer.
3. The accounting records your company submitted are incorrect. You have obviously miscalculated the figures.	3. After reviewing your company's accounting records, please recalculate the numbers to ensure that they correspond to the new X44 tax laws (enclosed).
4. Your letter suggesting an improvement for the system has been rejected. The reconfigurations you suggest are too large for the area specifications. We need you to resubmit if you can solve your problem.	4. Thank you for your suggestions. Though you offer excellent ideas, the configurations you suggest are too large for our area specifications. Please resubmit your proposal based on the figures provided online.
5. You have not paid your bill yet. Failure to do so will result in termination of services.	5. Prompt payment of bills ensures continued service.

TABLE 6.5 Differences Among Adjustment Letters

	100% Yes	100% No	Partial Adjustment
Introduction	State the good news.	Begin with a buffer, a comment agreeable to both reader and writer.	State the good news.
Discussion	Explain what happened and what the reader should do and/or what the company plans to do next.	Explain what happened, state the bad news, and provide possible alternatives.	Explain what happened, state the bad news, and provide possible alternatives—what the reader and/or company should do next.
Conclusion	End upbeat and positive.	Resell (provide discounts, coupons, follow-up contact names and numbers, etc.) to maintain goodwill.	Resell (provide discounts, coupons, etc.) to maintain goodwill.

- 100% yes—you could agree 100% with the writer of the complaint letter.
- 100% no—you could disagree 100% with the writer of the complaint letter.
- Partial adjustment—you could agree with some of the writer's complaints but disagree with other aspects of the complaint.

Writing a 100% yes response to a complaint is easy. You are telling your audience what they want to hear. The challenge, in contrast, is writing a 100% no response or a partial adjustment. In these letters, you must convey bad news, but you do not want to convey bad news too abruptly. Doing so might offend, anger, or cause hurt feelings. Using a **buffer statement** delays bad news in written communication and gives you an opportunity to explain your position.

Buffers to Cushion the Blow. Use the following techniques to buffer the bad news:

- **Establish rapport with the audience through positive words to create a pleasant tone.** Instead of writing, "We received your complaint," be positive and say, "We always appreciate hearing from customers."
- **Sway your reader to accept the bad news to come with persuasive facts.** "In the last quarter, our productivity has decreased by 16%, necessitating cost-cutting measures."
- **Provide information that both you and your audience can agree upon.** "With the decline of dotcom jobs, many Information Technology positions have been lost."
- **Compliment your reader or show appreciation.** "Thank you for your June 9 letter commenting on fiscal year 2008."
- **Make your buffer concise, one to two sentences.** "Thank you for writing. Customer comments give us an opportunity to improve service."
- **Be sure your buffer leads logically to the explanation that follows. Consider mentioning the topic, as in the following example about billing practices.** "Several of our clients have noted changes in our corporate billing policies. Your letter was one that addressed this issue."
- **Avoid placing blame or offending the reader.** Rather than stating, "Your bookkeeping error cost us $9,890.00," write, "Mistakes happen in business. We are refining our bookkeeping policies to ensure accuracy."

See Figures 6.14, 6.15, and 6.16 for sample adjustment letters.

Check Online Resources

www.prenhall.com/gerson
For more information about sample adjustment letters, visit our companion Web site.

FIGURE 6.14 100% Yes Adjustment Letter, Complete with Letter Essentials

1101 21st Street
Galveston, TX 77001
(712) 451–1010

Gulfstream Auto

May 31, 2008

Mr. Carlos De La Torre
1234 18th Street
Galveston, TX 77001

Dear Mr. De La Torre:

Thank you for your recent letter. Gulfstream will replace your defective shock absorber according to the warranty agreement.

The Trailhandler Performance XT shock absorber that you purchased was discontinued in April 2008. Mr. Blanton, the mechanic to whom you spoke, incorrectly assumed that Gulfstream was no longer honoring the warranty on that product. Because we no longer carry that product, we either will replace it with a comparable model or refund the purchase price. Ask for Mrs. Cottrell at the automotive desk on your next visit to our store. She is expecting you and will handle the exchange.

We appreciate your business, Mr. De La Torre. I'm glad you brought this problem to my attention. If I can help you in the future, please contact me.

Sincerely,

Holbert Lang
Sales Manager

cc: Jordan Cottrell, Supervisor
    Jim Gaspar, CEO

Positive word usage ("Thank you") achieves audience rapport.

The introduction immediately states the good news.

The discussion explains what created the problem and provides an instruction telling the customer what to do next.

The conclusion ("we appreciate your business") resells to maintain customer

Bad-News Letters

DOT-COM UPDATES

For more information about delivering bad news, check out the following link:

- http://www.articles911. com/Communication/ Delivering_Bad_News/

Unfortunately, you occasionally will be required to write bad-news letters. These letters might reject a job applicant, tell a vendor that his or her company's proposal has not been accepted, or reject a customer's request for a refund. Maybe you will have to write to a corporation to report that its manufacturing is not meeting your company's specifications. Maybe you will need to write a letter to a union documenting a grievance. You might even have to write a bad-news letter to fire an employee.

In any of these instances, tact is required. You cannot berate a customer or client. You should not reject a job applicant offensively. Even your grievance must be worded carefully to avoid future problems.

FIGURE 6.15   100% No Adjustment Letter Beginning with a Buffer Statement

Gulfstream Auto

1101 21st Street
Galveston, TX 77001
(712) 451–1010

May 31, 2008

Mr. Carlos De La Torre
1234 18th Street
Galveston, TX 77001

Dear Mr. De La Torre:

Thank you for your May 10 letter. Gulfstream Auto always appreciates hearing from its customers.

The introduction begins with a buffer. The writer establishes rapport with the audience through positive words to create a pleasant tone.

The Trailhandler Performance XT shock absorber that you purchased was discontinued in April 2008. Mr. Blanton, the mechanic to whom you spoke, correctly stated that Gulfstream was no longer honoring the warranty on that product. Because we no longer carry that product, we cannot replace it with a comparable model or refund the purchase price. Although we cannot replace the shock absorber, we want to offer you a 10% discount off of a replacement.

The discussion explains the company's position, states the bad news, and offers an alternative.

We appreciate your business, Mr. De La Torre. I'm glad you brought this problem to my attention. If I can help you in the future, please contact me.

Sincerely,

Holbert Lang
Sales Manager

cc: Jordan Cottrell, Supervisor
    Jim Gaspar, CEO

angry readers will not necessarily go out of their way to help you, your best approach is diplomacy.

To strengthen your assertions, in the introduction, include supporting details, such as the following: serial numbers, dates of purchase, invoice numbers, check numbers, names of salespeople involved in the purchase, and/or receipts. When possible, include copies documenting your claims.

Discussion. In the discussion paragraph(s), explain in detail the problems experienced. This could include dates, contact names, information about shipping, breakage information, an itemized listing of defects, and poor service.

Be specific. Generalized information will not sway your readers to accept your point of view. In a complaint letter, you suffer the burden of proof. Help your audience understand the extent of the problem. After documenting your claims, state what you want done and why.

1101 21st Street
Galveston, TX 77001
(712) 451–1010
May 31, 2008

Gulfstream Auto

Mr. Carlos De La Torre
1234 18th Street
Galveston, TX 77001

Dear Mr. De La Torre:

Begin your letter with the good news. ———→ Thank you for your recent letter. Gulfstream will replace your defective shock absorber according to the warranty agreement.

Explain what happened, state the bad news, and provide a possible alternative. ——→ The Trailhandler Performance XT shock absorber that you purchased was discontinued in April 2008. Mr. Blanton, the mechanic to whom you spoke, incorrectly assumed that Gulfstream was no longer honoring the warranty on that product. However, we no longer carry that product. We will replace the shock absorber with a comparable model, but you will have to pay for installation.

We appreciate your business, Mr. De La Torre. I'm glad you brought this problem to my attention. If I can help you in the future, please contact me.

Sincerely,

Holbert Lang
Sales Manager

cc: Jordan Cottrell, Supervisor
      Jim Gaspar, CEO

Conclusion. If you end your bad-news letter with the bad news, then you leave your reader feeling defeated and without hope. You want to maintain a good customer–client, supervisor–subordinate, or employer–employee relationship. Therefore, you need to conclude your letter by giving your readers an opportunity for future success. Provide your readers options which will allow them to get back in your good graces, seek employment in the future, or reapply for the refund you have denied. Then, to leave your readers feeling as happy as possible, given the circumstances, end upbeat and positively.

Organizational Modes
See Chapter 3 for additional information.

The example in Figure 6.17, from an insurance company canceling an insurance policy, shows how a bad-news letter should be constructed.

FIGURE 6.17 Bad-News Letter

*S*tate of *M*ind *I*nsurance
11031 Bellbrook Drive
Stamford, CT 23091
213-333-8989

September 7, 2008

John Chavez
4249 Uvalde
Stamford, CT 23091

Dear Mr. Chavez:

Thank you for letting us provide you and your family with insurance for the last 10 years. We have appreciated your business.

However, according to our records, you have filed three claims in the past three years:

1. Damage to your basement due to a failed sump pump
2. Flooring damage due to a broken water seal in your second-floor bathroom
3. Fender repair coverage for a no-fault automobile accident

SMI's policy stipulates that no more than two claims may be filed by a client within a three-year period. Therefore, we must cancel your policy as of October 15, 2008.

If you have any questions about this cancellation, please call our 24-hour assistance line (913-482-0000). Our transition team can help you find new coverage. Thank you for your patronage and your understanding.

Sincerely,

Darryl Kennedy

Darryl Kennedy

> Begin with a buffer statement in a bad-news letter.

> In the body, explain what happened and then state the bad news.

> In the conclusion, give your reader follow-up options.

The letters checklist will give you the opportunity for self-assessment and peer evaluation of your writing. Getting input from peers who also have to write letters will allow you to consider aspects of the letter that you might have ignored.

LETTERS CHECKLIST

_____ 1. **Letter Essentials:** Does your letter include the eight essential components (writer's address, date, recipient's address, salutation, text, complimentary close, writer's signed name, and writer's typed name)?

_____ 2. **Introduction:** Does the introduction state _what_ you are writing about and _why_ you are writing?

_____ 3. **Discussion:** Does your discussion clearly state the details of your topic depending on the type of letter?

_____ 4. **Highlighting/Page Layout:** Is your text accessible? To achieve reader-friendly ease of access, use headings, boldface, italics, bullets, numbers, underlining, or graphics (tables and figures). These add interest and help your readers navigate your letter.

_____ 5. **Organization:** Have you helped your readers follow your train of thought by using appropriate modes of organization? These include chronology, importance, problem/solution, or comparison/contrast.

_____ 6. **Conclusion:** Does your conclusion give directive action (tell what you want the reader to do next and when) and end positively?

_____ 7. **Clarity:** Is your letter clear, answering reporter's questions and providing specific details that inform, instruct, or persuade?

_____ 8. **Conciseness:** Have you limited the length of your words, sentences, and paragraphs?

_____ 9. **Audience Recognition:** Have you written appropriately to your audience? This includes avoiding biased language, considering the multicultural/cross-cultural nature of your readers, and your audience's role (supervisors, subordinates, coworkers, customers, or vendors). Have you created a positive tone to build rapport?

_____ 10. **Correctness:** Is your text grammatically correct? Errors will hurt your professionalism. See Appendix for grammar rules and exercises.

E-MAIL

Why Is E-mail Important?

Unique Characteristics of the "E-Reader"

See Chapter 13 for additional information.

E-mail (electronic mail) is different from hard-copy memos and letters in many ways. The primary difference is that people read online text differently than they read hard-copy text. E-mail has become a predominant means of routine correspondence. Many companies are

Check Online Resources

www.prenhall.com/gerson

For more information about sample E-mail messages, visit our companion Web site.

FAQs: Professionalism in E-Mail Messages

Q: E-mail messages are just casual communication, right? So writing an e-mail message is easy, isn't it, since you don't have to worry about grammar or correct style?

A: Nothing could be further from the truth. E-mail might be your major means of communication in the workplace. Therefore, you must pay special attention to correctness.

Listen to what managers at an engineering company say about e-mail messages:

- "Most business communication is now via e-mail. Business e-mail needs to be almost as formal and as carefully written as a letter because it is a formal and legal document. Never send an e-mail that you would not be comfortable seeing on the front page of a newspaper, because some day you may."

- "I see more and more new hires wanting to rely on e-mail. It is a totally ineffective way to resolve many issues on an engineering project. But they seem to feel it is OK for almost any communication. I suspect the general acceptance by their peers for this form of communication has led them to mistakenly assume the same is true for a business setting."

- "Many young people tend to be very 'social' in e-mails. Your employer owns your e-mails written on your work computers. They are NOT private. They can be used not only against you, but against your firm in court. For example, if I send an e-mail to a coworker that states in it somewhere what a lousy job Frank is doing on the such-and-such project and that project goes bad, it is possible that e-mail could end up in court and be used against my employer. In my mind all I was doing was venting my frustrations to an understanding friend and co-worker. But, in reality, I am creating a permanent record of anything I say."

(Gerson, et al. 2004 "Core Competencies")

"geared to operate with e-mail," creating what the Harvard Business School calls "e-mail cultures" for the following reasons ("The Transition to General Management" 2002).

Time. "Everything is driven by time. You have to use what is most efficient" (Miller, et al. 1996, 10). The primary driving force behind e-mail's prominence is time. E-mail is quick. Whereas a posted letter might take several days to deliver, e-mail messages can be delivered within seconds.

Convenience. With wireless communication, you can send e-mail from notebooks to handhelds. Current communication systems combine a voice phone, personal digital assistant, and e-mail into a package that you can slip into a pocket or purse. Then, you can access your e-mail messages anywhere, anytime.

Internal/External. E-mail allows you to communicate internally to coworkers and externally to customers and vendors. Traditional communication channels, like letters and memos, have more limited uses. Generally, letters are external correspondence written from one company to another company; memos are internal correspondence transmitted within a company.

Cost. E-mail is cost effective because it is paper free. With an ability to attach files, you can send many kinds of documentation without paying shipping fees. This is especially valuable when considering international business.

Documentation. E-mail provides an additional value when it comes to documentation. Because so many writers merely respond to earlier e-mail messages, what you end up with is a "virtual paper trail" (Miller, et al. 1996, 15). When e-mail is printed out, often the printout will contain dozens of e-mail messages, representing an entire string of dialogue. This provides a company an extensive record for future reference. In addition, most companies archive e-mail messages in backup files.

Reasons for Writing E-mail Messages

E-mail is used to convey many types of information in business and industry. You can write an e-mail message to accomplish any of the following purposes.

- **Directive**—inform a subordinate or a team of employees to complete a task.
- **Cover/transmittal**—inform a reader or readers that you have attached a document, and list the key points that are included in the attachment.
- **Documentation**—report on expenses, incidents, accidents, problems encountered, projected costs, study findings, hiring, firings, and reallocations of staff or equipment.
- **Confirmation**—inform a reader about a meeting agenda, date, time, and location; decisions to purchase or sell; topics for discussion at upcoming teleconferences; conclusions arrived at; fees, costs, or expenditures.
- **Procedures**—explain how to set up accounts, research on the company intranet, operate new machinery, use new software, apply online for job opportunities through the company intranet, create a new company Web site, or solve a problem.
- **Recommendations**—provide reasons to purchase new equipment, fire or hire personnel, contract with new providers, merge with other companies, revise current practices, or renew contracts.
- **Feasibility**—study the possibility of changes in the workplace (practices, procedures, locations, staffing, or equipment).
- **Status**—provide a daily, weekly, monthly, quarterly, biannual, or yearly report about where you, the department, or the company is regarding a topic of your choice (class project, sales, staffing, travel, practices, procedures, or finances).
- **Inquiry**—ask questions about upcoming processes, procedures, or assignments.

Check Online Resources

www.prenhall.com/gerson
For more information about E-mail, visit our companion Web site.

How is e-mail used and misused in business today?

Michael Smith, PE, President of George Butler Associates, Inc., an architectural-engineering consulting firm, supervises 270 employees, located in Kansas, Missouri, and Illinois. These include architects, engineers, scientists, and administrative support staff such as human resources, marketing, information systems, and accounting.

Mike receives about 40 "meaningful e-mail" a day—and about 60 spam e-mail, attempting to tell him how to lose weight, buy gifts for his loved ones, make cost-effective travel arrangements, and so on. Mike deletes the spam, which are obvious at a glance in his e-mail inbox, and responds to the business-related correspondence.

He says that e-mail has been essential at GBA for at least two reasons.

The Good News:

1. **E-mail is fast and efficient.** For example, Mike finds that e-mail is ideal for setting up meetings with groups of individuals, including coworkers and clients. Before e-mail, he had to make multiple calls to his staff or customers, usually missing them while they were off-site. Or, he'd have to walk down the halls and up the stairs to leave messages on desks. Instead, with e-mail, Mike knows that the message will arrive at its intended location and that the recipients will respond.

2. **E-mail provides an "electronic record."** When GBA used to rely more heavily on the telephone, people had trouble remembering to document conversations. With e-mail, you have a communication trail: multiple e-mail messages reporting all communication that has occurred regarding a topic.

Unfortunately, Mike tells us that e-mail also has a few downsides.

The Bad News:

1. **E-mail lacks privacy.** Mike says that though phone calls made record keeping hard, they ensured privacy. All you needed to do was shut your office door or speak quietly. E-mail, in contrast, inadvertently can be sent to many people.

2. **E-mail can lead to inaccurate communication.** "Didn't I read somewhere that 65% of communication is body language?" Mike asks. This leads to a second problem. With e-mail, the people involved can't see each other's eyes, hand motions, shrugs, smiles, or frowns, nor can they hear the grunts, groans, or laughs. Much of this body language is lost with electronic communication.

3. **E-mail can be depersonalized.** Body language not only helps communication, but also it personalizes. E-mail, Mike notes, can be impersonal. In some instances, especially those regarding "contentious situations," Mike knows that a face-to-face discussion is the best way to solve a problem. In fact, Mike concludes that some people who are "conflict resistant" use e-mail as a way of avoiding person-to-person communication. To maintain good business relationships (with personnel as well as with customers), face-to-face talks often are needed.

Still, for speed and efficiency, e-mail is hard to beat.

Techniques for Writing Effective E-mail Messages

To convey your messages effectively and to ensure that your e-mail messages reflect professionalism, follow these tips for writing e-mail.

Recognize Your Audience.   E-mail messages can be sent to managers, coworkers, subordinates, vendors, and customers, among other audiences. Your e-mail readers will be high tech, low tech, and lay people. Thus, you must factor in levels of knowledge.

If an e-mail message is sent internationally, you also might have to consider your readers' language. Remember that abbreviations and acronyms are not universal. Dates, times, measurements, and monetary figures differ from country to country. Audience concerns are discussed in detail in Chapter 4. In addition, your reader's e-mail system might not have the same features or capabilities that you have. Hard-copy text will look the same to all readers. E-mail platforms, such as in AOL, Outlook, Juno, HotMail, and Yahoo, display text differently. To communicate effectively, recognize your audience's level of knowledge, unique language, and technology needs.

Identify Yourself.   Identify yourself by name, affiliation, or title. You can accomplish this either in the "From" line of your e-mail or by creating a signature file or .sig file. This .sig file acts like an online business card. Once this identification is complete, readers will be able to open your e-mail without fear of corrupting their computer systems.

Provide an Effective Subject Line.   Readers are unwilling to open unsolicited or unknown e-mail, due to fear of spam and viruses. In addition, corporate employees receive approximately 50 e-mail messages each day. They might not want to read every message sent to them. To ensure that your e-mail messages are read, avoid uninformative subject lines, such as "Hi," "What's New," or "Important Message." Instead, include an effective subject line, such as "Subject: Meeting Dates for Tech Prep Conference."

Audience

See Chapter 4 for additonal information.

Keep Your E-mail Message Brief.   E-readers skim and scan. To help them access information quickly, "Apply the 'top of the screen' test. Assume that your readers will look at the first screen of your message only" (Munter et al. 2003, 31). Limit your message to one screen (if possible).

Online Communication

See Chapter 13 for additional information on the characteristics of online communication.

Organize Your E-mail Message.   Successful writing usually contains an introductory paragraph, a discussion paragraph or paragraphs, and a conclusion. Although many e-mail messages are brief, only a few sentences, you can use the introductory sentences to tell the reader why you are writing and what you are writing about. In the discussion, clarify your points thoroughly. Use the concluding sentences to tell the reader what is next, possibly explaining when a follow-up is required and why that date is important.

Use Highlighting Techniques Sparingly.   Many e-mail packages will let you use highlighting techniques, such as boldface, italics, underlining, computer-generated bullets and numbers, centering, font color highlighting, and font color changes. Many other e-mail platforms will not display such visual enhancements. To avoid having parts of the message distorted, limit your highlighting to asterisks (*), numbers, double spacing, and all-cap headings.

Highlighting Techniques

See Chapter 8 for additional information on highlighting techniques.

Proofread Your E-mail Message.   Errors will undermine your professionalism and your company's credibility. Recheck your facts, dates, addresses, and numerical information before you send the message. Try these tips to help you proofread an e-mail message.

- Type your text first in a word-processing package, such as Microsoft Word.
- Print it out. Sometimes it is easier to read hard-copy text than text online. Also, your word-processing package, with its spell check and/or grammar check, will help you proofread your writing.

Once you have completed these two steps (writing in Word or WordPerfect and printing out the hard-copy text), copy and paste the text from your word-processing file into your e-mail.

Make Hard Copies for Future Reference.   Making hard copies of all e-mail messages is not necessary because most companies archive e-mail. However, in some instances, you might want to keep a hard copy for future reference. These instances could include transmissions of good news. For example, you have received compliments about your work and want to save this record for your annual job review. You also might save a hard copy of an e-mail message regarding flight, hotel, car rental, or conference arrangements for business-related travel.

Be Careful When Sending Attachments. When you send attachments, tell your reader within the body of the e-mail message that you have attached a file; specify the file name of your attachment and the software application that you have used (HTML, PowerPoint, PDF, RTF [rich text format], Word, or Works); and use compression (Zip) files to limit your attachment size. Zip files are necessary only if an attachment is quite large.

Practice Netiquette. When you write your e-mail messages, observe the rules of "netiquette."

- **Be courteous.** Do not let the instantaneous quality of e-mail negate your need to be calm, cool, deliberate, and professional.
- **Be professional.** Occasionally, e-mail writers compose excessively casual e-mail messages. They will lowercase a pronoun like "i," use ellipses (. . .) or dashes instead of more traditional punctuation, use instant messaging shorthand language such as "LOL" or "BRB," and depend on emoticons ☺ ☹. These e-mail techniques might not be appropriate in all instances. Don't forget that your e-mail messages represent your company's professionalism. Write according to the audience and communication goal.
- **Avoid abusive, angry e-mail messages.** Because of its quick turnaround abilities, e-mail can lead to negative correspondence called flaming. Flaming is sending angry e-mail, often TYPED IN ALL CAPS.

Many e-mail messages only require a sentence or two. If you need to convey more information than can be accomplished in only a few sentences, use the all-purpose template for memos, letters, and e-mail provided earlier in this chapter in Figure 6.2 (page 159).

Figure 6.18 is a flawed e-mail message. Figure 6.19 is an example of a well-written e-mail message.

In Figure 6.20 an Information Technology Supervisor writes an e-mail to be distributed companywide to many different employees with varied levels of technical expertise. Note how the writer uses a simple style of writing, carefully omitting any technical terms. Only the information necessary to assure audience comfort with the new identification system is included.

In contrast to Figure 6.20 designed for an audience with multiple levels of expertise, Figure 16.21 is written for a high-tech audience. This high-tech audience is familiar with the technical terms included in the e-mail so the writer does not have to define such terminology. In addition, the writer includes sophisticated details and content which the technically proficient audience can understand easily.

The e-mail checklist will give you the opportunity for self-assessment and peer evaluation of your writing.

FIGURE 6.18 Flawed E-mail Message

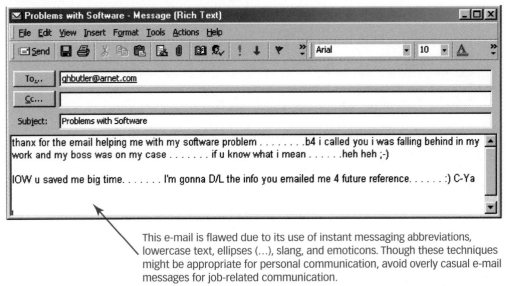

This e-mail is flawed due to its use of instant messaging abbreviations, lowercase text, ellipses (…), slang, and emoticons. Though these techniques might be appropriate for personal communication, avoid overly casual e-mail messages for job-related communication.

FIGURE 6.19 Successful E-mail Message about the Status of a Company's Web Site

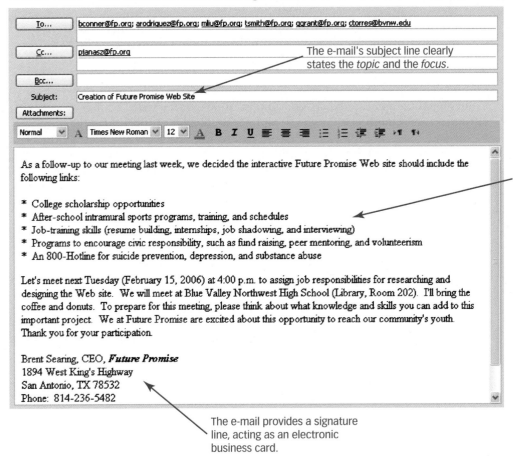

The e-mail's subject line clearly states the *topic* and the *focus*.

This e-mail message correctly uses asterisks for bullets so that the message will be compatible on all electronic platforms.

The e-mail provides a signature line, acting as an electronic business card.

FIGURE 6.20 E-mail for Multiple Audience Levels

To communicate effectively to multiple audiences, this writer defines high-tech terms like "IT" and "PIN." The e-mail uses a conversational tone and simple instructions to make this challenging topic more friendly and informative. The emphasis is on helpfulness vs. depth of detail.

Date: May 31, 2008
To: ncomlistserv
Cc: lhughes@ncomlistserv
From: vpatel@ncom.net, Information Technology Supervisor
Subject: Rollout of New Computer Security System

Our company is rolling out a new computer security system next week. This is an important step for the company for at least three reasons: to deter hackers from breaking into our company's computers, to minimize spam and viruses, and to ensure that our company's intellectual assets are secure from terrorists. This week, our information technology (IT) specialists will install the appropriate hardware and software needed to achieve this important security. Your workflow, however, will not be affected. Following are the key steps to this procedure:

1. To authenticate each computer user, we'll ask you to type in your 5 PIN (Personal Identification Number). If you have forgotten this number, just give us a call at 1-help-055.
2. Once the computer system verifies your identity, our new firewall system kicks in. This software will give you access to company data while denying the same access to others outside the company.
3. Finally comes protection from spam and viruses—"about time," you're probably saying. You'll never have to worry about this step. It's all taken care of for you by our network administrators.

Though our new system security should be seamless for end users, we're offering quick and easy hands-on demonstrations for interested employees. To make this convenient for you, look for our computer specialists in the company cafeteria from 11:00 a.m.–1:00 p.m. The staff wearing yellow shirts and sitting at the "FAQ" tables will answer your questions and walk you through the new procedures. As always, if you have any questions, just e-mail me (vapatel@ncom.net). We're here to help.

Veejay Patel
Supervisor, Information Technology
nCom
2134 Industrial Way
Sacramento, CA 22109
vpatel@ncom.net

FIGURE 6.21 E-mail for High-Tech Audience

Date: May 31, 2008
To: lhughes@ncom.net, Director, Network Systems
From: vpatel@ncom.net, Supervisor, Information Technology
Subject: Rollout of New Computer Security System

Lorette, as a follow-up to our meeting last week, below is an overview of the new computer security system we plan to implement:

1. nCom's network security begins with user authentifications. Starting next week, we will ask each company employee to use a PIN. These 5-digit numbers, ranging from 00000-99999, will give our employees 100,000 possible numbers to choose from, thus essentially negating chances of duplication. In addition, the nCom PIN verification system will allow employees only three attempts at logging in. This safeguards our system from intrusion, since hackers have only a 1/30,000 chance at guessing the correct PIN before access is blocked.

2. Authentification through PINs protects access, but it does not deter malware. That's where our firewall system kicks in. Our new Aponex firewall uses a two-pronged, doubleback security system to deflect adware, spyware, viruses, worms, and trojan horses. First, packet filtering will allow the passage of pre-defined data to pass through the system while discarding suspicious data. Next, as a backup filter, our software also uses stateful inspection. This allows us to match data against predefined characteristics, set by our network administrators. Together, these IPS systems detect anomalies to deny access and protect corporate privacy.

3. Finally, we have installed surveillance honeypots as network-accessible decoys. The honeypots, essentially, create a shadow network which we can monitor against unauthorized access attempts.

My IT staff is working this weekend to install all required hardware and software. In doing so, we will have the system up and running by Monday of next week, without interfering with regular workplace operations. IT staff will also be accessible to employees next week to help in the transition. I have arranged for FAQ tabling during lunch hours. I want to take this opportunity to thank you, Lorette, for your support and my staff for their expertise during this network security implementation.

Veejay Patel
Supervisor, Information Technology
nCom
2134 Industrial Way
Sacramento, CA 22109
vpatel@ncom.net

For a more high-tech audience, this e-mail does not define terms, such as "IPS" and "PIN." The e-mail contains detailed content and technical information about system security that the audience both needs and will understand. The tone of this e-mail is more businesslike, since the writer is a subordinate writing to management.

E-MAIL CHECKLIST

_____ 1. **Does the e-mail use the correct address?**

_____ 2. **Have you identified yourself?** Provide a "sig" (signature) line.

_____ 3. **Did you provide an effective subject line?** Include a _topic_ and a _focus_.

_____ 4. **Have you effectively organized your e-mail?** Consider including the following:

- Opening sentence(s) telling _why_ you are writing and _what_ you are writing about
- Discussion unit with itemized points telling _what exactly_ the e-mail is discussing
- Concluding sentence(s), _summing up_ your e-mail message or telling your audience what to do next

_____ 5. **Have you used highlighting techniques sparingly?**

- Avoid boldface, italics, color, or underlining.
- Use asterisks (*) for bullets, numbers, and double spacing for access.

_____ 6. **Did you practice netiquette?**

- Be polite, courteous, and professional.
- Don't flame.

_____ 7. **Is the e-mail concise?**

_____ 8. **Did you identify and limit the size of attachments?**

- Tell your reader(s) if you have attached files and what types of files are attached (PPt, PDF, RTF, Word, etc.).
- Limit the files to 750 K.

_____ 9. **Does the memo recognize audience?**

- Define acronyms or abbreviations where necessary.
- Consider a diverse audience (factoring in issues, such as multiculturalism or gender).

_____ 10. **Did you avoid grammatical errors?**

INSTANT MESSAGING

E-mail could be too slow for today's fast-paced workplace. Instant messaging (IM) could replace e-mail in the workplace within the next five years. Studies suggest IM pop-ups are already providing businesses many benefits.

Benefits of Instant Messaging

Following are benefits of instant messaging.

- Increased speed of communication.
- Improved efficiency for geographically dispersed workgroups.
- Collaboration by multiple users in different locations.
- Communication with colleagues and customers at a distance in real time, such as the telephone.
- Avoidance of costly long distance telephone rates (Note: Voice-over IP (VoIP) services, which allow companies to use the Internet for telephone calls, could be more cost efficient than IM.)
- More "personal" link than e-mail.
- Communication channel that is less intrusive than telephone calls.
- Communication channel that allows for multitasking (With IM, you can speak to a customer on the telephone or via an e-mail message and _simultaneously_ receive product updates from a colleague via IM.)
- Quick way to find out who is in the office, out of the office, available for conversation, or unavailable due to other activities. (Hoffman 2004; Shinder 2005)

See Figure 6.22 for an example of an instant message.

FIGURE 6.22 Instant Message

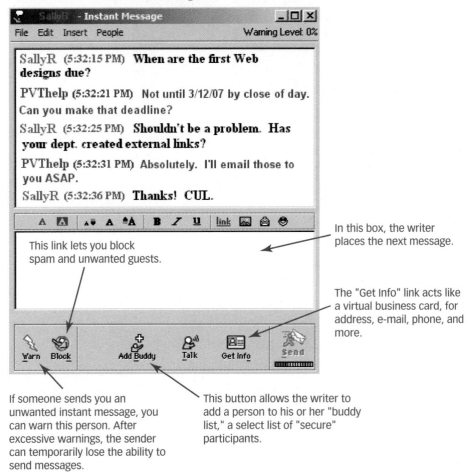

In this box, the writer places the next message.

This link lets you block spam and unwanted guests.

The "Get Info" link acts like a virtual business card, for address, e-mail, phone, and more.

If someone sends you an unwanted instant message, you can warn this person. After excessive warnings, the sender can temporarily lose the ability to send messages.

This button allows the writer to add a person to his or her "buddy list," a select list of "secure" participants.

Challenges of Instant Messaging

For business purposes, IM is so new that corporate standards have not been formalized. Software companies have not yet redesigned IM home versions for the workplace. This leads to numerous potential problems, including security, archiving, monitoring, and the following (Hoffman 2004; Shinder 2005):

- **Security issues.** This is the biggest concern. IM users are vulnerable to hackers, electronic identity theft, and uncontrolled transfer of documents. With unsecured IM, a company could lose confidential documents, internal users could download copyrighted software, or external users could send virus-infected files.
- **Lost productivity.** Use of IM on the job can lead to job downtime. First, we tend to type more slowly than we talk. Next, the conversational nature of IM leads to "chattiness." If employees are not careful, or monitored, a brief IM conversation can lead to hours of lost productivity.
- **Employee abuse.** IM can lead to personal messages rather than job-related communication with coworkers or customers.
- **Distraction.** With IM, a bored colleague easily can distract you with personal messages, online chats, and unimportant updates.

- **Netiquette.** As with e-mail, due to the casual nature of IM, people tend to relax their professionalism and forget about the rules of polite communication. IM can lead to rudeness or just pointless conversations.
- **Spim.** IM lends itself to "spim," instant messaging spam—unwanted advertisements, pornography, pop-ups, and viruses.

Techniques for Successful Instant Messaging

To solve potential problems, consider these 10 suggestions.

1. **Choose the correct communication channel.** Use IM for speed and convenience. If you need length and detail, other options—e-mail messages, memos, reports, letters—are better choices. In addition, sensitive topics or bad news should never be handled through IM. These deserve the personal attention provided by telephone calls or face-to-face meetings.

2. **Document important information.** For future reference, you must archive key text. IM does not allow for this. Therefore, you will need to copy and paste IM text into a word-processing tool for long-term documentation.

3. **Summarize decisions.** IM is great for collaboration; however, all team members might not be online when decisions are made. Once conclusions have been reached that affect the entire team, the designated team leader should e-mail everyone involved. In this e-mail, the team leader can summarize key points, editorial decisions, timetables, and responsibilities.

4. **Tune in, or turn off.** The moment you log on, IM software tells everyone who is active online. Immediately, your IM buddies can start sending messages. IM pop-ups can be distracting. Sometimes, in order to get your work done, you might need to turn off your IM system. Your IM product might give you status options, such as "on the phone," "away from my desk," or "busy." Turning on IM could infringe upon your privacy and time. Turning off might be the answer.

5. **Limit personal use.** Your company owns the instant messaging in the workplace. IM should be used for business purposes only.

6. **Create "buddy" lists.** Create limited lists of IM users, including legitimate business contacts (colleagues, customers, and vendors).

7. **Avoid public directories.** This will help ensure that your IM contacts are secure and business related.

8. **Disallow corporate IM users from installing their own IM software.** A company should require standardized IM software for safety and control.

Teamwork

See Chapter 1 for additional information on collaboration.

9. **Never use IM for confidential communication.** Use another communication channel if your content requires security.

10. **Use IM software that allows you to archive and record IM communications.** As with e-mail, IM programs can let systems administrators log and review IM conversations. Some programs create reports that summarize archived information and let users search for text by keywords or phrases. These systems are perfect for future reference (Hoffman 2004; Shinder 2005).

THE WRITING PROCESS AT WORK

the writing process

Prewriting	Writing	Rewriting

• Decide whether you are writing to inform, instruct, persuade, or build rapport.
• Determine whether your audience is internal, external, high-tech, low-tech, lay, or multiple.
• Choose the correct formal or informal tone for your audience and purpose.
• Choose the correct communication channel (e-mail, memo format, letter, or IM) for your audience and purpose.
	• Organize your content using modes such as problem/solution, cause/effect, comparison, argument/persuasion, analysis, chronology.
	• Use headings, bullets, and numbered lists for access.
		• Revise your draft by
		  • Adding details
		  • Deleting wordiness
		  • Simplifying words
		  • Enhancing the tone
		  • Reformatting your text
		  • Proofreading and correcting errors
		• Let a peer or colleague read your draft and provide input.

Effective writing follows a process of prewriting, writing, and rewriting. Each of these steps is sequential and yet continuous. The writing process is dynamic, with the three steps frequently overlapping. To clarify the importance of the writing process, look at how Jim Goodwin, the CEO of **CompuMed**, used prewriting, writing, and rewriting to write a memo to his employees.

Writing Process

See Chapter 2 for additional information.

Prewriting

No single method of prewriting is more effective than another. Throughout this textbook, you will learn many different types of prewriting techniques, geared uniquely for different types of communication. The goal of all prewriting is to help you overcome the blank page syndrome (writer's block). Prewriting will allow you to spend time before writing your memo or letter, gathering as much information as you can about your subject matter. In addition, prewriting lets you determine your objectives. Jim used mind mapping/clustering to gather data and determine objectives (Figure 6.23).

FIGURE 6.23   Mind Mapping/Clustering to Gather Data

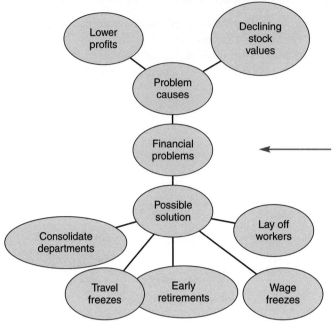

Clustering helps you see the different parts of any subject. Clustering helps you develop ideas and see the relationship between these concepts. Because clustering is less restrictive and less structured than outlining, this prewriting technique might allow you to gather data more creatively.

Different Types of Prewriting Techniques

See Chapter 2 for additional information.

Writing

Once you have gathered your data and determined your objectives in prewriting, your next step is to draft your memo or letter. In doing so, consider the following techniques.

Clarify Your Audience.   Before writing the draft, consider your audience. Are you writing laterally or vertically? Is your audience composed of management, subordinates, or colleagues? Is the audience multiple?

Organize Your Ideas.   If your supporting details are presented randomly, your audience will be confused. As a writer, develop your content logically. When you draft your memo or letter, choose a method of organization that will help readers understand your objectives. This could include comparison–contrast, problem–solution, chronology, cause–effect, and more.

Jim drafted a memo, focusing on the information he discovered in prewriting (Figure 6.24).

Rewriting

To edit and revise your memo or letter and make it as professional as it can be, follow these techniques.

Clarity and Conciseness

See Chapter 3 for additional information.

Add New Detail for Clarity.   Reread your draft. If you have omitted any information stemming from the reporter's questions (who, what, when, why, where, how), insert answers to these questions.

Delete Dead Words and Phrases for Conciseness.   Chapter 3 will help you understand how to achieve conciseness. Review techniques for revising unnecessarily long words, sentences, and paragraphs.

Move Information for Emphasis.   Place your most important information first, or move information around to maintain an effective chronological order.

Reformat for Access.   Use highlighting techniques (white space, bullets, boldface, italics, numbers, or headings) for reader-friendly appeal.

FIGURE 6.24   Rough Draft Memo

Date: October 14, 2008
To: CompuMed Employees
From: Jim Goodwin
Subject: Problems

As you know, we are experiencing some problems at CompuMed. These include lower profits and stock value declines. We have alot of unhappy stockholders. Its up to me to help everyone figure out how to solve our problems.

I have some ideas I want to share with you. I'm happy to have you share your ideas with me too. Here are my ideas: we need to consider consolidating departments and laying off some employees. We also might need to freeze wages and certainly its time to freeze travel.

The best idea I have is for some of you to take early retirement. If all of you who have over twenty years vested in the company would retire, that would save us around 2.1 million dollars over the next fiscal year. And, you know, saving money is good for all of us in the long run.

FIGURE 6.25   Revision Suggestions

Date: October 14, 2008
To: CompuMed Employees
From: Jim Goodwin
Subject: Problems

<u>As you know</u>, we are experiencing <u>some</u> problems at CompuMed. These include lower profits and stock value declines. We have <u>alot</u> of unhappy stockholders. <u>Its</u> up to me to help everyone figure out how to solve our problems.

I have <u>some</u> ideas I want to share with you. I'm happy to have you share your ideas with me too. Here are my ideas: we need to consider consolidating departments and laying off some employees. We also might need to freeze wages and certainly its time to freeze travel.

The best idea I have is for <u>some</u> of you to take early retirement. If all of you who have over twenty years vested in the company would retire, that would save us around 2.1 million dollars over the next fiscal year. And, you know, saving money is good for all of us in the long run.

Add a focus to the subject line, such as "Problems with . . . ."

Consider removing words such as "as you know," "some," and "a lot." Replace them with stronger words. Also, "a lot" and "it's" are spelled wrong.

List these problems and solutions to make them more accessible. Also, could you add more details?

Jim, I think you need to alter the tone of this memo. Is there some way to avoid talking about saving money by firing people?

Correct the Tone of Your Correspondence. Based on audience, achieve an appropriate tone.

Edit for Accuracy. Check your memo or letter for punctuation, grammar, and spelling errors. You must proofread to be professional.

After drafting this memo, Jim asked his administrative assistant to help revise the text (Figure 6.25).

Grammar and Mechanics
See Appendix for additional information.

FIGURE 6.26 Bad-News, Problem-Solution Memo Incorporating Revision Suggestions

Date: October 14, 2008
To: CompuMed Employees
From: Jim Goodwin
Subject: Suggestions for Improving Corporate Finances

CompuMed is experiencing lower profits and declining stock value. Consequently, stockholders are displeased with company performance. I have been meeting with the Board of Directors and division managers to determine the best course of action. Here are ideas to improve our company's financial situation.

1. Consolidating departments—by merging our marketing and advertising departments, for example, we can reduce redundancies. This could save CompuMed approximately $275,000 over a six-month period.
2. Reducing staff—we need to cut back employees by 15 percent. This does not necessarily mean that layoffs are inevitable. One way, for instance, to reduce staff is through voluntary retirements. We will be encouraging employees with over 20 years vested in the company to take our generous early-retirement package.
3. Freezing wages—for the next fiscal quarter, no raise increases will go into effect. Internal auditors will review the possibility of reestablishing raises after the first quarter.
4. Freezing travel—conference attendance will be stopped for six months.

I encourage you to visit with me and your division managers with questions or suggestions. CompuMed is a strong company and will bounce back with your help. Thank you for your patience and understanding.

Jim factored in his administrative assistant's suggestions and rewrote the memo. See Figure 6.26 for the finished product.

CHAPTER HIGHLIGHTS

1. Memos, letters, e-mail, and instant messages are an important part of your interpersonal communication on the job.
2. Memos, letters, and e-mail differ in destination, purpose, format, audience, tone, delivery time, and security.
3. Use an effective subject line including a topic and a focus.
4. Follow the all-purpose template for memos, letters, and e-mail including an introduction, a body or discussion section, and a conclusion.
5. The introduction states what you want and why you are writing.
6. In the discussion section, you state the details.
7. Conclude by telling readers what you plan to do next or what you expect them to do next. You can also date this action, and thank readers for their time.
8. Consider the audience when you are writing a memo, letter, or e-mail.
9. Wizards allow you to format your memos and letters but can be somewhat restrictive.
10. Follow techniques from this chapter to create effective instant messages.

CASE STUDIES

After reading the following case studies, write the appropriate correspondence required for each assignment.

1. As director of human resources at CompuMed biotechnology company, Andrew McWard helps employees create and implement their individual development plans (IDPs). Employees attend 360-Degree Assessment workshops where they learn how to get feedback on their job performance from their supervisors, coworkers, and subordinates. They also provide self-evaluations.

   Once the 360-Degree Assessments are complete, employees submit them to Andrew, who, with the help of his staff, develops IDPs. Andrew sends the IDPs to the employees, prefaced by a cover letter. In this cover letter, he tells them why he is writing and what he is writing about. In the letter's body, he focuses their attention on the attachment's contents: supervisor's development profile, schedule of activities which helps employees implement their plans, courses designed to increase their productivity, costs of each program, and guides to long-term professional development.

   In the cover letter's conclusion, Andrew ends upbeat by emphasizing how the employees' IDP help them resolve conflicts and make better decisions.

   Based on the information provided, write this cover letter for Andrew McWard. He is sending the letter to Sharon Baker, Account Executive, 1092 Turtle Hill Road, Evening Star, GA 20091.

2. ITCom is committed to increasing the diversity of its workforce and its clientele. ITCom realizes that a diverse population of employees and customers (in terms of gender, age, race, and religion) makes good business sense.

   To ensure that ITCom achieves diversity awareness, the company plans to do the following:

   - Develop a diversity committee
   - Focus on ways the company can be a responsible member of the community's diverse constituency
   - Hire a diverse workforce
   - Train employees to respect diversity
   - Write corporate communication (Web sites, e-mail, letters, corporate reports) that accommodate the unique needs of a diverse audience

   As CEO of ITCom, you want buy-in throughout the company. You want the company to realize that diversity is good for society and good for business.

   Write an e-mail message to your employees. In this e-mail, explain the company's diversity goals, highlight what the company plans to do to accomplish these goals, and ask the audience to participate in upcoming diversity workshops. Follow the criteria for effective e-mail provided in the textbook.

3. Mark Shabbot works for Apex, Inc., at 1919 W. 23rd Street, Denver, CO 80204. Apex, a retailer of computer hardware, wants to purchase 125 new flat-screen monitors from a vendor, Omnico, located at 30467 Sheraton, Phoenix, AZ 85023. The monitors will be sold to Northwest Hills Educational Cooperative. However, before Apex purchases these monitors, Mark needs information regarding bulk rates, shipping schedules, maintenance agreements, equipment specifications, and technician certifications. Northwest Hills needs this equipment before the new term (August 15). Write a letter of inquiry for Mr. Shabbot based on the preceding information.

4. Gregory Peña (121 Mockingbird Lane, San Marcos, TX 77037) has written a letter of complaint to Donya Kahlili, the manager of TechnoRad (4236 Silicon Dr., San Marcos,

TX 77044). Mr. Peña purchased a computer from a TechnoRad outlet in San Marcos. The *San Marcos Tattler* advertised that the computer "came loaded with all the software you'll need to write effective letters and perform basic accounting functions." (Mr. Peña has a copy of this advertisement.) When Mr. Peña booted up his computer, he expected to access word-processing software, multiple fonts, a graphics package, a grammar check, and a spreadsheet. All he got was a word-processing package and a spreadsheet. Mr. Peña wants Ms. Kahlili to upgrade his software to include fonts, graphics, and a grammar check; wants a computer technician from TechnoRad to load the software on his computer; and wants TechnoRad to reimburse him $400 (the full price of the software) for his trouble.

Ms. Kahlili agrees that the advertisement is misleading and will provide Mr. Peña software including the fonts, graphics, and grammar check (complete with instructions for loading the software). Write Ms. Kahlili's 100% yes adjustment to Mr. Peña based on the information provided.

5. Bob Ward, an account manager at HomeCare Health Equipment, has not received the raise that he thinks he deserves. When Bob met with his boss, Helene Koren, last Thursday for his annual evaluation, she told him that he had missed too many days of work (eight days during the year); was unwilling to work beyond his 40-hour workweek to complete rush jobs; and had not attended two mandatory training sessions on the company's new computerized inventory system.

Bob agrees that he missed the training sessions, but he was out of town on a job-related assignment for one of those sessions. He missed eight days of work, but he was allowed five days of sick leave as part of his contract. The other three days missed were due to his having to stay home to take care of his children when they were sick. He believed that these absences were covered by the company's parental leave policy. Finally, he does not agree that employees should be required to work beyond their contractual 40 hours.

Write a memo to Helene Koren, stating Bob's case.

INDIVIDUAL AND TEAM PROJECTS

1. Write a letter of inquiry. You might want to write to a college or university requesting information about a degree program, or to a manufacturer for information about a product or service. Whatever the subject matter, be specific in your request.

2. Write a cover letter. Perhaps your cover letter will preface a report you are working on in school, a report you are writing at work, or documentation you will need to send to a client.

3. Write a letter of complaint. You might want to write to a retail store, manufacturing company, restaurant, or governmental organization. Whatever the subject matter, be specific in your complaint.

4. Write an adjustment letter. Envision that a client has written a complaint letter about a problem he or she has encountered with your product or service. Write a 100 percent yes letter in response to the complaint.

5. Write a memo requesting office equipment. Your company plans to purchase new office equipment. Your memo will explain your office's needs. Specifically state what equipment and furniture you want and why these purchases are important.

PROBLEM-SOLVING THINK PIECES

1. Northwest Regional Governmental Training Consortium (GTC) provides educational workshops for elected and appointed officials, as well as employees of city and state governmental offices.

One seminar participant, Mary Bloom, supervisor of the North Platte County Planning and Zoning Department, attended a GTC seminar entitled, "Developing Leadership Skills," on February 12, 2008. Unfortunately, she was disappointed in the workshop and the facilitator. On February 16, Mary wrote a complaint letter to GTC's director, Leonard Liss, stating her dissatisfaction. She said that the training facilitator's presentation skills were poor. According to Mary, Doug Aaron, the trainer, exhibited the following problems.

- Late arrival at the workshop
- Too few handouts for the participants
- Incorrect cables for his computer, so he could not use his planned PowerPoint presentation
- Old, smudged transparencies as a backup to the PowerPoint slides

Mary also noted that the seminar did not meet the majority of the seminar participants' expectations. She and the other government employees had expected a hands-on workshop with breakout sessions. Instead, Mr. Aaron lectured the entire time. In addition, his information seemed dated and ignored the cross-cultural challenges facing today's supervisors.

Neither Leonard nor his employees had ever attended this workshop. They offered the seminar based on the seemingly reliable recommendation of another state agency, the State Data Collection Department. From Doug Aaron's course objectives and resume, he appeared to be qualified and current in his field.

However, Mary Bloom deserves consideration. Not only are her complaints justified by others' comments, but also she is a valued constituent. The GTC wants to ensure her continued involvement in their training program.

Assignments

- Leonard needs to write a 100% yes adjustment letter. In this letter, Leonard wants to accommodate the dissatisfied client. How should he recognize Mary's concern, explain what might have gone wrong, and offer satisfaction?
- Leonard needs to write an internal memo to his staff. In this memo, he will provide standards for hiring future trainers. What should the standards include?

2. You are the manager of a human resources department. You are planning a quarterly meeting with your staff (training facilitators, benefits employees, personnel directors, and company counselors). The staff works in three different cities and twelve different offices. You need to accomplish four goals: get their input regarding agenda items, find out what progress they have made on various projects, invite them to the meeting, and provide the final agenda.

How should you communicate with them? Should you write an e-mail, instant message, letter, or memo? Write an e-mail message to your instructor explaining your answer. Be sure to give reasons for and against each option.

WEB WORKSHOP

1. Many state governments provide guidelines for writing letters either to government officials or for government employees. In addition, you can go online, type an entry such as "how to write state government letters" in a search engine, and find sites with instructions and sample letters for writing to state officials or state agencies.

Research the Internet and find letter samples and guidelines. Once you have done so, continue with the following:

- Review sample letters to or from governmental agencies.
- Determine whether these letters are successful, based on the criteria provided in this chapter.
- Write either a letter or memo to your instructor explaining how and why the letters succeeded.
- Rewrite letters that can be improved, using the guidelines provided in this chapter.

2. Most city, county, and state governments have written guidelines for their employees' e-mail. For example, see "Electronic and Voice Mail: Connecticut's Management and Retention Guide for State and Municipal Government Agencies" (http://www.cslib.org/email.htm). In addition, see the e-mail policy provided by Hennepin County, Minnesota (http://www.hennepin.lib. mn.us/extranet/admin_policy/E-Mail.htm). Similarly, most companies and college/universities

have written guidelines for their employees' and students' e-mail use. For example, see what Florida Gulf Coast University says about e-mail use (http://admin.fgcu.edu/compservices/policies3.htm).

Research your city, county, and/or state e-mail policies. Research your city or college or university's e-mail policies. Determine what they say, what they have in common, and how they differ.

Write an e-mail to your instructor reporting your findings.

QUIZ QUESTIONS

1. What are five distinctions among memos, letters, and e-mail messages?
2. Why is a subject line in a letter, memo, or e-mail important, and what are the components of an effective subject line?
3. What do you hope to accomplish in the introduction to a memo, letter, or e-mail?
4. Why should you be concerned with document design in a memo, letter, or e-mail?
5. Why should you be careful when using highlighting techniques in e-mail?
6. What should you include in a conclusion to a memo, letter, or e-mail?
7. How do the audiences differ among memos, letters, and e-mail?
8. How can clustering or mind mapping be valuable to you as a form of prewriting a memo or e-mail?
9. What are three benefits of using instant messaging in the workplace?
10. What are three challenges of using instant messaging in the workplace?

CHAPTER 7

Employment Communication

In this scenario, a business owner interviews potential job applicants.

The job search involves at least two people—the applicant and the individual making the hiring decision. Usually more than two people are involved, however, because companies typically hire based on a committee's decision. That is the case at **DiskServe**. This St. Louis–based company is hoping to hire a customer service representative for its computer technology department. DiskServe is eager to hire a new employee because one of its best workers has just advanced to a new position in the company. DiskServe asked applicants to apply using e-mail (the quickest means of communication). Thus, the applicants submitted an e-mail letter of application and an attached scannable resume.

DiskServe advertised this opening in the career placement centers at local colleges, in the city newspaper, and online at its Web site: http://www.DiskServe.com.

In addition to DiskServe's chief executive officer (CEO), Sarah Beske, the hiring committee will consist of two managers from other DiskServe departments, the former employee whose job is being filled, and two coworkers in the computer technology department.

Ten candidates were considered for the position. All candidates first had teleconference interviews. While Sarah talked with the candidates, the other hiring committee members listened on a speaker phone. After the telephone interviews, four candidates were invited to DiskServe's worksite for personal interviews—Macy Heart, Aaron Brown, Rosemary Lopez, and Robin Scott.

Sarah Beske, who has worked hard to create a family-oriented environment at DiskServe, values three traits in her employees: technology know-how, teamwork, and a positive attitude toward customers and

Objectives

When you complete this chapter, you will be able to

1. Search for job openings applicable to your interests, education, and experience.

2. Compose effective letters of application that gain attention and are persuasive.

3. Choose either to write a functional or reverse chronological resume.

4. Write effective resumes consisting of your objectives, summary of qualifications, work experience, education, and professional skills.

5. Decide on the correct method of delivery of your resume, either through the mail, as an e-mail attachment, or scannable.

6. Understand effective interview techniques that demonstrate your professionalism.

7. Write appropriate follow-up correspondence to restate how you can benefit the company.

8. Use the job search checklists to evaluate your resume, letter of application, interview, and follow-up letter.

coworkers. When the candidates arrived at DiskServe, Sarah gave them a tour of the facilities, introducing them to many employees. Then the interviews began.

Each job candidate was asked a series of questions that included the following:

- What is your greatest strength? Give an example of how this reveals itself on the job.
- What did you like most about your past job?
- How have you handled customer complaints in the past?
- Where do you see yourself in five years?

Then, each candidate was taken to the computer repair lab and confronted with an actual hardware or software problem. The candidates were asked to solve the problem, and their work was timed. Finally, the applicants were allowed to ask questions about DiskServe and their job responsibilities.

Sarah Beske is a stickler for good manners and business protocol. She waited 48 hours after the final interview to make her hiring decision. The wait time allowed her to check references. More important, she wanted to see which of the candidates wrote follow-up thank-you notes, and she planned to assess the quality of their communication.

Sarah Beske takes the hiring process seriously. She wants to hire the best people because she hopes those employees will stay with the company a long time. Hiring well is a good corporate investment.

Check Online Resources

www.prenhall.com/gerson
For more information about the job search, visit our companion Web site.

Check out our quarterly newsletters TechCom E-Notes at www.prenhall.com/gerson for dot.com updates, new case studies, insights from business professionals, grammar exercises, and facts about technical communication.

HOW TO FIND JOB OPENINGS

When it's time for you to look for a job, how will you begin your search? You know you can't just wander up and down the street, knocking on doors randomly. That would be time consuming, exhausting, and counterproductive. Instead, you must approach the job search more systematically.

Visit Your College or University Job Placement Center

The school's job placement center is an excellent place to begin a job search, for several reasons.

- Your school's job placement service might have job counselors who will counsel you regarding your skills and job options.
- Your job placement center can give you helpful hints on preparing resumes, letters of application, and follow-up letters.
- The center will post job openings within your community and possibly in other cities.
- The center will be able to tell you when companies will visit campus for job recruiting.
- The service can keep on file your letters of recommendation or portfolio. The job placement center will send these out to interested companies upon your request.

Attend a Job Fair

Many colleges and universities host job fairs. A job fair will allow you to research job openings, make contacts for internships, or submit resumes for job openings. If you attend a job fair, treat it like an interview. Dress professionally and take copies of your resume and letters of recommendation.

Talk to Your Instructors

Whether in your major field or not, instructors can be excellent job sources. They will have contacts in business, industry, and education. They may know of job openings or people who might be helpful in your job search. Furthermore, because your instructors

obviously know a great deal about you (having spent a semester or more working closely with you), they will know which types of jobs or work environments might best suit you.

Network with Friends and Previous Employers

A *Smart Money* magazine article reports that 62 percent of job searchers find employment through "face-to-face networking" (Bloch 2003, 12). A study performed by Drake Beam Morin confirms this, stating that "64 percent of . . . almost 7,500 people surveyed said they found their jobs through socializing and meeting people" (Drakeley 2003, 5). Tell friends and acquaintances that you are looking for a job. They might know someone for you to call. Why is networking so important? It is simple math. Your friends and past employers know people. Those people know people. By visiting acquaintances, you can network with numerous individuals. The more people you talk to about your job quest, the more job opportunities you will discover.

Get Involved in Your Community

There are many ways to network. In addition to talking to your family or past employers, you can network by getting involved in your community. Consider volunteering for a community committee, pursuing religious affiliations, joining community clubs, or participating in fundraising events. Join Toastmasters. Take classes in accounting, HTML, RoboHelp, or Flash at your local community college. Each instance not only teaches you a new skill but also provides an opportunity to meet new people. Then, "when a job comes open, you'll be on their radar screen" (Drakeley 2003, 6).

Check Your Professional Affiliations and Publications

If you belong to a professional organization, this could be a source of employment in three ways. First, your organization's sponsors or board members might be aware of job openings. Second, your organization might publish a listing of job openings. Finally, you can gain professional experience by serving on boards or being an officer in an organization. These experiences can be included in your resume under professional skills, work experience, and affiliations.

Read the Want Ads

Check the classified sections of your local newspapers or in newspapers in cities where you might like to live and work. These want ads list job openings, requirements, and salary ranges. In addition, by reading the classifieds, you can learn keywords in your industry and incorporate them in your resume.

Read Newspapers or Business Journals to Find Growing Businesses and Business Sectors

Which companies are receiving grants, building new sites, winning awards, or creating new service or product lines? Which companies have just gained new clients or received expanded contracts? Newspapers and journals report this kind of news, and a growing company or business sector might be good news for you. If a company is expanding, this means more job opportunities for you to pursue.

Take a "Temp" Job

Temporary ("temp") jobs, accessed through staffing agencies, pay you while you look for a job, help you acquire new skills, allow you to network, and can lead to full-time employment.

Get an Internship

Internships provide you outstanding job preparedness skills, help you meet new people for networking, and improve your resume. An unpaid internship in your preferred work area might lead to full-time employment. An internship "gives you the opportunity to show your skills, work ethic, positive attitude, and passion for your work." By interning, you can prove that you should be "the next employee the company hires" (Drakeley 2003, 7). In fact, the National Association of Colleges and Employers lists internships as one of the top ten skills employers want ("Top Ten Qualities" 2005).

Job Shadow

Job shadowing allows you to visit a worksite and follow employees through work activities for a few hours or days. This allows you to learn about job responsibilities in a certain field or work environment. In addition, job shadowing also helps you find out if a company or industry is hiring, allows you to make new contacts, and places you in a favorable position for future employment at that company.

Set Up an Informational Interview

In an informational interview, you talk with people currently working in a field, asking them questions about career opportunities and contacts. Informational interviews allow you to

- Explore careers and clarify your career goal.
- Expand your professional network.
- Build confidence for your job interviews.
- Access the most up-to-date career information.
- Identify your professional strengths and weaknesses.

Research the Internet

In the mid-to-late 1990s, the Internet was the preferred means by which job seekers found employment. That has changed drastically. *Newsweek* magazine calls the Internet "a time waster." Quoting the head of a Chicago outplacement firm, *Newsweek* writes that some of the Internet's popular job sites are "big black holes"—your resume goes in, but you never hear from anyone again (Stern 2003, 67).

Others are equally pessimistic about the Internet's value as a source for jobs. Only 10 percent of technical or computer-related jobs are found from electronic job searches, "about 13 percent of interviews for managerial-level jobs result from responding to an online posting," and a mere 4 percent of jobs overall are found through the Internet (Bloch 2003, 12). Nonetheless, you should make the Internet part of your job search strategy. If it is not the best place to find a job, the Internet still provides numerous benefits. Internet job search engines provide excellent job search resources. These sites and many more provide the following career information resources.

- **Resumes**—explaining the difference between resumes and curriculum vitae (CV), how to address gaps in your career history, avoiding the top 10 resume mistakes, and 15 tips for writing winning resumes.
- **Interviews**—interviewing to get the job and handling illegal questions.

- **Cover letters and thank-you letters**—sample cover letter techniques and 10 ways to build a better thank-you letter.
- **Job search tips**—leaping to a twenty-first-century technology career and employing the correct netiquette.

See Table 7.1 for several online job search links.

TABLE 7.1 Online Job Search Links

- http://online.onetcenter.org/—O *Net Online bills itself as "the nation's primary source of occupational information, providing comprehensive information on key attributes and characteristics of workers and occupations."
- http://www.Monster.com—lets you post resumes and search for jobs, and provides career advice.
- http://www.CareerBuilder.com—lets you search for jobs by company, industry, and job type.
- http://www.CollegeRecruiter.com/—lists the latest job postings, "coolest career resources, and most helpful employment information."
- http://www.CareerJournal.com—the *Wall Street Journal's* career search site; provides salary and hiring information, a resume database, and job hunting advice.
- http://www.FlipDog.com—provides national job listings in diverse fields by job title, location, and date of listing.
- http://www.HotJobs.yahoo.com—lets you search for jobs by keyword, city, and state.
- http://www.WantedJobs.com—lists jobs in the United States and Canada.
- http://www.WetFeet.com—gives company and industry profiles, resume help, city profiles, and international job sites.
- http://www.CareerLab.com—provides a cover letter library.
- http://www.CareerCity.com—offers a detailed discussion and samples of functional versus chronological resumes.
- http://www.JobOptions.com—lists job opportunities in the fields of accounting, customer service, engineering, human resources, sales, and technology.
- http://www.CareerMag.com—allows for online job searches by keyword.
- http://www.CareerShop.com—helps you post and edit resumes, offers career advice, and lists hiring employers.
- http://www.ajb.dni.us/—America's Job Bank posts new jobs and provides resources to help you develop credentials, acquire educational financial aid, upgrade and measure your job-related skills, and help you start you own business.
- http://www.dice.com/—Dice focuses on technology careers.
- http://www.JobWeb.com/—helps college students, seniors, new college graduates, and alumni with career development and the job search.
- http://www.Job-Hunt.org/—called by *PC Magazine* and *Forbes* the Internet's best Web site for job hunting and resources.
- http://www.FedWorld.gov/jobs/jobsearch.html—is a site dedicated to government job searches and advice.

CRITERIA FOR EFFECTIVE RESUMES

Once you have found a job that interests you, it is time to apply. Your job application will start when you send the prospective employer your resume. Resumes are usually the first impression you make on a prospective employer. If your resume is effective, you have opened the door to possible employment—you have given yourself the opportunity to sell your skills during an interview. If, in contrast, you write an ineffective resume, you have closed the door to opportunity.

Your resume should present an objective, easily accessible, detailed biographical sketch. However, do not try to include your entire history. Because the primary goal of your resume, together with your letter of application, is to get an interview, you can use your interview to explain in more detail any pertinent information that does not appear on your resume. When writing a resume, you have two optional approaches. You can write either a reverse chronological resume or a functional resume.

Reverse Chronological Resume

Write a reverse chronological resume if you

- Are a traditional job applicant (a recent high school or college graduate, aged 18 to 25).
- Hope to enter the profession in which you have received college training or certification.
- Have made steady progress in one profession (promotions or salary increases).
- Plan to stay in your present profession.

Functional Resume

Write a functional resume if you

- Are a nontraditional job applicant (are returning to the workforce after a lengthy absence, are older, are not a recent high school or college graduate).
- Plan to enter a profession in which you have not received formal college training or certification.
- Have changed jobs frequently.
- Plan to enter a new profession.

Key Resume Components

Whether you write a reverse chronological or a functional resume, include the following key components.

Identification.   Begin your resume with the following:

- **Your name (full first name, middle initial, and last name).**   Your name can be in boldface and printed in a larger type size (14 point, 16 point, etc.).
- **Contact information.**   Include your street address, your city, state (use the correct two-letter abbreviation), and zip code. If you are attending college or serving in the armed forces, you might also want to include a permanent address. By including alternative addresses, you will enable your prospective employer to contact you more easily.
- **Your area code and phone numbers.**   As with addresses, you can provide alternative phone numbers, such as cell phone numbers. However, limit yourself to two phone numbers, and don't provide a work phone. Having prospective employers call you at your present job is not wise. First, your current employer

will not appreciate your receiving this sort of personal call; second, your future employer might believe that you often receive personal calls at work—and will continue to do so if he or she hires you.

- **Your e-mail or Web site address and fax number.** Be sure that your e-mail address is professional sounding. An e-mail address such as "ILuvDaBears," "Hotrodder," or "HeavyMetalDude" is not likely to inspire a company to interview you.

Career Objectives. A career objectives line is not mandatory. Some employers like you to state your career objectives; others do not. You will have to decide whether you want to include this section in your resume. The career objectives line is like a subject line in a memo or report. Just as the subject line clarifies your memo or report's intent, your career objective informs the reader of your resume's focus. Be sure your career objective is precise. Too often, career objectives are so generic that their vagueness does more harm than good.

BEFORE

Flawed Career Objective

Career Objective: Seeking employment in a business environment offering an opportunity for professional growth.

This poorly constructed career objective provides no focus. What kind of business? What kind of opportunities for professional growth? Employers don't want to hire people who have only vague notions about their skills and objectives.

AFTER

Better Career Objective

Career Objectives: To market financial planning programs and provide financial counseling to ensure positive client relations.

This improved career objective not only specifies which job the applicant is seeking but also how he or she will benefit the company.

Summary of Qualifications. As an option to beginning your resume with career objectives, you might want to consider starting with a summary of qualifications. According to Monster.Com, "resumes normally get less than a 15-second glance at the first screening" (Isaacs 2006). A summary of qualifications allows the employer an immediate opportunity to see how you can add value to the company.

A summary of qualifications should include the following:

- An overview of your skills, abilities, accomplishments, and attributes
- Your strengths in relation to the position for which you are applying
- How you will meet the employer's goals

To write an effective summary of qualifications, list your top three to seven most marketable credentials.

example

Summary of Qualifications

- Over four years combined experience in marketing and business
- Developed winning bid package for promotional brochures
- Promoted to Shift Manager in less than two years
- Maintained database of customers, special ordered merchandise, and tracked inventory

A: Do not worry about limiting yourself to the traditional one-page resume. Conciseness is important in all technical communication, but if your education, work experience, and professional skills merit more than a page, you must show your accomplishments. In addition, if you submit your resume as part of an e-mail message, readers will scroll. However, don't pad the resume. Limit yourself to jobs within the last ten years and skills that fit the job you're seeking.

Q: In my resume, which should I list first, my work experience or my education?

A: It's all about location, location, location. You should present your most important section first. If education is your strength and will help you get the job, lead with education. If, in contrast, your work experience is stronger, begin your resume with work.

Q: Can I omit jobs that I didn't like?

A: Yes and no. You cannot have any large gaps in your resume, such as a missing year or more. If you have any large gaps, you must either explain the gap or fill it with other activities (education, volunteerism, or childrearing, for instance). However, a missing month or so is not a problem. If you worked a job for a month, left that job, and then found other employment, you do not need to list the short-term job.

Employment. The employment section lists the jobs you've held. This information must be presented in reverse chronological order (your current job listed first, previous jobs listed next). This section must include the following:

- Your job title (if you have or had one)
- The name of the company you worked for
- The location of this company (either street address, city, and state or just the city and state)
- The time period during which you worked at this job
- Your job duties, responsibilities, and accomplishments

This last consideration is important. This is your chance to sell yourself. Merely stating where you worked and when you worked there will not get you a job. Instead, what did you achieve on the job? In this part of the resume, you should detail how you met deadlines, trained employees, cut expenses, exceeded sales expectations, decreased overage, and so forth. Plus, you want to quantify your accomplishments.

---

example

Listing your job title, company name, location, and dates of employment merely shows where you were in a given period of time. To prove your contributions to the company, provide specific details highlighting achievements.

Assistant Manager

McConnel Oil Change, Beauxdroit, LA

2005 to present

- Tracked and maintained over $25,000 inventory
- Trained a minimum of four new employees quarterly
- Achieved a 10 percent growth in customer car count for three consecutive years
- Developed a written manual for hazardous waste disposal, earning a "Citizen's Recognition Award" from the Beauxdroit City Council

---

Education. In addition to work experience, you must include your education. Document your educational experiences in reverse chronological order (most recent education first;

previous schools, colleges, universities, military courses, and training seminars next). When listing your education, provide the following information.

- Degree. If you have not yet received your degree, you can write "Anticipated date of graduation June 2008" or "Degree expected in 2008."
- Area of specialization.
- School attended. Do not abbreviate. Although you might assume that everyone knows what *UT* means, your readers won't understand this abbreviation. Is UT the University of Texas, the University of Tennessee, the University of Tulsa, or the University of Toledo?
- Location. Include the city and state.
- Year of graduation or years attended.

As you can see, this information is just the facts and nothing else. Many people might have the same educational history as you. For instance, just imagine how many of your current classmates will graduate from your school, in the same year, with the same degree. Why are you more hirable than they are? The only way you can differentiate yourself from other job candidates with similar degrees is by highlighting your unique educational accomplishments. These might include any or all of the following:

Grade point average	Academic honors, scholarships, and awards
Academic club memberships and leadership offices held	Fraternity or sorority leadership offices held
Unique coursework	Percentage of your college education costs you paid for while attending school
Special class projects	Technical equipment you can operate

Please note a key concern regarding your work experience and education. You should have no chronological gaps when all of your work and education are listed. You can't omit a year without a very good explanation. (A missing month or so is not a problem.)

Professional Skills.   If you are changing professions or reentering the workforce after a long absence, you will want to write a functional resume. Therefore, rather than beginning with education or work experience, which won't necessarily help you get a job, focus your reader's attention on your unique skills. These could include the following:

Proficiency with computer hardware and software	New techniques you have invented or implemented
Procedures you can perform	Numbers of and types of people you have managed
Special accomplishments and awards you have earned	Publications you have created or worked on
On-the-job training you have received	Certifications you have earned
Training you have provided	Languages you speak, read, and write

These professional skills are important because they help show how you are different from all other applicants. In addition, they show that although you have not been trained in the job for which you are applying, you can still be a valuable employee.

Professional Skills

- Proficient in Microsoft Word, Excel, PowerPoint, and FrontPage
- Knowledge of HTML, Java, Visual Basic, and C++
- Certified OSHA Hazardous Management Safety Trainer
- Fluent in Spanish and English
- Completed Second Shift Administration Certificate

Military Experience. If you served for several years in the military, you might want to describe this service in a separate section. You would state the following:

Rank	Discharge status
Service branch	Special clearances
Location (city, state, country, ship, etc.)	Achievements and professional skills
Years in service	Training seminars attended and education received

Professional Affiliations. If you belong to regional, national, or international clubs or have professional affiliations, you might want to mention these. Such memberships might include the Rotarians, Lions Club, Big Brothers, Campfire Girls, or Junior League. Maybe you belong to the Society for Technical Communication, the Institute of Electrical and Electronic Engineers, the National Office Machine Dealers Association, or the American Helicopter Society. Listing such associations emphasizes your social consciousness and your professional sincerity.

Optional Resume Components

Portfolios. As an optional component for your job search, consider using a portfolio. If you are in fashion merchandising, heating/ventilation/air conditioning (HVAC), engineering, drafting, architecture, or graphic design, for example, you might want to provide samples of your work. These samples could include schematics, drawings, photographs, or certifications.

However, avoid sending an unsolicited portfolio electronically. Many companies are reluctant to open unknown attachments. Make hard copies of your work or a CD-ROM. When you have an interview with a prospective employer, you can give the employer the hard copy or CD-ROM.

References. Avoid a reference line that reads, "Supplied on request," "Available on request," or "Furnished on request." Every employer knows that you will provide references if asked. Instead of wasting valuable space on your resume with unnecessary text, use this space to develop your summary of qualifications, education, work experience, or professional skills more thoroughly. Create a second page for references, and bring this reference page to your interview. On the reference page, list three or four colleagues, supervisors, teachers, or community individuals who will recommend you for employment. Provide their names, titles, addresses, and phone numbers. By bringing the reference page to your interview, you will show your prospective employer that you are proactive and organized.

Personal Data. Do not include any of the following information: birth date, race, gender, height, weight, religious affiliations, marital status, or pictures of yourself. Equal opportunity laws disallow employers from making decisions based on these factors.

Effective Resume Style

The preceding information suggests *what* you should include in your resume. Your next consideration is *how* this information should be presented. As mentioned throughout this textbook, format or page layout is essential for effective technical communication. The same holds true for your resume.

Choose Appropriate Font Types and Sizes.   As with most technical communication, the best font types are Times New Roman and Arial. These are readable and professional looking. Avoid "designer fonts," such as Comic Sans, and cursive fonts, such as Shelley Vollante. In addition, use a 10- to 12-point font for your text. Smaller font sizes are hard to read; larger font sizes look unprofessional. Headings can be boldface and 12- to 14-point font size. Limit your resume to no more than two font types.

Avoid Sentences.   Sentences create three problems in a resume. First, if you use sentences, the majority of them will begin with the first-person pronoun *I*. You'll write, "I have . . . ," "I graduated . . . ," or "I worked. . . ." Such sentences are repetitious and egocentric. Second, if you choose to use sentences you'll run the risk of committing grammatical errors: run-ons, dangling modifiers, agreement errors, and so forth. Third, sentences will take up room in your resume, making it longer than necessary.

Format Your Resume for Reader-Friendly Ease of Access.   Instead of sentences, highlight your resume with easily accessible lists. Set apart your achievements by bulletizing your accomplishments, awards, unique skills, and so on. In addition to bullets, make your resume accessible by capitalizing and boldfacing headings, indenting subheadings to create white space, and italicizing subheadings and major achievements. Avoid underlining headings. Studies show that most people find underlined text hard to read (Vogt 2007).

Begin Your Lists with Verbs.   To convey a positive, assertive tone, use verbs when describing your achievements.
Table 7.2 provides a list of verbs you might use.

Quantify Your Achievements.   Your resume should not tell your readers how great you are; it should prove your worth. To do so, quantify by precisely explaining your achievements. (See Before and After boxes on p. 216.)

Make It Perfect.   You cannot afford to make an error in your resume. Remember, your resume is the first impression you'll make on your prospective employer. Errors in your resume will make a poor first impression.

TABLE 7.2 Active Verbs to Highlight Achievement

Accomplished	Designed	Initiated	Planned
Achieved	Developed	Installed	Prepared
Analyzed	Diagnosed	Led	Presented
Awarded	Directed	Made	Programmed
Built	Earned	Maintained	Reduced
Completed	Established	Managed	Resolved
Conducted	Expanded	Manufactured	Reviewed
Coordinated	Gained	Negotiated	Sold
Created	Implemented	Ordered	Supervised
Customized	Improved	Organized	Trained

BEFORE	AFTER
Maintained positive customer relations with numerous clients.	Maintained positive customer relations with 5,000 retail and 90 wholesale clients.
Improved field representative efficiency through effective training.	Improved field representative efficiency by writing corporate manuals for policies and procedures.
Achieved production goals.	Achieved 95 percent production, surpassing the company's desired goal of 90 percent.

TECHNOLOGY TIP

Using Resume Templates in Microsoft Word 2007

Microsoft Word 2007 provides you a resume template if you want help getting started.

1. Click on the **Office Button** located on the top left of your toolbar and scroll to **New.**

The following **New Document** window will pop up.

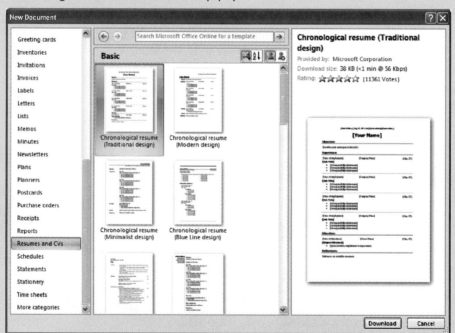

2. Click on the **Resumes and CVs** tab to find 24 optional resume layouts ("Traditional" chronological resume design, "Modern" chronological resume design, "Minimalist" chronological resume design, etc.).

These resume templates provide benefits as well as create a few problems. On the positive side, the templates are great reminders of what to include in your resume, such as objectives, work experience, education, and skills. In contrast, the templates also limit you and perhaps suggest that you include information that isn't needed. First, the templates mandate font sizes and page layout. Second, a few of the templates suggest that you include information about "Interests" and "References." Rarely should you include "Interests," such as hobbies. Furthermore, most experts suggest that you omit the "References" line, saving valuable resume space for more important information. You can include references on a separate page, especially for interviews. More important, if you use the same templates that everyone else does, then how will your resume stand out as unique? A good compromise is to review the templates for ideas and then create your own resume with your unique layout.

Check Online Resources

www.prenhall.com/gerson

For more information on sample resumes, visit our companion Web site.

METHODS OF DELIVERY

When writing either a reverse chronological or a functional resume, you can deliver your document in several ways.

Mail Version

The traditional way to deliver a resume is to insert it into an envelope and mail it. This resume can be highly designed, using bullets, boldface, horizontal rules, indentations, and different font sizes. Because this document will be a hard copy, what the reader sees will be exactly what you mail. Do not be tempted to overdesign your resume, however. For example, avoid decorative fonts, clip art, borders, or photos. Do not print your resume on unusual colors, like salmon, baby blue, tangerine, or yellow. Instead, stick to heavy white paper and standard fonts, like Times New Roman and Arial.

Figures 7.1 and 7.2 are excellent examples of traditional resumes, ready to be mailed.

E-mail Resume

Delivering your resume by U.S. mail can take several days. The quickest way to get your resume to the prospective employer is as an e-mail attachment. Speed isn't the only issue. "Hiring managers and recruiters have become as addicted to e-mail as everyone else. More than one-third of human-resource professionals reported a preference for e-mailed resumes, according to . . . the Society for Human Resource Management" (Dixson 2001).

If you choose to submit the resume as an attachment, be sure to write a brief e-mail cover message (we discuss this later in the chapter as well, under letters of application). In addition, you must clarify what software you have used for this attachment—Microsoft Word, Works, WordPerfect, or Rich Text Format (rtf), for example.

Scannable Resume

Many companies use computers to screen resumes with a technique called electronic applicant tracking. The company's computer program scans resumes as raster (or bitmap) images. Next, the software uses artificial intelligence to read the text, scanning for keywords. If your resume contains a sufficient number of these keywords, the resume will then be given to someone in the human resources department for follow-up.

A scannable resume can be e-mailed or sent through the U.S. mail. To create a scannable resume, type your text using Notepad for Windows, Simpletext for Macintosh, or Note Tab, which is available as freeware. You also could type your

FIGURE 7.1 Chronological Resume

Sharon J. Barenblatt

1901 Rosebud Avenue
Boston, MA 12987
Cell phone: 202-555-2121
E-mail: sharonbb@juno.com

You could replace the "Objective" line with a "Summary of Qualifications," as follows:

Summary of Qualifications

- Over five years customer service experience
- Work experience in public relations
- Proven record of written and interpersonal communication abilities
- Outstanding leadership skills
- Fluent in Spanish

OBJECTIVE:

Employment as an Account Manager in public relations, using my education, work experience, and interpersonal communication skills to generate business.

EDUCATION:

BS, Business. Boston College. Boston, MA 2005

- 3.2 GPA
- Social Justice Chair, Sigma Delta Tau, 2004
- Study Abroad Program, Madrid, Spain 2003
- Internship, Ace Public Relations, Boston, MA 2002

Frederick Douglas High School. Newcastle, MA 2001

- 3.5 GPA
- Member, Honor Society
- Captain, Frederick Douglas High School tennis team

List your education and work experience in reverse chronological order.

WORK EXPERIENCE:

Salesperson/assistant department manager. Jessica McClintock Clothing Store. Boston, MA 2004 to present.

- Prepare nightly deposits, input daily receipts
- Open and close the store
- Provide customer service
- Trained six new employees

List current jobs using present tense verbs and past jobs using past tense verbs.

Salesperson. GAP Clothing. Newcastle, MA 1999 to 2001

- Assisted customers
- Stocked shelves

Do not only list where you worked and when you worked there. Also include your job responsibilities.

PROFESSIONAL SKILLS:

- Made oral presentations to the Pan-Hellenic Council to advertise sorority philanthropic activities
- Helped plan community-wide "Paul Revere's Ride Day"
- Created advertising brochures and fliers

---

resume using Microsoft Word and save the document as a text file, with a *.txt* extension (Dikel 1999, 3).

To create a successful scannable resume, try the following techniques.

- Use high-quality paper (send an original), optimum contrast (black print on white paper), and paper without wrinkles or folds (if you are mailing the resume).
- Use a Courier, Helvetica, or Arial typeface (10- to 14-point type).

FIGURE 7.2 Functional Resume

Jody R. Seacrest
1944 W. 112th Street
Salem, OR 64925
(513) 451-4978   jseacrest12@hotmail.com

PROFESSIONAL SKILLS

- Operated sporting goods/sportswear mail-order house. Business began as home-based but experienced 125% growth and was purchased by a national retail sporting goods chain.
- Managed a retail design studio producing over $500,000 annually.
- Hired, trained, and supervised an administrative staff of 15 employees for a financial planning institution.
- Sold copiers through on-site demonstrations. Exceeded corporate sales goals by 10% annually.
- Provided purchaser training for office equipment, reducing labor costs by 25%.
- Acquired modern management skills through continuing education courses.

WORK EXPERIENCE

Office manager, Simcoe Designs, Salem, OR 2004 to present.
Sales representative, Hi-Tech Office Systems, Salem, OR 2000 to 2004.
Office manager, Lueck Finances, Portland, OR 1998 to 2000.
President, Good Sports, Inc., Portland, OR 1996 to 1998.

COMPUTER PROFICIENCY

Microsoft Office XP, Visual Basic 6, C++, Oracle, Microsoft SQL Server, Network Administration

MILITARY EXPERIENCE

Corporal, U.S. Army, Fort Lewis, WA 1992 to 1996. Honorably Discharged.
- Served as Company network administrator.
- Planned and budgeted all IT purchases.

EDUCATION

BA, Communication Studies, Portland State University, Portland, OR 1992.

In a functional resume, emphasize skills you have acquired which relate to the advertised position. Also quantify your accomplishments.

In a functional resume, you still should list education and work experience in a reverse chronological order.

A functional resume is organized by importance. Begin with the skills or accomplishments that will get you the job. Place less important information lower in the resume.

- Place your name at the top of the page. "Scanners assume that whatever is at the top is your name. If your resume has two pages, place your name and a 'page two' designation on the second page, and attach with a paper clip—no staples" (Kendall 2003).
- Avoid italics, underlining, colors, horizontal and vertical bars, and iconic bullets.
- White space is still important, but do not use your Tab key for spacing. Tabs will be interpreted differently in different computer environments. Use your space bar instead.
- Avoid organizing information in columns.
- Do not center text.
- Limit your screen length to approximately 60 characters per line.

- Hit a hard return at the end of each line instead of using word wrap.
- Use all-cap headings and place your text below the headings, spacing for visual appeal.
- Create bullets using an asterisk (*) or a hyphen (-).
- Create horizontal rules by using the equal sign (=), underline (_), or hyphen (-).
- Use *keywords* in your summary of qualifications, work experience, and professional skills.

Keywords are the most important feature of scannable resumes. OCR searches focus on keywords and phrases specifically related to the job opening. The keywords include job titles, skills and responsibilities, corporate buzzwords, acronyms and abbreviations related to hardware and software, academic degrees, and certifications.

You can find which keywords to focus on by carefully reading the following:

- Job advertisements
- Your prospective employer's Web site
- Government job descriptions
- Industry-specific Web sites
- The *Occupational Outlook Handbook* (found online at http://www.bls.gov/oco/)
- Career-related discussion groups or blogs
- Sample resumes found online

When using keywords, be specific; avoid vague words and phrases.

BEFORE	AFTER
Knowledge of various software products	Can create online help using AuthorIT and have expertise with PageMaker and Quark

BEFORE	AFTER
Familiar with computer technology	Proficient in multimedia, HTML, and Windows and Macintosh platforms

Other industry keywords are shown in Table 7.3.

TABLE 7.3 Hardware and Software Keywords

NT	FrontPage
DOS	AutoCAD
Microsoft Office	LAN/WAN
Corel Office Suite	IBM Client Access
Lotus Notes	CAD/CAM
Novell GroupWise	Windows XP
Internet Explorer	Visual Basic
Netscape Communicator	IBM AS/400
Flow Chart	C++
Excel	Dreamweaver
PowerPoint	Flash

In addition to hardware and software skills, OCR also scans for "soft skills," as shown in Table 7.4.

Figure 7.3 shows an excellent example of a scannable resume.

TABLE 7.4 Employment Skills Keywords

Oral presentations	Customer service
Oral communication	Telemarketing
Effective writing skills	Marketing
Interpersonal communication	Product information
Teamwork	Self-motivated
Flexibility	Organization
Time management	Innovation
Management	Ethical
Web design	Quality assurance
Project management	Training
Supervision	Problem solving

FIGURE 7.3 Scannable Resume

Rochelle J. Kroft
1101 Ave. L
Tuscaloosa, AL 89403
Home: (313) 690-4530
Cell: (313) 900-6767
E-mail: rkroft90@aol.com

Place your name at the top of a scannable resume and avoid centering text.

OBJECTIVES
To use HAZARDOUS WASTE MANAGEMENT experience and knowledge to ensure company compliance and employee safety.

SUMMARY OF QUALIFICATIONS
* HAZARDOUS WASTE MANAGEMENT with skills in teamwork, end-user support, quality assurance, problem solving, and written documentation.
* Five years' experience working with international and national businesses and regulatory agencies.
* Skilled in assessing environmental needs and implementing hazardous waste improvement projects.
* Able to communicate effectively with multinational, cross-cultural teams, consisting of clients, vendors, coworkers, and local and regional stakeholders.
* Excellent customer service, using strong problem-solving techniques.
* Effective project management skills, able to multitask.

Use key words to summarize your accomplishments.

COMPUTER PROFICIENCY
Microsoft Windows XP, FrontPage, PowerPoint, C++, Visual Basic, Java, CAD/CAM

(continued)

FIGURE 7.3 Continued

Use hyphens for design elements such as line dividers and asterisks (*) for bullets.

Type your scannable resume in Courier, Arial, Verdana, or Helvetica. Avoid "designer" fonts like Comic Sans, Lucida, or Corsiva.

EXPERIENCE

Hazardous Waste Manager
Shallenberger Industries, Tuscaloosa, AL (2000 to present)

*   Assess client needs for root cause analysis and recommend strategic actions.
*   Oversee waste management improvements, using project management skills.
*   Conduct and document follow-up quality assurance testing on all newly developed applications to ensure compliance.
*   Develop training manuals to ensure team and stakeholder safety. Shallenberger has had NO injuries throughout my management.
*   Manage a staff of 25 employees.
*   Achieved "Citizen's Recognition" Award from Tuscaloosa City Council for safety compliance record.

Hazardous Waste Technician
CleanAir, Montgomery, AL (1998–2000)

*   Developed innovative solutions to improve community safety.
*   Created new procedure manuals to ensure regulatory compliance.

EDUCATION

B.S. Biological Sciences, University of Alabama, Tuscaloosa, AL (1997)

*   Biotechnology Honor Society, President (1996)
*   Golden Key National Honor Society

AFFILIATIONS

Member, Hazardous Waste Society International

CRITERIA FOR EFFECTIVE LETTERS OF APPLICATION

Your resume, whether hard copy or electronic, will be prefaced by a letter of application (or cover e-mail if you submit the resume as an e-mail attachment). These two components of your job package serve different purposes.

The resume is generic. You'll write one resume and use it over and over again when applying for numerous jobs. In contrast, the letter of application is specific. Each letter of application will be different, customized specifically for each job. Criteria for an effective letter of application include the following.

Letter Formats

See Chapter 6 for more discussion of letter formats.

Letter Essentials

Letters contain certain mandatory components: your address, the date, your reader's address, a salutation, the letter's body, a complimentary close, your signed name, your typed name, and an enclosure notation if applicable. If you are submitting an electronic resume along with an e-mail cover message, you will not need these letter essentials. We discuss the e-mail cover message later in this chapter.

Introduction. In your introductory paragraph, include the following:

*   Tell where you discovered the job opening. You might write, "In response to your advertisement in the May 31, 2008, *Lubbock Avalanche Journal* . . ." or "Bob Ward, manager of human resources, informed me that . . ."

- State which specific job you are applying for. Often, the classified section of your newspaper will advertise several jobs at one company. You must clarify which of those jobs you're interested in. For example, you could write, "Your advertisement for a computer maintenance technician is just what I have been looking for."
- Sum up your best credentials. "My BS in chemistry and five years of experience working in a hazardous materials lab qualify me for the position."

Discussion. In the discussion paragraph(s), sell your skills. To do so, describe your work experience, your education, and your professional skills. This section of your letter of application, however, is not meant to be merely a replication of your resume. The resume is generic; the letter of application is specifically geared toward your reader's needs. Therefore, in the discussion, follow these guidelines.

- Focus on your assets uniquely applicable to the advertised position. Select only those skills from your resume that relate to the advertisement and which will benefit the prospective employer.
- Don't explain how the job will make you happy: "I will benefit from this job because it will teach me valuable skills." Instead, using the pronouns *you* and *your*, show reader benefit: "My work with governmental agencies has provided me a wide variety of skills from which your company will benefit."
- Quantify your abilities. Don't just say you're great ("I am always looking for ways to improve my job performance"). Instead, prove your assertions with quantifiable facts: "I won the 2007 award for new ideas saving the company money."

Conclusion. Your final paragraph should be a call to action. You could say, "I hope to hear from you soon" or "I am looking forward to discussing my application with you in greater detail." You could tell your reader how to get in touch with you: "I can be reached at 913-469-8500." If you are more daring, you could write, "I will contact you within the next two weeks to make an appointment at your convenience. At that time I would be happy to discuss my credentials more thoroughly."

In addition to these suggestions, you should mention that you have enclosed a resume. You can do this either in the introduction, discussion, or conclusion. Select the place that best lends itself to doing so.

E-mail Cover Message

If you submit the resume as an attachment to e-mail, you should write a brief e-mail cover message. This e-mail message will serve the same purpose as a hard-copy letter of application. Therefore, you want to include an introductory paragraph, body, and conclusion.

E-Mail Messages

See Chapter 6 for more discussion of e-mail messages.

Introduction—Tell the reader which job you are applying for and where you learned of this position.

Body—State that you have attached a resume. Tell the reader which software you have used to write your resume.

I have attached a resume for your review. To open this document, you will need Microsoft Word.

I have saved the resume as an RTF (rich text file).

example

Briefly explain why you are the right person for the job. You can do this in a short paragraph or by briefly listing three to five key assets.

Conclusion—Sum up your e-mail message pleasantly. Tell your reader that you would enjoy meeting him or her and that you look forward to an interview.

Online Application Etiquette

Check Online Resources

www.prenhall.com/gerson

For more information on sample e-mail cover messages, visit our companion Web site.

If you send your resume as an attachment to an e-mail message, be sure to follow online etiquette.

- **Do not use your current employer's e-mail system**—That clearly will tell your prospective employer that you misuse company equipment and company time.
- **Avoid unprofessional e-mail addresses**—Addresses such as Mustang65@aol.com, Hangglider@yahoo.com, or HotWheels@juno.com are inappropriate for business use. When you use e-mail to apply for a job, it is time to change your old e-mail address and become more professional. Use your initials or your name instead.
- **Send one e-mail at a time to one prospective employer**—Do not mass mail resumes. No employer wants to believe that he or she is just one of hundreds to whom you are writing.
- **Include a clear subject line**—Announce your intentions or the contents of the e-mail: "Resume—Vanessa Diaz" or "Response to Accountant Job Opening."

Figure 7.4 shows an effective e-mail cover message prefacing an attached resume. Figure 7.5 is an example of an effective letter of application.

FIGURE 7.4 E-mail Cover Message

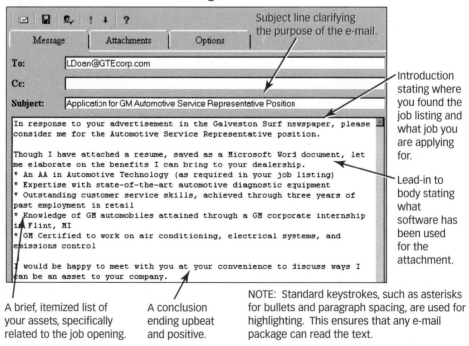

Subject line clarifying the purpose of the e-mail.

To: LDoan@GTEcorp.com

Cc:

Subject: Application for GM Automotive Service Representative Position

In response to your advertisement in the Galveston Surf newspaper, please consider me for the Automotive Service Representative position.

Though I have attached a resume, saved as a Microsoft Word document, let me elaborate on the benefits I can bring to your dealership.
* An AA in Automotive Technology (as required in your job listing)
* Expertise with state-of-the-art automotive diagnostic equipment
* Outstanding customer service skills, achieved through three years of past employment in retail
* Knowledge of GM automobiles attained through a GM corporate internship in Flint, MI
* GM Certified to work on air conditioning, electrical systems, and emissions control

I would be happy to meet with you at your convenience to discuss ways I can be an asset to your company.

Introduction stating where you found the job listing and what job you are applying for.

Lead-in to body stating what software has been used for the attachment.

A brief, itemized list of your assets, specifically related to the job opening.

A conclusion ending upbeat and positive.

NOTE: Standard keystrokes, such as asterisks for bullets and paragraph spacing, are used for highlighting. This ensures that any e-mail package can read the text.

FIGURE 7.5 Cover Letter

11944 West 112th Street
St. Louis, MO 66221

December 11, 2008

Ms. Sarah Beske:

CEO
DiskServe
9659 W. 157th St.
St. Louis, MO 78580

Dear Ms. Beske:

I am responding to your advertisement in the November 24, 2008, issue of *The St. Louis Courier* for a job in your computer technology department. Because of my two years experience in computer information systems, I believe I have the skills you require.

Although I have enclosed a resume, let me elaborate on my achievements. While working as computer technician for Radio Shack, I troubleshot motherboards, worked the help desk, and made service calls to businesses and residential customers. Your advertisement listed the importance of customer service and technical skills. I have expertise in both areas.

Your job description also mentioned the importance of working in a team environment. At Radio Shack, I frequently made service calls with a team of technicians. Together, we provided quality service.

I would like to meet with you and discuss employment possibilities at your company. Please call me at (913) 469-8500 so that we can set up an interview at your convenience. I appreciate your consideration.

Sincerely,

Macy G. Heart

The introduction mentions where the writer found the job posting, which job he is applying for, and key qualifications for the position.

The letter's body quantifies the writer's abilities. The body also shows how the writer meets the job requirement.

The conclusion ends positively, using words and phrases like "convenience," "Please," and "appreciate." It also provides follow-up information.

FAQs

Q: Do prospective employers really care about how well I write? Aren't they more concerned about my area of expertise?

A: According to a report from The National Commission on Writing, employers care deeply about the quality of your writing.

- Writing is a basic consideration for hiring and promotion. More than 75 percent of respondents report that they take writing into consideration in hiring and promoting professional employees.
- State agencies frequently require writing samples from job applicants. Fully 91 percent of respondents that "almost always" take writing into account also require a writing sample from prospective "professional" employees.
- Poorly written applications are likely to doom candidates' chances of employment. Four of five respondents agree that poorly written materials would count against "professional" job applicants either "frequently" or "almost always."

("Writing: A Powerful Message from State Government" 2005)

TECHNIQUES FOR INTERVIEWING EFFECTIVELY

The goal of writing an effective resume and letter of application is to get an interview. The resume and letter of application open the door; only a successful interview will win you the job. In fact, some sources suggest that the interview is *the most important stage of your job search*. The Society for Human Resource Management states that 95 percent of respondents to a 2003 survey ranked "interview performance" as a "very influential" factor when deciding to hire an employee. "Interview performance [is] more influential than 17 other criteria, including years of relevant work experience, resume quality, education levels, test scores or references" (Stafford 2003, L1).

Dress Professionally

Professionalism starts with your appearance. The key to successful dressing is to wear clean, conservative clothing. No one expects you to spend money on high-fashion, stylish clothes, but everyone expects you to look neat and acceptable. Business suits are still best for both men and women.

Be on Time

Plan to arrive at your interview at least 20 to 30 minutes ahead of schedule.

Watch Your Body Language

To make the best impression, don't slouch, chew your fingernails, play with your hair or jewelry, or check your watch. These actions will make you look edgy and impatient. Sit straight in your chair, even leaning forward a little to show your enthusiasm and energy. Look your interviewer in the eye. Smile and shake your interviewer's hand firmly.

Don't Chew Gum, Smoke, or Drink Beverages during the Interview

The gum might distort your speech; the cigarette will probably offend the interviewer, particularly if he or she is a non-smoker; and you might spill the beverage.

Turn Off Your Cell Phone

Today, cell phones are commonplace. However, the interview room is one place where cell phones must be avoided. Taking a call while you are being interviewed is rude and will ensure that you will not be hired.

Watch What You Say and How You Say It

Speak slowly, focus on the conversation, and don't ramble. Once you have answered the questions satisfactorily, stop.

Bring Supporting Documents to the Interview

Supporting documents can include extra copies of your resume, a list of references, letters of recommendation, employer performance appraisals, a portfolio (paper copies and electronic version) of technical documents you have written, and transcripts.

Research the Company

Show the interviewer that you are sincerely interested in and knowledgeable about the company. Dr. Judith Evans, vice president of Right Management Consultants of New York, says that the most successful job candidates show interviewers that they "know the company inside and out" (Kallick 2003, D1).

Be Familiar with Typical Interview Questions

You want to anticipate questions you will be asked and be ready with answers. Some typical questions include the following:

What are your strengths and weaknesses?	Can you travel?
Why do you want to work for this company?	Will you relocate?
Why are you leaving your present employment?	What do you want to be doing in five years, ten years?
What did you like least about your last job?	How would you handle this (hypothetical) situation?
What computer hardware are you familiar with, and what computer languages do you know?	What was your biggest accomplishment in your last job or while in college?
What machines can you use?	What about this job appealed to you?
What special techniques do you know, or what special skills do you have?	What starting salary would you expect?
What did you like most about your last job?	How do you get along with colleagues and with management?

When Answering Questions, Focus on the Company's Specific Need

For example, if the interviewer asks if you have experience using FrontPage, explain your expertise in that area, focusing on recent experiences or achievements. Be specific. In fact, you might want to tell a brief story to explain your knowledge. This is called "behavioral description interviewing" (Ralston et al. 2003, 9). It allows an interviewer to learn about your speaking abilities, organization, and relevant job skills. To respond to a behavioral description interview question, answer questions as follows (Ralston et al. 2003, 11):

- Organize your story chronologically.
- Tell who did what, when, why, and how.
- Explain what came of your actions (the result of the activity).
- Depict scenes, people, and actions.
- Make sure your story relates exactly to the interviewer's needs.
- Stop when you are through—do not ramble on. Get to the point, develop it, and conclude. If, however, you do not have the knowledge required, then "explain

How Do Human Resource Professionals Conduct a Job Search?

Maria Levit, Human Resources Director, says that for each open staff position at DeVry University, Kansas City, over 100 people apply. How do you find the best applicant out of so many possibilities?

Maria follows a step-by-step approach, which includes

- Reviewing the resumes
- Following with a prescreen telephone call to the top 10 to 15 applicants
- Inviting between 6 and 8 individuals to face-to-face interviews
- Calling references

Ms. Levit looks for two strengths above all others in the resumes. First, she wants to see "evidence of skills and credentials applicable to the job." Next, Maria needs proof. "I look for evidence of success. I want concrete examples that prove the applicant's accomplishments. For example, I want a resume to read something like, 'reduced turnover by a specific percent within such and such a time.'"

If an applicant's resume meets the job's criteria, Maria conducts the phone prescreen. In these phone calls, she wants evidence of good communication skills and a positive attitude. Maria wants to hear that the applicant "cares about people, wants to make a positive impact on students and coworkers, and has a passion" for the job. Most importantly, Maria uses the phone prescreen to detect "red flags." She is wary of applicants who make negative comments about current coworkers, bosses, or work environment.

In the face-to-face interview, Maria looks for applicants who come prepared with extra copies of resumes, names and telephone numbers of references, or transcripts. In contrast, if they "have poor posture, speak in a monotone, don't make eye contact, and dress inappropriately," then their general day-to-day job preparedness might be lacking as well.

Maria's most challenging activity is the follow-up reference call. Current and past employers are cautious about discussing employees. To overcome this reticence, Maria begins with a nonthreatening question: "What was John's basic job description?" Next she moves on to more challenging probes, like "identify two to three of John's strengths and weaknesses."

Whatever Maria does, it must be working. Many faculty and staff have worked at the Kansas City DeVry for over ten years. Hiring the right person for the right job leads to continuity in the workplace.

how you can apply the experience you *do* have" (Hartman 2003, 24). You could say, "Although I've never used FrontPage, I have built Web sites using Netscape Composer and HTML coding. Plus, I'm a quick learner. I was able to learn RoboHelp well enough to create online help screens in only a week. Our customer was very happy with the results." This will show that you understand the job and can adapt to any task you might be given.

CRITERIA FOR EFFECTIVE FOLLOW-UP CORRESPONDENCE

Once you have interviewed, don't just sit back and wait, hoping that you will be offered the job. Write a follow-up letter or e-mail message. This follow-up accomplishes three primary things: It thanks your interviewers for their time, keeps your name fresh in their memories, and gives you an opportunity to introduce new reasons for hiring you.

A follow-up letter or e-mail message contains an introduction, discussion, and conclusion.

1. **Introduction.** Tell the readers how much you appreciated meeting them. Be sure to state the date on which you met and the job for which you applied.

2. **Discussion.** In this paragraph, emphasize or add important information concerning your suitability for the job. Add details that you forgot to mention during the interview, clarify details that you covered insufficiently, and highlight your skills that match the job requirements. In any case, sell yourself one last time.

3. **Conclusion.** Thank the readers for their consideration, or remind them how they can get in touch with you for further information. Don't, however, give them any deadlines for making a decision.

See Figure 7.6 for a sample follow-up correspondence.

This letter succeeds for several reasons. First, it is short, merely reminding the reader of the writer's interest, instead of overwhelming him or her with too much new information. Second, the letter is positive, using words such as *enjoyed, ability, welcome, opportunity, exciting,* and *thank you.* Finally, the letter provides the reader an e-mail address for easy follow-up.

FIGURE 7.6   Follow-up Correspondence

Thank you for allowing me to interview with Acme Corporation on July 8. I enjoyed meeting you and the other members of the team to discuss the position of Account Representative.

You stated in the interview that Acme is planning to expand into international marketing. With my Spanish speaking ability and my study-abroad experience, I would welcome the opportunity to become involved in this exciting expansion.

Again, thank you for your time and consideration. I look forward to hearing from you. Please e-mail me at gfiefer21@aol.com.

For an e-mail follow-up, you would include your reader's e-mail address and a subject line, such as "Thank you for the July 8 Interview" or "Follow-up to July 8 Interview."

For a hard-copy follow-up letter, you would include all letter components: writer's address, date, reader's address, salutation, address, salutation, complimentary close, and signature.

CHECKLISTS

Seeking a job is not easy. However, it can be a manageable activity if you approach it as a series of separate but equal tasks. To get the job you want, you must search for an appropriate position, write an effective resume and letter of application, interview successfully, and write a follow-up letter.

Job Search Checklist

Job Openings

_____ 1. Did you visit your college or university job placement center?

_____ 2. Did you talk to your professors about job openings?

_____ 3. Have you networked with friends or past employers?

_____ 4. Have you checked with your professional affiliations or looked for job openings in trade journals?

_____ 5. Did you search the Internet for job openings?

Resume

_____ 1. Are your name, address, and phone number correct?

_____ 2. Is your job objective specific?

_____ 3. Have you included a summary of qualifications?

_____ 4. Is all information within your education, work experience, and military experience sections accurate?

_____ 5. Have you used lists beginning with verbs?

_____ 6. Have you quantified each of your achievements?

_____ 7. Have you avoided using sentences and the word *I*?

_____ 8. Does your resume use highlighting techniques to make it reader friendly?

_____ 9. Have you proofread your resume to find grammatical and mechanical errors?

_____ 10. Have you decided whether you should write a reverse chronological resume or a functional resume?

Letter of Application

_____ 1. Have you included all of the letter essentials?

_____ 2. Does your introductory paragraph state where you learned of the job, which job you are applying for, and your interest in the position?

_____ 3. Does your letter's discussion unit pinpoint the ways in which you will benefit the company?

_____ 4. Does your letter's concluding paragraph end cordially and explain what you will do next or what you hope your reader will do next?

_____ 5. Is your letter free of all errors?

Interview

_____ 1. Will you dress appropriately?

_____ 2. Will you arrive ahead of time?

_____ 3. Have you practiced answering potential questions?

_____ 4. Have you researched the company so you can ask informed questions?

_____ 5. Will you bring to the interview additional examples of your work or copies of your resume?

Follow-up Letter

_____ 1. Have you included all the letter essentials?

_____ 2. Does your introductory paragraph remind the readers when you interviewed and what position you interviewed for?

_____ 3. Does the discussion unit highlight additional ways in which you might benefit the company?

_____ 4. Does the concluding paragraph thank the readers for their time and consideration?

_____ 5. Does your letter avoid all errors?

CHAPTER HIGHLIGHTS

1. Use many different resources to locate possible jobs, such as college placement centers, instructors, friends, professional affiliations, want ads, and the Internet.

2. Use either a reverse chronological resume or a functional resume.

3. Write a traditional hard-copy resume or a scannable resume.

4. Indicate a specific career objective on your resume.

5. Write your letter of application so that it targets a specific job.

6. Prepare before your interview so you can anticipate possible questions.

7. Your follow-up letter or e-mail message will impress the interviewer and remind him or her of your strengths.

8. Over 60 percent of jobs are found through networking.

9. Relying only on the Internet for your job search is a mistake. Less than 4 percent of new jobs are obtained this way.

10. On your resume, place education first if that is your strongest asset, or begin with work experience if this will help you get the job.

CASE STUDIES

1. DiskServe, a St. Louis–based company, is hoping to hire a customer service representative for its computer technology department. In addition to DiskServe's CEO, Sarah Beske, the hiring committee will consist of two managers from other DiskServe departments and two coworkers in the computer technology department.

2. Rewrite the flawed resume on page 232. In doing so, revise the errors and create three different types of resumes for Macy G. Heart—a chronological resume, a functional resume, and a scannable resume.

   The position requires a bachelor's degree in information technology (or a comparable degree) and/or four years of experience working with computer technology. Candidates must have knowledge of C++, Visual Basic (VB), SQL, Oracle, and Microsoft Office applications. In addition, customer service skills are mandatory. Four candidates were invited to DiskServe's worksite for personal interviews—Macy Heart, Aaron Brown, Rosemary Lopez, and Robin Scott.

   Macy has a bachelor's degree in computer information systems. He has worked two years part time in his college's technology lab helping faculty and students with computer hardware and software applications, including Microsoft Office and VB. He worked for two years at a computer hardware/software store as a salesperson. His supervisor considers Macy to be an outstanding young man who works hard to please his supervisors and to meet customer needs. According to the supervisor, Macy's greatest strength is customer service, because he is patient, knowledgeable, and respectful.

   Aaron has an information technology certificate from Microsoft, where he has worked for five years. Aaron began his career at Microsoft as a temporary office support assistant, but progressed to a full-time salesperson. When asked where he saw himself in five years, Aaron stated, "The sky's the limit." References proved Aaron's lofty goals by calling him "a self-starter, very motivated, hardworking, and someone with excellent customer service skills." He is taking programming courses at night from the local community college, focusing on C++, Visual Basic, and SQL.

   Rosemary has an associate's degree in information technology. She has five years of experience as the supervisor of Oracle application. Prior to that, Rosemary worked with C++, VB, and SQL. She also has extensive knowledge of Microsoft Office. Rosemary was asked, "How have you handled customer complaints in the past?" and responded, "I rarely handle customer complaints. In my past job, I assigned that work to my subordinates."

   Robin has a bachelor's degree in information technology. To complete her degree, Robin took courses in C++, VB, Oracle, and SQL. She is very familiar with Microsoft Office. Since Robin just graduated from college, she has no full-time experience in the computer industry. However, she worked in various retail jobs (food services, clothing stores, and book stores) during high school, summers, and in her senior year. She excelled in customer service, winning the "Red Dragon Employee of the Month Award" from her last job as a server in a Chinese restaurant.

Assignment

Who would you hire? Give an oral presentation or write an e-mail or memo to Sarah Beske, DiskServe's CEO, explaining which of the candidates she should hire. Explain your choice.

1890 Arrowhead Dr.
Utica, MO 51246
710-235-9999

Resume of Macy G. Heart

Objective: Seeking a position in Computer Technology Customer Service where I can use my many technology and people-person skills. I want to help troubleshoot software and hardware problems and work in a progressive company which will give me an opportunity for advancement and personal growth.

Work Experience

Jan. 2006 to now Aramco.net   St. Louis, Missouri   Tech Support Specialists Primarily I provide customer support for customer problems with C ++, Visual Basic, Java, Networking, and Databases (Access, Oracle, SQL). I provide solutions to software and hardware problems and respond to e-mail queries in a timely manner.

Oct. 2004 to Dec. 22, 2005   DocuHelp   Chesterfield, MO Computer Consultant Provide technical support for PC's and Mac's. I also trained new PC and Mac users in hardware applications. When business was slow, I repaired computer problems, using my many technology skills.

May 2002 to Oct. 2003   Ram-on-the-Run   East St. Louis, Illinois   Computer Salesman Sold laptops, PCs, printers, and other computer accessories to men and women. Answered customer questions. Won "Salesman of the Month Award" three months in a row due to exceeding sales quotas.

Jan. 2002 to May 2002   Carbondale High School   Carbondale, IL   Lab Tech Worked in the school's computer lab, helping Mr. Jones with computer-related classwork. This included fixing computer problems and tutoring new students having trouble with assignments.

Education

Aug. 2006   Bachelor's Degree, Information Technology, Carbondale Institute
    of Technology, Carbondale, IL
Concentration: Database/Programming Applications
Relevant classes: Business Information Systems, Hardware Maintenance, Database Management, Visual Basic, Systems Analysis and Design, C++, Web Design

May 2002 Graduate Carbondale High School
Member of the Computer Technology Club
Member of FFA and DECA
Principal's Honor Roll, senior year

Computer Expertise

Cisco Certified, knowledge of C++, VB, Microsoft Office Suite XP, HTML, Java, SML

Additional Information

A good team player, who works well with others
Made all A's in my college major classes
Built my own computer from scratch in high school
Starting football player in high school Junior Varsity tight end

INDIVIDUAL AND TEAM PROJECTS

1. Practice a job search. To do so, find examples of job openings in newspapers, professional journals, at your college or university's job placement service, and online. Bring these job possibilities to class for group discussions. From this job search, you and your peers will get a better understanding of what employers want in new hires.

2. An informational interview can help you learn about the realities of a specific job or work environment. Interview a person currently working in your field of interest. You can find such employees as follows:

   • **Alumni Office**—ask your college or university for a list of alums who are willing to speak to students about careers.
   • **Career Center**—your school's career center might give you names of people to contact.
   • **Human Resources**—visit a company in your field of interest and ask the human resources staff for help.
   • **Networking**—do you have friends or family who know of employees in your field of interest?

   Once you find an employee willing to help, visit with him or her and ascertain the following:

   • What job opportunities exist in your field?
   • What does a job in your field require, in terms of writing, education, interpersonal communication skills, teamwork, and so on, as well as the primary job responsibilities?

   After gathering this information, write a thank-you letter to the employee who helped you. Then, write a report documenting your findings and give an oral presentation to your classmates.

3. Write a resume. To do so, follow the suggestions provided in this chapter. Once you have constructed this resume, bring it to class for peer review. In small groups, discuss each resume's successes and areas needing improvement.

4. Write a letter of application. To do so, find a job advertisement in your newspaper's classified section, in your school's career planning and placement office, at your worksite's personnel office, or in a trade journal. Then construct the letter according to the suggestions provided in this chapter. Next, in small groups, review your letter of application for suggested improvements.

5. Practice a job interview in small groups, designating one student as the job applicant and other students as the interview committee. Ask the applicant the sample interview questions provided in this chapter or any others you consider valid. This will give you and your peers a feel for the interviewing process.

PROBLEM-SOLVING THINK PIECES

1. You need to submit a resume for a job opening. However, you have problems with your work history, such as the following:
   • You have had no jobs.
   • You have worked as a baby sitter, or you have cut grass in your neighborhood.
   • You have been fired from a job.
   • You have had five (or more) jobs in one year.
   Consider how you would meet the challenges of your job history.

2. You have found a job that you want to apply for. The job requires a bachelor's degree in a specific field. Though you had been enrolled in that specific degree program for three years, you never completed the degree.

What should you say in your resume and letter of application to apply for this position, even though you do not meet the degree requirement? Which type of resume (reverse chronological or functional) should you write? Explain your answer.

3. You are in the middle of an interview. Though the interview had been scheduled for 1:00 P.M. to 1:45 P.M., it is running late. You had planned to pick up your son from daycare at 2:00 P.M. How do you handle this problem?

4. You have just completed the interview process for a job in your field. You have impeccable credentials, meeting and exceeding all of the requirements. Your interview went very well. At the close of the interview, one of the interviewers says, "Thank you for interviewing with us. You did a great job answering our questions, and your credentials are truly excellent. However, I think you would find the job unchallenging, maybe even boring. You are overqualified." How should you handle this situation?

5. During an interview, you are asked to "describe a problem you encountered at work and explain how you handled that challenge." How do you answer this typical question, but avoid giving an answer that paints a negative picture of a boss, coworker, or your work environment?

WEB WORKSHOP

1. Access any Internet search engine to find information about the job search. You can go online to research resumes, cover letters, and follow-up thank-you notes. For example, check out JobsMart.org and Monster.com for resume guidelines, samples, resume makeovers, and do's and don'ts. Research CareerJournal.com for up-to-date articles about resume writing (tips, samples, whether you should pay to have someone write a resume for you, and case studies about "red-flag" resumes). CareerPerfect.com provides guidelines for electronic resume submission and techniques for creating scannable versions of your resume. Check out JobSearchTech to learn why and how to write follow-up thank-you letters, and to see samples.

   Once you research any of these sites, analyze your findings. How do the letters and resumes compare to those discussed in this textbook? What new information have you learned from the online articles?
   a. Report your findings, either in an oral presentation or in writing (e-mail, memo, letter, or report).
   b. If you find sample letters or resumes that you dislike, rewrite them according to the criteria provided in this textbook.

2. Using an Internet search engine, find job openings in your area of interest. Which companies are hiring, what skills do they want from prospective employees, and what keywords are used to describe preferred skills in this work field? Report your findings either to your instructor by writing a memo or e-mail, or give an oral presentation to your class about the job market in your field.

QUIZ QUESTIONS

1. What are four places you can search for a job?
2. When would you write a reverse chronological resume?
3. In what order should you list jobs you have held?
4. What should you include when you list your education?
5. Why would you write a functional resume?
6. What types of personal data should you exclude?
7. Why should you avoid sentences in a resume?
8. What is the main difference between a resume and a letter of application?
9. What are five things you can do to interview effectively?
10. What are three things you can accomplish in a follow-up letter?

CHAPTER 8

Document Design

COMMUNICATION *at work*

In this scenario, Roger Traver relies on meeting minutes to record information for his company.

DesignGlobal Incorporated (DGI), an engineering consulting company holds monthly meetings at the Carriage Club, a privately owned banquet facility. To communicate to its 3,750 employees located internationally, these meetings are simulcast through videoconference technologies. Roger Traver, CEO of DGI, invites speakers to make presentations related to news of interest to his employees. Last month, Roger invited George Smith, a university chancellor, to speak about the connection between industry and academia. George hopes to attract more professionals such as the engineers at DGI to mentor professors and students.

Figure 8.1 shows you the meeting minutes taken at the monthly meeting where George Smith spoke. These minutes are clear, answering reporter's questions such as who, what, when, where, why, and how. The minutes also are concise in terms of word usage and sentence length. However, the minutes are unsuccessful technical communication. The wall-to-wall words not only are visually unappealing but also disallow easy access of information. An improved document design would help readers understand the meeting minutes.

Objectives

When you complete this chapter, you will be able to

1. Understand the importance of document design in technical communication.

2. Improve the organization of your technical communication through chunking.

3. Prioritize your technical information by ordering ideas.

4. Use headings and talking headings effectively.

5. Help your audience to access information through highlighting techniques.

6. Add variety to your technical communication through effective layout.

7. Evaluate your document design through the checklist.

FIGURE 8.1 Flawed Document Design

MINUTES

The meeting at the Carriage Club was attended by thirty members and guests. After the dinner, Roger Traver introduced the guest speaker, George Smith, university chancellor, and noted his accomplishments and experiences prior to education—U.S. Navy commander, Oak Ridge Laboratory researcher, and politician. Dr. Smith's talk, "Industry and Education Collaboration," was very interesting and included a history of special projects enjoyed by both academics and corporate heads. Dr. Smith suggested that we engineers could work with education to (1) provide training seminars, (2) help in urban development, and (3) provide intern opportunities. Recent industry–education collaborations include training seminars in computers, fiber optics, and human resource options. The chancellor's primary thrust was a request for $100,000 in financial aid for urban development. He said money had already been donated from three sources: a large realty firm, Capital Homes, had given $20,000; a philanthropic group, We Care, had donated a matching $20,000; Dr. Smith's university gave a matching $20,000. The remaining $40,000, Dr. Smith hoped, would come from industry donations. Finally, the chancellor noted that industry could help itself, as well as the community, by providing internships for university undergraduate majors. These internships could either be semester- or year-long arrangements, whereby students would work for minimum wage to learn more about the day-to-day aspects of their chosen fields. The chancellor said that these internships would not only increase the students' theoretical knowledge of engineering by giving them hands-on experience but also make them better future employees for the host engineering companies. Everyone would benefit. Dr. Smith noted that the students would receive a grade and credit for their work. After the speech, out VP introduced new business, calling for nominations for next year's officers; gave us the agenda for our next meeting; and adjourned the meeting.

Check Online Resources

www.prenhall.com/gerson
For more information about document design, visit our companion Web site.

Check out our quarterly newsletters TechCom E-Notes at www.prenhall.com/gerson for dot.com updates, new case studies, insights from business professionals, grammar exercises, and facts about technical communication.

IMPORTANCE OF DOCUMENT DESIGN

In technical communication, words are not your only concern. What you write is important, but how the text looks on the page is equally important. If you give your readers excessively long paragraphs, or pages full of wall-to-wall words, you have made a mistake. Ugly blocks of unappealing text negatively impact readability. In contrast, effective technical communication allows readers rapid access to the information, highlights important information, and graphically expresses your company's identity.

The Technical Communication Context

Why do people read correspondence? Although individuals read poetry, short stories, novels, and drama for enjoyment, few people read memos, letters, reports, or instructions for fun. They read these types of technical communication for information about a product or service. They read this correspondence while they talk on the telephone, while they commute to work, or while they walk to meetings.

Given these contexts, your readers want you to provide them information quickly, information they can understand at a glance. Reading word after word, paragraph after paragraph takes time and effort, which most readers cannot spare. Therefore, if your technical communication is visually unappealing, your audience might not even read your words. Readers will either give up before they have begun or be unable to remember what they have read. You cannot assume they will labor over your text to uncover its worth. Good technical communication allows readers rapid access to information.

Damages and Dangers

If your intended readers fail to read your text because it is visually inaccessible, imagine the possible repercussions. They could damage equipment by not recognizing important information which you have buried in dense blocks of text. The readers could give up on the text and call your company's toll-free hotline for assistance. This wastes your readers' and your coworkers' time and energy. Worse, your readers might hurt themselves and sue your company for failing to highlight potential dangers in a user manual.

Corporate Identity

Your document—whether a memo, letter, report, instruction, Web site, or brochure—is a visual representation of your company, graphically expressing your company's identity. It might be the only way you meet your clients. If your text is unappealing, that is the corporate image your company conveys to the customer. If your text is not reader friendly, that is how your company will appear to your client. Visually unappealing and inaccessible correspondence can negatively affect your company's sales and reputation. In today's competitive workplace, any leverage you can provide your company is a plus. Document design is one way to appeal to a client.

Time and Money

Successful technical communication involves the audience. Your goal is to help the readers understand the organization of your text, recognize its order, and access information at a glance. You also want to use varied types of communication, including graphics, for readers who absorb information better visually, rather than through words. What's the payoff? Studies tell us that effective document design saves money and time. One international customs department, after revising the design of its lost-baggage forms, reduced its error rate by over 50 percent. When a utilities company changed the look of its billing statements, customers asked fewer questions, saving the company approximately $250,000 per year. The U.S. Department of Commerce, Office of Consumer Affairs, reported that when several companies improved the visual appeal of their documents, the companies increased business and reduced customer complaints (Schriver 1993, 250–51). Document design isn't a costly frill. Effective document design is good for your company's business.

To clarify how important document design is, look at the inaccessible meeting minutes in this chapter's opening "Communication at Work" scenario (see Figure 8.1). You are given so much data in such an unappealing format that your first response upon seeing the correspondence probably is to say, "I'm not going to read that."

How can you make these minutes more inviting? How can you break up the wall-to-wall words and make key points more accessible? To achieve effective document design, you should provide your readers visual

- Organization
- Order
- Access
- Variety

ORGANIZATION

The easiest way to organize your document's design is to break text into smaller chunks of information, a technique called *chunking*. When you use chunking to separate blocks of text, you help your readers understand the overall organization of your correspondence. They can see which topics go together and which are distinct.

Chunking to organize your text is accomplished by using any of the following techniques.

- *White space* (horizontal spacing between paragraphs, created by double or triple spacing)
- *Rules* (horizontal lines typed or drawn across the page to separate units of information)
- *Section dividers and tabs* (used in longer reports to create smaller units)
- *Headings and talking headings*

DOT-COM UPDATES

For more information about chunking, check out the following link:

- http://www.webstyleguide.com/site/chunk.html

Headings and Talking Headings

To improve your page layout and make content accessible, use headings and talking headings. Headings—words or phrases such as "Introduction," "Discussion,"

"Conclusion," "Problems with Employees," or "Background Information"—highlight the content in a particular section of a document. When you begin a new section, you should use a new heading. In addition, use subheadings if you have a long section under one heading. This will help you break up a topic into smaller, more readable units of text.

Talking headings, in contrast, are more informative than headings. A heading helps your readers navigate the text by guiding them to key parts of a document. However, headings such as "Introduction," "Discussion," and Conclusion" do not tell the readers what content is included in the section. Talking headings, such as "Human Resources Committee Reviews 2008 Benefits Packages," informatively clarify the content that follows. One way to create a talking heading is to use a subject (someone or something performing the action), a verb (the action), and an object (something acted upon).

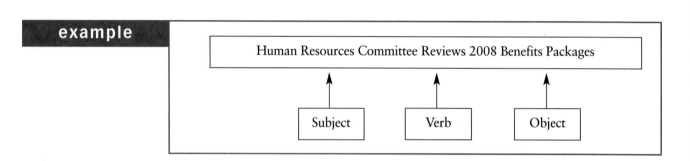

Another way to create talking headings is to use informative phrases, such as "Problems Leading to Employee Dissatisfaction," "Uses of Company Cars for Personal Errands," and "Cost Analysis of Technology Options for the Accounting Department."

Table 8.1 provides examples of informative talking headings.

Notice how using some of these techniques improves the document design in Figure 8.2.

TABLE 8.1 Talking Headings

Sentences Used as Talking Heads
Rude Customer Service Leads to Sales Losses
Accounting Department Requests Feedback on Benefits Package
Corporate Profit Sharing Decreases to 27 Percent
Parking Lot Congestion Angers Employees
Phrases Used as Talking Heads
Frozen Budget Increases until 2008
Outsourced Workers Leading to Corporate Layoffs
Harlan Cisneros—New Departmental Supervisor
EEOC: Questions about Company Hiring Practices

FIGURE 8.2 Document Design Using Chunking to Organize the Information

MINUTES — First-level heading

The meeting at the Carriage Club was attended by 30 members and guests. After the dinner, Roger Traver introduced the guest speaker, George Smith, university chancellor, and noted his accomplishments and experiences prior to education—U.S. Navy commander, Oak Ridge Laboratory researcher, and politician. Dr. Smith's talk, "Industry and Education Collaboration" was very interesting and included a history of special projects enjoyed by both academics and corporate heads. Dr. Smith suggested that we engineers could work with education to accomplish three goals.

— Horizontal rule

Training Seminars

Recent industry–education collaborations include training seminars in computers, fiber optics, and human resource options.

Urban Development — Second-level heading

The chancellor's primary thrust was a request for $100,000 in financial aid for urban development. He said money had already been donated from three sources: a large realty firm, Capital Homes, had given $20,000; a philanthropic group, We Care, had donated a matching $20,000; Dr. Smith's university gave a matching $20,000. The remaining $40,000, Dr. Smith hoped, would come from industry donations.

Internships

The chancellor noted that industry could help itself, as well as the community, by providing internships for university undergraduate majors. These internships could either be semester- or year-long arrangements, whereby students would work for minimum wage to learn more about the day-to-day aspects of their chosen fields. The chancellor said that these internships would not only increase the students' theoretical knowledge of engineering by giving them hands-on experience but also make them better future employees for the host engineering companies. Everyone would benefit. Dr. Smith noted that the students would receive a grade and credit for their work.

Conclusion

After the speech, our VP introduced new business, calling for nominations for next year's officers; gave us the agenda for our next meeting; and adjourned the meeting.

ORDER

Once a wall of unbroken words has been separated through chunking to help the reader understand the text's organization, the next thing a reader wants from your text is a sense of order. What's most important on the page? What's less important? What's least important? You can help your audience prioritize information by ordering—or *queuing*—ideas. The primary way to accomplish this goal is through a hierarchy of headings set apart from each other through various techniques.

- **Typeface**—There are many different *typefaces* (or *fonts*), including Times New Roman, `Courier`, Verdana, Helvetica, Arial, Bauhaus 93, Comic Sans MS, *Lucida Calligraphy*, **Cooper Black,** and STENCIL. Whichever typeface you choose, it will either be a *serif* or *sans serif* typeface. Serif type has "feet" or decorative strokes at the edges of each letter. This typeface is commonly used in text because it is easy to read, allowing the reader's eyes to glide across the page.

Serif ← decorative feet

Sans serif is a block typeface that omits the feet or decorative lines. This typeface is best used for headings.

Sans Serif ← no decorative feet

DOT-COM UPDATES

To access hundreds of designer fonts, check out the following link:

- http://flamingtext.com/

   Though you have many font typefaces to choose from, all are not appropriate for every technical document. Times New Roman and Arial are best to use for letters, memos, reports, resumes, and proposals, because these font types are most professional looking and are easiest to read. Arial and Verdana are considered best for Web sites since these fonts are very readable online. If you want to use "designer fonts," limit them to brochures and sales letters, for example.

- **Type size**—Another way of queuing for your readers is through the size of your type. A primary, first-level heading should be larger than subsequent, less important headings: second level, third level, and so forth. For example, a first-level heading could be in 18-point type. The second-level heading would then be set in 16-point type, the third-level heading in 14-point type, and the fourth-level heading in 12-point type.
   Figure 8.3 shows examples of different typefaces and type sizes.

- **Density**—The *weight* of the type also prioritizes your text. Type *density* is created by bold-facing or double-striking words.

FIGURE 8.3 Examples of Typefaces and Type Sizes

Sans Serif Typefaces	Serif Typefaces
Avant Garde 12 point	Courier 12 point
Avant Garde 14 point	Courier 14 point
Avant Garde 18 point	Courier 18 point
Futura 12 point	Bookman 12 point
Futura 14 point	Bookman 14 point
Futura 18 point	Bookman 18 point
Helvetica 12 point	Goudy 12 point
Helvetica 14 point	Goudy 14 point
Helvetica 18 point	Goudy 18 point

FIGURE 8.4 Outdented and Indented Headings

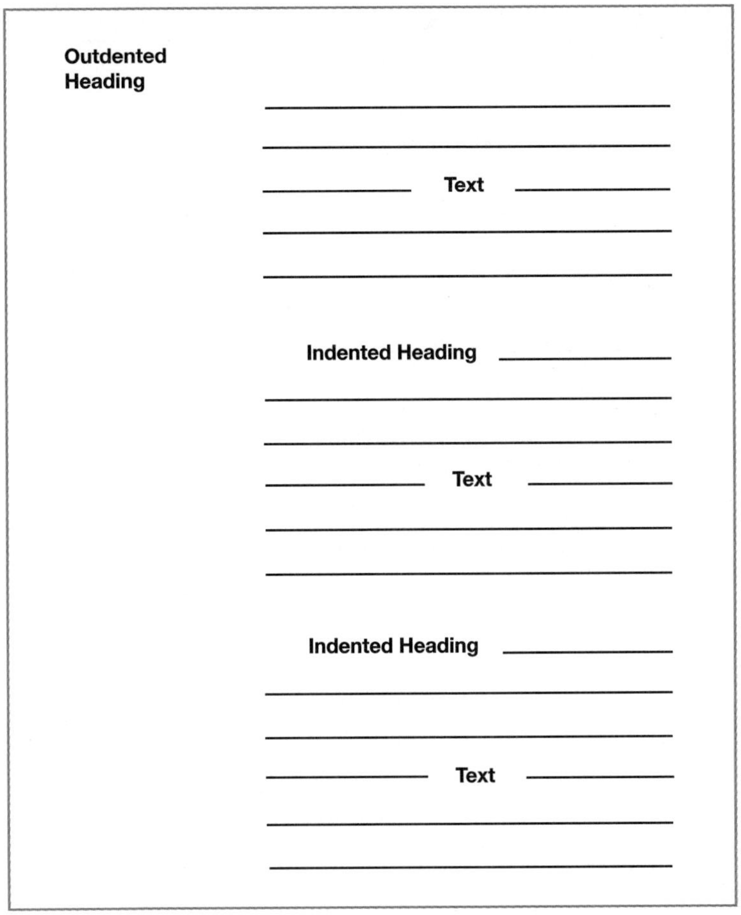

• **Spacing**—Another queuing technique to help your readers order their thoughts is the *amount of horizontal space* used after each heading. The first-level heading should have more space following it than the second-level heading, and so forth.

• **Position**—Your headings can be centered, aligned with the left margin, indented, or outdented (*hung heads*). No one approach is more valuable or more correct than another. The key is consistency. If you center your first-level heading, for example, and then place subsequent heads at the left margin, this should be your model for all chapters or sections of that report.

DOT-COM UPDATES

For more information about typography (links and news about font selection), check out the following link:

• http://www.microsoft.com/typography/default.mspx

Figure 8.4 shows an *outdented* first-level heading with *indented* subsequent headings. Figure 8.5 shows a *centered* heading with subsequent headings aligned with the left margin.

Figure 8.6 reformats the meeting minutes seen in Figure 8.1 and uses queuing to order the hierarchy of ideas. The outdented first-level heading is set in a 12-point bold sans serif typeface, all caps. The second-level heading is set in a 10-point bold serif typeface and is separated from the preceding text by horizontal white space. The third-level heading is set

FIGURE 8.5 Centered and Left-Margin Aligned Headings

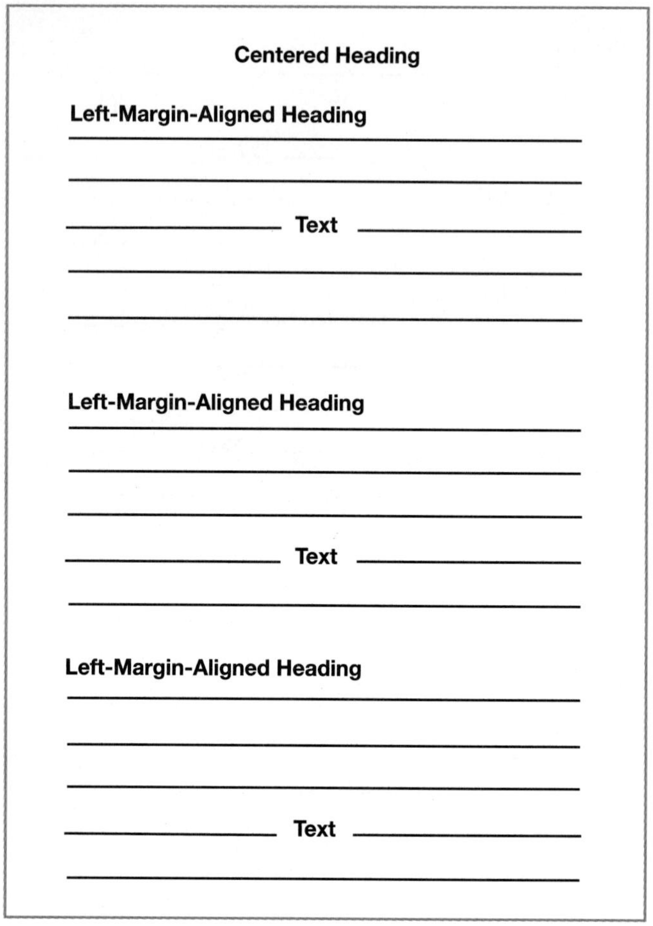

in a 10-point bold serif typeface and is separated from the preceding text by double-spacing. It is also set on the same line as the following text. Hierarchical heading levels shown in Figure 8.6 allow the readers to visualize the order of information to see clearly how the writer has prioritized text.

ACCESS

Chunking helps the reader see which ideas go together, and a hierarchy of headings helps the reader understand the relative importance of each unit of information. Nonetheless, the document design in Figure 8.6 needs improvement. The reader still must read every word carefully to see the key points within each chunk of text. Readers are not that generous with their time. As the writer, you should make your reader's task easier.

FIGURE 8.6 Document Design Using a Hierarchy of Heading Levels to Order the Information

MINUTES   The meeting at the Carriage Club was attended by 30 members and guests. After the dinner, Roger Traver introduced the guest speaker, George Smith, university chancellor, and noted his accomplishments and experiences prior to education—U.S. Navy commander, Oak Ridge Laboratory researcher, and politician. Dr. Smith's talk, "Industry and Education Collaboration," was very interesting and included a history of special projects enjoyed by both academics and corporate heads. Dr. Smith suggested that we engineers could work with education to accomplish three goals.

This hanging head is set in all caps and a sans serif font.

Urban Development
The chancellor's primary thrust was a request for $100,000 in financial aid for urban development. He said money had already been donated from three sources: a large realty firm, Capital Homes, had given $20,000; a philanthropic group, We Care, had donated a matching $20,000; Dr. Smith's university gave a matching $20,000. The remaining $40,000, Dr. Smith hoped, would come from industry donations.

Create a hierarchy of headings by changing font size and style.

Internships
The chancellor noted that industry could help itself, as well as the community, by providing internships for university undergraduate majors. These internships could either be semester- or year-long arrangements, whereby students would work for minimum wage to learn more about the day-to-day aspects of their chosen fields. The chancellor said that these internships would not only increase the students' theoretical knowledge of engineering by giving them hands-on experience but also make them better future employees for the host engineering companies. Everyone would benefit. Dr. Smith noted that the students would receive a grade and credit for their work.

Training Seminars. Recent industry–education collaborations include training seminars in computers, fiber optics, and human resource options.

Headings create accessible content.

Conclusion
After the speech, our VP introduced new business, calling for nominations for next year's officers, gave us the agenda for our

A third way to assist your audience is by helping them *access* information rapidly—at a glance. You can use any of the following highlighting techniques to help the readers filter out extraneous or tangential information and focus on key ideas.

- **White space**—In addition to horizontal space, created by double or triple spacing, you also can create *vertical space* by indenting. This vertical white space breaks up the monotony of wall-to-wall words and gives your readers breathing room. White space invites your readers into the text and helps the audience focus on the indented points you want to emphasize.
- **Bullets**—Bullets, used to emphasize items within an indented list, are created by using asterisks (*), hyphens (-), a lowercase *o*, degree signs (°), typographic symbols (■, ❏, ●, or ◆), or iconic dingbats (☞, ☺, or ✔).
- **Numbering**—*Enumeration* creates itemized lists that can show sequence or importance and allow for easy reference.
- **Boldface**—Boldface text emphasizes a keyword or phrase.
- **All caps**—The technique of capitalizing text is an excellent way to highlight a WARNING, DANGER, CAUTION, or NOTE. However, capitalizing other types of information is not suggested because reading lowercase words is easier for your audience. All caps creates a block of letters in which individual letters aren't easily distinguished from each other.
- **Underlining**—Underlining should be used cautiously. If you underline too frequently, none of your information will be emphatic. One underlined word or phrase will call attention to itself and achieve reader access. Several underlined words or phrases will overwhelm your readers.
- **Italics**—Italics and underlining are used similarly as highlighting techniques.
- **Text boxes**—Place key points in a text box for emphasis. Here is an example also italicizing for emphasis.

note	Be sure to *hand-tighten* the nuts at this point. Once you have completed the installation, go back and securely tighten all nuts.

- **Fills**—You can further highlight text boxes through fills (lines, patterns, waves, bricks, gradients, and shadings).
- **Inverse type**—You can also help readers access information by using inverse type—printing white on black, versus the usual black on white.
- **Color**—Another way to make keywords and phrases leap off the page is to color them. *Danger* would be red, for example, *Warning* orange, and *Caution* yellow. You can also use color to help a reader access the first-level heading, a header, or a footer. (*Headers* contain information placed along the top margin of text; *footers* contain information placed along the bottom margin of text.)

  For instance, if your text is typed in a black font color, then headings typed in a blue font would stand out more effectively. However, as with all highlighting techniques, a little bit goes a long way. Do not overuse color. Do not type several headings in different colors. Doing so could produce an unprofessional impression.

Effective Use of Colored Headings

Committee Action
The Budget and Personnel Committee will vote to approve the audit report at the July meeting.

Recommendation
The committee will recommend that a proposal be submitted to improve roadway construction.

Staff Contacts
Mel Henderson
Sean Thomson

Ineffective Use of Colored Headings

Committee Action
The Budget and Personnel Committee will vote to approve the audit report at the July meeting.

Recommendation
The committee will recommend that a proposal be submitted to improve roadway construction.

Staff Contacts
Mel Henderson
Sean Thomson

All colors are not equal in visual value. Generally, dark-colored fonts provide the most contrast against light-colored backgrounds, or vice versa. For example, a black font on a white background (or a white font on a black background) creates optimum contrast. On the other hand, a light-colored font on a light background does not improve access.

Good Contrast Helps Access

Committee Action
The Budget and Personnel Committee will vote to approve the audit report at the July meeting.

Recommendation
The committee will recommend that a proposal be submitted to improve roadway construction.

Staff Contacts
Mel Henderson
Sean Thomson

← Inverse print (white on black) creates optimum contrast.

Bad Contrast Hurts Access

Committee Action
The Budget and Personnel Committee will vote to approve the audit report at the July meeting.

Recommendation
The committee will recommend that a proposal be submitted to improve roadway construction.

Staff Contacts
Mel Henderson
Sean Thomson

← Light-colored text on a light-colored background harms readability.

You also should use colors tastefully. Avoid garish color combinations in graphics or backgrounds (red and orange, pink and green, or purple and yellow). Colors that clash will distract the reader more than aid access.

Clashing Color Combinations that Hurt Access

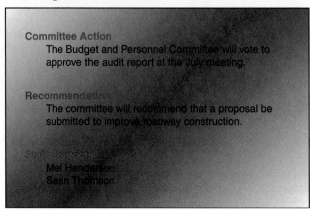

Here is a very important consideration: When it comes to using highlighting techniques, *more is not better*. A few highlighting techniques help your readers filter out background data and focus on key points. Too many highlighting techniques are a distraction and clutter the document design. Be careful not to overdo a good thing. Figure 8.7 gives examples of several highlighting methods. Notice how Figure 8.8 uses highlighting techniques to help the readers access information in the meeting minutes.

VARIETY

You can print your document using portrait orientation as seen in Figure 8.9. This figure uses one column and is printed vertically on a traditional $8\frac{1}{2} \times 11$ inch page. This is not your only option. Your reader might profit from more variety. For example, you might want to use a smaller or larger size paper, vary the weight of your paper (for example, 10 pound, 12 pound, or heavier card stock), or even print your text on colored paper.

More important, you can vary the document design as follows:

- **Print horizontally**—Rather than print your text vertically—$8\frac{1}{2} \times 11$ inch portrait—you could print *horizontally*, as an $11 \times 8\frac{1}{2}$ inch landscape.
- **Use more columns**—Provide your reader two to five columns of text.
- **Vary gutter width**—Columns of text are separated by vertical white space called the *gutter*.
- **Use ragged-right margins**—Some text is fully justified (both right and left margins are aligned). Once this was considered professional, giving the text a clean look. Now, however, studies confirm that right-margin-justified text is harder for the audience to read. It's too rigid. In contrast, *ragged-right* type (the right margin is not justified) is easier to read and more pleasing to the eye. You can use this method to vary page layout.

FIGURE 8.7 Highlighting Techniques for Access

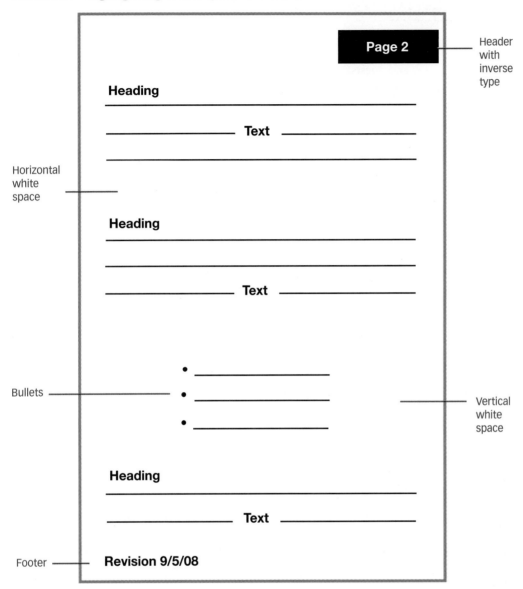

Figure 8.10 shows how you can use columns, landscape orientation, and ragged-right margins to vary your document design.

Although you can vary your document design by printing horizontally and by using multiple columns, the audience is still confronted by words, words, and more words. The majority of readers do not want to wade through text. Luckily, words are not your only means of communication. You can reach a larger audience with different learning styles by varying your method of communication. Graphics are an excellent alternative. Many people are more comfortable grasping information visually than verbally. Although it's a cliché, a picture is often worth a thousand words.

To clarify our point about the value of variety, see Figure 8.11, which adds a graphic to the meeting minutes.

Communication Channels

See Chapter 9 for more discussion of tables and figures.

FIGURE 8.8 Document Design Using Highlighting Techniques to Help Readers Access Key Ideas

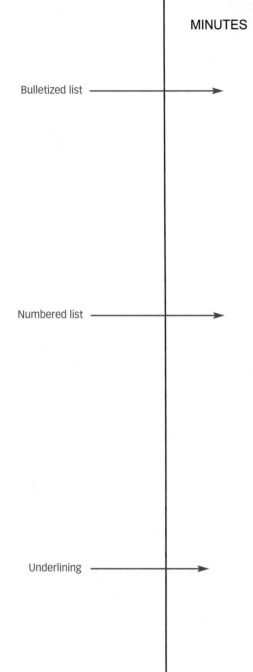

MINUTES   The meeting at the Carriage Club was attended by 30 members and guests. After the dinner, Roger Traver introduced the guest speaker, George Smith, university chancellor, and noted his accomplishments and experiences prior to education:

- U.S. Navy commander
- Oak Ridge Laboratory researcher
- Politician

Dr. Smith's talk, "Industry and Education Collaboration," was very interesting and included a history of special projects enjoyed by both academics and corporate heads. Dr. Smith suggested that we engineers could work with education to accomplish three goals.

Urban Development

The chancellor's primary thrust was a request for $100,000 in financial aid for urban development. He said money had already been donated from three sources:

1. A large realty firm, Capital Homes, had given $20,000.
2. A philanthropic group, We Care, had donated a matching $20,000.
3. Dr. Smith's university also gave $20,000.

The remaining $40,000 would come from industry.

Internships

The chancellor noted that industry could help itself, as well as the community, by providing internships for university undergraduate majors:

1. Semester-long internships
2. Year-long internships

Students would work for minimum wage to learn more about the day-to-day aspects of their chosen fields. The chancellor said that these internships would not only increase the students' theoretical knowledge of engineering by giving them hands-on experience but also make them better future employees for the host engineering companies. <u>Everyone would benefit.</u> Dr. Smith noted that the students would receive a grade and credit for their work.

Training Seminars. Recent industry–education collaborations include training seminars in computers, fiber optics, and human resource options.

Conclusion

After the speech, our VP introduced new business, calling for nominations for next year's officers, gave us the agenda for our next meeting, and adjourned the meeting.

FIGURE 8.9 Portrait with One Column Fully Justified

FIGURE 8.10 Landscape with Two Columns and Ragged-Right Margins

FIGURE 8.11 Document Design Using a Pie Chart to Vary the Communication

MINUTES   The meeting at the Carriage Club was attended by 30 members and guests. After the dinner, Roger Traver introduced the guest speaker, George Smith, university chancellor, and noted his accomplishments and experiences prior to education:

- U.S. Navy commander
- Oak Ridge Laboratory researcher
- Politician

Bulleted list for accessibility

Dr. Smith's talk, "Industry and Education Collaboration," was very interesting and included a history of special projects enjoyed by both academics and corporate heads. Dr. Smith suggested that we engineers could work with education to accomplish three goals.

Urban Development
The chancellor's primary thrust was a request for $100,000 in financial aid for urban development. He said money had already been donated from three sources, but business and industry can still help significantly. The following pie chart clarifies what money has been encumbered and how industry donations are still needed.

Internships
The chancellor noted that industry could help itself, as well as the community, by providing internships for university undergraduate majors:

1. Semester-long internships
2. Year-long internships

Students would work for minimum wage to learn more about the day-to-day aspects of their chosen fields. The chancellor said that these internships would not only increase the students' theoretical knowledge of engineering by giving them hands-on experience but

FIGURE 8.11 Continued

Graphic for visual appeal and accessible data ⟶

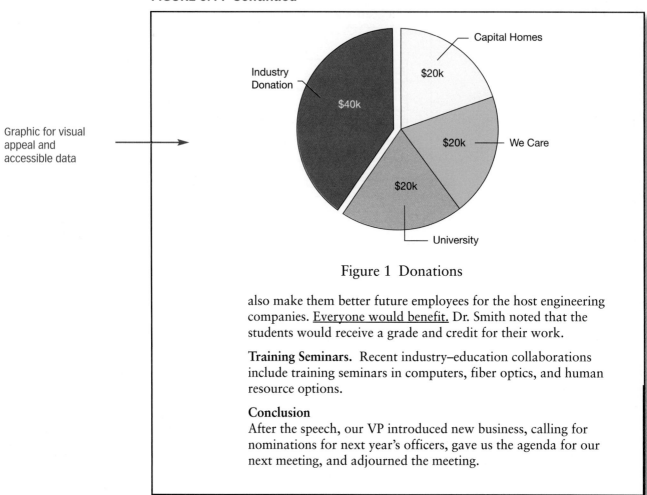

Figure 1 Donations

also make them better future employees for the host engineering companies. <u>Everyone would benefit.</u> Dr. Smith noted that the students would receive a grade and credit for their work.

Training Seminars. Recent industry–education collaborations include training seminars in computers, fiber optics, and human resource options.

Conclusion
After the speech, our VP introduced new business, calling for nominations for next year's officers, gave us the agenda for our next meeting, and adjourned the meeting.

TECHNOLOGY TIPS

Document Design Using Microsoft Word 2007
When it comes to document design, you have a world of options at your fingertips. Not only does your word-processing software offer you possibilities to enhance page layout, but also the Internet provides unlimited resources.

Word Processing
In a word-processing software program like Microsoft Word 2007, you enhance your document's design in many ways. From the **Home** tab ribbon, you can make changes to **Font**, **Paragraphs**, and **Styles**.

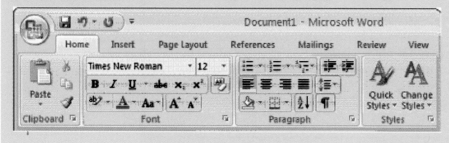

"**Font**" Category	"**Paragraph**" Category	"**Styles**" Category
• Choose your **FONT TYPE** and SIZE Times New Roman ▾ 12 ▾ • **Boldface**, *italicize,* and <u>underline</u> the **B** *I* <u>U</u> ▾ text • ~~Strikethrough~~, create _{subscripts} and ^{superscripts} abc X₂ X² • Highlight the color of text, change the font color, or change the case of selected text to uppercase or ab▾ A▾ Aa▾ lowercase • Increase or decrease the font size A A	• Create bulleted, numbered, or multilevel lists (as with outlines) • Decrease or increase an indentation • Change the margins (ragged right, centered, block right, or full block) • Change the spacing between lines • Color background behind selected text, create borders, alphabetize, and show or hide paragraphing	• Format text ("Quick Styles") and change colors and fonts throughout a document ("Change Styles") Quick Change Styles ▾ Styles ▾ Styles

From the **Insert** tab ribbon shown below, you can add **Tables, Illustrations, Links, Headers & Footers**, enhance **Text**, and include mathematical **Symbols**.

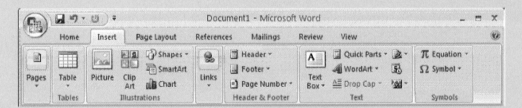

The **Illustrations** category lets you insert pictures, clip art, and charts (pie charts, bar charts, line charts, etc.). **SmartArt** provides access to more complex charts such as Venn diagrams and organizational charts. The **Shapes** pull-down menu lets you insert lines, arrows, flowchart symbols, callouts, and stars and banners.

The **Text** category lets you insert the following:

• 36 different kinds of text boxes including sidebars Text boxes can be enhanced further by changing the color of the lines or fill. You can also create shadows, as shown in this example.	• WordArt **This is an example of WordArt**

DOCUMENT DESIGN CHECKLIST

_____ 1. Have you broken up the text in your document into smaller chunks (units of information)?

_____ 2. Have you added white space to enhance readability of the text?

_____ 3. Did you use headings or talking headings to improve page layout and make content accessible?

_____ 4. Did you prioritize information by ordering (queuing) ideas?

_____ 5. Did you create a hierarchy of headings by relying on different typefaces and sizes, boldface, spacing, and varied positions?

_____ 6. Have you relied on a variety of highlighting techniques (indented text, bullets, numbers, boldface text, all caps, underlining or italicizing, textboxes, or color) to make content accessible?

_____ 7. Have you avoided overuse of highlighting techniques?

_____ 8. Have you added variety to your technical document through landscape versus portrait layout, columns, margin alignments, or graphics?

_____ 9. Have you considered using a different weight, size, or color paper?

_____ 10. Have you used color sparingly, and avoided overloading the document with excessive variety?

CHAPTER HIGHLIGHTS

1. Breaking your text into smaller chunks of information will help you create a more readable document.

2. When you create an order among the items in a document through use of typeface and type size, your audience can more easily prioritize the information.

3. Your reader should be able to glance at the document and easily pick out the key ideas. Highlighting techniques will help accomplish this goal.

4. Vary the appearance of your document by using columns, varying gutter widths, and printing in portrait or landscape orientation.

5. Your audience will access the content easily if you use white space, bullets, numbering, underlining, and text boxes.

6. Effective technical communication allows readers rapid access to information, highlights important information, and graphically expresses a company's identity.

7. A hierarchy of headings helps the reader understand the relative importance of each unit of information.

8. A variety of highlighting techniques includes white space, bullets, numbers, boldface, all caps, underlining, italics, text boxes, fills, inverse type, and color.

9. Dark-colored fonts provide the most contrast against light-colored backgrounds, and vice versa.

10. A few highlighting techniques helps your readers filter out background data and focus on key points.

CASE STUDIES

1. In teams of three to five class members, reformat the following summary to improve its document design.

> **SUMMARY**
>
> The City of Waluska wants to provide its community with a safe and reliable water treatment facility. The goal is to protect Waluska's environmental resources and to ensure community values.
>
> To achieve these goals, the city has issued a request for proposal to update the Loon Lake Water Treatment Plant (LLWTP). The city recognizes that meeting its community's water treatment needs requires overcoming numerous challenges. These challenges include managing changing regulations and protection standards, developing financially responsible treatment services, planning land use for community expansion, and upholding community values.
>
> For all of the above reasons, DesignGlobal, Inc. (DGI) Engineering is your best choice. We understand the project scope and recognize your community's needs.
>
> We have worked successfully with your community for a decade, creating feasibility studies for Loon Lake toxic control, developing odor-abatement procedures for your streams and creeks, and assessing your water treatment plant's ability to meet regulatory standards.
>
> DGI personnel are not just engineering experts. We are members of your community. Our dynamic project team has a close working relationship with your community's regulatory agencies. Our Partner in Charge, Julie Schopper, has experience with similar projects worldwide, demonstrated leadership, and the ability to communicate effectively with clients.
>
> DGI offers the City of Waluska an integrated program that addresses all your community's needs. We believe that DGI is your best choice to ensure that your community receives a water treatment plant ready to meet the challenges of the twenty-first century.

2. In teams of three to five class members, reformat the following memo to improve its document design.

DATE: November 30, 2008
TO: Jan Hunt
FROM: Tom Langford
SUBJECT: CLEANING PROCEDURES FOR MANUFACTURING WALK-IN OVENS #98731, #98732, AND #98733

The above-mentioned ovens need extensive cleaning. To do so, vacuum and wipe all doors, walls, roofs, and floors. All vents/dampers need to be removed, and a tack cloth must be used to remove all loose dust and dirt. Also, all filters need to be replaced.

I am requesting this because loose particles of dust/dirt are blown onto wet parts when placed in the air-circulating ovens to dry. This causes extensive rework. Please perform this procedure twice per week to ensure clean production.

3. The following text is visually unappealing and inaccessible. Using the techniques discussed in this chapter (organization, order, access, and variety), improve the document's design.

In 1998, the Transportation Equity Act for the 21st Century (TEA-21) was passed and authorized Federal programs for roadway safety.

To limit its focus, the TEA-21 discussed safety as it was affected by facility planning, roadway design, and maintenance.

The TEA-21 was concerned about roadway safety for pedestrians, motorized vehicles, and non-motorized vehicles. Motorized vehicles included cars, vans, trucks, SUV's and motorcycles. Non-motorized vehicles include bicycles, roller skates, rollerblades, and scooters.

The TEA-21's planning guidelines consisted of many goals. These included communicating revised safety programs to pedestrians, non-motorized vehicle users, and motor vehicle users; organizing safety data for analysis by highway and city road departments; strategies for reducing fatalities for pedestrians and vehicle users; and developing city safety blueprints based on statewide prototypes.

One prototypical model that the TEA-21 recommends following is Operation Green Light. This model uses a traffic signal coordination system to maintain a steady flow of traffic. Steady traffic flow has been proven to reduce fatalities by 15 percent.

4. The following short report is poorly formatted. The text is so dense that readers would have difficulty understanding the content easily. Improve the document's design to aid access. Use highlighting techniques discussed in this chapter to revise the text.

DATE: May 18, 2008
TO: Martha Collins
FROM: Richard Davis
SUBJECT: 2008 Switch Port Carriers

Attached are the supplemental 2008 Switch Port Carriers that are required to support this year's growth patterns. As we have discussed in previous phone conversations, the May numbers show a decrease in traffic, but forecasts still suggest increased traffic. Therefore, we are issuing plans for this contingency. If the June forecasts prove to be accurate, the ports being placed in the network via these plans will support our future growth except for areas where growth can not be predicted. Some areas, for example, are too densely populated for forecasting because the company did not hire enough survey personal to do a thorough job.

Following is an update of our suggested Port Additions. For Port 12ABR, add 16 Ports in Austin. For Port 13RgX, add 27 Ports in Houston. For Port 981D, add 35 Ports in San Antonio. For Port 720CT, add 18 Ports in Dallas. The total Port Additions will equal 96 and cost $3,590,625.

After working long hours on these suggestions, Port Additions should be considered mandatory. However, follow-up forecasts are probably needed due to the short time we were provided to do these studies. If you are going to perform follow-up forecasts, do so before September 1. The survey teams, if you want a successful forecast, need at least three months. Twenty-five team members should be sufficient.

5. Review the following PowerPoint screen. First, explain how the screen is visually flawed. Second, revise it for improved access using the techniques suggested in this chapter.

Recommendations

Table 1 **outlines the investments recommended for** *80 percent of the fiscal year 2007 funds.*
The table focuses on 80 percent of the available funds. The remaining *20 percent are being withheld until fiscal year 2008.*

Company	Costs
Harness	$2,300
MarTT3	$4,500
Notary Rex	$5,200
Mobile CRT	$2,750

INDIVIDUAL AND TEAM PROJECTS

1. Bring samples of technical communication to class. These could include letters, brochures, newsletters, instructions, reports, or advertisements from magazines or newspapers.

   In groups of three to five class members, assess the document design of each sample. Determine which samples have successful document designs and which samples have poor document designs. Base your decisions on the criteria provided in this chapter: organization, order, access, and variety. Either orally or in writing, share your findings with other teams in the class.

2. In teams of three to five class members, take one of the less successful samples from team project 1 and reformat it to improve its document design. Focus on improving the sample's organization, order, access, and variety.

   Once your team has completed reformatting the text, make photocopies or transparencies of your work. Then review the various team projects, determining which team created the best document design. Select a winner, and explain why this text now has a successful document design.

3. Read the following headings and make them more accessible by creating a hierarchy using different font sizes and font types.

   > Meeting Minutes
   > Agenda
   > Discussion of Ongoing Projects
   > Recommendations
   > Pricing
   > Cost of Equipment
   > Cost of Facilities Update
   > Cost of Insurance Benefits
   > New Hiring Policies
   > Job Requirements
   > Employee Credentials
   > Licensing

4. Change the following headings into "talking headings" by changing them to phrases or complete sentences. To do so, add any information you choose to clarify your content.

   - Computer Problems
   - WiFi Compatibility
   - Biotechnology Advances
   - Accounting Regulations
   - Facilities Update

PROBLEM-SOLVING THINK PIECES

1. The following two lists need to separate and highlight the information more clearly. But which highlighting technique should you use for each list—bullets or numbers? Explain why you would use bullets versus numbers, or vice versa for both lists.

List 1	List 2
To access your online course, follow these steps:	To choose the right car for your family and business needs, consider these factors:
Turn on the computer.	Price
Double-click the Internet icon on your desktop.	Options
Type in the following URL:	Fuel economy
http://webct.acc.edu.	Cost of repairs
Type your username and password.	Availability of dealerships
Click on the online course of your choice.	Financing
Complete assignment 1.	Capacity

2. Look at the following two lists with headings. Both use Comic Sans and colored headings, but should they? Assessing these two lists, explain when and why it's okay to use designer fonts and color or when and why you shouldn't?

Example 1 Resume

Objective	Use my information technology skills to improve a company's network capabilities, computer security, and troubleshooting.
Qualifications	• Four years experience in information technology • BS in Computer Science • Security certified
Work History	Information Technologist Pantheon Corp. St. Louis, Mo 2005-present • Networked 24 computer stations • Created computer passwords for all employees • Installed Spam and Virus protection on all systems

Example 2 Brochure

Prices

- Buy one, get one free
- Guaranteed lowest in the market
- Last year's prices—*today!*

Options

- Sizes—4″ × 2″, 3″ × 2″, and 2″ × 2″
- Colors—red, green, black, and silver
- WiFi compatible

Service

- 24/7
- On site or online help

3. Which is the best way to show a comparison/contrast to your audience: a list with headings or a table? Assess the following two choices, decide which document design is best, and justify your answer.

Option 1 List with Headings

Renner Road

- Two lanes

- No sidewalk

- Limited lighting

Bannister Road

- Four lanes

- Sidewalks on both sides of the road

- Street lighting on one side of the road

Shawnee Road

- Four lanes

- Sidewalks on both sides of the road

- Street lighting in the center esplanade

Option 2 Table

Renner Road	Bannister Road	Shawnee Road
• Two lanes	• Four lanes	• Four lanes
• No sidewalk	• Sidewalks on both sides of the road	• Sidewalks on both sides of the road
• Limited lighting	• Street lighting on one side of the road	• Street lighting in the center esplanade

WEB WORKSHOP

On the Internet, access ten corporate Web sites. Study them and make a list of the techniques used for visual communication. Which Web sites are successful, and why? Which Web sites are unsuccessful, and why? How would you redesign the less successful Web sites to achieve better visual communication?

QUIZ QUESTIONS

1. What are three reasons for designing a technical document effectively?
2. What is chunking?
3. What do you achieve by chunking?
4. In what four ways can you accomplish chunking?
5. What is the purpose of creating order or queuing your ideas?
6. How can you achieve a hierarchy of headings in your technical document?
7. What is the difference between headings and talking headings?
8. How can you help your readers access your information, filtering out extraneous or tangential information?
9. What can happen if you overuse highlighting techniques?
10. What are four ways in which you can vary your document design?

CHAPTER 9

Graphics

To help his clients understand complex figures, Bert Lang includes visual aids in his proposals.

Bert Lang is an investment banker at Country Commercial Bank. He is writing a proposal to a potential client, Sylvia Light, a retired public

health nurse. Sylvia is 68 years old and worked for the Texas Public Health Department for 36 years. She has earned her State of Texas retirement and Social Security benefits. She now has $315,500 allocated as follows: $78,000 in an individual retirement account (IRA), $234,000 in a low-earning certificate of deposit (CD), and $3,500 in her checking account.

Sylvia contacted Bert, asking him to help her organize her portfolio for a comfortable retirement. Bert has studied Sylvia's various accounts and considered her lifestyle and expenditures. Now, he is ready to write the proposal.

In this proposal, Bert wants to show Sylvia how to reallocate her funds. She should keep some ready money available and invest a portion of capital for long-term returns. Currently, too much of her money is tied up in a CD earning 3.7%. Bert plans to propose that Sylvia could reallocate her funds as follows:

- $110,000 in an annuity
- $78,000 in an individual retirement account (IRA)

Objectives

When you complete this chapter, you will be able to

1. Recognize the benefits of visual aids in your technical communication.

2. Consider the use of color or three-dimensional graphics to enhance technical communication.

3. Understand the criteria for creating effective tables and figures.

4. Know how to use a variety of visual aids.

5. Evaluate your visual aids with the checklist.

- $45,000 in municipal bonds
- $37,000 in a stock fund
- $42,000 in certificates of deposit (CDs)
- $3,500 in a checking account

Like most people, Sylvia is unfamiliar with financial planning. Though she was an expert in her health field, tuberculosis treatment, money matters confuse her. Numbers alone will not explain Bert's vision for her money management.

To make this proposal visually appealing and more readily understandable to Sylvia, Bert will use visual aids. He will provide Sylvia a pie chart to show how he wants to invest her money. Bert will use a line graph to predict how much more money she can earn by reallocating her assets. Finally, he will create a table to clarify the types of investments, the amount in each investment, the interest to be earned, and the fees.

Although Sylvia has always been fiscally conservative, Bert hopes that he can explain the need for growth of capital even in retirement. The graphic aids will visually enhance his written explanation.

Check out our quarterly newsletters TechCom E-Notes at www.prenhall.com/gerson for dot.com updates, new case studies, insights from business professionals, grammar exercises, and facts about technical communication.

THE BENEFITS OF VISUAL AIDS

Although your writing may have no grammatical or mechanical errors and you may present valuable information, you won't communicate effectively if your information is inaccessible. Consider the following paragraph.

> In January 2008, the actual rainfall was 1.50", but the average for that month was 2.00". In February 2008, the actual rainfall was 1.50", but the annual average had been 2.50". In March 2008, the actual rainfall was 1.00", but the yearly average was 2.50". In April 2008, the actual rainfall was 1.00", but annual averages were 2.50". The May 2008 actual rainfall was 0.50", whereas the annual average had been 1.50". No rainfall was recorded in June 2008. Annually, the average had been 0.50". In July 2008, only 0.25" rain fell. Usually, July had 0.50" rain. In August 2008, again no rain fell, whereas the annual August rainfall measured 0.25". In September and October 2008, the actual rainfall (0.50") matched the annual average. Similarly, the November actual rainfall matched the annual average of 1.50". Finally, in December 2008, 2.00" rain fell, compared to the annual average of 1.50".

If you read the preceding paragraph in its entirety, you are an unusually dedicated reader. Such wall-to-wall words mixed with statistics do not create easily readable writing. The goal of effective technical communication is accessible information. The example paragraph fails to meet this goal. No reader can digest the data easily or see clearly the comparative changes from one month's precipitation to the next. To present large blocks of data or reveal comparisons, you can supplement, if not replace, your text with graphics. In technical communication, visual aids accomplish several goals. Graphics (whether hand drawn, photographed, or computer generated) will help you achieve conciseness, clarity, and cosmetic appeal.

Conciseness

Visual aids allow you to provide large amounts of information in a small space. Words used to convey data (such as in the example paragraph) double, triple, or even quadruple the space needed to report information. By using graphics, you can also delete many unnecessary words and phrases.

Clarity

Visual aids can clarify complex information. Graphics help readers see the following:

- **Trends**—Certain trends, such as increasing or decreasing sales figures, are most evident in line graphs.
- **Comparisons between like components**—Comparisons such as actual monthly versus average rainfalls can be seen in grouped bar charts.
- **Percentages**—Pie charts help readers discern these.
- **Facts and figures**—A table states statistics/numbers more clearly than a wordy paragraph.

Cosmetic Appeal

Visual aids help you break up the monotony of wall-to-wall words. If you only give unbroken text, your reader will tire, lose interest, and overlook key concerns. Graphics help you sustain your reader's interest. Let's face it, readers like to look at pictures. The two types of graphics important for technical communication are tables and figures. This chapter helps you correctly use both.

COLOR

All graphics look best in color, don't they? Not necessarily. Without a doubt, a graphic depicted in vivid colors will attract your reader's attention. However, the colors might not aid communication. For example, colored graphics could have the following drawbacks (Reynolds and Marchetta 1998, 5–7).

1. The colors might be distracting (glaring orange, red, and yellow combinations on a bar chart would do more harm than good).
2. Colors that look good today might go out of style in time.
3. Colored graphics increase production costs.
4. Colored graphics consume more disk space and computer memory than black-and-white graphics.
5. The colors you use might not look the same to all readers.

Let's expand on this last point. Just because you see the colors one way does not mean your readers will see them the same. We are talking about what happens to your color graphics when someone reproduces them as black-and-white copies. We are also talking about computer monitor variations. The color on a computer monitor depends on its resolution (the number of pixels displayed) and the monitor's RGB values (how much red, green, and blue light is displayed). Because all monitors do not display these same values, what you see on your monitor will not necessarily be the same as what your reader sees. To solve this problem, test your graphics on several monitors. Also, limit your choices to primary colors instead of the infinite array of other color possibilities. More important, use patterns to distinguish your information so that the color becomes secondary to the design.

THREE-DIMENSIONAL GRAPHICS

Many people are attracted to three-dimensional (3-D) graphics. After all, they have obvious appeal. Three-dimensional graphics are more interesting and vivid than flat, one-dimensional graphics. However, 3-D graphics have drawbacks. A 3-D graphic is visually appealing, but it does not convey information quantifiably. A word of caution: Use 3-D graphics sparingly. Better yet, use the 3-D graphic to create an impression; then include a table to quantify your data.

CRITERIA FOR EFFECTIVE GRAPHICS

Figure 9.1 is an example of a cosmetically appealing, clear, and concise graphic. At a glance, the reader can pinpoint the comparative prices per barrel of crude oil between 2001 and 2007. Thus, the line graph is clear and concise. In addition, the writer has included an interesting artistic touch. The oil gushing out of the tower shades parts of the graph to emphasize the dollar amounts. Envision this graph without the shading. Only the lines would exist. The shading provides the right touch of artistry to enhance the information communicated.

The graph shown in Figure 9.1 includes the traits common to effective visual aids. Whether hand drawn or computer generated, successful tables and figures have the following characteristics.

1. They are integrated with the text (i.e., the graphic complements the text; the text explains the graphic).
2. They are appropriately located (preferably immediately following the text referring to the graphic and not a page or pages later).
3. They add to the material explained in the text (without being redundant).
4. They communicate important information that could not be conveyed easily in a paragraph or longer text.

FIGURE 9.1 Line Graph with Shading

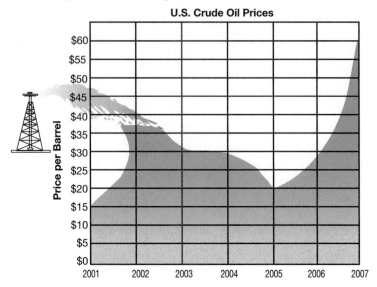

5. They do not contain details that detract from, rather than enhance, the information.
6. They are an effective size (not too small or too large).
7. They are neatly printed to be readable.
8. They are correctly labeled (with legends, headings, and titles).
9. They follow the style of other figures or tables in the text.
10. They are well conceived and carefully executed.

TYPES OF GRAPHICS

Graphics can be of two basic types: tables and figures. Tables provide columns and rows of information. You should use a table to make factual information—such as numbers, percentages, and monetary amounts—easily accessible and understandable. Figures, in contrast, are varied and include bar charts, line graphs, photographs, pie charts, schematics, line drawings, and more.

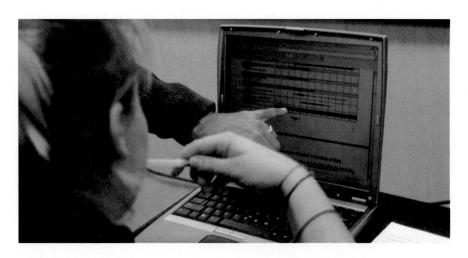

Tables

Let's tabulate the information about rainfall in 2008 presented on page 264. Because effective technical communication integrates text and graphic, you will want to provide an introductory sentence prefacing Table 9.1, as follows:

> Table 9.1 reveals the actual amount of rainfall for each month in 2008 versus the average documented rainfall for those same months.

Table 9.1 has advantages for both the writer and the reader. First, the headings eliminate needless repetition of words, thereby making the text more readable. Second, the audience can see easily the comparison between the actual amount of rainfall and the monthly averages. Thus, the table highlights the content's significant differences. Third, the table allows for easy future reference. Tables could be created for each year. Then the reader could compare quickly the changes in annual precipitation. Finally, if this information is included in a report, the writer will reference the table in the table of contents' list of illustrations. This creates ease of access for the reader.

Criteria for Effective Tables. To construct tables correctly, do the following:

1. Number tables in order of presentation (i.e., Table 1, Table 2, Table 3).
2. Title every table. In your writing, refer to the table by its number, not its title. Simply say, "Table 1 shows . . . ," "As seen in Table 1," or "The information in Table 1 reveals . . . "
3. Present the table as soon as possible after you have mentioned it in your text. Preferably, place the table on the same page as the appropriate text, not on a subsequent, unrelated page or in an appendix.
4. Don't present the table until you have mentioned it.
5. Use an introductory sentence or two to lead into the table.
6. After you have presented the table, explain its significance. You might write, "Thus, the average rainfall in both March and April exceeded the actual rainfall by 1.50 inches, reminding us of how dry the spring has been."

TABLE 9.1 2008 Monthly Rainfall versus Average Rainfall (All Figures in Inches)

Month	2008 Rainfall	Average Rainfall
January	1.50	2.00
February	1.50	2.50
March	1.00	2.50
April	1.00	2.50
May	0.50	1.50
June	0.00	0.50
July	0.25	0.50
August	0.00	0.25
September	0.50	0.50
October	0.50	0.50
November	1.50	1.50
December	2.00	1.50

7. Write headings for each column. Choose terms that summarize the information in the columns. For example, you could write "% of Error," "Length in Ft.," or "Amount in $."

8. Because the size of columns is determined by the width of the data or headings, you may want to abbreviate terms (as shown in item 7). If you use abbreviations, however, be sure your audience understands your terminology.

9. Center tables between right and left margins. Don't crowd them on the page.

10. Separate columns with ample white space, vertical lines, or dashes.

11. Show that you have omitted information by printing two or three periods or a hyphen or dash in an empty column.

12. Be consistent when using numbers. Use either decimals or numerators and denominators for fractions. You could write 3 1/4 and 3 3/4 or 3.25 and 3.75. If you use decimal points for some numbers but other numbers are whole, include zeroes. For example, write 9.00 for 9.

13. If you do not conclude a table on one page, on the second page write *Continued* in parentheses after the number of the table and the table's title.

Table 9.2 is an excellent example of a correctly prepared table.

TABLE 9.2   Student Headcount Enrollment by Age Group and Student Status, Fall 2008

Age Group	New Students	Continuing Students	Readmitted	Other	Total
15–17	453	33	2	2	490
18–20	1,404	1,125	132	—	2,661
21–23	339	819	269	—	1,427
24–26	263	596	213	—	1,072
27–29	250	436	134	—	820
30–39	524	1,168	372	—	2,064
40–49	271	510	186	—	967
50–59	76	121	54	—	251
60+	19	48	16	—	83
Unknown	109	92	27	2	230
Total	3,708	4,948	1,405	4	10,065

TECHNOLOGY TIPS

Creating Graphics in Microsoft Word 2007 (Pie Charts, Bar Charts, Line Graphs, etc.)

You can create customized graphics in Microsoft Word 2007 as follows:

1. Click on **Insert** on the Menu bar.

2. Click on **Chart**. Word will open a datasheet and bar chart template that you can customize by inserting your own text and numbers.

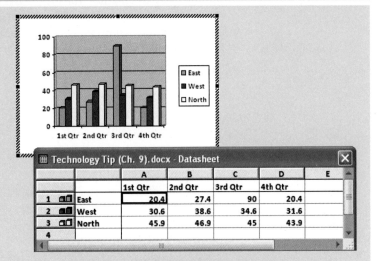

		A	B	C	D	E
		1st Qtr	2nd Qtr	3rd Qtr	4th Qtr	
1	East	20.4	27.4	90	20.4	
2	West	30.6	38.6	34.6	31.6	
3	North	45.9	46.9	45	43.9	
4						

Once you have opened Word's graphic's datasheet and template, you can customize the graphic further as follows:

3. Choose the type of graphic you want by clicking on **Chart** on the Menu bar and scrolling to and selecting **Chart Type**. The **Chart Type** dialog box displays, allowing you to select a chart type.

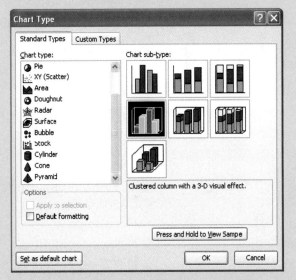

4. To add figure numbers, figure titles, legends, gridlines, data labels, and data tables, click on **Chart** and scroll to and select **Chart Options**.

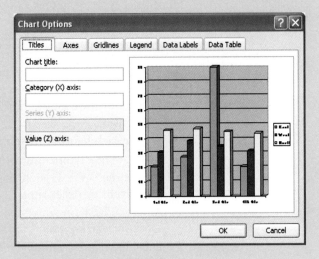

Figures

Another way to enhance your technical communication is to use figures. Whereas tables eliminate needless repetition of words, figures highlight and supplement important points in your writing. Like tables, figures help you communicate with your reader. Types of figures include the following:

- Bar charts
  - Grouped bar charts
  - 3-D (tower) bar charts
  - Pictographs
  - Gantt charts
  - 3-D topographical charts
- Pie charts
- Line charts
  - Broken line charts
  - Curved line charts
- Combination charts
- Flowcharts
- Organizational charts
- Schematics
- Geologic maps
- Line drawings
  - Exploded views
  - Cutaway views
  - Super comic book look
  - Renderings
  - Virtual reality drawings
- CAD drawings
- Photographs
- Icons
- Internet downloadable graphics

All of these types of figures can be computer generated using an assortment of computer programs. The program you use depends on your preference and hardware.

Criteria for Effective Figures. To construct figures correctly, do the following:

1. Number figures in order of presentation (i.e., Figure 1, Figure 2, Figure 3).
2. Title each figure. When you refer to the figure, use its number rather than its title: "Figure 1 shows the relation between the average price for houses and the actual sales prices."
3. Preface each figure with an introductory sentence.
4. Avoid using a figure until you have mentioned it in the text.
5. Present the figure as soon as possible after mentioning it instead of several paragraphs or pages later.
6. After you have presented the figure, explain its significance. Don't let the figure speak for itself. Remind the reader of the important facts you want to highlight.
7. Label the figure's components. For example, if you are using a bar or line chart, label the x- and y-axes clearly. If you're using line drawings, pie charts, or

photographs, use clear *callouts* (names or numbers that indicate particular parts) to label each component.

8. When necessary, provide a legend or key at the bottom of the figure to explain information. For example, a key in a bar or line chart will explain what each differently colored line or bar means. In line drawings and photographs, you can use numbered callouts in place of names. If you do so, you will need a legend at the bottom of the figure explaining what each number means.

9. If you abbreviate any labels, define these in a footnote. Place an asterisk (*) or a superscript number (^{1}, ^{2}, ^{3}) after the term and then at the bottom of the figure where you explain your terminology.

10. If you have drawn information from another source, note this at the bottom of the figure.

11. Frame the figure. Center it between the left and right margins or place it in a text box.

12. Size figures appropriately. Don't make them too small or too large.

Bar Charts   Bar charts show either vertical bars (as in Figure 9.2) or horizontal bars (as in Figure 9.3). These bars are scaled to reveal quantities and comparative values. You can shade,

FIGURE 9.2 Vertical Bar Chart

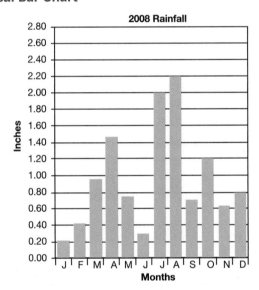

FIGURE 9.3 Horizontal Bar Chart for High-tech Readers

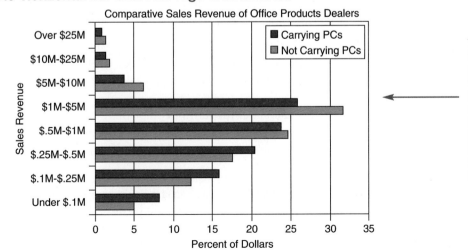

This bar chart, geared toward office supply sales managers, is as factual as the pictograph in Figure 9.4. However, Figure 9.3, which omits the drawings of the PCs, keeps the graphic more businesslike for the intended audience of high-tech peers.

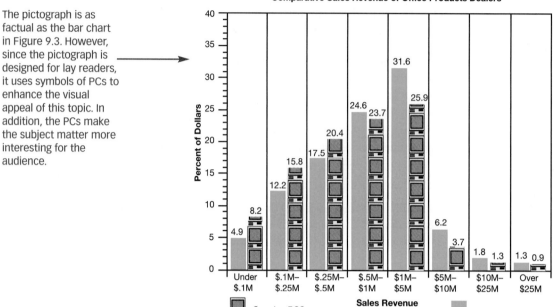

FIGURE 9.4 Pictograph for a Lay Audience

The pictograph is as factual as the bar chart in Figure 9.3. However, since the pictograph is designed for lay readers, it uses symbols of PCs to enhance the visual appeal of this topic. In addition, the PCs make the subject matter more interesting for the audience.

Comparative Sales Revenue of Office Products Dealers

Carrying PCS

Not Carrying PCS

color, or crosshatch the bars to emphasize the contrasts. If you do so, include a key explaining what each bar represents, as in Figure 9.3. *Pictographs* (as in Figure 9.4) use picture symbols instead of bars to show quantities. To create effective pictographs, do the following:

1. The picture should be representative of the topic discussed.
2. Each symbol equals a unit of measurement. The size of the units depends on your value selection as noted in the key or on the *x*- and *y*-axes.
3. Use more symbols of the same size to indicate a higher quantity; do not use larger symbols.

Gantt Charts   *Gantt charts,* or *schedule charts* (as in Figure 9.5), use bars to show chronological activities. For example, your goal might be to show a client phases of a project. This could include planned start dates, planned reporting milestones, planned completion dates, actual progress made toward completing the project, and work remaining. Gantt charts are an excellent way to represent these activities visually. They are often included in proposals to project schedules or in reports to show work completed. To create successful Gantt charts, do the following:

1. Label your *x*- and *y*-axes. For example, the *y*-axis represents the various activities scheduled, then the *x*-axis represents time (either days, weeks, months, or years).
2. Provide gridlines (either horizontal or vertical) to help your readers pinpoint the time accurately.
3. Label your bars with exact dates for start or completion.
4. Quantify the percentages of work accomplished and work remaining.
5. Provide a legend or key to differentiate between planned activities and actual progress.

3-D Topographical Charts   Three-dimensional contour representations are not limited to land elevations. A three-dimensional surface chart could be used to represent many different forms of data. These 3-D "topos" (as shown in Figure 9.6) are used in industries as varied as aerospace, defense, education, research, oil, gas, and water. Applications include CAD/CAM, statistical analysis, and architectural design.

FIGURE 9.5 Gantt Chart

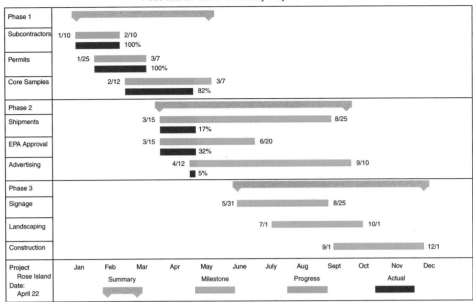

FIGURE 9.6 3-D Topographical Chart

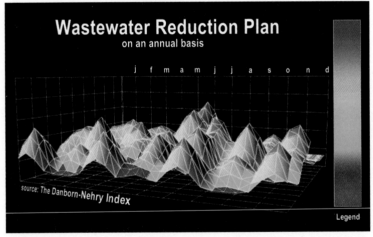

Courtesy of Brandon Henry

Pie Charts Use pie charts (as in Figure 9.7) to illustrate portions of a whole. The pie chart represents information as pie-shaped parts of a circle. The entire circle equals 100 percent or 360 degrees. The pie pieces (the wedges) show the various divisions of the whole.

To create effective pie charts, do the following:

1. Be sure that the complete circle equals 100 percent or 360 degrees.
2. Begin spacing wedges at the 12 o'clock position.
3. Use shading, color, or crosshatching to emphasize wedge distributions.
4. Use horizontal writing to label wedges.
5. If you don't have enough room for a label within each wedge, provide a key defining what each shade, color, or crosshatching symbolizes.
6. Provide percentages for wedges when possible.
7. Do not use too many wedges—this would crowd the chart and confuse readers.
8. Make sure that different sizes of wedges are fairly large and dramatic.

FIGURE 9.7 Pie Chart

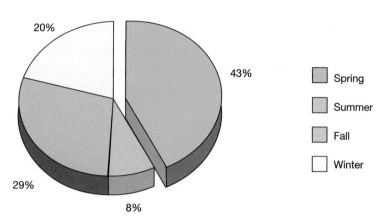

Rainfall Amounts for 2008

20%

43%

29%

8%

■ Spring

■ Summer

■ Fall

□ Winter

FIGURE 9.8 3-D Line Chart

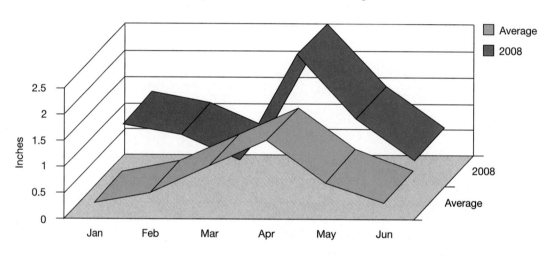

Comparison of 2008 Rainfall to Average

■ Average

■ 2008

Line Charts.   Line charts reveal relationships between sets of figures. To make a line chart, plot sets of numbers and connect the sets with lines. These lines create a picture showing the upward and downward movement of quantities. Line charts of more than one line (see Figure 9.8) are useful in showing comparisons between two sets of values. However, avoid creating line charts with too many lines, which will confuse your readers.

Combination Charts.   A combination chart reveals relationships between two sets of figures. To do so, it uses a combination of figure styles, such as a bar chart, line chart, and table (as shown in Figure 9.9). The value of a combination chart is that it adds interest and distinguishes the two sets of figures by depicting them differently.

Flowcharts.   You can show chronological sequence of activities using a flowchart. Flowcharts are especially useful for writing technical instructions. When using a flowchart, remember that ovals represent starts and stops, rectangles represent steps, and diamonds equal decisions (see Figure 9.10).

Organizational Charts.   The chart in Figure 9.11 shows the chain of command in an organization. You can use boxes around the information or use white space to distinguish among levels in the chart. An organizational chart helps your readers see where individuals work within a business and their relation to other workers.

Instructions and Flowcharts

See Chapter 12 for more discussion of instructions and flowcharts.

FIGURE 9.9 Combination Chart (Bar, Line, Table)

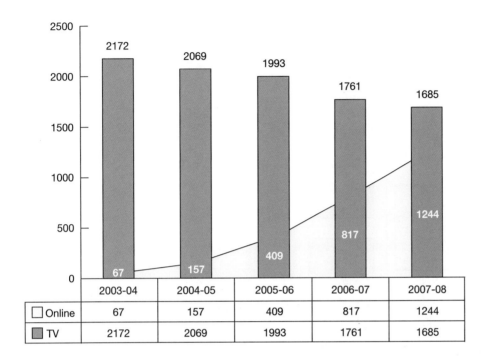

**The Relative Growth of Distance Learning
Class Enrollment, 2003–2008**

	2003-04	2004-05	2005-06	2006-07	2007-08
☐ Online	67	157	409	817	1244
◼ TV	2172	2069	1993	1761	1685

FIGURE 9.10 Flowchart

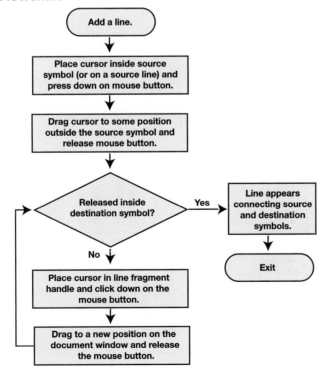

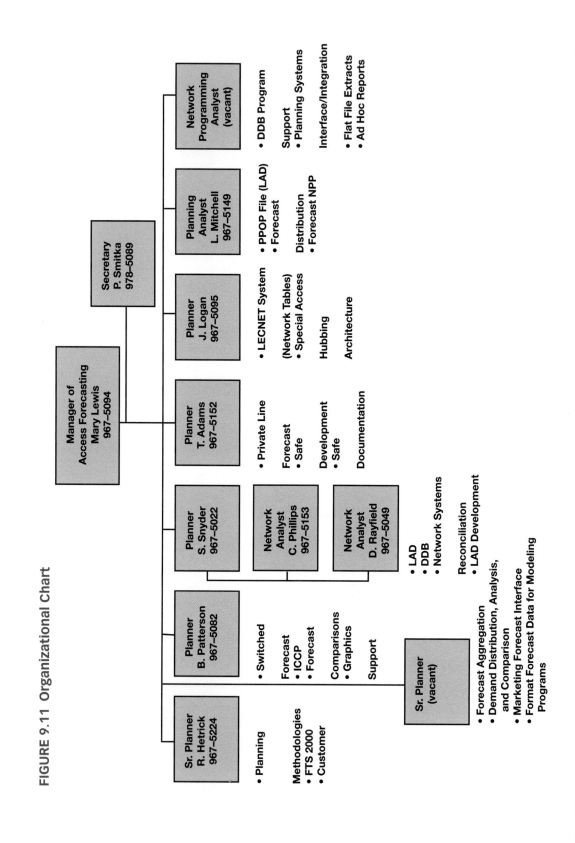

FIGURE 9.11 Organizational Chart

Schematics. Schematics are useful for presenting abstract information in technical fields such as electronics and engineering. A schematic diagrams the relationships among the parts of something such as an electrical circuit. The diagram uses symbols and abbreviations familiar to highly technical readers. The schematic in Figure 9.12 shows various electronic parts (resistors, diodes, condensors) in a radio.

Geologic Maps. Maps of any sort help us understand locations. Usually these show cities, streets, roads, highways, rivers, lakes, mountains, and so forth. Geologic maps do more. They also show terrain, contours, heat ranges, the surface features of a place or object, or an analysis of an area. Often, to help readers orient themselves, these maps are printed on top of a regular map (called a base map). The base map is black and white. The geologic map, in contrast, uses colors, contact and fold lines, and special symbols to reveal the geology of an area. These features are then defined on a map legend or key (as in Figure 9.13).

Line Drawings. Use line drawings to show the important parts of a mechanism or to enhance your text cosmetically. To create line drawings, do the following:

1. Maintain correct proportions in relation to each part of the object drawn.
2. If a sequence of drawings illustrates steps in a process, place the drawings in left-to-right or top-to-bottom order.

FIGURE 9.12 Schematic of a Radio

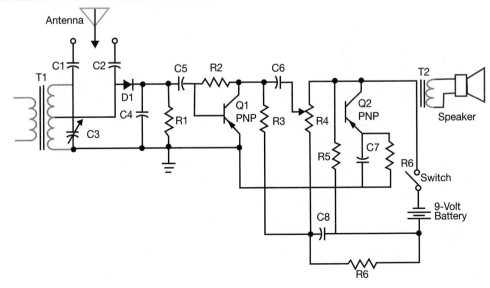

3. Using callouts to name parts, label the components of the object drawn (see Figure 9.14).

4. If there are numerous components, use a letter or number to refer to each part. Then reference this letter or number in a key (see Figure 9.15).

5. Use exploded views (Figure 9.15) or cutaways (Figures 9.16, 9.17, and 9.18) to highlight a particular part of the drawing.

FIGURE 9.13 Geologic Map

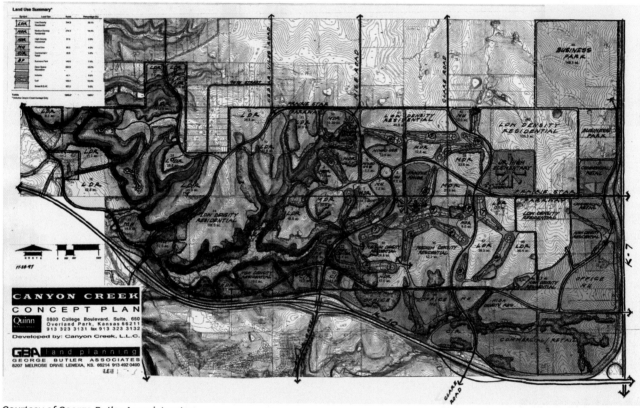

Courtesy of George Butler Associates, Inc.

Renderings and Virtual Reality Drawings. Two different types of line drawings are renderings and virtual reality views. Both offer 3-D representations of buildings, sites, or objects. Often used in the architectural/engineering industry, these 3-D drawings (as shown in Figures 9.19 and 9.20) help clients get a visual idea of what services your company can provide. Renderings and virtual reality drawings add lighting, materials, and shadow and reflection mapping to mimic the real world and allow customers to see what a building or site will look like in a photorealistic setting.

FIGURE 9.14 Line Drawing of Ventilator (Exploded View with Call-outs)

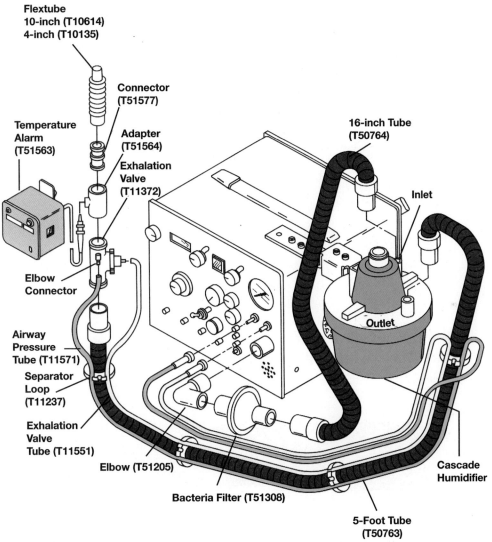

Courtesy of Nellcor Puritan Bennett Corp.

FIGURE 9.15 Line Drawing of Exhalation Valve (Exploded View with Key)

Item	Part Number	Description
\multicolumn{3}{c}{Exhalation Valve Parts List}		
1	000723	Nut
2	003248	Cap
3	T50924	Diaphragm
4	Reference	Valve Body
5	Reference	Elbow Connector
—	T11372	Exhalation Valve

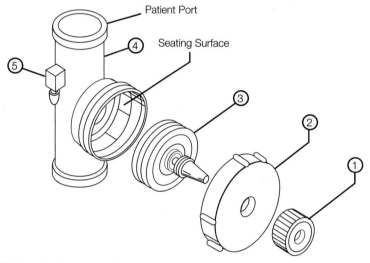

Courtesy of Nellcor Puritan Bennett Corp.

FIGURE 9.16 Line Drawing of Cable (Cutaway View)

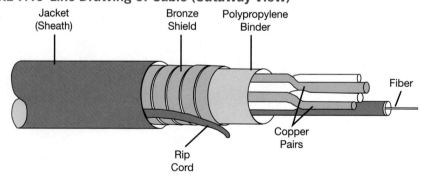

Courtesy of Nellcor Puritan Bennett Corp.

FIGURE 9.17 Comparison of Planets with Cutaway View

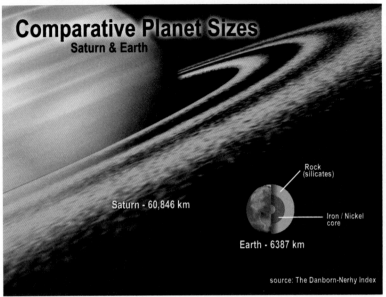

Courtesy of Brandon Henry

FIGURE 9.18 Cutaway View of a Railcar Braking System

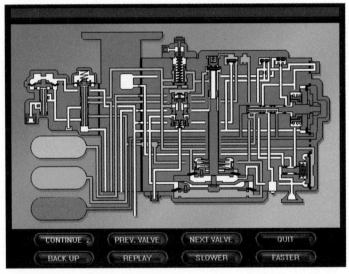

Courtesy of Burlington Northern Santa Fe Railroad

CAD Drawings.   Computer-aided design (CAD) drawings, such as the floor plan in Figure 9.21, use geometric shapes and symbols to provide a customer with a graphic view of a setting drawn to a particular scale. CAD drawings include *notations* to define scale and a *title block*. The title block gives the date of completion, name of the draftsperson, company, client, and project name.

Photographs.   A photograph can illustrate your text effectively. Like a line drawing, a photograph can show the components of a mechanism. If you use a photo for this purpose, you will need to label (name), number, or letter parts and provide a key. Photographs are

FIGURE 9.19 Architectural Rendering

Courtesy of George Butler Associates, Inc.

FIGURE 9.20 3-D Drawing

Courtesy of Johnson County Community College

excellent visual aids because they emphasize all parts equally. Their primary advantage is that they show something as it truly is. Photographs have one disadvantage, however. They are difficult to reproduce. Whereas line drawings photocopy well, photographs do not. See Figure 9.22.

Icons.   Approximately 23 percent of America's population is functionally illiterate. In today's global economy, consumers speak diverse languages. Given these two facts, how can technical writers communicate to people who cannot read and to people who speak different languages? Icons offer one solution. Icons (as in Figures 9.23 to 9.26) are visual representations of a capability, a danger, a direction, an acceptable behavior, or an unacceptable behavior.

FIGURE 9.21 Computer-Aided Design Drawing

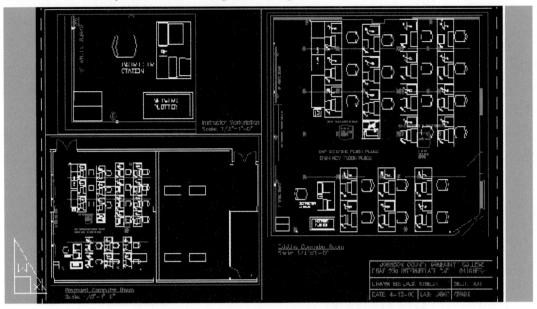

FIGURE 9.22 Photograph of Mechanical Piping

Courtesy of George Butler Associates, Inc.

FIGURE 9.23 An Icon of Explosives

FIGURE 9.24 An Icon of Dangerous Machinery

FIGURE 9.25 An Icon of Electric Shock

FIGURE 9.26 An Icon of Corrosive Material

For example, the computer industry uses icons—open manila folders to represent computer files. In manuals, a jagged lightning stroke iconically represents the danger of electrocution. On streets, an arrow represents the direction we should travel; on computers, the arrow shows us which direction to scroll. Universally depicted stick figures of men and women greet us on restroom doors to show us which rooms we can enter and which rooms we must avoid.

When used correctly, icons can save space, communicate rapidly, and help readers with language problems understand the writer's intent. To create effective icons, follow these suggestions.

1. **Keep it simple**—You should try to communicate a single idea. Icons are not appropriate for long discourse.
2. **Create a realistic image**—This could be accomplished by representing the idea as a photograph, drawing, caricature, outline, or silhouette.
3. **Make the image recognizable**—A top view of a telephone or computer terminal is confusing. A side view of a playing card is completely unrecognizable. Select the view of the object that best communicates your intent.
4. **Avoid cultural and gender stereotyping**—For example, if you are drawing a hand, you should avoid showing any skin color, and you should stylize the hand so it is neither clearly male nor female.
5. **Strive for universality**—Stick figures of men and women are recognizable worldwide. In contrast, letters—such as *P* for *parking*—will mean very little in China, Africa, or Europe. Even colors can cause trouble. In North America, red represents danger, but red is a joyous color in China. Yellow calls for caution in North America, but this color equals happiness and prosperity in the Arab culture (Horton 1993, 682–93).

Internet Downloadable Graphics

Website Design

See Chapter 13 for more discussion of Web site design.

More and more, you will be writing online as the Internet, intranets, and extranets become prominent in technical communication. You can create graphics for your Web site in three ways.

Download Existing Online Graphics.   The Internet contains thousands of Web sites that provide online clip art. These graphics include photographs, line drawings, cartoons, icons, animated images, arrows, buttons, horizontal lines, balls, letters, bullets, and hazard signs. In fact, you can download any image from any Web site. Many of these images are freeware, which you can download without cost and without infringing on copyright laws.

To download these images, just place your cursor on the graphic you want, then right-click on the mouse. A pop-up menu will appear. Click on Save Picture As. Once you have done this, a new menu will appear. You can save your image in the file of your choice, either on the hard drive or on your disk. The images from the Internet will already be GIF (graphics interchange format) or JPEG (joint photographic experts group) files. Thus, you will not have to convert them for use in your Web site.

Modify and Customize Existing Online Graphics.   If you plan to use an existing online graphic as your company's logo, for example, you will need to modify or customize the graphic. You will want to do this for at least two reasons: to avoid infringing on copyright laws and to make the graphic uniquely yours.

To modify and customize graphics, you can download them in two ways. First, you can print the screen by pressing the Print Screen key (usually found on the upper right of your keyboard). This captures the entire screen image in a clipboard. Then you can open a graphics program and paste the captured image. Second, you can save the image in a file (as discussed) and then open the graphic in a graphics package. Most graphics programs will allow you to customize a graphic. Popular programs include Paint, Paint Shop Pro, PhotoShop, Corel Draw, Adobe Illustrator, Freehand, and Lview Pro. In these graphics programs, you can manipulate the images by changing colors, adding text, reversing the images, cropping, resizing, redimensioning, rotating, retouching, deleting or erasing parts of the images, overlaying multiple images, joining multiple images, and so forth. After you make substantial changes, the new image becomes your property.

You could also take any existing graphic from hard-copy text (magazines, journals, books, brochures, user manuals, reports, etc.), scan the image, crop and retouch it, save it, and then reopen this saved file in one of the graphics programs for further manipulation. Some graphics programs, such as Paint, save an image only as a BMP, a bitmap image. Once the image has been altered, you will need to convert your bitmap file to a GIF or JPEG format for use in your Web site. Doing so is important because the Internet will not read BMP images.

Create New Graphics.   A final option is to create your own graphic. If you are artistic, draw your graphic in a graphics program, save the image as a GIF or JPEG file, and then load the image into your Web site. This option might be more challenging and time consuming. However, creating your own graphic gives you more control over the finished product, provides a graphic precisely suited to your company's needs, and helps avoid infringement of copyright laws.

VISUAL AID CHECKLIST

_____ 1. Will a visual aid add to your technical communication and make the document concise, clear, and add cosmetic appeal?

_____ 2. Should you use color or three-dimensional graphics in your visual aids?

_____ 3. Have you considered the following criteria when you created your tables or figures?

    a. Are visuals integrated with the text?

    b. Do the visuals add to the text and enhance it without being redundant?

    c. Do they communicate information visually that could not easily be conveyed in text?

    d. Are the visuals the correct size, labeled, readable, and similar in style?

_____ 4. Did you use a table when you presented factual information, such as numbers, percentages, or monetary amounts?

_____ 5. Have you used a figure to highlight and supplement important points in your writing?

CHAPTER HIGHLIGHTS

1. Using graphics can allow you to create a more concise document.
2. Graphics add variety to your text, breaking up wall-to-wall words.
3. Color and 3-D graphics can be effective. However, these two design elements also can cause problems. Your color choices might not be reproduced exactly as you planned, and a 3-D graphic could be misleading rather than informative.
4. Tables are effective for presenting numbers, dates, and columns of figures.
5. You can often communicate more easily with your audience when you use figures, such as bar charts, pie charts, line charts, flowcharts, and organizational charts, to highlight and supplement important parts of your text.

APPLY YOUR KNOWLEDGE

CASE STUDIES

1. Bert Lang is an investment banker at Country Commercial Bank, as noted in this chapter's beginning scenario. He is writing a proposal to a potential client, Sylvia Light, a retired public health nurse. She now has $315,500 in savings, allocated as follows: $78,000 in an individual retirement account (IRA), $234,000 in a low-earning certificate of deposit (CD), and $3,500 in her checking account. Sylvia contacted Bert, asking him to help her organize her portfolio for a comfortable retirement. Bert has studied Sylvia's various accounts and considered her lifestyle and expenditures. Now, he is ready to write the proposal. Bert plans to propose that Sylvia could reallocate her funds as follows:

   - $110,000 in an annuity
   - $78,000 in an individual retirement account (IRA)
   - $45,000 in municipal bonds
   - $37,000 in a stock fund
   - $42,000 in certificates of deposit (CDs)
   - $3,500 in a checking account

   To make this proposal visually appealing and more readily understandable to Sylvia, Bert will use visual aids.

Assignment

   - Create a table to show how he wants to invest her money.
   - Create a bar chart comparing her current allocations versus his proposed allocations.

2. Jim Goodwin owns an insurance agency, Goodwin and Associates Insurance (GAI). Letters are a major part of his technical communication with vendors, clients, and his insurance company's home office. However, many of Jim's employees fail to recognize the importance of business communication. To prove how important their communication is to the company, Jim plans to summarize the amount of writing they do on the job. He has determined that they write the following each week.

   - **57 inquiry letters**—Insurance coverage changes constantly. To clarify these changes for customers, employees write inquiries, asking home office questions about new insurance laws, levels of coverage, coverage options, and rate changes.
   - **253 response letters**—A key to GAI's success is acquiring new customers. When potential customers call, e-mail, or write letters asking for insurance quotes, Jim and his employees write response letters.

- **43 cover (transmittal) letters**—Once customers purchase insurance policies, Jim and his coworkers mail these policies, prefaced by cover letters. The transmittal letters clarify key points within the documents.
- **12 good news letters**—Jim's clients receive discounts when they add new cars to an existing policy or include both car and home coverage. GAI writes good news letters to convey this information.
- **25 adjustment letters**—When accidents occur or when losses take place, GAI writes adjustment letters stating that the claim is covered.
- **2 bad news letters**—Like any business, GAI needs to communicate bad news to clients or vendors.
- **7,000 e-mail messages**—Jim and his 20 employees send and receive a minimum of 50 e-mail messages each day.

Assignment

To show his employees the importance of technical communication, Jim wants to create a visual aid depicting this data. Create the appropriate graphic for GAI's employees.

INDIVIDUAL AND TEAM PROJECTS

1. Present the following information in a pie chart, bar chart, and table.

    In 2007, the Interstate Telephone Company bought and installed 100,000 relays. It used these for long-range testing programs that assessed failure rates. It purchased 40,000 Nestor 221s; 20,000 VanCourt 1200s; 20,000 Macro R40s; 10,000 Camrose Series 8s; and 10,000 Hardy SP6s.

2. Using the information presented in activity 1 and the following revised data, show the comparison between 2007 and 2008 purchases through two pie charts, a grouped bar chart, and a table.

    In 2008, after assessing the success and failure of the relays, the Interstate Telephone Company made new purchases of 200,000 relays. It bought 90,000 VanCourt 1200s; 50,000 Macro R40s; 30,000 Camrose Series 8s; and 30,000 Hardy SP6s. No Nestors were purchased.

3. Create a line chart. To do so, select any topic you like. The subject matter, however, must include varying values. For example, present a line chart of your grades in one class, your salary increases (or decreases) at work, the week's temperature ranges, your weight gain or loss throughout the year, the miles you've run during the week or month, amounts of money you've spent on junk food during the week, and so forth.

4. Analyze the following graphics and explain which ones succeed and which ones fail.

    a. Vertical bar chart

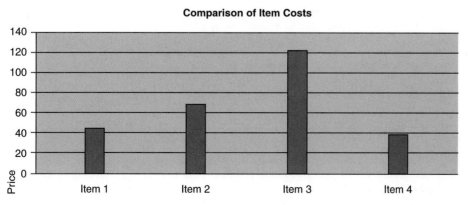

b. Pie chart

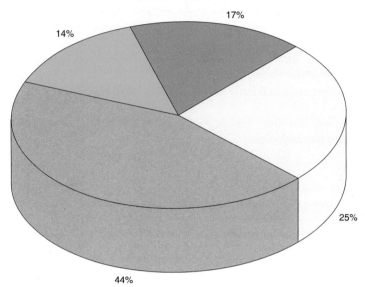

Comparison of Item Costs

17%

14%

25%

44%

c. Horizontal bar chart

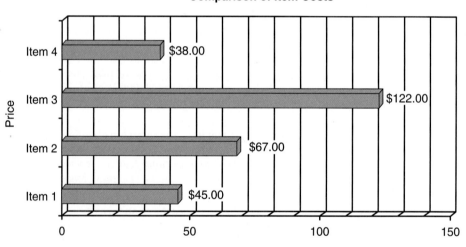

Comparison of Item Costs

Item 4 $38.00

Item 3 $122.00

Item 2 $67.00

Item 1 $45.00

Price

0   50   100   150

d. Table (Table 9.3)

TABLE 9.3   Comparison of Item Costs

	Price	Percentage
Item 1	$ 45	17%
Item 2	$ 67	25%
Item 3	$122	44%
Item 4	$ 38	14%
Total	$272	100%

5. Based on the criteria provided in this chapter, revise any of the graphics from assignment 4. Add any additional information necessary to improve the graphics.

PROBLEM-SOLVING THINK PIECES

1. Angel Guerrero, computer information systems technologist at HeartHome Insurance, was responsible for making an inventory of his company's hardware. He learned the following: The company had 75 laptops, 159 PCs, 27 printers, 10 scanners, 59 handheld computers (PDAs), 238 cell phones with e-mail capabilities, and 46 digital cameras.

   To write his inventory report, Angel needs to chart the above data. Which type of visual aid should Angel use? Explain your answer, based on the information provided in this chapter. Create the appropriate visual aid.

2. Minh Tran works in the marketing department at Thrill-a-Minute Entertainment Theme Park (TET). Minh and her project team need to study entry prices, ride prices, food and beverage prices, and attendance of their park versus their primary competitor, Carnival Towne (CT).

   Minh and her team have found that TET charges $16.50 admission, while CT charges $24.95. Most of TET's rides are included in the entry price, but special rides (the Horror, the Bomber, the Avenger, and Peter Pan's Train) cost $2.50. At CT, the entry fee covers many rides, excluding Alice's Teacup, Top-of-the-World Ferris Wheel, and the Zinger, which cost $2.00 each. Food and beverages at TET cost $1.95 for a hot dog, $2.50 for a burger, and $3.95 for nachos, and $1.50 to $2.50 for drinks. At CT, food and beverages cost $1.75 for hot dogs, $2.75 for burgers, $2.50 for nachos, and $1.50 to $2.50 for drinks. Attendance at TET last year was 250,000, while attendance at CT was 272,000.

   What type of visual aid should Minh and her team use to convey this information? Explain your decision, based on the criteria for graphics provided in this chapter. Create the appropriate visual aid.

3. Toby Hebert is human resource manager at Crab Bayou Industries (Crab Bayou, Louisiana), the world's largest wholesaler of frozen Cajun food. Toby and her management are concerned about the company's hiring trends. A prospective employee complained about discriminatory hiring practices at Crab Bayou.

   To prove that the company has not practiced discriminatory hiring practices, Toby has studied the last 10 years' hires by age. She found that in 1997, the average age per employee was 48; in 1998, the average age was 51; in 1999, the average age was 47; in 2000, the average age was 52; in 2001, the average age was 45; in 2002, the average age per employee was 47; in 2003, the average age was 42; in 2004, the average age was 39; in 2005, the average age rose to 42; in 2006 and 2007, the average age fell to 29 and 30, respectively (due to a large number of early retirements).

   What type of visual aid should Toby use to convey this information? Explain your decision, based on the criteria for graphics provided in this chapter. Create the appropriate visual aid.

4. Yasser El-Akiba is a member of his college's International Students Club. Yasser is a native of Israel. Other members of the club are from other countries: three from Australia, two from Ecuador, eight from Mexico, five from Africa, two from England, three from Canada, four from the Dominican Republic, and nine from China.

   What kind of visual aid could Yasser create to show his club members' homelands? Defend your decision based on criteria for graphics in this chapter. Create the appropriate visual aid.

WEB WORKSHOP

Using an Internet search engine, type in phrases such as "automobile sales+line graph," "population distribution by age+pie chart," or "California+organizational chart." Create similar phrases for bar charts, pictographs, flowcharts, tables, and so forth. Open several links from your Web search and study the examples you have found. Which examples of graphics are successful and why? Which examples of graphics are unsuccessful and why? Explain your reasoning, based on this chapter's criteria.

QUIZ QUESTIONS

1. How can you effectively present large blocks of data or reveal comparisons?
2. How do graphics allow you to be concise?
3. What are four things graphics help readers to see?
4. What can happen when you use only unbroken text in a document?
5. When should you avoid using color in a graphic?
6. What are drawbacks to 3-D graphics?
7. When you use graphics, what are five criteria to follow to achieve successful graphics?
8. What are two things you achieve by using a table?
9. After a table, what should you provide in your text?
10. In a table, should you use decimals or numerators and denominators for fractions?
11. Figures accomplish what two things in a document?
12. What are five types of figures?
13. What are two types of bar charts?
14. What is a pictograph?
15. What are three ways to achieve an effective pictograph?
16. When would you use a Gantt chart?
17. What do pie charts illustrate?
18. What type of sequence can you reveal in a flowchart?
19. How are schematics useful?
20. What are two reasons for using icons?

CHAPTER 10

Communicating to Persuade

In this scenario, TechToolshop relies on persuasive writing to expand business.

TechToolshop provides sales, service, maintenance, installation, and data recovery for PC and Mac hardware and software tools. Located in Big Springs, Iowa, this business of 1,200 employees began in 2002. Now, however, the company wants to expand its business.

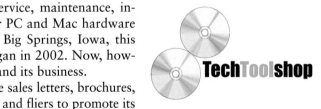

TechToolshop will use sales letters, brochures, and fliers to promote its expansion plans. To inform its current customers about new service offerings, the company will write a sales letter. To attract new customers to the store, TechToolshop is producing three, full-color brochures that focus on the company's primary functions: printers, desktops, mainframes, mini-towers, software, accessories, and encryption devices.

In contrast to the three brochures, the company will produce many fliers throughout the year. The full-color brochures are costly to write and publish, but the single-sided, postcard-sized fliers are more cost effective. These fliers, which can be produced quickly, focus on individual topics where quick turnaround is

Objectives

When you complete this chapter, you will be able to

1. Understand the importance of argument and persuasion in technical communication.

2. Recognize the traditional methods of argumentation.

3. Use the ARGU technique to organize persuasive technical communication.

4. Avoid logical fallacies in persuasive communication.

5. Evaluate persuasive documents using the checklist.

important: what's on sale this month, recent computer industry news that affects business owners, innovative software, and technology updates.

As Director of Marketing Communications at TechToolshop, Amanda Carroll's job is to facilitate transmission of information between key internal and external stakeholders. Primarily she communicates information to the public about product offerings, company product-line changes, socially responsible products, and warranty changes. Internally she disseminates information about the products, customer service, and many other companywide issues affecting the 1,200 employees. Many of Amanda's written documents and oral presentations depend on persuasion. To be persuasive, she relies on the traditional methods of argument by appealing to the emotions, being logical, and maintaining ethical standards. She wants to communicate effectively to ensure that information contributes to "product branding" and, in turn, enhances the profitability of the company. This is good business for both management and employees.

One of Amanda's major goals is maintaining excellent customer relations. Hard work and client referrals obviously help contribute to this goal. However, Amanda has found that continuous communication is another way to accomplish client rapport. Though TechToolshop's business is increasing, it wants to keep adding customers. Outstanding technical communication through persuasive sales letters, brochures, and fliers equals more business for TechToolshop.

Check Online Resources

www.prenhall.com/gerson
For more information about persuasive communication, visit our companion website.

Check out our quarterly newsletters TechCom E-Notes at www.prenhall.com/gerson for dot.com updates, new case studies, insights from business professionals, grammar exercises, and facts about technical communication.

THE IMPORTANCE OF ARGUMENT AND PERSUASION IN TECHNICAL COMMUNICATION

Communication Goals

See Chapter 1 for more discussion of communication goals such as informing, instructing, and building rapport.

When communicating technical information, you will write and speak for many reasons. You might write a memo, letter, or e-mail message to *inform* about an upcoming meeting, a job opportunity, a new product release, or a facilities change. You might give an oral presentation to local businesses, government, or educational organizations in which you hope to *build rapport*. In writing a user manual, your goal will be to *instruct*. When you write a proposal, your goal is to *recommend* changes. If you write a technical specification, you will *analyze* components of a piece of equipment.

Informing, Analyzing, and Recommending

See Chapter 16 for more discussion of informing, analyzing, and recommending in long, formal reports.

In addition to informing, building rapport, instructing, recommending, and analyzing, you also will need to communicate persuasively. Let's say you are a customer who has purchased a faulty product. You might want to write a letter of complaint to the manufacturer of this product. To make your case strongly, you will need to convince your audience, clarifying how the product failed. If your argument is effective, then you will persuade the company to give you a refund or new product.

Professionally, you will need to use argument and persuasion daily. As a manager, you might need to argue the merits of a company policy to an unhappy customer. If your colleagues have decided that the department should pursue a course of action, you might need to persuade your boss to act accordingly. Maybe you are asking your boss for a raise or promotion, for office improvements, or for changes to the work schedule. Your task is to persuade the boss to accept your suggestions. In these instances, you will communicate persuasively using any of the following communication channels: memos, e-mail and instant messages, letters, reports, proposals, or oral presentations.

Figure 10.1 is a persuasive memo from a subordinate to a boss, documenting a problem and suggesting a course of action.

Frequently, you will write e-mail messages to supervisors and colleagues, persuading these readers to accept your point of view. Topics could range from requests for promotion, equipment needs, days off from work, assistance with projects, or financial assistance for job-related travel.

The "Before" and "After" e-mail messages seek to persuade a colleague to attend a work-related conference instead of the e-mail writer. The first "Before" sample (Figure 10.2) is not effectively persuasive. It reads more like a command and fails to consider the audience's reaction. The second "After" sample (Figure 10.3) is more persuasive.

In addition to memos, e-mail, or reports, you might write persuasively for the following reasons.

Routine Correspondence

See Chapter 6 for more discussion of routine correspondence.

- **Automotives:** You are starting a new business, geared toward automobile service and sales. Now, you need to communicate this new venture to the public. To do so, you will write a sales letter, persuading the public to visit your service center and purchase your automobile products.
- **Electrical/HVAC:** A new mall is being built in your community. Every store in the mall will require electrical work, including heating, ventilation, and air conditioning. To showcase your product, services, and credentials, you plan to send electronic fliers to potential clients.
- **Healthcare:** Your doctor's office or care center provides clients with brochures that market health services. Patients need easy-to-read literature to persuade them that they have chosen the correct physician, hospital, or laboratory.

FIGURE 10.1 Persuasive Memo

DATE: March 22, 2008
FROM: Bob Ward
TO: Lynn Richards
SUBJECT: MAINTENANCE OF PHOTO LAB MACHINE

On March 21, our one-hour photo employee Brian Syoung reported that the photo lab machine was malfunctioning. It was cropping photos out of specification and producing blurred images. A service technician examined the machine and estimated that repairs would cost $1,000. Just as a reminder, Lynn, in one week, we were scheduled to have this machine replaced with an upgrade.

The problem is whether to repair the machine or wait a week for the replacement. Here are some facts that I've collected to help us decide.

- The lab makes $850 in profits each week. If we spend $1,000 to repair the machine, we will lose money.
- Closing the lab will save the company around $200 in supplies and electricity.
- If the photo lab is closed, then our higher-paid photo specialists will have to be assigned other jobs. We will lose approximately $2.00 an hour paying them to do work usually assigned to lower-paid employees.
- Closing the photo lab for a week to save repair expenses will inconvenience our repeat customers. In fact, they might shop elsewhere, which would impact us negatively long term.

Lynn, though repairing the machine costs $1,000, we will more than make up for this cost in customer satisfaction. We should repair the machine for the sake of our customers. Since this is a difficult decision, let's meet later this afternoon at 3:00 p.m. in your office. We need to resolve this issue today before we lose more money.

The body objectively recognizes both sides of an issue and alternative approaches. The first two bullets focus on delaying the repair. The last two bullets explain why repair delays are bad for business.

The body develops the argument persuasively by providing facts and figures. In doing so, the writer reveals his knowledge of the subject matter.

The conclusion emphasizes persuasively the urgency of action.

BEFORE

FIGURE 10.2

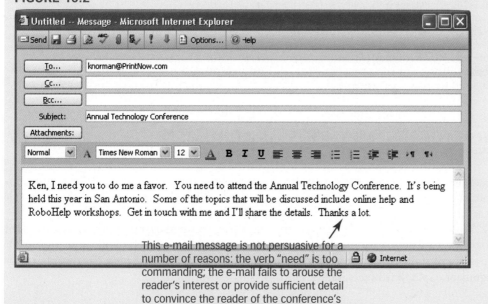

This e-mail message is not persuasive for a number of reasons: the verb "need" is too commanding; the e-mail fails to arouse the reader's interest or provide sufficient detail to convince the reader of the conference's worth; it also fails to urge the reader to action.

FIGURE 10.3

The introductory paragraph arouses reader interest with an anecdote and a question.

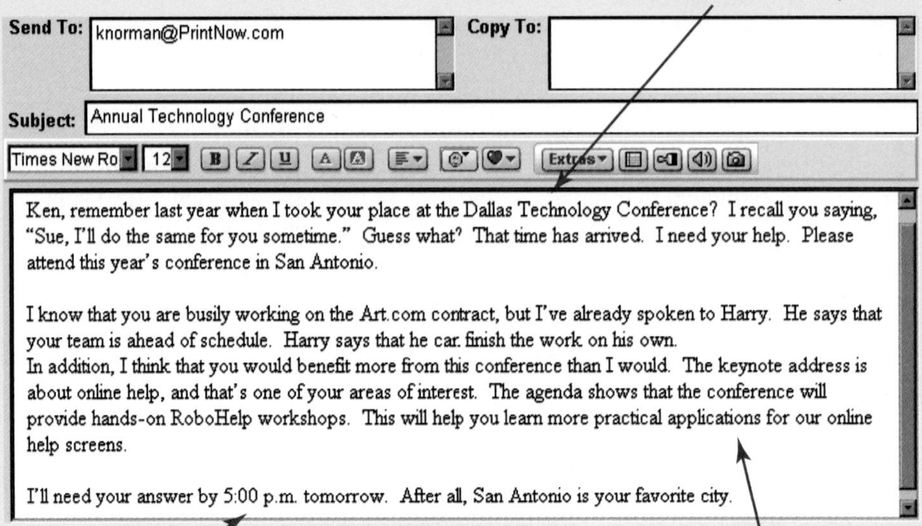

Send To: knorman@PrintNow.com   **Copy To:**

Subject: Annual Technology Conference

Times New Ro 12 **B** *I* U A A E▾ ☺▾ ♥▾ Extras▾ ▢ ☒ ◁) ◙

Ken, remember last year when I took your place at the Dallas Technology Conference? I recall you saying, "Sue, I'll do the same for you sometime." Guess what? That time has arrived. I need your help. Please attend this year's conference in San Antonio.

I know that you are busily working on the Art.com contract, but I've already spoken to Harry. He says that your team is ahead of schedule. Harry says that he can finish the work on his own.
In addition, I think that you would benefit more from this conference than I would. The keynote address is about online help, and that's one of your areas of interest. The agenda shows that the conference will provide hands-on RoboHelp workshops. This will help you learn more practical applications for our online help screens.

I'll need your answer by 5:00 p.m. tomorrow. After all, San Antonio is your favorite city.

The conclusion urges action by giving a due date and highlighting reader enjoyment.

The e-mail body begins by refuting any objections. Then, following sentences show audience benefit.

DOT-COM UPDATES

For more information about the importance of persuasive marketing in technical communication, check out the following links.

- de la Giroday, Maryse. "MarCom and the technical writer or why do people keep calling me a liar?" *STCWestCoast*. Nov/Dec. 2003. http://www.stcwestcoast. ca/04_news/articles/ novdec_200301.asp.
- Graham, Gordon. "Tech writers as sales reps?" 8 Mar. 2005. http://www. softwareceo.com/attachm ents/stc/com030805.php.
- "Marketing Collateral." *InfoPros*. 2005. http://www.infopros.com/ productsservices/ ps_marketing.html.
- "Technical writers turn to marketing to survive." *Ottawa Business Journal Staff*. 8 Sep. 2003. http://www. ottawabusinessjournal. com/284317458401581. php.

FAQs: How Does Marketing Fit in with Technical Communication?

A: Technical communication is more than hard, cold memos, letters, reports, user manuals, and technical descriptions. Technical communication also has a soft side—*marketing*. The bottom line is every company is in business to make money. Thus, all employees should perceive themselves as marketing personnel.

Examples: George Butler Associates (GBA), an architectural and engineering company in Kansas, Missouri, and Illinois, asks every one of its employees to take marketing classes on site. The goal is to help GBA employees work well with clients, vendors, and partners.

The Society for Technical Communication also recognizes the importance of marketing for technical writers. STC's Web site (stc.org) provides links to special interest groups. One such group is entitled, "Marketing Communication." In addition, two recent links for general news include articles about the importance of marketing for technical writers: "Tech writers as sales reps? Interface Software's award-winning docs boost brand, revenues, and customer satisfaction" and "Technical writers turn to marketing to survive." Persuasive marketing materials are essential to technical communication.

TRADITIONAL METHODS OF ARGUMENT AND PERSUASION

To argue a point persuasively, you can use any of the traditional methods of argumentation: *ethos* (ethical), *pathos* (emotional), and *logos* (logical) appeals. These three appeals to an audience are called "the rhetorical triangle," as shown in Figure 10.4.

This equilateral triangle suggests that each part of a persuasive appeal is as important as any other part. In addition, it emphasizes the need for balance of all three appeals or types of proof. Excess emotion, for example, might detract from the logical appeal of your argument. Cold, hard facts might fail to persuade your audience.

FIGURE 10.4 The Rhetorical Triangle

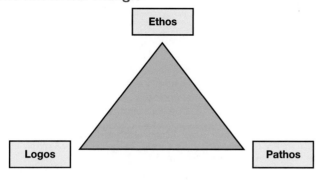

Ethical Argumentation

Arguments based on ethics (ethos) depend on personal character. If you make arguments based on personal experiences, then you must appear to be trustworthy and credible as a writer or speaker. To accomplish this goal, present information that is unbiased, reliable, and evenhanded.

Tiger Woods and Mia Hamm are classic examples of corporate spokespeople hired for their character. When Tiger promotes golf equipment, you trust him. He is a reliable voice for these products because of his successes on the links. The same holds true for Mia. When she speaks about soccer, you know she is an expert. In fact, Tiger and Mia are so trustworthy in the eyes of the public that they also can sell products outside of their primary areas of expertise, such as cars and healthcare products.

Former New York City Mayor Rudy Guiliani has written a book entitled, *Leadership*. Oprah Winfrey writes and speaks about health, education, entertainment, books, money, philanthropy, and more. When these individuals make arguments drawn from personal experiences, audiences respond favorably. Given their reputations for excellence and success, both are trustworthy and reliable experts.

Emotional Argumentation

Arguments based on emotion (pathos) seek to change an audience's attitudes and actions by focusing on feelings. If you want to move an audience emotionally, you would appeal to passion. You can do this either positively or negatively.

To sway an audience positively, you would focus on positive concepts like joy, hope, honor, pleasure, happiness, success, and achievement. You would use positive words to create an appealing message. In contrast, you also can appeal to emotions negatively. Fear, horror, anger, and unhappiness can be powerful tools in an argument.

Notice how the National Center for Environmental Health warns about the dangers of carbon monoxide: "Carbon monoxide is a silent killer. This colorless, odorless, poisonous gas kills nearly 500 U.S. residents each year, five times as many as West Nile virus" ("CDC and CPSC Warn of Winter Home Heating Hazards" 2007). To highlight the dangers associated with carbon monoxide, the Center for Disease Control uses emotional words such as *killer* and *poisonous*, and compares this problem to the frightening "West Nile virus."

Logical Argumentation

Argumentation based on logic (logos) depends on rationality, reason, and proof. You can persuade people logically when you provide them the following:

- Facts—statistics, evidence, data, and research
- Testimony—citing customer or colleague comments, expert authorities, and results of interviews

FIGURE 10.5 E-mail Message Effectively Using Ethos, Pathos, and Logos

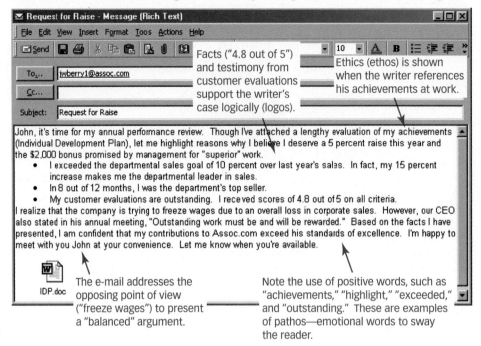

- Examples—anecdotes, instances, and personal experiences
- Strong, clear claims—including warrantees and guarantees
- Acknowledgment of the opposing points of view to ensure that information is "balanced"

Figure 10.5 is an e-mail message that argues a case for a raise by focusing on **ethics.** The writer refers to his work-related achievements to support his credibility. He uses **logic**—facts and testimony. Also note how the e-mail factors in an opposing point of view. In addition, the e-mail uses **emotion** to sway the reader through positive words.

ARGU TO ORGANIZE YOUR PERSUASION

Effective persuasive communication entails ethical, logical, and emotional appeals. Understanding the importance of this rhetorical triangle is only part of your challenge as a persuasive communicator. The next step is deciding how best to present your argument. Using our *ARGU* approach will help you organize your argument, as follows:

- *A*rouse audience involvement—grab the audience's attention in the introduction of your communication.
- *R*efute opposing points of view in the body of your communication.
- *G*ive proof to develop your thoughts in the body of your communication.
- *U*rge action—motivate your audience in the conclusion.

Arouse Audience Involvement

You have only about five to eight seconds to grab your readers' attention in a sales letter, persuasive e-mail message, marketing brochure, speech, or any persuasive communication. You must arouse the audience's interest imaginatively in the first few sentences of your document or oral presentation. Try any of the following attention grabbers in the **introduction** of your persuasive message.

- **Use an anecdote—a brief, dramatic story relating to the topic.** Stories engage your audience. You can involve your readers or listeners by recounting an interesting story to which they can relate. The story should be specific in time, place, person, and action. The drama should highlight an event. However, this must be a short scenario, since your time is limited. If you do not capture their interest quickly, the audience might lose interest.

E-mail Message Persuading a Manager to Take Action

Sam, last week, a customer fell down in our parking lot, cutting her knee, tearing her slacks, and requiring medical attention. We can't let this happen again. Please consider hiring a maintenance crew to salt and sand our lot during icy weather.

example

This anecdote is specific in time, place, person, and action.

- **Start with a question to interest your audience.** By asking a question, you involve your audience. Questions imply the need for an answer. A question can make readers or listeners ask themselves, "How would I answer that?" This encourages their participation.

Sales Letter from a Financial Planner

"Where will I get money for my kid's college education?" "How can I afford to retire?" "Will my insurance cover all medical bills?" You have asked yourself these questions. Our estate planning video has the answers.

example

By asking questions, you involve your audience.

- **Begin with a quotation to give your communication the credibility of authority.** By quoting specialists in a field or famous people, you enhance your credibility and support your assertions. A quote from Warren Buffett, world-renowned investor and businessman, about economics, for example, gives credence to a financial topic. Similar quotes from Bill Gates, founder of Microsoft, about the computer industry are trustworthy.

An Oral Presentation about Stock Performance to Shareholders

As Warren Buffett says, "Our favorite holding period is forever." Though our stock prices are down now, don't panic and sell. The company will rebound.

example

Use a quote from an authority to lend credence to your argument.

- **Let facts and figures enhance your credibility.** Factual information (percentages, sales figures, amounts, or dates) catches the audiences' attention and lays the groundwork for your persuasion.

Sales Letter about Computer Technology

Eighty-seven percent of all college students own a computer. Don't go off to college unprepared. Buy your laptop or PC at *CompuRam* today.

example

Facts and figures help persuade your reader to your point of view.

- **Appeal to the senses.** You can involve your audience by letting them hear, taste, smell, feel, or visualize a product.

Marketing Flier Advertising a New Restaurant

Hickory-smoked goodness and fire-flamed Grade A beef—let Roscoe's BBQ bring the taste of the South to your neighborhood.

- **Use comparison/contrast to highlight your message.** Comparison/contrast lets you make your point more persuasively. For example, you might want to show a client that your computer software is superior to the competition. You might want to show a boss that following one plan for corporate restructuring is better than an alternative plan. You could justify why an employee has not received a raise by comparing/contrasting his or her performance to the departmental norm. Comparison/contrast is useful in many different arguments for a variety of audiences.

E-mail from Management to Subordinates Proposing Changes in Procedures

Last year our department fell short of corporate sales goals. This year we must surpass expectations. To do so, I propose a seven-step procedure.

- **Begin with poetic devices.** Advertising has long used alliteration (the repetition of sounds), similes (comparisons), and metaphors to create audience interest. Poetic devices are memorable, clever, fun, and catchy.

Sales Letter from a Bank

Looking for a **low-loan lease?** Let USB Bank take care of your car leases. Our 2 percent loans **beat the best.**

- **Create a feeling of comfort, ease, or well-being.** To welcome your audience, making them feel calm and peaceful, invoke nostalgia or good times.

Slogan from a Mortgage Company Brochure

Come home again to Countryside Mortgage. We treat our customers like neighbors.

- **Create a feeling of discomfort, fear, or anxiety.** Another way to involve your audience is through stress. Beginning communication by highlighting a problem allows you to persuade the audience that you offer the solution.

Sales Letter from an Apartment Complex

Why pay too much? If you paid over $300 a month for your apartment last semester, you got robbed. We'll charge 20 percent less—guaranteed.

Refute Opposing Points of View

You can strengthen your argument by considering opposing opinions. Doing so shows your audience that you have considered your topic thoroughly. Rather than looking at the subject from only one perspective, you have considered alternatives and discarded them as lacking in merit. In addition, by refuting opposing points of view, you anticipate negative comments an audience might make and defuse their argument.

To refute opposing points of view in the **discussion** (**body**) of your communication, follow these steps.

- Recognize and admit conflicting views.
- Let the audience know that you understand their concerns.
- Provide evidence.
- Allow for alternatives.

Figure 10.6 shows a letter of application that successfully uses refutation as part of its persuasion.

FIGURE 10.6 Letter of Application Using Refutation

1901 West King's Highway
San Antonio, TX 77910

February 27, 2008

Marissa Verona
Manager, Human Resources
PMBR Marketing
20944 Wildrose Dr.

San Antonio, TX 78213

Dear Ms. Verona:

In response to your advertisement in the *San Antonio Daily Register,* please consider me for the position of Marketing Accounts Associate. I have enclosed a resume elaborating on ways I can benefit your company.

Your advertisement requires someone with a BA in Marketing. Though I do not have this degree, my experience and skills prepare me for this position:

- Ten years experience in Marketing.
- Prepared press releases and created public service announcements.
- Created 30-second spots for local radio and television stations.
- Participated in press conferences providing corporate information about declining stock values and company layoffs.
- Created PowerPoint presentations for local governmental agencies.
- Currently enrolled in a Marketing class at the University.

My experience and abilities will make me an asset in your Marketing Department. I would be happy to meet with you to discuss ways in which I can benefit PMBR. Please contact me at jobrien22@hotmail.com.

Sincerely,

Joan O'Brien

Attachment: Resume

The writer's refutation anticipates negative comments the audience might have and diffuses the argument through proof and alternatives.

Admits an opposing point of view, the BA requirement.

The first five bullets provide evidence to show that the writer has skills appropriate for the job versus a degree.

The current enrollment provides an alternative to the degree requirements.

TABLE 10.1 Techniques for Supporting an Argument

Provide facts and figures to document your assertions.	*Eighty-five percent of the homeowners contend that...* or *Seven out of ten buyers said they would...*
Persuade through graphics.	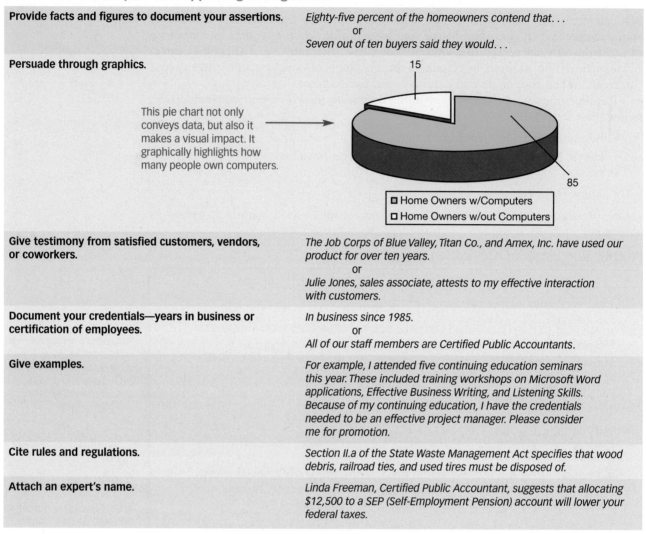
Give testimony from satisfied customers, vendors, or coworkers.	*The Job Corps of Blue Valley, Titan Co., and Amex, Inc. have used our product for over ten years.* or *Julie Jones, sales associate, attests to my effective interaction with customers.*
Document your credentials—years in business or certification of employees.	*In business since 1985.* or *All of our staff members are Certified Public Accountants.*
Give examples.	*For example, I attended five continuing education seminars this year. These included training workshops on Microsoft Word applications, Effective Business Writing, and Listening Skills. Because of my continuing education, I have the credentials needed to be an effective project manager. Please consider me for promotion.*
Cite rules and regulations.	*Section II.a of the State Waste Management Act specifies that wood debris, railroad ties, and used tires must be disposed of.*
Attach an expert's name.	*Linda Freeman, Certified Public Accountant, suggests that allocating $12,500 to a SEP (Self-Employment Pension) account will lower your federal taxes.*

The pie chart text within the graphics cell reads: "This pie chart not only conveys data, but also it makes a visual impact. It graphically highlights how many people own computers." Values shown: 15, 85. Legend: Home Owners w/Computers, Home Owners w/out Computers.

Give Proof to Develop Your Thoughts

In the **body** of your communication, develop your argument with proof. Arousing an audience's interest and refuting opposing points of view will not necessarily persuade the audience. Most people require details and supporting evidence before making decisions. You can provide specific details to support your argument using any of the techniques in Table 10.1.

Urge Action—Motivate Your Audience

Throughout your communication, you have worked to persuade the audience to accept your point of view. In the **conclusion**, you need to motivate the audience to action. This could include any of the following: attend a meeting, purchase new equipment, invite you to interview for a position, vote on a proposition, promote you, give you a raise, allow you to work a flexible schedule, or change a company policy.

Writing Persuasively in Business and Industry

Logos, ethos, and *pathos*—these argumentation devices are ancient, but they're everyday communication tools in modern marketing. Here's what Ron Dubrov says about persuasive communication.

"Where words appeal primarily to rationality, the graphics and sound help to create emotional impact." To build trust, faith, and credibility in a product or service, Ron believes that words provide descriptive detail and supportive facts, while graphics, music, rhyme, and rhythm add interest.

Ron is CEO of RonDubrovCreative, an advertising/marketing company that utilizes creative copywriting, audio, video, and graphic design techniques to help clients develop their corporate identity and achieve their marketing communications objectives.

Words Provide Rational Proof

For example, in developing the Kansas Department of Transportation's "Kansas Clicks"© campaign to raise awareness and emphasize the importance of wearing seat belts, Ron created messages for radio, television, billboards, and print media based on the use of persuasive logic. To arouse the audience's interest, he used a variety of traditional techniques:

- **Questions**   ("Hear it click? Your seat belt is trying to tell you something. Listen. It's the first sound of a responsible driver.")
- **Facts**   ("If you want your kids to be safe, let them hear your seat belt. Because when you click, they click!")
- **Startling images**   (headline superimposed over a shattered windshield: "What's going to stop you from not wearing a seat belt?")

Statistical facts also served as the basis of the seat belt campaign message directed toward drivers of pickup trucks. Research showed that over half the people in Kansas who drive pickup trucks fail to wear their seat belt—the one safety feature that could save their life.

To urge action on the part of his audience, Ron uses "classic retail techniques"—reward for action, such as gifts or incentives. These include "limited time offers and coupon redemption." In the "Kansas Clicks" and "Click It or Ticket" campaigns, the rewards or benefits of taking positive action become even more emphatic—avoiding fines, being able to keep your license and continue to drive, and ensuring you and your family's safety by surviving a car crash.

Graphics and Sound Create Emotional Context

To add interest and build emotional impact in his radio advertisements, Ron appeals to the "mind's eye" and memorable rhythm. For the "Kansas Clicks" campaign, these included a custom audio track with seat belt clicking sound; street/traffic sounds; driving on rough, gravel roads; automobile doors opening and slamming shut; and screeching tires and crashes.

Many of the radio commercials were written in Spanish and English, and the audio effects helped to maintain a cohesive campaign as well as maintain emotional impact. "It's a powerful tool," Ron says. "Sound transcends language."

Valuing the Audience

How does Ron judge the success of his persuasive documents? He offers value to his clients by being "the voice" of the customer, by reacting as the customer would, and by searching for and promoting the "real truth and value to the end user." These are attributes that not only abide by the Society for Technical Communication's ethical guidelines but also contribute to Ron's success as a professional.

TABLE 10.2   Techniques for Motivating an Audience to Action

Give due dates.	*Please respond by January 15.*
Explain why a date is important.	*Your response by January 15 will give me time to prepare a quarterly review and meet with you if I have additional questions.*
Provide contact information for follow-up.	*Please submit your proposal to Hank Green, Project Director. You can e-mail him at hgreen@modernco.com.*
Suggest the next course of action.	*We need to plan our presentation before the next City Council meeting, so please attend Tuesday's meeting at 9:00 a.m.*
Show negative consequences.	*You must repair your sidewalk within 30 days to comply with city laws regarding pedestrian safety. Failure to do so will result in a $150 fine.*
Reward people for following through.	*Following these ten simple steps will help you load the software easily and effectively.*

To urge the audience to action, consider the techniques in Table 10.2.

The following paragraph concludes a persuasive letter from a governmental agency to a business owner.

example

This conclusion provides a due date, explains why the date is important, shows negative consequences, and tells the reader what to do next.

> Your business has not been in compliance for over 16 months. We have written three letters to give you guidance and information necessary to get the facility into compliance. You have not responded to our efforts, with the exception of last week's phone call. The deadline for the pending abatement order is 30 days. You must reply to this letter in a timely manner to avoid fines. We are open to having a meeting at our regional office to discuss the corrective measures you can take to address the pending order. Thank you for considering your options.

AVOIDING LOGICAL FALLACIES

In a corporate environment, you must persuade your audience not only logically but also ethically. Your persuasive communication must be honest and reasonable. Honesty demands that you avoid the following logical fallacies.

Inaccurate Information

Facts and figures must be accurate. One way in which dishonest appeals are made in communication is through inaccurate graphics. Look at Figures 10.7 and 10.8.

Figure 10.8 shows a $25,000 deficit in the third quarter. So does Figure 10.7. However, the size of the bar in Figure 10.8 is misleading, visually suggesting that the loss is less. This is an inaccurate depiction of information.

Unreliable Sources

Being an expert in one field does not mean you are an expert in all fields. For example, quoting a certified public accountant to support an important healthcare issue is illogical. Not all specialists are reliable in all situations.

Sweeping Generalizations

Avoid exaggerating. Allow for exceptions. For instance, it is illogical to say that "*All* marketing experts believe that newsletters are effective." Either qualify this with a word like *some,* or quantify with specific percentages.

FIGURE 10.7 Annual Income

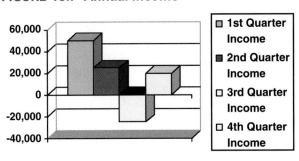

FIGURE 10.8 Annual Income

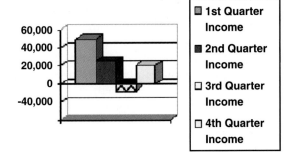

Either . . . Or

Suggesting that a reader has only two options is deceitful if other options exist. Allow for other possibilities. It is wrong to write, "Either all employees must come to work on time, or they will be fired." This blanket statement excludes alternatives or exceptions. Is the "either . . . or" statement true if an employee has a car accident, if an employee's child is sick, or if the employee is caught in heavy traffic due to a snow storm?

Circular Reasoning (Begging the Questions)

"Some accountants are ambitious because they wish to succeed." This statement is illogical and uses circular reasoning because it states the same thing twice. The words *ambitious* and *succeed* are essentially synonyms. The writer fails to prove the assertion.

Inaccurate Conclusions

When communicating persuasively, consider all possible causes and effects. Exact causes of events often are difficult to determine.

- A condition that <u>precedes</u> another is not necessarily the <u>cause</u> of it. This error is called *post hoc, ergo propter hoc.*

Example: The contractor lost the bid, so he cannot expect to have increased revenues this fiscal year.

Yes, the contractor lost a bid, but this does not necessarily mean that the contractor won't increase revenues in some other way. To say so is to make a hasty conclusion based on too little information.

- A condition that <u>follows</u> is not necessarily the <u>effect</u> of another. This is called a *non sequitur.*

Example: Because the manager is inexperienced, the report will be badly written.

Again, this is a hasty conclusion. Lack of experience in one area does not necessarily lead to a lack of ability in another area.

Red Herrings

If you focus on an irrelevant issue to draw attention from a central issue, this is called a red herring. For instance, you have failed to pay fines following citations for the mishandling of hazardous wastes. You contact the state environmental agency and complain about state taxes being too high. This is an irrelevant issue. By focusing on high taxes, you are merely avoiding the central issue. To write effective persuasive documents, you must develop your assertions correctly, avoiding logical fallacies.

TYPES OF PERSUASIVE DOCUMENTS

You have manufactured a new product (a mini digital camera, a paper-thin plasma computer monitor, a fiber optic cable, or a microfiber rain jacket). Perhaps you have just created a new service (home visitation healthcare, a mobile accounting business, a Web design consulting firm, or a gourmet food preparation and delivery service). Congratulations! However, if your product sits in your basement gathering dust or your service exists only in your imagination, what have you accomplished? To benefit from your labors, you must market your product or service.

To convince potential customers to purchase your merchandise, you could write any of the following persuasive documents.

- Sales letters
- Fliers
- Brochures

Check Online Resources

www.prenhall.com/gerson
For more information about sales letters, figures, and brochures, visit our companion Web site.

Routine Correspondence
See Chapter 6 for more discussion of routine correspondence.

Sales Letters

To write your sales letter, follow the format for letters discussed in Chapter 6. Include the letter essentials (letterhead address, date, reader's address, salutation, text, complimentary close, and signatures). Your sales letter should accomplish the following objectives, relating to effective persuasion.

Arouse Reader Interest.   The introductory paragraph of your sales letter tells your readers why you are writing (you want to increase their happiness or reduce their anxieties, for example). Your introduction should highlight a reader problem, need, or desire. If the readers do not need your services, then they will not be motivated to purchase your merchandise. The introductory sentences also should mention the product or service you are marketing, stating that this is the solution to their problems. Arouse your readers' interest with anecdotes, questions, quotations, or facts.

Refute Opposing Points of View.   By mentioning competitors and undercutting their worth, you can emphasize your product's value.

Give Proof to Develop Your Thoughts.   In the discussion paragraph(s), specify exactly what you offer to benefit your audience or how you will solve your readers' problems. You can do this in a traditional paragraph. In contrast, you might want to itemize your suggestions in a numbered or bulleted list. Whichever option you choose, the discussion should provide data to document your assertions, give testimony from satisfied customers, or document your credentials.

Urge Action.   Make readers act. If your conclusion says, "We hope to hear from you soon," you have made a mistake. The concluding paragraph of a sales letter should motivate the reader to act. Conclude your sales letter in any of the following ways.

- Give directions (with a map) to your business location.
- Provide a tear-out to send back for further information.
- Supply a self-addressed, stamped envelope for customer response.
- Offer a discount if the customer responds within a given period of time.
- Give your name or a customer-contact name and a phone number (toll free if possible).

Figure 10.9 provides a sample sales letter using the ARGU method of persuasion.

FIGURE 10.9 Sales Letter

4520 Shawnee Dr. Tulsa, OK 86221 721-555-2121

November 12, 2008

Bill Schneider
Office Manager
REM Technologies
2198 Silicon Way
Tulsa, OK 86112

Dear Mr. Schneider:

Are hardware and software upgrades making your profits plummet? Would you like to reduce your company's computer purchase and maintenance costs? Do computer breakdowns hurt your business productivity? Don't let technology breakdowns harm your bottom line. Many companies have taken advantage of **Office Station's** computer prices, service guarantees, and certified technicians.

Office Station, located in your neighborhood, offers you the following benefits:

 Purchase prices at least 10 percent lower than our competitors.

 IBM-trained technicians, available on a yearly contract or per-call basis.

 An average response time to service calls of under two hours.

 Repair loaners to keep your business up and running.

 Over 5,000 satisfied customers, like IBM, Ford, Chevrolet, and Boeing.

 All-inclusive agreements that cover travel, expenses, parts, and shop work.

 State-of-the-art technologies, featuring the latest hardware and software.

Our service is prompt, our technicians are courteous, and our prices are unbeatable. For further information and a written proposal, please call us at **721-555-2121** or e-mail your sales contact, Steve Hudson (shudson@os.com). He's waiting to hear from you. Take advantage of our *Holiday Season Discounts!*

Sincerely,

Rachel Adams,
Sales Manager

Office Station
Authorized Sales and Service for
Gateway   3M   Microsoft   HP   Apple   Dell   Swingline

The introduction arouses reader interest by using alliteration ("profits plummet") and asking questions. The questions highlight reader problems: *profits, costs, productivity, and breakdowns.*

The last sentence shows how Office Station will solve the problems.

The letter uses positive words to persuade: *advantage, guarantees, certified, benefits, satisfied, prompt,* and *courteous.*

The body provides specific proof to sway the reader: 10 percent lower, IBM-trained technicians, two-hour response, and renowned satisfied customers.

The conclusion urges action by giving contact names and numbers and seasonal discounts.

Fliers

Corporations and companies, educational institutions, religious organizations, museums, zoos, amusement parks, cities and states—they all need to communicate with their constituencies (clients, citizens, members). An effective way for these organizations to communicate persuasively is with a one- or two-page flier.

Fliers provide the following benefits.

- **Cost effective**—A flier costs less than an expensive advertising campaign and can be produced in-house.
- **Time efficient**—Creating a flier can take only a few hours of work or less by the company's employees.
- **Responsive to immediate needs**—Different fliers can be created for different audiences and purposes to meet unique, emerging needs.
- **Personalized**—Fliers can be created with a specific market or client in mind. Then, these fliers either can be mailed to that client or hand delivered for more personalization.
- **Persuasive**—In a compact format, fliers concisely communicate audience benefit.

Flier Criteria. When writing your flier, follow these criteria.

Keep the Flier Short. Though one page might be preferable, you could create a two-page flier, using the front and back of an 8 1/2 × 11 inch piece of paper. If you keep your flier to one page (front only), then you can save money by folding the flier in thirds, stapling it, and using the blank side for mailing purposes (addresses and stamp). A flier even could be smaller, the size of a postcard, for example. Many companies create electronic fliers, transmitted via their Web sites. This is even more cost effective than a hard-copy flier, since no postage is needed.

Focus on One Idea, Topic, or Theme Per Flier. A flier should make one key point. This is how you make the flier's content relevant to your audience, fulfilling that audience's unique needs. For example, if your company's focus is automotive parts, avoid writing a flier covering every car accessory. Write the flier with one accessory in mind, such as windshield wipers, batteries, or custom rims.

Use a Title at the Top of Your Flier to Identify Its Theme. The title can be one or two words long; you could use a phrase; or you might want to write an entire sentence at the top of the flier. An effective persuasive approach is to begin your flier with a question to immediately arouse reader interest: "Software giving you a headache?" or "How usable is your Web site?"

Limit Your Text. Using few words, provide reader benefit, involve the audience, and motivate them to act. The action could be to purchase a product, attend an event, or contact you for additional information. By limiting your text, you avoid overwhelming either the flier's appearance or the audience's attention span. Getting to the point in a flier is a key concern. Limiting your text helps you achieve this goal.

Increase Font Size. In a flier you can use a 16-point font and up for text, and a 20-point font and up for titles. This will make the text more readable and dramatic. Your heading must be eye catching. To accomplish this goal, make sure the heading's font size and style are emphatic—at a glance, even from across a room.

Use Graphics. One graphic, at least, will emphasize your theme and visually make your point memorable. Another graphic could include your company logo (for corporate identity and namesake recognition). In addition, the logo should be accompanied by a street address, e-mail address, Web site URL, fax number, or phone number so clients can contact you or visit your site.

Use Color for Audience Appeal. Pick one dominant color to emphasize key points. Use a color in your company logo to remind your reader of your company's identity. However, don't overuse color. Excess will distract your reader.

Use Highlighting Techniques. Bullets, white space, tables, boldface, italics, headings, or subheadings will help your reader access information. A little highlighting goes a long way, especially on a one-page flier. Too much makes a jumble of your text and distracts from the message.

Find the Phrase. Select a catchy phrase, which you can use continually in all your fliers, to personify your company's primary focus. For example, McDonald's repeats the line, "We love to make you smile"; Pepsi uses, "The Joy of Cola"; Gateway writes, "Get more out of the box"; and Ford states, "Quality is Job #1." What phrase captures your company's personality?

Recognize Your Audience. You want to show the readers how your product or service will benefit them. Understand your audience's needs and direct the flier to meet those concerns. In addition, you want to engage the reader. To do this, use *pronouns* which speak to the reader on a personal level and *positive words* which motivate the reader to action. Remember to speak at the reader's level of understanding, defining terms as necessary.

Figure 10.10 provides a sample persuasive flier.

FIGURE 10.10 Persuasive Flier

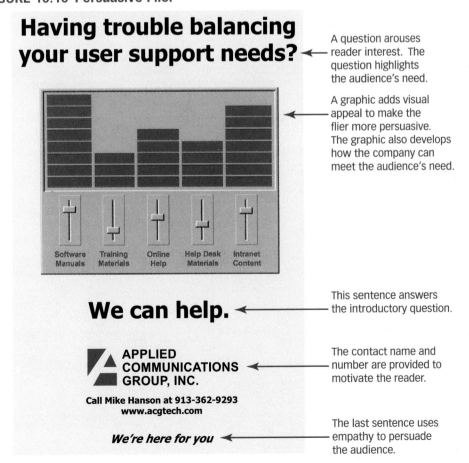

Using Word 2007 for Flier and Brochure Design Templates

You can access many design templates for fliers and brochures using your word-processing software. For example, in Microsoft Word 2007, try this approach:

1. On the Menu bar, click on the **Office Button** and scroll to **New**. You will see a **New Document/Templates** pop up.

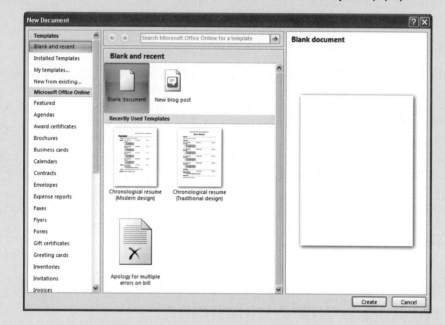

2. Click on the document of your choice (**Fliers** or **Brochures**). Word 2007 gives you 14 different brochure options for events, travel, recruiting, professional services, real estate, and financial. These include templates for 8 1/2 x 11 landscape with three folds, 8 1/2 × 14 landscape with three and four folds, and 8 1/2 × 11 portrait letter folds. When you click on fliers, you get templates for 25 different "Events" fliers, 13 "Marketing" options, 3 "Real Estate" choices, and 5 "Other Fliers."

NOTE: The templates will not tell you what to write in each panel. Instead, Word 2007 provides you a design layout, a shell for you to fill with your content.

Creating Your Own Brochure Layout

Instead of using a predetermined template, you can create your own as follows:

1. Change from portrait to landscape by clicking on the **Page Layout** tab and then **Orientation.**

2. Create three panels for your brochure by clicking on the **Columns** icon.

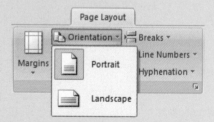

Brochures

A flier *must* be short—one or two pages. If you have more information to convey than can fit on a one-page flyer, then a brochure might be a good option. Brochures offer you a detailed overview of products, services, options, and opportunities, complete with photographs, maps, or charts. Brochures are persuasive for the following reasons.

- Create awareness of your company, product, or service
- Increase understanding of a product, service, or your company's mission
- Advertise new aspects about your company, product, or service
- Change negative attitudes
- Show ways in which your company, product, or service surpasses your competition
- Increase frequency of use, visit, or purchase
- Increase market share

Criteria for Writing Brochures. Brochures come in many shapes and sizes. They can range from a simple front and back, four-panel, 8 1/2 × 5 1/2 inch brochure (one landscape 8 1/2 × 11 inch page folded in half vertically) to six-, eight-, or even twelve-panel brochures printed on any size paper you choose (see Figure 10.11). Brochures, like fliers, also can be transmitted electronically as Web pages. Your topic and the amount of information you are delivering will determine your brochure's size and means of transmission.

To determine what you will write in each panel of your brochure and how the brochure should look, follow these criteria for writing an effective persuasive brochure.

Title Page (Front Panel). Usually, the title page includes at least three components.

- **Topic**—In the top third of the title panel, name the topic. This includes a product name, a service, a location, or the subject of your brochure.
- **Graphic**—In the middle third of the panel, include a graphic to appeal to your reader's need for a visual representation of your topic. The graphic will sell the value of your subject (its beauty, usefulness, location, or significance) or visually represent the focus of your brochure.
- **Contact Information**—In the bottom third of the panel, place contact information. Include your name, your company's name, street address, city, state, zip code, telephone number, fax number, or e-mail address. See Figure 10.12 for a sample brochure.

FIGURE 10.11 Four- and Six-Panel Brochures

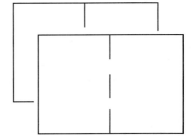
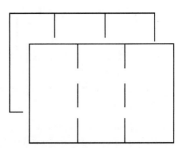

FIGURE 10.12 Sample Brochure

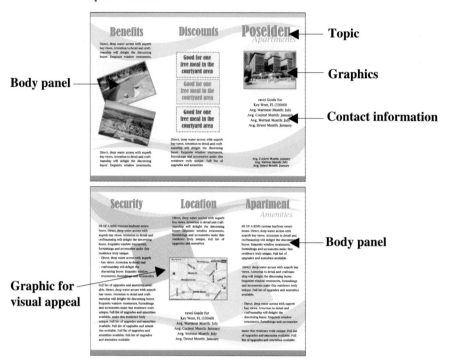

Back Panel. The back panel could include the following:

- **Conclusion**—Summarize your brochure's content. Restate the highlights of your topic or suggest a next step for your readers to pursue.
- **Mailing**—The back panel could be used like the face of an envelope. On this panel, when left blank, you could provide your address, a place for a stamp or paid postage, and your reader's address.
- **Coupons**—As a tear-out, this panel could be an incentive for your readers to visit your site or use your service. Here you could urge action by providing discounts or complimentary tickets.
- **Location**—A final consideration would be to provide your reader with your address, hours of operation, phone numbers, e-mail, and a map to help them locate you.

Body Panels (Fold-in and Inside). Following are suggestions for creating the brochure's text.

- **Provide headings and subheadings.** These act as navigational tools to guide your readers, direct their attention, and help them find the information they need. The headings and subheadings should follow a consistent pattern of font type and size. First-level headings should be larger and more emphatic than your second-level subheadings. The headings must be parallel to each other grammatically.

  For example, if your first heading is entitled, "Introduction," a noun, all subsequent headings must be nouns, like "Location," "Times," "Payment Options," and "Technical Specifications." If your first heading is a complete sentence, like "This is where it all began," then your subsequent headings must also be complete sentences: "It's still beautiful," "Here's how to find us," and "Prices are affordable."

- **Use graphics.** Use photographs, maps, line drawings, tables, or figures to vary the page layout, add visual appeal, and enhance your text.

- **Develop your ideas.** Consider including locations, options, prices, credentials, company history, personnel biographies, employment opportunities, testimonials from satisfied customers, specifications, features, uses of the product or service, payment schedules, or payment plans.
- **Persuade your audience.** Review the tips provided in this chapter for persuasive arguments. Use ethics, logic, and emotion to sway your reader.

Document Design. Visual appeal helps to interest and persuade an audience. Compelling graphics, for example, can help to convince an audience. Use pie charts, bar charts, tables, or photographs to highlight key concerns. In addition to graphics, make your brochure visually appealing by doing the following:

- Limit sentence length to 10 to 12 words and paragraph length to 4 to 6 lines. When you divide paper into panels, text can become cramped very easily. Long sentences and long paragraphs then become difficult to read. By limiting the length of your text, you will help your readers access the information.
- Use white space instead of wall-to-wall words. Indent and itemize information so readers won't have to wade through too much detail.
- Use color for interest, variety, and emphasis. For example, you can use a consistent color for your headings and subheadings.
- Bulletize key points.
- Boldface or underline key ideas.
- Do not trap yourself within one panel. For variety and visual appeal, let text and graphics overlap two or more of the panels.
- Place graphics at angles (occasionally) or alternate their placement at either the center, right, or left margin of a panel. Panels can become rigid if all text and graphics are square. Find creative ways to achieve variety.

See Figure 10.13 to get a better idea of what a typical brochure looks like.

DOT-COM UPDATES

For more information about creating brochures, check out the following links.

- http://www.hp.com/ sbso/productivity/howto/ marketing_main/ marketing_brochure/
- www.smallbusinessnotes. com/operating/marketing/ brochures.html

Persuasive Communication Checklist

_____ 1. **Ethical Argumentation:** Have you made an ethical argument based on character? You must be trustworthy and credible as a writer or speaker.

_____ 2. **Emotional Argumentation:** Have you used emotion to change an audience's attitudes? You can appeal to an audience's emotions either positively or negatively.

_____ 3. **Logical Argumentation:** Have you developed your persuasion by depending on rationality, reason, and proof? You can persuade people logically by providing facts, testimony, and examples.

_____ 4. **Arouse audience interest:** Have you used questions, quotes, anecdotes, comparison/contrast, poetic language, or an appeal to senses to interest your audience?

_____ 5. **Refute opposing points of view:** Have you presented a balanced argument? To do so, recognize and admit conflicting views, let the audience know that you understand their concerns, and allow for alternatives.

_____ 6. **Give proof:** Have you provided evidence to prove your point?

_____ 7. **Urge to action:** Have you motivated your audience to act? To do so, provide incentives, give discounts, mention warranties, provide contact information, or suggest a follow-up action.

_____ 8. **Highlighting/Page Layout:** Is your text accessible? To achieve reader-friendly ease of access, use headings, boldface, italics, bullets, numbers, underlining, or graphics (tables and figures). These add interest and help your readers navigate your text.

_____ 9. **Conciseness:** Have you limited the length of your sentences, words, and paragraphs?

_____ 10. **Audience Recognition:** Have you written appropriately to your audience? This includes avoiding biased language, considering the multicultural/cross-cultural nature of your readers, and your audience's role (supervisors, subordinates, coworkers, customers, or vendors).

_____ 11. **Correctness:** Is your text grammatically correct? Errors will hurt your professionalism.

FIGURE 10.13 Brochure

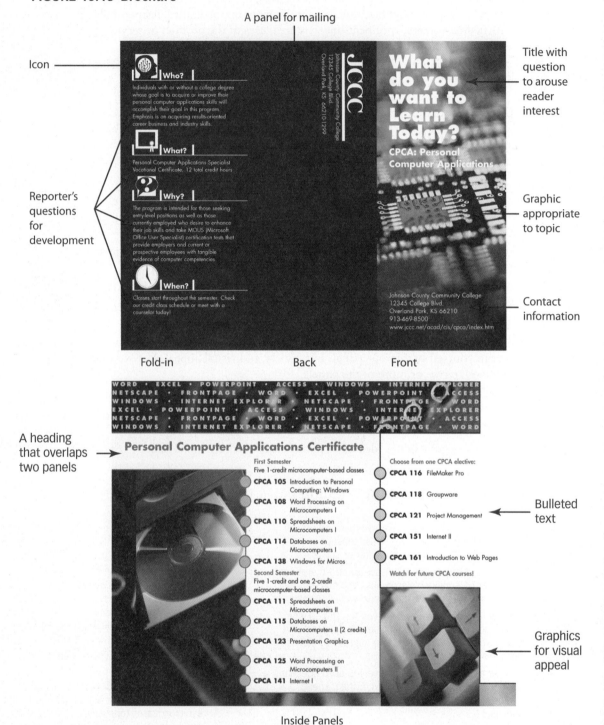

A panel for mailing

Icon

Reporter's questions for development

Fold-in

Back

Front

Title with question to arouse reader interest

Graphic appropriate to topic

Contact information

A heading that overlaps two panels

Bulleted text

Graphics for visual appeal

Inside Panels

THE WRITING PROCESS AT WORK

As shown throughout this book, the best way to approach any writing activity is as a process. First, prewrite to gather data, then write a rough draft, and finally rewrite by revising your text. Remember that the writing process is dynamic, with the three steps frequently overlapping.

The Writing Process

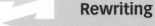

Prewriting	Writing	Rewriting

• Decide which communication channel to use: sales letter, brochure, flier, memo, e-mail, etc.
• Determine how you can use ethical, emotional, and logical arguments.
• Gather details for your content.
	• Organize your content by arousing interest, refuting opposing points of view, giving proof, and urging action.
	• Determine where graphics will enhance your persuasion.
		• Revise your draft by
		  • adding details
		  • deleting wordiness
		  • simplifying words
		  • enhancing the tone
		  • reformatting your text
		  • proofreading and correcting errors
		  • avoiding logical fallacies

FIGURE 10.14 Cubing

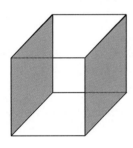
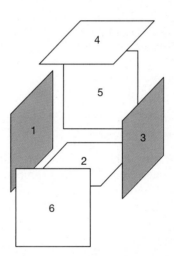

Prewriting

We have suggested different types of prewriting techniques to help you gather ideas and organize your information. Some techniques work best for specific types of correspondence. For example, reporter's questions are an excellent way to prewrite for sales letters. A new prewriting technique especially effective for brochures is *cubing*.

Cubing. Visualize a three-dimensional cube, with six sides (see Figure 10.14). Then try looking at a topic from the following six angles or perspectives.

1. **Describe your topic**—What does it look like? Consider focusing on all five senses. We realize that sight, sound, smell, taste, and touch might not be appropriate for every topic. Just use whichever senses apply to your subject matter and disregard the rest.

2. **Compare/contrast**—Compare the subject to other topics to discover the similarities and dissimilarities. Make a list of several similarities and differences. Your topic should compare favorably, of course, to your competition.

3. **Associate your topic**—What does your subject remind you of? Create analogies, similes, and metaphors. People often get a better understanding of a topic when you help them envision what it's similar to. For example, one could say that Las Vegas is *an oasis* in the Nevada desert. That's "sort of" true because the literal definition of "oasis" is a small watering hole surrounded by palm trees and dunes. To call Las Vegas an oasis is more figuratively metaphorical than actual.

4. **Analyze it**—Tell how it's made. What are its components, its various facets or aspects? Decide how it will benefit the reader or meet the readers' needs.

5. **Apply it**—What can you do with your topic? How can it be used on the job, in the home, for educational purposes, for pleasure, and so on?

6. **Argue the pros and cons**—What's good about your topic? Conversely, what's bad? Again, answer these questions with a specific audience in mind. Who is your client, for example? How will your topic help him or her (that represents the pros)? How might your topic not help the audience (that would represent the cons)? The reason for focusing on negatives, as well as positives, is that by knowing one's complaints beforehand, you can refute potential criticisms.

Cubing allows you to analyze your topic from multiple perspectives and thus gather a great deal of information. Some information you will use; some you won't. The key to prewriting is to analyze a topic and to gather data—perhaps more than you will need. The time for deleting information and for sharpening your focus comes later, during the rewriting phase. In addition to helping you develop your ideas thoroughly, cubing helps you organize your thoughts. Each side of the cube might represent a panel in your brochure.

Writing

Now that you have gathered information, the next step is to start constructing your brochure. This is the drafting stage, when words are placed on the individual panels. Follow this procedure:

Create the Panels.   Technical writing differs from other forms of communication because it is limited by space. For example, a novel can be 100, 500, or 1,000 pages in length. Technical writing, in contrast, must "fit in the box." A car's user manual has to fit in the glove compartment; a brochure's content must fit inside its panels. You might as well set the parameters of your brochure now. If you envision a four-panel brochure, set your page layout to "landscape" and create two columns. If you envision a six-panel brochure, set the page layout to "landscape" and create three columns (or use an existing template, present in your word-processing software). Doing so allows you to work within the boundaries mandated by a brochure format.

Title Your Brochure.   In the top third of the right-hand panel of your brochure, write a title. For example, you could name your company or product, plus give an accompanying descriptive phrase: "Moody Gardens . . . a World of Wonders"; "17 Historic Sites: Place Yourself in Texas History." The title should be relatively short, attract your reader's attention, and clearly inform your reader of your brochure's primary focus.

Select a Graphic.   On the title page, include a graphic that pictorially represents your topic and entices the reader to open the brochure for more information.

Subdivide Your Topic.   What are you planning to discuss about your topic—benefits, prices, locations, uses, specifications, options, warranties, and so forth? Make this decision based on your prewriting. These components—subdivisions of your topic—become the content for your panels.

Use Headings for Your Panels.   Once you have topics for discussion, give each panel a heading. The heading serves two purposes. It keeps you on track as the writer, helping you to maintain your focus, ensuring that you do not wander off the topic. Of more importance, the heading acts as a signpost for your reader, like street signs along a roadway. The heading gives your reader direction, guiding your audience through the brochure.

Write the Text.   Writing the text, of course, is your most important job. Your ultimate goal is to communicate information about your topic. To do so, develop your ideas by answering reporter's questions (*who, what, when, where, why*, and *how*), drawing from your prewriting. What have you discovered while cubing that will help your reader better understand your topic? Add this information to your brochure.

Rewriting

To help you revise your rough draft, consider the following effective brochure usability checklist. What is "usability"? Usability helps you determine whether your reader can "use" the brochure effectively and whether your brochure meets your user's needs and expectations. That is, does your brochure work? The reader not only wants to find information that helps him or her better understand the topic, but the audience also wants the text to be readable, accurate, up to date, and easy to access. Thus, usability focuses on four key factors (Dorazio 2000).

1. **Retrievability**—Can the user find specific information quickly and easily?
2. **Readability**—Can the user read and comprehend information quickly and easily?
3. **Accuracy**—Is the information complete and correct?
4. **User satisfaction**—Does the brochure present information in a way that is easy to learn and remember?

EFFECTIVE BROCHURE USABILITY CHECKLIST

First Panel

_____ 1. Does the first panel name the product, company, or service?

_____ 2. Does the first panel provide a graphic to attract the reader's attention and pictorially represent the topic?

_____ 3. Does the first panel provide contact information, such as address, phone number, e-mail address, fax number, and so on?

Body Panels

_____ 1. Are headings presented similarly, maintaining parallelism?

_____ 2. Are graphics effectively used for interest and to clarify ideas?

_____ 3. Does the brochure vary font sizes and type to emphasize key points and add visual appeal?

_____ 4. Are bullets and numbers used to itemize ideas for better access?

_____ 5. Has ample white space been used to help the reader access information and to make reading easier?

_____ 6. Has color been used effectively for visual appeal?

Content

_____ 1. Are all unfamiliar terms defined?

_____ 2. Is the technical content correct, verified by peer review?

_____ 3. Is the brochure grammatically correct?

_____ 4. Is the text clear, answering reporter's questions (*who, what, when, where, why*, and *how*)?

_____ 5. Is the text concise, using short words, short sentences, and short paragraphs?

_____ 6. Does the brochure meet the writer's goals: to create awareness of the company, product, or service; to increase understanding; to advertise new aspects; to change negative attitudes; to show ways in which the topic surpasses the competition?

_____ 7. Are maps used to help the reader find the writer's location?

_____ 8. Are discounts or promotional incentives offered to ensure reader participation?

Audience

_____ 1. Is the level of writing appropriate for the audience (high tech, low tech, lay, multiple readers)?

_____ 2. Has the appropriate tone been achieved through positive words and personalized pronouns?

Once you have answered the questions posed in the brochure usability checklist, revise accordingly.

- Add missing information for clarity, such as a missing e-mail address, phone number, building address, maps, and so on.
- Delete unnecessary words, phrases, and content for conciseness.
- Simplify words and phrases, or define acronyms and abbreviations to better communicate with your audience.
- Move information within the brochure. What is the best place to locate a map? Where should you list your credentials? What should be on the first panel your reader sees? You will have to assess these issues individually. However, remember that the most important information should probably be the first thing your reader sees, not something buried on the last panel of the brochure.
- Reformat for access. Now is the time to add the bullets, italics, or boldfacing. Maybe you need to reconsider your color choice. Is the background you've used too dark for readability? Is your font selection too hard to read? If your paragraphs are too long, cut them in half, and add more white space.
- Add more graphics.
- Enhance the tone of your brochure. Add more pronouns and contractions to make the brochure more friendly and casual.
- Correct errors. Nothing will destroy your professionalism more quickly than typographical errors or errors in content. Imagine how your business will be hurt if you have typed the wrong e-mail address or phone number.

CHAPTER HIGHLIGHTS

1. Persuasive technical communication consists of a combination of ethos, logos, and pathos.
2. Use the ARGU technique to create persuasive fliers, brochures, or sales letters.
3. Avoid logical fallacies when communicating persuasively.
4. Sales letters market services and products.
5. Fliers should be short, focused on one idea, titled, have limited text, and be visually appealing.
6. Recognize your audience and their needs when you write a flier, brochure, or sales letter.
7. The front panel of a brochure contains a title and a graphic that depicts the topic.
8. The interior panels of a brochure contain information about the topic.
9. The back panel of a brochure can be used for mailing purposes.
10. Document design is important when you create a flier, brochure, or sales letter.

CASE STUDY

A company has just expanded into your city. Having been in business in the Midwest since 2002, TechToolshop provides sales, service, maintenance, installation, and data recovery for PC and Mac hardware and software tools.

They install, repair, and maintain workstations, servers, printers, and peripherals. Through an online catalog and storefront site, they sell printers, desktops, mainframes, minitowers, software, accessories, and encryption devices. TechToolshop has proudly worked with many large companies, including McDonald's, Pepsi, Ford, Texaco, Sprint, Transamerica, GE, JCPenney, and Chase Manhattan Bank. For their latest venture, this company will help you create your own corporate Web page and maintain the site, fees determined by number of screens, plug-ins, and graphics.

TechToolshop's home office is in Big Springs, Iowa, at 11324 Elm, where over 1,200 employees work. The phone number at this site is 212-345-6666, and the e-mail address is ToolHelp@TechTools.com. This company's Web site can be found at www.TechTools.com. TechToolshop's new local address in your city is 5110 Nueces Avenue. The phone number is 345-782-8776.

TechToolshop is most proud of its product support and performance guarantees. The firm offers free product support, 24 hours a day at 1-800-TechHelp. By moving to your city, TechToolshop also can guarantee arrival at your site within two hours of any service emergency call. Plus—the company's greatest innovation—TechToolshop has installed service kiosks in every mall, library, and bank in your city. By keying in your personal identification number (obtained by paying a small monthly fee), you can have answers to technical questions at your convenience. Of course, you can get help via the Web site also. In addition, the company warrants all products and services—money back—for 90 days, covering defects in material and workmanship.

The company is owned by James Wilcox Oleander (president) and his brother Harold Robert Oleander (CEO). They started this company after graduating from Midwestern State University, with degrees in IT, IS, and CPCA. Their first store had only three employees, but through hard work, their business grew 45 percent in the first two years of operation. Expert recruiting of the best State U. graduates increased their workforce, as did the Oleander's philosophy of "employee ownership." TechToolshop's workers take pride in their work, since their success increases their stock options (TechToolshop's stock at NASDAQ opened in 2003 at $18 a share but has listed as high as $45). The Oleanders plan to open at least 12 new stores each year throughout the United States, as well as pursue franchise options. They have high hopes for their future success.

Write either a persuasive sales letter, flier, or brochure about TechToolshop, based on the information provided. Follow the criteria provided in this chapter.

INDIVIDUAL OR TEAM PROJECTS

ARGU

For the ARGU method of organization, you *A*rouse reader interest, *R*efute opposing arguments, *G*ive proof to support your argument, and *U*rge reader action. Read the following situations and complete the assignments.

- You plan to sell a flash drive that is small enough to fit on a key chain. For a sales letter, write five different introductions to <u>arouse</u> reader interest. Use any of the options provided in the chapter for arousing reader interest.
- Write a body that <u>refutes</u> opposing arguments (too expensive, easily damaged, easily lost, etc.) and that <u>gives proof</u> to support your product claims.
- In the sales letter for the portable computer zip drive, conclude by <u>urging</u> reader action. Use at least two of the methods discussed in this chapter.

Analysis of Persuasive Writing

Find examples of persuasive writing (sales letters, fliers, or brochures). Bring these to class, and in small groups or individually, accomplish the following:

1. Decide which methods of persuasion have been used. Where in the documents do the writers appeal to logic, emotion, and ethics? Give examples and explain your reasoning, either in writing or orally.

2. Have the writers aroused reader interest? Give examples and explain your reasoning, either in writing or orally. If the writers have not aroused reader interest, should they have? Explain why. Rewrite the introductions using any of the techniques discussed in this chapter to arouse reader interest.

3. Have the writers refuted opposing points of view? Give examples and explain your reasoning, either in writing or orally. If the writers have not negated opposing points of view, should they have? Explain why. Rewrite the text using any of the techniques discussed in this chapter to negate opposing points of view.

4. Have the writers developed their arguments persuasively? Give examples and explain your reasoning, either in writing or orally. If the writers have not provided persuasive proof, rewrite the text using any of the techniques discussed in this chapter to improve the arguments.

5. Do any of the examples you have found use logical fallacies to persuade the readers? Give examples and explain your reasoning, either in writing or orally. If the writers have used logical fallacies, rewrite the text using any of the techniques discussed in this chapter to improve the arguments.

PROBLEM-SOLVING THINK PIECES

Logical, Emotional, and Ethical Appeals

Read the following situations and determine whether the persuasive arguments appeal to logic, emotion, or ethics. These argumentation techniques can overlap. Explain your answers.

1. If you purchase this product, you can benefit from a healthier, happier, and longer life!
2. Seventy-two percent of SUV owners say that high gas prices will influence their next car purchases.
3. CEO Jim Snyder, an expert in the field of sports management, says, "Building a downtown sports arena enhances a city's image."
4. Style-tone Hair Gel improves your hair quality by preventing split ends, generating new hair growth, and inhibiting "frizzies."
5. Failure to recycle will cause 52 percent more dangerous hydrofluoric carbons to be released into the atmosphere, leading to harmful decreases in the ozone layer.

Logical Fallacies

Read the following logical fallacies and revise them, ensuring that the sentences provide logical, ethical, and correct argumentation.

1. All marketing experts believe that newsletters are effective.
2. Either all employees must come to work on time, or they will be fired.
3. Some accountants are ambitious because they wish to succeed.
4. The contractor lost the bid, so he cannot expect to have increased revenues this fiscal year.
5. Because the manager is inexperienced, the report will be badly written.

WEB WORKSHOP

You can find electronic brochures and fliers on the Internet. Using a search engine of your choice, type in phrases like "online brochure," "e-brochure," "online flier," "e-flier," "electronic brochure," or "electronic flier." Once you find examples of these online persuasive documents, do the following:

- Compare and contrast the electronic documents with hard-copy brochures and fliers.
- Compare and contrast the electronic documents with the criteria provided in this chapter.
- Decide how the online communication is similar to and different from the criteria and from hard-copy versions.
- If you decide that the online versions can be improved, print them out and revise them.

QUIZ QUESTIONS

1. What are the traditional methods of argumentation?
2. Which type of argumentation depends on your character?
3. Which type of argumentation focuses on feelings?
4. Which type of argumentation depends on rationality, reason, and proof?
5. What are four ways to arouse reader interest?
6. Which steps do you follow to refute opposing points of view in the body of your argument?
7. How can you urge your audience to action?
8. What are three types of logical fallacies?
9. Why are brochures effective marketing tools?
10. Why are fliers effective for marketing?

CHAPTER 11

Technical Descriptions and Process Analyses

COMMUNICATION *at work*

The information technology scenario shows the importance of technical descriptions and process analyses in the engineering profession.

Alpha Beta Consulting (A/B/C)

ABC ALPHA BETA CONSULTING

specializes in creating software for public service engineering purposes, such as storm water, water distribution, transportation and traffic control, parks and recreation, facility and building management, and inventorying of fleets and equipment.

The city of Maple Valley, Texas, is interested in purchasing one of A/B/C's software suites related to storm water maintenance. The suite focuses on accurate and easy-to-use inventory and inspection tools for maintaining and improving storm water systems. A/B/C's software will help Maple Valley inspect and test all components of the city's storm water system including conduits, structures, detention basins, pump stations, and pumps.

Objectives

When you complete this chapter, you will be able to

1. Distinguish between technical descriptions and process analyses.

2. Follow criteria for writing an effective technical description or process analysis.

3. Use graphics with callouts effectively in your description or process analysis.

4. Follow a process approach—prewriting, writing, and rewriting—to write your description or process.

5. Evaluate your description or process by using the checklist.

Before A/B/C can sell the software to Maple Valley, the company's information technology director, Neeha Patel, must do the following:

- Determine the city's unique needs.
- Research software options.
- Create or adapt software to meet the city's technical specifications.
- Write technical descriptions of the software.
- Provide process analyses, explaining how the software works to meet the city's needs.

Once Neeha creates this documentation, she can provide it to the Maple Valley Planning Commission, who will decide whether to purchase the software and work with A/B/C.

DEFINING TECHNICAL DESCRIPTIONS

A *technical description* is a part-by-part depiction of the components of a mechanism, tool, or piece of equipment. Technical descriptions are important features in several types of technical communication.

TYPES OF TECHNICAL DESCRIPTIONS

Operations Manuals

Manufacturers often include an *operations manual* in the packaging of a mechanism, tool, or piece of equipment. This manual helps the end user construct, install, operate, and service the equipment. Operations manuals often include technical descriptions.

Technical descriptions provide the end user with information about the mechanism's features or capabilities. For example, this information may tell the user which components are enclosed in the shipping package, clarify the quality of these components, specify what function these components serve in the mechanism, or allow the user to reorder any missing or flawed components. Following is a brief technical description found in an operations manual.

example

> The Modern Electronics Tone Test Tracer, Model 77A, is housed in a yellow, high-impact plastic case that measures 1¼ inch × 2 inch × 2¼ inch, weighs 4 ounces, and is powered by a 1604 battery. Red and black test leads are provided. The 77A has a standard four-conductor cord, a three-position toggle switch, and an LED for line polarity testing. A tone selector switch located inside the test set provides either solid tone or dual alternating tone. The Tracer is compatible with the EXX, SetUp, and Crossbow models.

Product Demand Specifications

Sometimes a company needs a piece of equipment that does not currently exist. To acquire this equipment, the company writes a *product demand* specifying its exact needs, as follows:

example

This product demand specification is written for a high-tech audience. It assumes knowledge on the part of the reader, such as definitions for "high-speed" and "deep-hole," which a lay reader would not understand. In addition, the high-tech abbreviation "AISI" is not defined.

> Subject: Pricing for EDM Microdrills
>
> Please provide us with pricing information for the construction of 50 EDM Microdrills capable of meeting the following specifications:
>
> - Designed for high-speed, deep-hole drilling
> - Capable of drilling to depths of 100 times the diameter using 0.012-inch to 0.030-inch diameter electrodes
> - Able to produce a hole through a 1.000-inch-thick piece of AISI D2 or A6 tool steel in 1.5 minutes, using a 0.020-inch diameter electrode
>
> We need your response by January 13, 2008.

Study Reports Provided by Consulting Firms

Companies hire a consulting engineering firm to study a problem and provide a descriptive analysis. The resulting *study report* is used as the basis for a product demand specification requesting a solution to the problem. One firm, when asked to study crumbling cement walkways, provided the following technical description in its study report.

<table>
<tr>
<td>

The slab construction consists of a wearing slab over a ½ -inch thick waterproofing membrane. The wearing slab ranges in thickness from 3½ inches to 8½ inches, and several sections have been patched and replaced repeatedly in the past. The structural slab varies in thickness from 5½ inches to 9 inches with as little as 2 inches over the top of the steel beams. The removable slab section, which has been replaced since original construction, is badly deteriorated and should be replaced. Refer to Appendix A, Photo 9, and Appendix C for shoring installed to support the framing prior to replacement.

</td>
<td>

example

</td>
</tr>
</table>

Sales Literature

Companies want to make money. One way to market equipment or services is to describe the product. Such descriptions are common in sales letters, proposals, and on Web sites. Figure 11.1 is a technical description from Hewlett-Packard.

Instructions

See Chapter 12 for more discussion of instructions.

FIGURE 11.1   Technical Description for a Lay Audience

The title identifies the equipment to be described.

HP Officejet Pro K550 series color printer

Impressionistic words such as "quick access," "easy troubleshooting," and "User-friendly control panel" are used for sales purposes to the lay audience.

HP Officejet Pro K550dtwn color printer shown

Numbered callouts refer to the text and help the readers locate parts of the equipment.

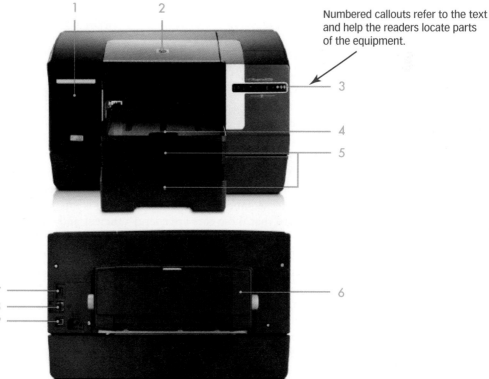

1 Ink cartridge cover provides quick access to four snap-in ink cartridges.

2 Top cover flips open for easy troubleshooting or maintenance.

3 User-friendly control panel includes graphical icons on the buttons and lights. It provides printer and wireless networking status at-a-glance and simplifies wireless configuration with SecureEasySetup™.

4 150-sheet output tray with extender.

5 250-sheet and 350-sheet input trays for high-volume printing.

6 Automatic two-sided printing accessory for professional, two-sided documents.

7 Built-in (Full Speed) USB host connector enables wireless configuration through Windows® Connect Now.

8 Built-in wired Ethernet networking.

9 Built-in Hi-Speed USB 2.0 port enables fast and easy direct connections.

Courtesy of Hewlett-Packard

DEFINING PROCESS ANALYSIS

What is a process analysis? Let us define a process analysis by comparing it to an instruction. Instructions provide a step-by-step explanation of how to do something. For example, in an instruction, you explain how to change the oil in a car, how to connect a DVD player to a computer, how to make polenta, or how to put together a child's toy. In an instruction, the audience wants to know how to do a job. A process analysis, in contrast, focuses not on how to do something, but on how something works. For instance, a process analysis might explain how viruses attack our bodies, how airbags save lives, how metal detectors work, or how e-mail messages are transmitted.

EXAMPLES OF PROCESS ANALYSIS

A discussion of *process* is a common part of many technical descriptions.

- **Engineering**—Dubai, second largest emirate of the United Arab Emirates, plans to develop several islands off its coast. To increase tourism, these islands will have both private residences and hotels. Your engineering company has been hired to build bridges connecting the islands to the coastline. Your options include beam bridges, arch bridges, and suspension bridges. To clarify which type of bridge will best meet your client's needs, you provide a PowerPoint presentation explaining how each bridge works.

- **Automotive Sales**—A potential customer wants to buy a new car. This customer is concerned about the environment and is interested in buying a hybrid, having heard that these cars are energy efficient and ecologically friendly. However, the customer doesn't understand how hybrids work. Your job is to explain the process of how gasoline-electric hybrid cars work.

- **Biomedical Technology**—You and a team of scientists have been working on new treatments for diabetes, a health problem in the United States that affects approximately 20.8 million Americans. One possible treatment is "pancreatic islet transplantation." Islets are injected through a catheter into a diabetic's liver. In time, the islets begin releasing insulin. This treatment has great potential for diabetics, but research is ongoing. You are writing a proposal requesting additional funding for research. In this proposal, you provide a process description of how diabetes affects a patient and how the pancreatic islet transplantation will work to treat this disease.

Why Write Process Analyses?

Stacy Gerson is a technical writer for GBA Master Series, a software company that creates computerized maintenance management systems (CMMS) for public works and utilities organizations. Their clients include street, water, and sewer departments for cities, counties, and states, helping these constituencies track their inventory and assets. As a documentation specialist, Stacy writes online user manuals and training guides for end users such as office workers, dispatchers, maintenance supervisors, field workers, engineers, and employees of citizen complaint call centers.

These manuals help her clients use GBA Master Series software related to public works preventative maintenance scheduling, work orders, asset inventory, asset inspections, employee timesheets, and administrative modules for city, county, and state planning and budgeting.

As part of these instructions, Stacy writes process analyses, complete with technical descriptions. She includes these process analyses for at least two reasons:

- **FYIs.** Many of Stacy's clients will e-mail her company asking, "How did that work?" Her process analyses give these inquiring minds a view "behind the scenes."
- **Troubleshooting.** More importantly, process analyses help her clients complete a task. The software asks clients to input data regarding their asset management. If the clients get an error reading, knowing how the software works can give them a better understanding of how a problem might have occurred and how to fix it.

Here's a good example of one of Stacy's process analyses. The GBA Master Series "Accident Manager" software allows a city's public works department to determine if an intersection or stretch of road needs improvement. If many traffic accidents are occurring in one location, perhaps the city needs to add more signals, remove roadway obstructions, add speed bumps, change traffic patterns, and so on.

Stacy writes, "The screens in this ['Accident Manager'] are fully customizable, allowing you to tailor the module to fit your accident or case report forms. Additionally, the accident records are validated by the system to create an accurate collision diagram report."

To help her clients better understand the software and use it effectively, Stacy then writes, "For additional information on the validation performed by the system, consult the Accident Validation Process help guide." This process analysis says the following:

"The GBA Accident Manager™ Data Management module allows you to record details of your traffic accidents. Based on the information found in your accident records, the system conducts a validation process to verify the location of the accident, direction of the accident, and the type of accident that took place. These validated results are then compiled in an Intersection Collision Diagram. This diagram allows you to accurately project accident patterns and make recommendations for street improvements. Listed below are the steps taken by the system to validate the user-entered accident data.

1. The system checks to make sure there is at least one vehicle involvement record for the accident. This is found on the Involvement tab on the Data Management module.

   Note: You must have at least one vehicle record associated with the accident before it can be saved.

2. The system will then perform validations for the following: pedestrian involvement, type of accident (rear-end collision, head-on collision, sideswipe), pre-crash maneuver, and direction of vehicle travel."

In addition to her instructions and technical descriptions, a process analysis gives Stacy another way to meet her end-user's needs for effective technical communication.

CRITERIA FOR WRITING TECHNICAL DESCRIPTIONS AND PROCESS ANALYSES

As with any type of technical communication, there are certain criteria for writing technical descriptions.

Title

Preface your text with a title precisely stating the topic of your description. This could be the name of the mechanism, tool, or piece of equipment you're writing about.

Overall Organization

In the **introduction** specify and define what topic you are describing, explain the mechanism's *functions* or *capabilities,* and list its *major components.*

example	The Apex Latch (#12004), a mechanism used to secure core sample containers, is composed of three parts: the hinge, the swing arm, and the fastener.

example	The DX 56 DME (Distance Measuring Equipment) is a vital piece of aeronautical equipment. Designed for use at altitudes up to 30,000 feet, the DX 56 electronically converts elapsed time to distance by measuring the length of time between your transmission and the reply signal. The DX 56 DME contains the transmitter, receiver, power supply, and range and speed circuitry.

In the **discussion,** use highlighting techniques (itemization, headings, underlining, white space) to describe each of the mechanism's components and how the mechanism works—its process.

Your **conclusion** depends on your purpose in describing the topic. Some options are as follows:

Sales—"Implementation of this product will provide you and your company . . ."	Guarantees—"The XYZ carries a 15-year warranty against parts and labor."	Comparison/contrast—"Compared to our largest competitor, the XYZ has sold three times more . . ."
Uses—"After implementation, you will be able to use this XYZ to . . ."	Testimony—"Parishioners swear by the XYZ. Our satisfied customers include . . ."	Reiteration of introductory comments—"Thus, the XYZ is composed of six interchangeable parts."

Graphics

See Chapter 9 for more discussion of graphics.

Use *graphics* in your technical descriptions. Today, many companies use the *super comic book look*—large, easy-to-follow graphics that complement the text. You can use line drawings, photographs, clip art, exploded views, or sectional cutaway views of your topic, each accompanied by *callouts* (labels identifying key components of the mechanism).

Internal Organization

When describing your topic in the discussion portion of the technical description, itemize the topic's components in logical sequence. Components of a piece of equipment, tool, or product can be organized by importance.

However, *spatial* organization is better for technical descriptions. When a topic is spatially organized, you literally lay out the components as they are seen, as they exist in space. You describe the components as they are seen either from left to right, from right to left, from top to bottom, from bottom to top, from inside to outside, or from outside to inside. In contrast, when writing your process analysis, you will use *chronological* organization to show how the tool or mechanism works.

Development

To describe your topic clearly and accurately, detail the following:

Weight	Materials (composition)
Size (dimensions)	Identifying numbers
Color	Make/model
Shape	Texture
Density	Capacity

Word Usage

Your word usage, either photographic or impressionistic, depends on your purpose. For factual, objective technical descriptions, use photographic words. For subjective, sales-oriented descriptions, use impressionistic words. Photographic words are denotative, quantifiable, and specific. Impressionistic words are vague and connotative. Table 11.1 shows the difference.

TECHNOLOGY TIPS

How to Make Callouts Using Microsoft Word 2007

Callouts are a great way to label mechanism components in your technical description or process analyses. Follow these steps to create callouts.

1. Click on **Insert** in your Menu bar.

2. Click on **Shapes.** You will see a dropdown box.

3. Select the type of callout you want to use, such as the **Rounded Rectangular Callout** shown below.

4. When you choose the callout, a crosshair will appear. Use it to draw your callout, as shown.

Note: When you draw the callout, Word 2007 opens up the **Format** tab, as shown.

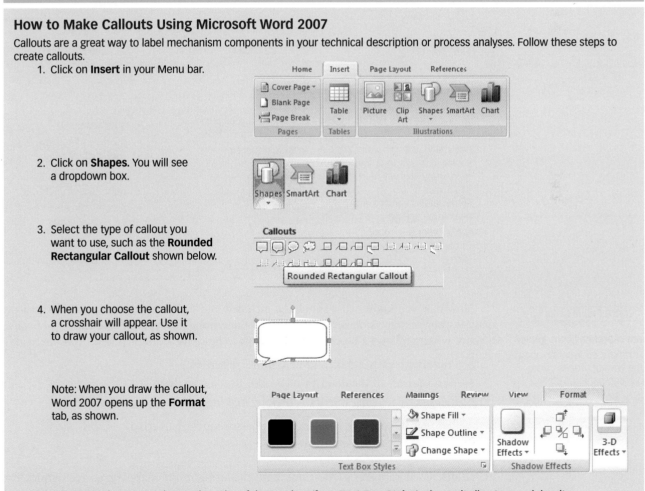

From the **Format** tab, you can change the color of the text box (from **Text Box Styles**), change its line type and density (from **Shape Outline**), create shadow effects, and 3-D effects, and add content.

TABLE 11.1 Photographic Versus Impressionistic Word Usage

Photographic	Impressionistic
6'9"	tall
350 lb	heavy
gold	precious metal
6,000 shares of United Can	major holdings
700 lumens	bright
0.030 mm	thin
1966 XKE Jaguar	impressive car

THE WRITING PROCESS AT WORK

The Writing Process

Prewriting	Writing	Rewriting

• Decide whether you are writing to inform, instruct, or persuade.
• Determine whether your audience is high tech, low tech, or lay. This will help you decide which abbreviations and acronyms need to be defined.
• Gather your information through brainstorming/listing.
	• Organize your description spatially and your process analysis chronologically.
	• Include visual aids and callouts.
	• Use headings and subheadings for easy navigation.
		• Revise your draft by
		  • adding details
		  • deleting wordiness
		  • simplifying words
		  • enhancing the tone
		  • reformatting your text
		  • proofreading and correcting errors

Now that you know what should be included in a technical description and/or process analysis, your next question is, "How do I go about writing this type of document?" As always, effective writing follows a process. To write your technical description and process analysis,

- *Prewrite* to gather data and determine objectives
- *Write* a draft of the description and/or process analysis
- *Rewrite* by revising the draft, thereby making it as perfect as possible

Remember that the parts of the writing process frequently overlap.

Prewriting

We've discussed reporter's questions and clustering/mind mapping as techniques for gathering data and determining objectives. Either of these two prewriting tools can be used for your technical description. However, another prewriting technique can be helpful: *brainstorming/listing*.

Brainstorming/listing, either individually or as a group activity, is a good, quick way to sketch out ideas for your description or process analysis. To brainstorm/list, do the following:

Title Your Activity: Write the title of your activity at the top of the page, to help you maintain your focus (for example, *Description: XYZ model 2267 Light Pen*).

List Any and All Ideas: Jot down any ideas or thoughts you might have on the topic. Don't editorialize. Avoid criticizing ideas at this point. Just randomly jot down as many aspects of the topic as possible, without making value judgments.

Edit Your List: Reread the list and evaluate it. To do so, (a) select the most promising features, (b) cross out any that don't fit, and (c) add any obvious omissions.

Writing

Once you've gathered your data and determined your objectives, the next step is writing your description or process analysis.

Review Your Prewriting: After a brief gestation period (an hour, a half-day, or a day) in which you wait to acquire a more objective view of your prewriting, go back to your brainstorming/listing and reread it. Have you omitted all unneeded items and added any important omissions? If not, do so.

Organize Your List: Make sure the items in the list are organized effectively. To communicate information to your readers, organize the data so readers can follow your train of thought. This is especially important in technical descriptions. You want your readers not only to understand the information but also to visualize the mechanism or component. This visualization can be achieved by organizing your data spatially.

Title Your Draft: At the top of the page, write the name of your piece of equipment, tool, mechanism, or process.

Write a Focus Statement: In one sentence, write the following:

- Name of your topic (plus any identifying numbers)
- Possible functions of your topic or your reason for writing the description and analysis
- Number of parts composing your topic

This will eventually act as the introduction. For now, however, it can help you organize your draft and maintain your focus.

Draft the Text: Use the sufficing technique discussed in earlier chapters. Write quickly without too much concern for grammatical or textual accuracy, and let what you write suffice for now. Just get the information on the page, focusing on overall spatial organization for the description, chronology for the process analysis, and a few highlighting techniques (headings, perhaps). The time to edit is later.

Sketch a Rough Graphic of Your Equipment, Component, or Mechanism: You can refine this rough graphic later during the rewriting stage.

Rewriting

Rewriting is the most important part of the process. This is the stage during which you fine-tune and polish your technical description and process analysis, making it as perfect as it can be. Perfected communication ensures your credibility. Imperfect communication makes you and your company look bad.

Revise your draft as follows:

Add Any Detail Required to Communicate More Effectively: Add information such as brand names, model numbers, and specificity of detail (size, material, density, dimensions, color, weight, etc.).

Delete Dead Words and Phrases for Conciseness: Refer to our discussion of effective technical communication style in Chapter 3.

Simplify Long-Winded, Old-Fashioned Words and Phrases for Easier Understanding: Chapter 3 will help you with this important aspect of your writing.

Move (Cut and Paste) Information to Ensure Effective Organization: You don't want to give your reader a distorted view of your mechanism or process analysis. To avoid doing so, maintain a spatial order—left to right, right to left, and so on for your description and a chronological order for the process analysis. For example, let's say you are describing a pencil, graphically depicted horizontally so the eraser end is to the left and the graphite tip is to the right. This pencil consists of the eraser, metal ferrule, wooden body, and sharpened point (that's the correct spatial order). If you describe the eraser first, then the tip, then the body, and then the ferrule, you've distorted the spatial order.

Reformat Your Text for Ease of Access through Highlighting Techniques: Use highlighting techniques such as headings, boldface type, font size and style, itemization, and so forth. Rewriting is also the time to perfect your graphics. When revising your description or process analysis, make sure that your graphics are effective, as follows:

- Add a figure number and title. Place these beneath the graphic (for example, "Figure 11.2 Exhalation Valve with Labeled Callouts").
- Draw, photograph, or reproduce graphics neatly.
- Place the graphics in an appropriate location, near where you first mention them in the text.
- Make sure that the graphics are an effective size. You don't want the graphics to be so small that your reader must squint to read them, nor do you want them to be so large that they overwhelm the text.
- Label the components. Use callouts to name each part, as in Figure 11.2.

FIGURE 11.2   Exhalation Valve with Labeled Callouts

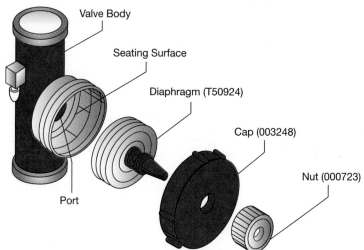

Courtesy of Nellcor Puritan Bennett Corp.

Enhance the Style of Your Text: Make sure that you have avoided impressionistic words. Instead, use photographic words, which are precise and specific. In addition, strive for a personalized tone. It is difficult to incorporate pronoun usage in technical descriptions. However, even when writing about a piece of equipment, remember that "people write to people." Although you are describing an inanimate object, you're writing to another human being. Therefore, use pronouns to personalize the text. (Examples later in this chapter clarify how you can accomplish this.)

Correct Your Draft: Proofread to correct grammar errors as well as errors in content: measurements, shapes, textures, materials, colors, and so on.

Avoid Biased Language: Remember that your text will be read by multiculturally diverse audiences as well as by readers of different ages and genders.

The following checklist will help you to write technical descriptions and process analyses.

Multicultural Audience

See Chapter 12 for more discussion of multicultural audience.

TECHNICAL DESCRIPTION AND PROCESS ANALYSIS CHECKLIST

_____ 1. Does the technical description or process analysis have a title noting your topic's name (and any identifying numbers)?

_____ 2. Does the technical description or process analysis's introduction (a) state the topic, (b) mention its functions or the purpose, and (c) list the components?

_____ 3. Does the technical description or process analysis's discussion use headings to itemize the components for reader-friendly ease of access?

_____ 4. Do you need to define the mechanism and its main parts?

_____ 5. Is the detail within the technical description or process analysis's discussion precise? That is, does the discussion portion of the description or process analysis specify the following?

Colors	Capacities
Sizes	Textures
Materials	Identifying numbers
Shapes	Weight
Density	Make/model

_____ 6. Are all of the calculations, measurements, or process timelines correct?

_____ 7. Do you sum up your discussion using any of the optional conclusions discussed in this chapter?

_____ 8. Does your technical description or process analysis provide graphics that are correctly labeled, appropriately placed, neatly drawn or reproduced, and appropriately sized?

_____ 9. Do you write using an effective technical style (low fog index) and a personalized tone?

_____10. Have you avoided biased language and grammatical and mechanical errors?

PROCESS EXAMPLE

The following student-written process log clarifies the way process is used in writing technical descriptions.

Prewriting

We asked students to describe a piece of equipment, tool, or mechanism of their choice. One student chose to describe a cash register pole display. To do so, he first had to gather data using brainstorming/listing. He provided the following list:

1. Three parts
2. Pole printed circuit board (PCB)
3. Case assembly
4. Filter
5. Plastic materials

FIGURE 11.3 Student's Brainstorming/Listing

Pole PCB

- length—15 mm
- width—5.1 mm
- tube length—10.8 mm
- face plate width—2.3 mm
- thick—1.7 mm
- PCB thickness—0.2 mm
- 10 inch stranded wire with female connectors
- fiberglass and copper construction

Pole Case Assembly

- long—15.5 mm
- bottom width—2.5 mm
- top width—0.9 mm
- mounting pole—5 mm high × 3.2 mm diameter
- tounge for mounting—3.1 mm
- lower mounting tounge—1.5 mm
- side mounting tounge—0.8 mm high
- high—6.1 mm
- almond-colored plastic

Filter

- long—15.6 mm
- high—6.2 mm
- thick—0.7 mm
- plastic, blue

Next, we asked students to go back to their initial lists and edit them. Because this student's list was so sketchy, he did not need to omit any information. Instead, he had to add omissions and missing detail, as in Figure 11.3.

Writing

After the prewriting activity, we asked students to draft a technical description. Focusing on overall organization, highlighting, detail, and a sketchy graphic, this student wrote a rough draft (Figure 11.4).

Rewriting

Once students completed their rough drafts, we used peer review groups to help each student revise his or her paper. In Figure 11.5, students

- Added new detail for clarity
- Corrected any errors
- Perfected graphics

FIGURE 11.4 Student's Rough Draft

The QL169 Customer Pole Display provides the viewing of all transaction data for the customer. The display consist of a printed circuit board, a case assembly, and a filter display.

Display Circuit Board

- length—15 mm
- width—5.1 mm
- tube length—10.8 mm
- face plate width—2.3 mm
- thick—1.7 mm
- PCB thickness—0.2 mm
- 10 inch stranded wire with female connectors
- fiberglass and copper construction

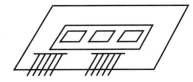

Display Case Assembly

- long—15.5 mm
- bottom width—2.5 mm
- top width—0.9 mm
- mounting pole—
  5 mm high × 3.2 mm diameter
- tounge for mounting—3.1 mm
- lower mounting tounge—1.5 mm wide
- side mounting tounge—0.8 mm high
- high—6.1 mm
- almond-colored plastic

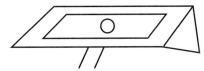

Display Filter

- long—15.6 mm
- high—6.2 mm
- thick—0.7 mm
- plastic, blue

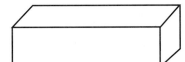

The student incorporated these suggestions and prepared the finished copy (Figure 11.6).

SAMPLE TECHNICAL DESCRIPTION AND PROCESS ANALYSIS

Figure 11.7 shows a professionally written technical description merging graphic and text. This is a highly effective technical description with these characteristics:

- Title
- Excellent introduction that lists topic, function, and components
- Effectively drawn and placed graphic

FIGURE 11.5 Student's Suggested Revisions

No Title — define — awkward —

The QL169 Customer Pole Display provides the viewing of ~~all transaction data~~ for the customer. The display consist of a printed circuit board, a case assembly, and a filter display. (SP)

Display Circuit Board
Make All Caps
- length—15 mm
- width—5.1 mm
- tube length—10.8 mm
- face plate width—2.3 mm
- thick—1.7 mm   thickness
- PCB thickness—0.2 mm
- 10 inch stranded wire with female connectors
- fiberglass and copper construction
  Too vague (specify)

Number your components

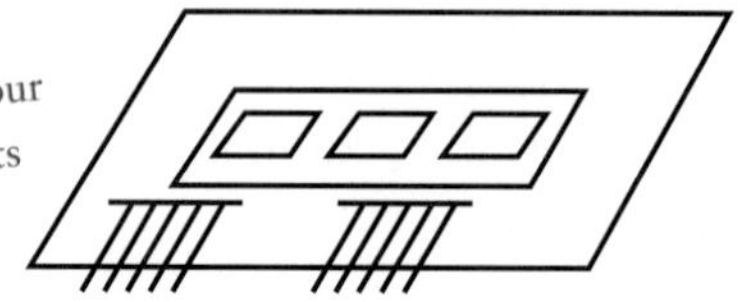

Display Case Assembly
Make All Caps   Number your components
- long—15.5 mm   length
- bottom width—2.5 mm
- top width—0.9 mm
- mounting pole—
  5 mm high × 3.2 mm diameter
- tounge for mounting—3.1 mm
- (SP) lower mounting tounge—1.5 mm wide
- side mounting tounge—0.8 mm high
- high—6.1 mm   height
- almond-colored plastic
  vague

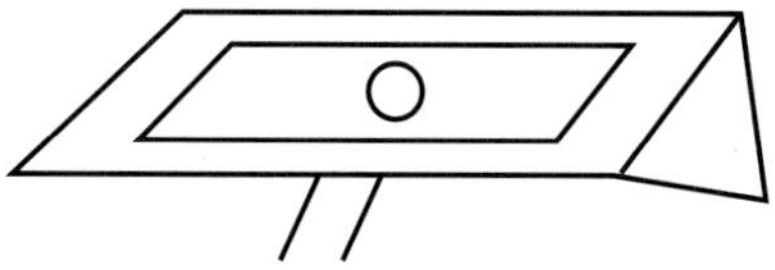

Display Filter
- long—15.6 mm   length
- high—6.2 mm   height
- thick—0.7 mm   thickness
- plastic, blue
  what

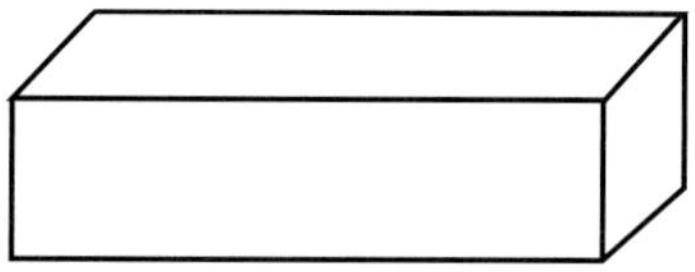

Add a conclusion

FIGURE 11.6 Student's Finished Copy

QL169 Pole Display

The QL169 pole is an electronic mechanism that provides an alphanumeric display for customer viewing of cash register sales. The display consists of a printed circuit board (PCB) assembly, a case assembly, and a display filter.

Include a definition in the introduction.

Item Description
1. PCB ASSEMBLY
2. CASE ASSEMBLY
3. DISPLAY FILTER

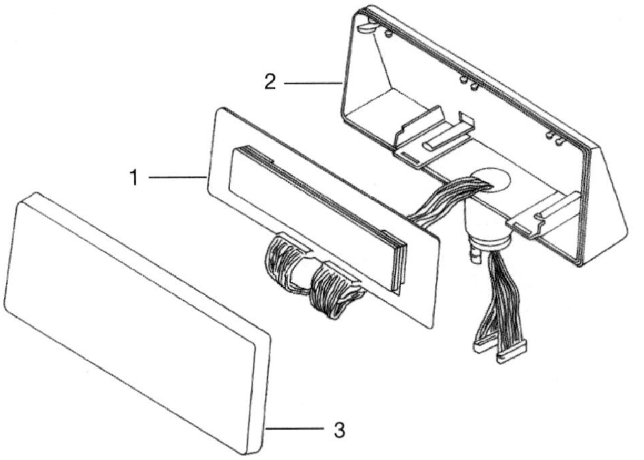

Figure 1. QL169

Exploded graphics help readers visualize the topic.

1.0 PCB Assembly

The printed circuit board, containing the display's electrical circuitry, is constructed of fiberglass with copper etchings.
The board consists of the following features:

1.1 Length—15 mm
1.2 Width—5.1 mm
1.3 Tube length—10.8 mm
1.4 Tube faceplate
    width—2.3 mm
1.5 Tube total width—2.8 mm
1.6 Tube thickness—1.7 mm
1.7 PCB thickness—0.2 mm
1.8 20 conductor 10" 22-gauge
    stranded wire with two AECC
    female connectors
    (AECC part #7214-001) 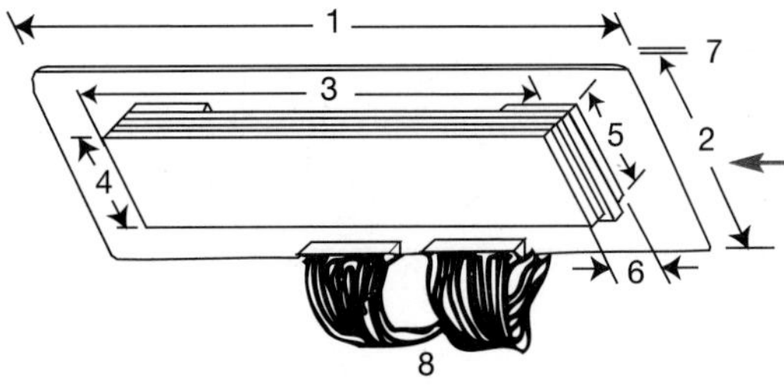
1.9 American Display Company blue phosphor display tube
    (ADC part #1172177)

Figure 2. PCB

Specificity of detail adds clarity.

continued

FIGURE 11.6 Continued

2.0 Case Assembly

Almond-colored ABS plastic, used to construct the case assembly, protects the PCB.

2.1 Length—15.5 mm

2.2 Bottom width—2.5 mm

2.3 Top width—0.9 mm

2.4 Mounting pole—5 mm high and 3.2 mm in diameter

2.5 Mounting tongue inside width—3.1 mm from side of assembly

2.6 Lower mounting tongue— 1.5 mm wide

2.7 Side mounting tongue— 0.8 mm high

2.8 Tongue thickness—0.2 mm

2.9 Height—6.1 mm

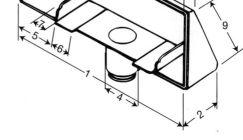

Figure 3. Case Assembly

3.0 Display Filter

3.1 Length—15.6 mm

3.2 Height—6.2 mm

3.3 Thickness—0.7 mm

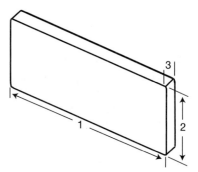

Figure 4. Display Filter

Transparent blue plastic, used to construct the display filter, allows the customer to view the readings. The QL169 pole provides easy viewing of clerk transactions and ensures cashier accuracy.

Conclusion focuses on end-user benefits.

- Reader-friendly discussion with headings exactly corresponding to the callouts in the drawing
- Precisely detailed discussion focusing on materials, color, dimensions, and so on
- Effective conclusion

Figure 11.8 is a process analysis of why, when, and how automobile air bags inflate.

FIGURE 11.7   Professionally Written Technical Description

A1 Feed Switch

The A1 Feed Switch is an electronic device that provides an automatic solid-state closure (normally open type) after sensing the lack of presence of food. The feed switch can detect dry concentrate, meal, rolled barley, and fine and coarse feeds, as well as special feeds such as soybean meal and high-moisture corn. Figure 1 shows the three principal parts of the feed switch.

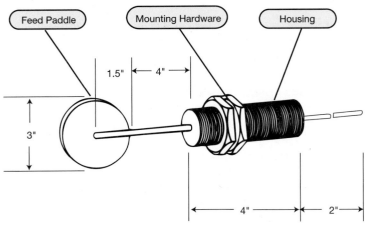

Figure 1. Electronic A1 Feed Switch

- HOUSING: The housing consists of a blue molded polycarbonate cylinder, 4″ in length and 1″ in diameter. Encircling the complete length of the housing are 1″–18 threads. At the middle of one end of the housing is a 2″ Belden 6906 cable. Attached to the middle of the other end of the housing is a feed paddle.
- MOUNTING HARDWARE: The mounting hardware consists of two white, flat, hexagonal, plastic Delryn locking nuts, Grade 2, 1″ ID, 1 1/4″ OD, 1/8″ thick, and 1″–18 EF threads per inch.
- FEED PADDLE: The feed paddle consists of a 3″ diameter, 16-gauge (0.060″) No. 304 galvanized steel, Finish 2D or better, welded to a No. 101 Finish 3D or better galvanized steel shaft 5 1/2″ in length and 1/4″ in diameter. Exposed threads 1/4″–20 and 1/2″ in length protrude from the free end of the shaft. Welded on three sides to the middle of the 3″ diameter steel is the 1 1/2″ of the unthreaded shaft end. The feed paddle is white Teflon coated to a thickness of 1/64″.

The Electronic A1 Feed Switch, with its five-year warranty, is reliable in dusty and dirty environments. The easy-to-install, short-circuit-protected switch is internally protected for transient voltage peaks to a maximum of 5 KV, for up to 10 msec duration.

FIGURE 11.8 Why, When, and How Does an Air Bag Inflate?

The introduction explains what the topic is, why it is important, and where the mechanism is located.

Air bags save lives. Driver and passenger air bags are designed to inflate in frontal or side crashes. Steering wheel, right front instrument panel, or side-panel air bags will not inflate on all occasions. If a car drives over a bump or if a crash is "minor," such as hitting an object while driving at a slow speed, an air bag will not deflate. However, when cars hit walls (or trucks or cars or trees), air bags inflate to minimize injury and to save people's lives.

Before Inflation

How does the air bag detect whether the car has hit a bump versus being involved in a collision?

The airbag system's crash sensor can differentiate between head-on collisions and simple bumps in the road as follows:

1. A steel ball slides inside a smooth bored cylinder.
2. The ball is secured by a magnet or by a stiff spring. This inhibits the ball's motion when the car encounters minor motion changes, such as bumps or potholes.
3. If the car comes to a dramatic and rapid stop, a force equal to running into a brick wall at about 10 to 15 miles per hour, such as in a head-on crash, the ball quickly moves forward. This closes a contact and completes an electrical circuit, which initiates the process of inflating the airbag.

Parts of an Air Bag

This technical description discusses the parts of the mechanism, the materials used, the location of these components, and the chemical compounds required for activation.

Air bag systems consist of the following:

- **Air bag:** made of thin nylon. A nontoxic powder (cornstarch or talcum powder) inside the air bag keeps it flexible and ensures the parts of the air bag do not stick together.
- **Air bag holding compartment:** the steering wheel, dashboard, seat, or door.
- **Sensor:** a device that tells the bag to inflate.
- **Inflator canister:** consisting of sodium azide (NaN_3), potassium nitrate (KNO_3), and silicon dioxide (SiO_2), which produce nitrogen gas (N_2).

During Inflation

In a severe impact, the air bag sensing system will deploy in milliseconds. The following occurs:

1. The air bag's crash sensor triggers a switch that energizes a wire, sending electricity into a heating element in the propellant that releases gas from the inflator canister.
2. A pellet of sodium azide (NaN_3) is ignited, generating nitrogren gas (N_2).
3. A nylon air bag, folded into the dashboard, steering wheel, and/or side panels of the door, inflates at a speed of 150–250 mph, taking only about 40 milliseconds (about 1/20 of a second) for the inflation to be complete.

Here, providing the mechanism's process, the text explains how an air bag works. Note the specificity of detail: 150–250 mph and 1/20 of a second.

FIGURE 11.8 *Continued*

4. The sodium azide inside the air bag produces sodium metal, which is highly reactive, potentially explosive, and harmful when in contact with eyes, nose, or mouth. To render this harmless, the sodium azide reacts with potassium nitrate (KNO_3) and silicon dioxide (SiO_2)—also inside the air bag—to produce silicate glass, a harmless and stable compound.

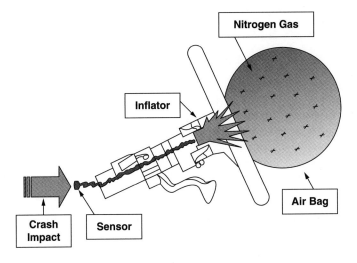

This illustration graphically depicts the process in action. When the crash sensor is triggered, it activates the inflator. Nitrogen gas explodes, inflating the air bag.

After Inflation

Air bag vents, minute holes in the bag, allow the deployed air bags to deflate immediately after impact. This ensures that the car's inhabitants do not smother.

Though air bags were first used in 1973, they have only been mandatory in all cars since 1998. Have air bags made a difference? Absolutely! The Bureau of Transportation Statistics says air bags saved an estimated 19,659 lives from 1975–2005.

The conclusion sums up the process description by quantifying the significance of air bags as a means of saving lives.

CHAPTER HIGHLIGHTS

1. Technical descriptions are often used in user manuals, product specifications, reports, and sales literature.
2. Process analyses tell how a tool, mechanism, or piece of equipment works.
3. The introduction of a technical description or process analysis explains the mechanism's function and major components.
4. Graphics with callouts make a technical description and process analysis comprehensible to the reader.
5. Follow a spatial pattern of organization in a technical description.
6. Follow a chronological order of organization when describing how a process works.
7. Word usage for descriptions and processes will be photographic and/or impressionistic.
8. Use headings and subheadings to organize the content of your description or process analysis.

9. Use personal pronouns to personalize your description or process analysis and achieve more effective technical communication.

10. Using descriptive titles will enhance your description or process analysis.

APPLY YOUR KNOWLEDGE

CASE STUDY

Cathode Ray Tube Monitors Versus Liquid Crystal Display Flat Panels

Nitrous Systems Biotechnology, Inc. plans to expand its company. Part of this expansion includes the need to hire 50 new biotechnicians. Each, of course, will need his or her own desktop computer, complete with monitor. But what kind of monitor should the company buy—CRTs or LCDs? Sure, LCD flat panels are sleek and take up less room, but CRTs are cheaper, right? To meet their computer technology needs, Nitrous has hired Alpha Beta Consulting (A/B/C). Nitrous wants a brief comparison/contrast written, complete with a product description and a process analysis explaining the differences between these two monitor types.

In addition, which monitor type provides the best resolution and viewing quality, refresh rates, and screen-viewable size options? Are CRTs cheaper, or has that price break changed? Even energy savings are a consideration when purchasing either CRTs or LCDs? Finally, how do CRTs work versus LCDs?

As a CIS/IT tech who works for A/B/C, you need to research this topic and write the technical description, complete with process analysis, to meet the client's request.

INDIVIDUAL OR TEAM PROJECTS

1. Write a technical description, either individually or as a team. To do so, first select a topic. You can describe any tool, mechanism, or piece of equipment. However, don't choose a topic too large to describe accurately. To provide a thorough and precise description, you will need to be exact and minutely detailed. A large topic, such as a computer, an oscilloscope, a respirator, or a Boeing airliner, would be too demanding for a two- to four-page description. On the other hand, do not choose a topic that is too small, such as a paper clip, a nail, or a shoestring. Choose a topic that provides you with a challenge but that is manageable. You might write about any of the following topics.

USB flash drive	Computer disk	DVD remote control
Wrench	Computer mouse	Mechanical pencil
Screwdriver	Light bulb	Ballpoint pen
Pliers	Calculator	Computer monitor
Wall outlet	Automobile tire	Cell phone

Once you or your team have chosen a topic, prewrite (listing the topic's components), write a draft (abiding by the criteria provided in this chapter), and rewrite (revising your draft).

2. Write a process analysis, either individually or as a team, on one or more of the following topics.
   - How does blood coagulate?
   - How does an x-ray machine work?
   - How are viruses spread?
   - How do Wi/Fi connections work?

- How do rotary engines work?
- How do fuel gauges work?
- How is metal welded?
- How does a metal detector work?
- How do browsers work?
- How do computer viruses work?

3. Find examples of professionally written technical descriptions or process analyses in technical books and textbooks, professional magazines and journals, or on the Internet. Bring these examples to class and discuss whether they are successful according to the criteria presented in this chapter. If the descriptions or process analyses are good, specify how and why. If they are flawed, state where and suggest ways to improve them. You might even want to rewrite the flawed descriptions or process analyses.

4. Select a simple topic for description, such as a pencil, coffee cup, toothbrush, or textbook (you can use brainstorming/listing to come up with additional topics). Describe this item without mentioning what it is or providing any graphics. Then, read your description to a group of students/peers and ask them to draw what you have described. If your verbal description is good, their drawings will resemble your topic. If their drawings are off base, you haven't succeeded in providing an effective description. This is a good test of your writing abilities.

5. In degree-specific teams, choose a topic from your area of interest and expertise to describe (including a process analysis). For example, students majoring in HVAC could describe a humidifier and explain how it works. Biotech and nursing students could describe and provide a process analysis for a nebulizer, blood pressure monitor, or glucometer. EMT students could write about defibrillators. Automotive technicians could describe and include a process analysis for jumpstarters, battery chargers, or hydraulic jacks. Welding students could write about MIG, TIG, or stick welding.

PROBLEM-SOLVING THINK PIECE

Read the following process analysis and reorganize the numbered sentences to achieve a clear, chronological order. If necessary, research this process to learn more about how blood clots.

How does blood clot?

1. In our bodies, blood can clot due to platelets and the thrombin system.

2. When bleeding occurs, chemical reactions make the surface of the platelet "sticky." These sticky platelets adhere to the wall of blood vessels where bleeding has occurred.

3. Platelets, tiny cells created in our bone marrow, travel in the bloodstream and wait for a bleeding problem.

4. Soon, a "white clot" is formed, so called because the clotted platelets look white.

5. Blood clot consists of both platelets and fibrin. The fibrin strands bind the platelets and make the clot stable.

6. In the thrombin system, several blood proteins become active when bleeding occurs. Clotting reactions produce fibrin, long, sticky strings. These sticky strands catch red blood cells and form a "red clot."

7. Primarily, arteries clot due to platelets, while veins clot due to our thrombin system.

WEB WORKSHOP

1. You are ready to purchase a product. This could include printers, monitors, digital cameras, scanners, PCs, laptops, speakers, cables, adapters, automotive engine hoists, generators, battery chargers, jacks, power tools, truck boxes, screws, bolts, nuts, rivets, hand tools, and more. A great place to shop is online. By going to an online search engine, you can find not only prices for your products but also technical descriptions or specifications. These will help you determine if the product has the size, shape, materials, and capacity you want.

   Go online to search for a product of your choice and review the technical description or specifications provided. Using the criteria in this chapter and your knowledge of effective technical writing techniques, analyze your findings.

   - How do the online technical descriptions compare to those discussed in this textbook?
   - Are graphics used to help you visualize the product?
   - Are callouts used to help you identify parts of the product?
   - Are high-tech terms defined?
   - Is the use of the product explained?
     a. Report your findings, either in an oral presentation or in writing (e-mail message, memo, letter, or report).
     b. Rewrite any of the technical descriptions that need improvement according to the criteria provided in this textbook.

2. HowStuffWorks (http://www.howstuffworks.com/index.htm) is an extensive online library of process descriptions. Revise a process description in this Web site by improving its layout, content, organization, and/or graphics.

QUIZ QUESTIONS

1. What is a technical description?
2. Why would you use a technical description?
3. What are the components of a technical description?
4. What is a process analysis?
5. Give three examples of topics for a process analysis.
6. What is a callout?
7. What is the best organizational pattern for a process analysis?
8. What is the best way to organize a technical description?
9. What information is included in a technical description?
10. What is the difference between photographic and impressionistic words?

CHAPTER 12

Instructions

COMMUNICATION *at work*

This scenario shows the importance of effective instructions in the biomedical industry and various communication channels used to convey step-by-step procedures.

PhlebotomyDR

As the baby boom generation ages, medical needs are expanding—sometimes faster than medical care facilities and medical professionals can manage. One area in which this has been felt most acutely is in medical laboratories. Thousands of medical technicians need to be trained to accommodate increased demand.

PhlebotomyDR is a medical consulting firm seeking to solve this problem. Its primary area of concern is training newly hired technicians responsible for performing blood collection. PhlebotomyDR facilitates training workshops to teach venipuncture standards and venipuncture procedures.

PhlebotomyDR focuses on the following venipuncture instructions.

- Proper patient identification procedures
- Proper equipment selection, sterilization, use, and cleaning
- Proper labeling procedures
- Order of phlebotomy draw
- Patient care before, during and following venipucture
- Safety and infection control procedures
- Procedures to follow when meeting quality assurance regulations

Objectives

When you complete this chapter, you will be able to

1. Understand when to write an instruction.

2. Follow criteria in order to write an effective instruction.

3. Recognize your audience when writing an instruction.

4. Include key components for an effective instruction.

5. Follow the writing process to create instructions.

6. Evaluate instructions using the effective instruction checklist and the usability checklist.

Each of the instructions mentioned requires numerous steps, complete with visual aids. PhlebotomyDR offers its audience various communication channels. Hospitals, labs, and treatment centers can access PhlebotomyDR's instruction as follows:

- Hard-copy instructional manuals
- Online instructions, accessible at *http://www.phlebotomydr.com*
- Videos showing step-by-step performances of blood collection, complete with case studies enacted by technicians, patients, and supervisors
- Computer-aided instruction (CAI) for individual tutorials
- One-on-one tutorials with trained phlebotomists
- Instructional workshops geared toward groups of seminar participants

PhlebotomyDR's outstanding staff realizes that trained technicians make an enormous difference. Training, achieved through instructional manuals, electronic aids, and individual facilitation, ensures the health and safety of patients. Of equal importance, excellent training also benefits many stakeholders. Untrained technicians make errors that cost us all. Medical errors create insurance problems, the need to redo procedures, increased medical bills, the potential involvement of regulators and legislators, and dangerous repercussions for patients.

In contrast, effective communication, achieved through successful instruction, saves lives, time and money.

WHY WRITE AN INSTRUCTION?

Almost every manufactured product comes complete with instructions. You will receive instructions for baking brownies, making pancakes, assembling children's toys, or changing a tire, the oil in your car, or the coolant in your engine. Instructions help people set up stereo systems, construct electronic equipment, maintain computers, and operate fighter planes.

There are even instructions for making chicken noodle soup. When one major soup manufacturer stopped printing instructions for soup preparation on its cans of soup, the company received thousands of calls from consumers asking how to prepare the soup. Responding to the customers' needs, the company reverted to printing these instructions. "Pour soup in pan. Slowly stir in one can of water. Heat to simmer, stirring occasionally."

Even pop-top cans of soda include instructions: "1. Lift up. 2. Pull back. 3. Push down." When such simple items as soup and soda need instructions, you can imagine how necessary instructions are for complex products and procedures.

We frequently see short instructions about computer-related problems. If you are at home or in the office, for example, and you can't figure out how to operate one of your computer's applications, you will dial a 1-800 hotline and speak to a computer technician. Following is an e-mail message we received in such an instance. Notice how this e-mail includes an introduction, a numbered body, a cautionary note, and a conclusion.

From: pscsupport@earthlink.net
Sent: Saturday, April 28, 2008, 7:23 P.M
To: Steve Gerson
Subject: Technical Support

Thank you for contacting us about your computer problem. Recently, we have encountered cases where an online Earthlink update modified existing profiles. This change can affect your e-mail. To solve this problem create a new profile in Earthlink 5.0. [Note: This will remove old e-mail and your address book.]

1. Click on the icon to open up the Earthlink 5.0 sign-on screen.
2. Click on Configure.
3. Click on New Profile.
4. Click on Manage profiles.
5. Click on Add an existing account to this computer.
6. Click in the user name field and type in your user name in lowercase letters.
7. Click in the password field and type in your password in lowercase letters.
8. Click Next.
9. When asked, "Do you want to configure Internet Explorer for Earthlink," click Yes.
10. When asked, "Do you want to configure Netscape Communicator for Earthlink," click Yes.
11. Print out your information.
12. Click Finish.

If you need further help, please let us know. You can always access our customer support site at http://support.earthlink.net or click on the support tab in your personal start page.

Include instructions or user manuals when your audience needs to know how to do the following:

Operate a mechanism	Restore a product
Install equipment	Correct a problem
Manufacture a product	Service equipment
Package a product	Troubleshoot a system
Unpack equipment	Care for a plant
Test components	Use software
Maintain equipment	Set up a product
Clean a product	Implement a procedure
Monitor a system	Construct anything
Repair a system	Assemble a product

DOT-COM UPDATES

For excellent examples of instructions about home improvement, automobiles, gardening, crafts, and more, check out the following Do It Yourself Network link.

- http://www.diynetwork.com/

CRITERIA FOR WRITING INSTRUCTIONS

Odds are good that you've read a badly written instruction. You've probably encountered an instruction that failed to define unfamiliar terminology, that didn't tell you what equipment you needed to perform the task, that didn't warn you about dangers, clarify how to perform certain steps, or provide you enough graphics to complete the job successfully. Usually, such poorly written instructions occur because writers fail to consider their audience's needs. Successful instructions, in contrast, start with audience recognition.

Audience Recognition

The instruction tells you to "place the belt on the motor pulley," but you don't know what a "motor pulley" is nor do you know how to "place the belt" correctly. Or, you are told to "discard the used liquid in a safe container," but you don't know what is safe for this specific type of liquid. You are told to "size the cutting according to regular use," but you have never regularly performed this activity before.

Here is a typical instruction written without considering the audience.

Audience Recognition

See Chapter 4 for more discussion of audience recognition.

> To overhaul the manual starter, proceed as follows: Remove the engine's top cover. Untie the starter rope at the anchor and allow the starter rope to slowly wind onto the pulley. Tie a knot on the end of the starter rope to prevent it from being pulled into the housing. Remove the pivot bolt and lift the manual starter assembly from the power head.

example

Although many high-tech readers might be able to follow these instructions, many more readers will be confused. How do you remove the engine's top cover? Where is the anchor? Where is the pivot bolt and how do you remove it? What is the power head?

The problem is caused by writers who assume that their readers have high-tech knowledge. This is a mistake for several reasons. First, even high-tech readers often need detailed information because technology changes daily. You cannot assume that every high-tech reader is up to date on these technical changes. Thus, you must clarify. Second, low-tech and lay readers—and that's most of us—carefully read each and every step, desperate for clear and thorough assistance.

As the writer, you should provide your readers with the clarity and thoroughness they require. To do this, recognize accurately who your readers are and give them what they want, whether that amounts to technical updates for high-tech readers or precise, even simple, information for low-tech or lay readers. The key to success as a writer of instructions is, "Don't assume anything. Spell it all out—clearly and thoroughly!"

Components of Instructions

Not every instruction will contain the same components. Some very short instructions will consist of nothing more than a few, numbered steps. Other instructions, however, will consist of the components shown in Figure 12.1.

Title Page

Preface your instruction with a title page that consists of the *topic* about which you are writing, the *purpose* of the instruction, and a *graphic* depicting your product or service. For example, to merely title your instruction "iPod" would be uninformative. This title names the product, but it does not explain why the instruction is being written. Will the text discuss operating instructions, troubleshooting, service, or maintenance? A better title would be "Operating Instructions for the iPod." The graphical representation lets your audience see what the finished product will look like and helps you market your product or service.

Hazards

You can place hazard notations anywhere throughout your text. If a particular step presents a danger to the reader, you might want to call attention to this hazard just before asking the reader to perform the step. Similarly, to help your audience complete an action, you might want to place a note before the step, suggesting the importance of using the correct tool, not overtightening a bolt, or wearing protective equipment.

In addition to placing hazard notations before a required step, consider prefacing your entire instructions with hazard notations. By doing so, you make the audience aware of possible dangers, warnings, cautions, or notes in advance of performing the instructions. This is important to avoid potentially harming an individual, damaging equipment, and avoiding costly lawsuits. Include the following in your hazard notations.

- **Access.** Make the hazard notations obvious. To do so, vary your typeface and type size, use white space to separate the warning or caution from surrounding text, box the warning or caution, and call attention to the hazards through graphics.
- **Definitions.** What does *caution* mean? How does it differ from *warning, danger,* or *note*?

FIGURE 12.1 Key Components of Instructional Manuals

Title Page Topic Graphic Purpose	Hazards	Table of Contents	Introduction	List of Required Tools / Equipment
Glossary of Terms	Steps 1. 2.	Steps 3. 4. Etc.	Additional Components	Corporate Contact Information

Four primary institutions that seek to provide a standardized definition of terms are American National Standards Institute (ANSI), the U.S. military (MILSPEC), the Occupational Safety and Health Administration (OSHA), and the MSDSSearch National Repository (material safety data sheets).

To avoid confusion, we suggest the following hierarchy of definitions, which clarifies the degree of hazard.

1. **Note:** Important information, necessary to perform a task effectively or to avoid loss of data or inconvenience
2. **Caution:** The potential for damage or destruction of equipment
3. **Warning:** The potential for serious personal injury
4. **Danger:** The potential for death

Our goal is to avoid confusion, which can occur if one writes, "Cautionary Warning Notice!"

- **Colors.** Another way to emphasize your hazard message is through a colored window or text box around the word. Usually, *Note* is printed in blue or black, *Caution* in yellow, *Warning* in orange, and *Danger* in red.
- **Text.** To further clarify your terminology, provide the readers text to accompany your hazard alert. Your text should have the following three parts.

  1. **A one- or two-word identification alerting the reader.** Words such as *High Voltage, Hot Equipment, Sharp Objects*, or *Magnetic Parts*, for example, will warn your reader of potential dangers, warnings, or cautions.
  2. **The consequences of the hazards, in three to five words.** Phrases like "Electrocution can kill," "Can cause burns," "Cuts can occur," or "Can lead to data loss," for example, will tell your readers the results stemming from the dangers, warnings, or cautions.
  3. **Avoidance steps.** In three to five words, tell the readers how to avoid the consequences noted: "Wear rubber shoes," "Don't touch until cool," "Wear protective gloves," or "Keep disks away."

- **Icons.** Equipment is manufactured and sold globally; people speak different languages. Your hazard alert should contain an icon—a picture of the potential consequence—to help everyone understand the caution, warning, or danger.

DOT-COM UPDATES

For samples of hazard icons, check out the following link.

- http//www.speedysigns.com/signs/osha_readymade.asp

Figure 12.2 shows an effective page layout and the necessary information to communicate hazard alerts.

Table of Contents

Your instruction might have several sections. In addition to the actual steps, the instructional manual could include technical specifications, warranties, guarantees, FAQs, troubleshooting tips, and customer service contact numbers. An effective table of contents will allow your readers to go to any of these sections individually on an as-needed basis.

Introduction

Companies need customers. A user manual might be the only contact a company has with its customer. Therefore, instructions often are reader friendly and seek to achieve audience recognition and audience involvement. The manuals try to reach out and touch the customer in a personalized way. Look at these introductions from two user manuals.

> Thank you for your purchase. Installation and operating procedures are contained in this booklet about your new color TV. The minutes you spend reading this book will contribute to hours of viewing pleasure.

example

FIGURE 12.2 Hazard Alert

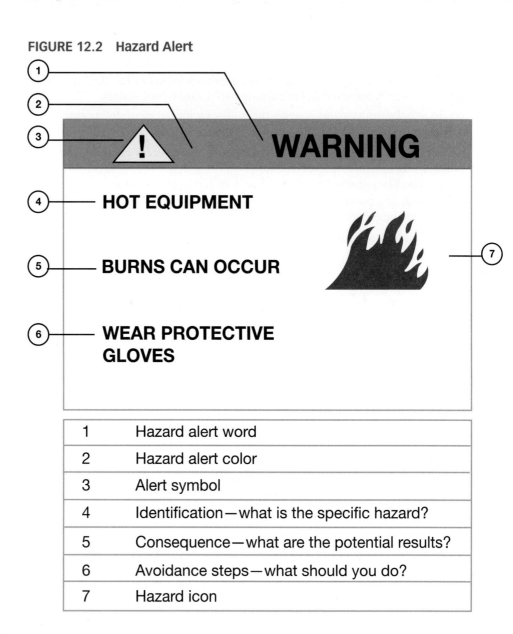

1	Hazard alert word
2	Hazard alert color
3	Alert symbol
4	Identification—what is the specific hazard?
5	Consequence—what are the potential results?
6	Avoidance steps—what should you do?
7	Hazard icon

example Welcome to the world of modern electronics. The cordless phone you have purchased is the product of advanced technology. This manual will help you install your new phone easily.

Each of these introductions uses pronouns (*you*, *your*, and *our*) to personalize the manual. The introductions also use positive words, such as *welcome*, *thank you*, and *pleasure* to achieve a positive customer contact. This is important because companies know that without customers, they're out of business. An effective introduction promotes good customer–company relationships.

Glossary

If your instruction uses the abbreviations *BDC*, *CCW*, or *CPR*, will the readers know that you are referring to "bottom dead center," "counterclockwise," or "continuing property records"?

If your audience is not familiar with your terminology, they might miss important information and perform an operation incorrectly.

To avoid this problem, define your abbreviations, acronyms, and symbols. You can define your terms early in the instruction, throughout the manual, or in a glossary located at the end of your manual. The following example defines terms alphabetically.

example

	This symbol designates an important operating or maintenance step.
BDC	Bottom dead center
CCW	Counterclockwise
Danger	This hazard alert designates the possibility of death. Be extremely careful when performing an operation.
RMS	Root mean square

Required Tools or Equipment

What tools or equipment will the audience need to perform the procedures? You don't want your audience to be in the middle of performing a step and suddenly realize that they need a missing piece of equipment. Provide this important information either through a simple list or graphics depicting the tools or equipment necessary to complete the tasks.

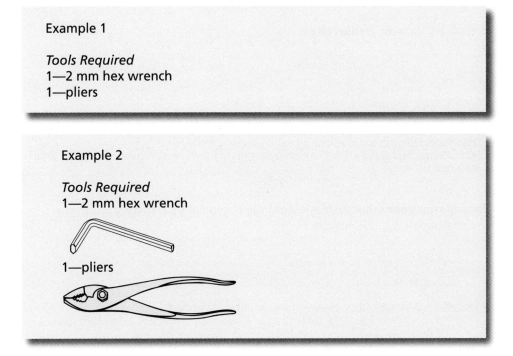

Example 1

Tools Required
1—2 mm hex wrench
1—pliers

Example 2

Tools Required
1—2 mm hex wrench

1—pliers

Instructional Steps

The instructional steps are the most important part of your manual—the actual actions required of the audience to complete a task. To successfully write your instruction, follow these steps.

1. **Organize the steps chronologically**—as a step-by-step sequence. You cannot tell your readers to do step 6, then go back to step 2, then accomplish step 12, then do step 4. Such a distorted sequence would fail to accomplish your goals. To operate machinery, monitor a system, or construct equipment, your readers must follow a chronological sequence. Be sure that your instruction is chronologically accurate.

2. **Number your steps**—Do not use bullets or the alphabet. Numbers, which you can never run out of, help your readers refer to the correct step. In contrast, if you used bullets, your readers would have to count to locate steps—seven bullets for step 7, and so on. If you used the alphabet, you'd be in trouble when you reached step 27.

3. **Use highlighting techniques**—Boldface, different font sizes and styles, text boxes, emphatic warning words, color, and italics call attention to special concerns. A danger, caution, warning, or specially required technique must be evident to your reader. If this special concern is buried in a block of unappealing text, it will not be read. This could be dangerous to your reader or costly to you and your company. To avoid lawsuits or to help your readers see what is important, call it out through formatting.

4. **Limit the information within each step**—Don't overload your reader by writing lengthy steps:

BEFORE

Overloaded Steps
1. Start the engine and run it to idling speed while opening the radiator cap and inserting the measuring gauge until the red ball within the glass tube floats either to the acceptable green range or to the dangerous red line.

Instead, separate the distinct steps.

AFTER

Separated Steps
1. Start the engine and run it to idling speed.
2. Open the radiator cap and insert the measuring gauge.
3. Note whether the red ball within the glass tube floats to the acceptable green range or up to the dangerous red line.

5. **Develop your points thoroughly**—Avoid vague content by clarifying directions, needed equipment, and cautions.

BEFORE

After rotating the discs correctly, grease each with an approved lubricant.

AFTER

This clarifies what is meant by correct rotating and approved lubricant. The steps are also separated for enhanced readability.

1. Rotate the disks clockwise so that the tabs on the outside edges align.
2. Lubricate the discs with 2 oz of XYZ grease.

6. **Use short words, short sentences, and short paragraphs**—Help your audience complete the task quickly and easily. People read instructions because they need help. It's hard to complete a task when they are unfamiliar with equipment, tools, dangers, and technical concepts. You don't want to compound their challenge with long words, sentences, and paragraphs.

7. **Begin your steps with verbs**—the imperative mood. Note that each of the numbered steps in the following example begins with a verb.

> **example**
>
> **Verbs Begin Steps**
>
> 1. *Number* your steps.
> 2. *Use* highlighting techniques.
> 3. *Limit* the information within each step.
> 4. *Develop* your points thoroughly.
> 5. *Use* short words and phrases.
> 6. *Begin* your steps with verbs.

8. **Personalize your text**—Remember, people write to people. Involve your readers in the instruction by using pronouns.
9. **Do not omit articles**—Articles such as *a, an*, and *the* are part of our language. Although you might see instructions that omit these articles, please don't do so yourself. Articles do not take up much room in your text, but they do make your sentences read more fluidly.

Audience Involvement

See Chapter 4 for more discussion of audience involvement through personalization.

Additional Components

Your instruction might include the following additional components:

Technical Descriptions.   In addition to the step-by-step instructions, many manuals contain technical descriptions of the product or system. A description could be a part-by-part explanation or labeling of a product or system's components. Such a description helps readers recognize parts when they are referred to in the instruction. For example, if the user manual tells the reader to lay shingles with the tabs pointing up, but the reader doesn't know what a "tab" is, the step cannot be performed. In contrast, if a description with appropriate callouts is provided, then the reader's job has been simplified.

Perhaps the user manual will contain a list of the product's specifications, such as its size, shape, capacity, capability, and materials of construction. Table 12.1 is an

TABLE 12.1   Cordless Telephone Specifications

Base Unit	Specifications
Transmit Frequency	46.6 to 46.9 MHz
Receive Frequency	49.6 to 49.9 MHz
Power Requirements	117 V, 60 Hz, 6 W
Size	84 mm (W) × 60 m (H) × 246 mm (D)
Weight	755 g
Handset	**Specifications**
Transmit Frequency	49.6 to 49.9 MHz
Receive Frequency	46.6 to 46.9 MHz
Power Requirements	Rechargeable nickel cadmium batteries
Size	60 mm (W) × 210 mm (H) × 44 mm (D)
Weight	345 g

Using Screen Captures to Visually Depict Steps

Here's a good, simple way to visually depict steps in your instruction—screen captures. A screen capture lets you copy any image present on your computer monitor. To make screen captures, follow these steps.

Capturing the Image

1. Find the graphic you want to include in your instructions.
2. Press the "Print Screen" key on your computer keyboard.

Cropping the Image

Pressing the "print screen" key captures the *entire* image seen on the monitor. This might include images that you do not want to include in your instruction. To crop your image, follow these steps.

1. Paste the "print screen" image into a graphics program like MS Paint.
2. Click on the Select icon.

3. Click and drag the Select tool to choose the part of the captured graphic you want to crop.

4. Copy and paste this cropped image into your user manual.

Note: When capturing screenshots, be careful to avoid infringing upon a company's copyright to the image. Make sure that your use of the image meets the principle of "fair use" (which permits the use of images for criticism, comment, news reporting, teaching, scholarship, or research—Copyright Act of 1976, 17 U.S.C. § 107).

example of a specification from a cordless telephone user manual. Such a specification allows the user to decide whether the product meets his or her needs. Does it state the desired frequency? Is it the preferred weight and size? The specification answers these questions.

Finally, a user manual might include a schematic, depicting the product or system's electrical layout. This would help the readers troubleshoot the mechanism.

Warranties. Warranties protect the customer and the manufacturer. Many warranties tell the customer, "This warranty gives you specific legal rights, and you may also have other rights that vary from state to state." A warranty protects the customer if a product malfunctions sooner than the manufacturer suggests it might: "This warrants your product against defects due to faulty material or installation." In such a case, the customer usually has a right to free repairs or a replacement of the product.

However, the warranty also protects the manufacturer. No product lasts forever, under all conditions. If the product malfunctions after a period of time designated by the manufacturer, then the customer is responsible for the cost of repairs or a replacement product. The designated period of time differs from product to product. Furthermore, many warranties tell the customers, "This warranty does not include damage to the product resulting from normal wear, accident, or misuse."

Disclaimers are another common part of warranties that protect the manufacturer. For example, some warranties include the following disclaimers.

- Note: Any changes or modifications to this system not expressly approved in this manual could void your warranty.
- Proof of purchase with a receipt clearly noting that this unit is under warranty must be presented.
- The warranty is only valid if the serial number appears on the product.
- The manufacturer of this product will not be liable for damages caused by failure to comply with the installation instructions enclosed within this manual.

Accessories. A company always tries to increase its income. One way to do so is by selling the customer additional equipment. A user manual promotes such equipment in an accessories list. This equipment isn't mandatory, required for the product's operation. Instead, accessories lists offer customers equipment such as extra-long cables or cords, carrying cases, long-life rechargeable batteries, extendable antennas, a modem, CD-ROM capabilities, and video instruction packages. Often the specifications are also provided for these accessories.

Frequently Asked Questions. Why take up your customer support employees' valuable time by having them answer the same questions over and over? By including a frequently asked questions (FAQs) page in the user manual, common consumer concerns can be addressed immediately. This will save your company time and money while improving customer relations.

Disclaimers. You can ensure that your company is legally protected by clearly stating what your company is *not* responsible for. This could include risks inherent to using the product. You also might want to clarify timeframes for legitimate repair requests, who can and should work on a product, equipment limitations, copyright laws applicable to the product or service, and explanations of the intended use of the product. Sometimes, before a customer can use a product, the company will require that the end user click on the "I Agree" dialogue box, therefore demanding that customer accept the disclaimer. This acts as a contract between the end user and the company.

Disclaimer

This operating guide to the Udell PQ 4454 Overhead Projector is included primarily for ease of customer use. Service and application for anything other than normal use or replacement parts must be performed by a trained and qualified technician.

Corporate Contact Information. The instructions not only help customers complete a task, but also allow the end user to purchase accessories, receive answers to FAQs, and register customer complaints. Therefore, end your manual by providing your company's street, city, and state address; 1-800 hotline telephone number; e-mail address; fax number; or other ways your audience could contact the company with questions or requests for more information.

Graphics

Clarify your points graphically. Use drawings, photographs, and screen captures that are big, simple, clear, keyed to the text, and labeled accurately. Not only do these graphics make your instructions more visually appealing, but also they help your readers and you. What the reader has difficulty understanding, or you have difficulty writing clearly, your graphic can help explain pictorially.

A company's use of graphics often depends on the audience. With a high-tech reader, the graphic might not be needed. However, with a low-tech reader, the graphic is used to help clarify. Look at the following examples of instructions for the same procedure. Figure 12.3 uses text with graphics to help the low-tech audience better understand the required steps; Table 12.2 uses text without graphics for the high-tech audience.

Note that Figure 12.3 uses large, bold graphics to supplement the text, clearly depicting what action must be taken. This is appropriate for a low-tech end user. On the other hand, Table 12.2, written for a manufacturer's technician (high tech), assumes greater knowledge and, therefore, deletes the graphic.

Check Online Resources

www.prenhall.com/gerson

For more information about the sample instructions, visit our companion website.

FIGURE 12.3 Instructions with Large, Bold Graphics for Low-Tech Audience

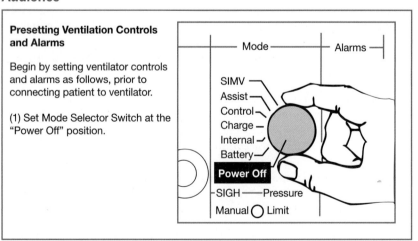

Presetting Ventilation Controls and Alarms

Begin by setting ventilator controls and alarms as follows, prior to connecting patient to ventilator.

(1) Set Mode Selector Switch at the "Power Off" position.

Courtesy of Puritan Bennett Corp.

TABLE 12.2 Text without Graphics for High-Tech Readers

Location Item	Action	Remarks
	Note For additional operating instructions, refer to Companion 2800 operator's manual.	
1. MODE switch	Set to POWER OFF.	
2. Rear panel	Ensure that instructions on all labels are observed.	
3. Patient tubing circuit	Assemble and connect to unit.	See Table 2, step 3.
4. Power cord	Connect to 120/220 V ac grounded outlet.	If power source is external battery, connect external battery cable to unit per Table 2, step 4.
5. Cascade I humidifier	Connect to unit.	See Table 2, step 2.

Courtesy of Puritan Bennett Corp.

SAMPLE INSTRUCTIONS

See Figures 12.4 (HP Officejet Pro Color Printer) and 12.5 (Crate&Barrel Elements Bookcase) for sample instructions.

FIGURE 12.4 HP Officejet Pro K550 Series Color Printer

Selecting print media

The printer is designed to work well with most types of office paper. It is best to test a variety of print media types before buying large quantities. Use HP media for optimum print quality. Visit HP website at www.hp.com for details on HP media.

- Load only one type of media at a time into a tray.
- For tray 1 and tray 2, load media print-side down and aligned against the right and back edges of the tray.

Loading media

This section provides instructions for loading media into the printer.

To load tray 1 or tray 2

1. Pull the tray out of the printer by grasping under the front of the tray.

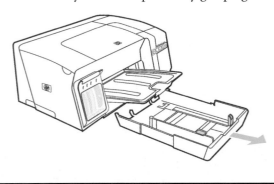

FIGURE 12.4 Continued

2. For paper longer than 11 inches (279 mm), lift the front cover of the tray (see shaded tray part) and lower the front of the tray.

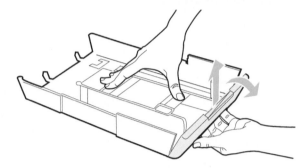

3. Insert the paper print-side down along the right of the tray. Make sure the stack of paper aligns with the right and back edges of the tray, and does not exceed the line marking in the tray.

 NOTE Tray 2 can be loaded only with plain paper.

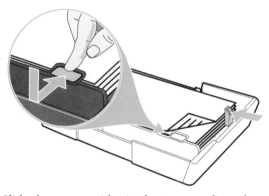

4. Slide the paper guides in the tray to adjust them for the size that you have loaded.
5. Gently reinsert the tray into the printer.

 CAUTION If you have loaded legal-size or longer media, keep the front of the tray lowered. Damage to the media or printer might result if you raise the front of the tray with this longer media loaded.

6. Pull out the extension on the output tray.

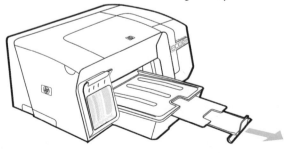

FIGURE 12.5   Elements Bookcase Instructions

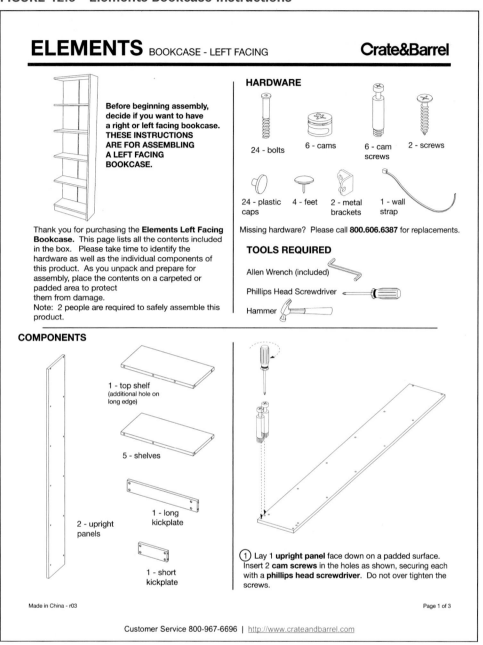

ELEMENTS BOOKCASE - LEFT FACING   Crate&Barrel

Before beginning assembly,
decide if you want to have
a right or left facing bookcase.
**THESE INSTRUCTIONS
ARE FOR ASSEMBLING
A LEFT FACING
BOOKCASE.**

Thank you for purchasing the **Elements Left Facing
Bookcase.** This page lists all the contents included
in the box.   Please take time to identify the
hardware as well as the individual components of
this product. As you unpack and prepare for
assembly, place the contents on a carpeted or
padded area to protect
them from damage.
Note: 2 people are required to safely assemble this
product.

HARDWARE

24 - bolts   6 - cams   6 - cam   2 - screws
                                 screws

24 - plastic   4 - feet   2 - metal   1 - wall
caps   brackets   strap

Missing hardware? Please call **800.606.6387** for replacements.

TOOLS REQUIRED

Allen Wrench (included)

Phillips Head Screwdriver

Hammer

COMPONENTS

1 - top shelf
(additional hole on
long edge)

5 - shelves

2 - upright
panels

1 - long
kickplate

1 - short
kickplate

(1) Lay 1 **upright panel** face down on a padded surface.
Insert 2 **cam screws** in the holes as shown, securing each
with a **phillips head screwdriver.** Do not over tighten the
screws.

Made in China - r03   Page 1 of 3

Courtesy of Crate&Barrel

FIGURE 12.5   Continued

ELEMENTS BOOKCASE - LEFT FACING Crate&Barrel

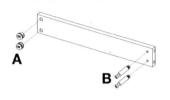

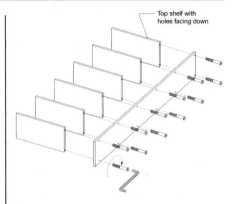

Top shelf with holes facing down

② A. Insert 2 **cams** into the large holes on the **long kickplate**. Make sure arrow on each cam points towards nearest edge with holes.
B. Insert 2 **cam screws** into the smaller holes of the long kickplate and secure with a screwdriver.
C. Insert 4 **cams** into the **short kickplate**, making sure each arrow points to the nearest edge with holes.

③ With the assistance of another adult, align holes in the **upright panel** (panel without cam screws) with holes in the side of each **shelf** and **top shelf**.   Attach each shelf with 2 **bolts**, fully tightening each bolt with the **allen wrench**.

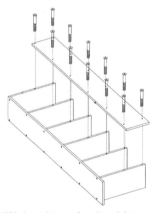

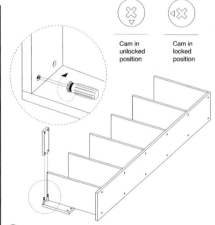

Cam in unlocked position

Cam in locked position

④ With the assistance of another adult, turn shelf assembly over. Attach remaining **upright panel** (panel with cam screws) with 12 **bolts**, fully tightening each bolt with the **allen wrench**. Make sure the cam screws face down.

⑤ Turn assembly back over. Fit **short kickplate** onto cam screws as shown. Secure by turning both **cams** clockwise with a **screwdriver** until the cam securely engages the cam screw.

Made in China - r03

Page 2 of 3

Customer Service 800-967-6696 | http://www.crateandbarrel.com

Courtesy of Crate&Barrel

FIGURE 12.5   Continued

ELEMENTS BOOKCASE - LEFT FACING   **Crate&Barrel**

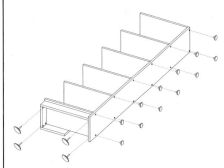

⑥ Fit cam screws of **long kickplate** into short kickplate and upright panel. Secure all 4 **cams** by turning each clockwise with a **screwdriver**.

⑦ A. Attach 4 **feet** to the bottom of the assembly as shown, gently tapping into place with a **hammer**.
B. Conceal all exposed bolt heads by inserting plastic caps. Assembly is now complete.

OPTIONAL WALL ATTACHMENT

No wall anchor hardware is included with this unit. For safe installation it is essential to use anchor hardware appropriate for your wall type. Mount to wood wall studs whenever possible. If you are unsure of what type of fasteners to use, consult your local hardware store or a qualified professional. This product is only deterrent, it is not a substitute for proper adult supervision.

Locate the pilot hole on the back of the shelf as shown.
A. Attach 1 **metal bracket** with 1 **screw** to the shelf.
B. With the assistance of another adult, place bookcase in desired location against the wall. Mark wall directly opposite of pilot hole in shelf.
C. Attach 1 **metal bracket** to the wall at the mark. Secure with 1 wall anchor (not included), preferably into a wall stud. If not attaching to a stud, be sure to use hardware appropriate to your wall type (drywall, masonry, etc).
D. Place bookcase back into position against the wall. Thread the end of the **wall strap** through each bracket and into the notched end of the strap. Pull until tight.

Pilot hole

Hole in Wall

CLEANING AND CARE

Clean surfaces with a dry or damp soft cloth. Do not use abrasive cleaners.

Made in China - r03

Page 3 of 3

Customer Service 800-967-6696 | http://www.crateandbarrel.com

Courtesy of Crate&Barrel

What Challenges Do Tax Specialists Face When Writing Instructions?

Gil Charney (CPA, CFP), Director of Tax Training at H&R Block, manages a staff of a dozen writers who create instructional manuals. Their mission is to "produce up-to-date, accurate, well-written tax training material and course textbooks for 80,000 tax professionals nationwide." That's a hard job!

Why? Gil and his staff must overcome numerous challenges to create their instructions, including the following:

Audience Recognition: H&R Block's tax preparation professionals are a broad audience. Some are CPAs and experienced tax preparers, but many do not have extensive tax backgrounds. They are all ages and come from all walks of life. To communicate effectively to this diverse audience, Gil and his staff try to write their documentation using common, easily understandable language to explain complex tax laws.

Complex Subject Matter: Gil's staff must stay up to date by researching our changing tax laws. In addition, Gil and his staff create over 50 different tax courses. These include instructions for documenting tax information on estates, employment taxes, farms, corporations and partnerships, gift taxation, foreign income taxation, and sole proprietorships, as well as tax information for individuals who are retired, in the military, or in the clergy.

Diverse Deliverables: As with most technical communicators today, Gil and his staff produce intranet training material. In addition to explaining tax law and providing common examples and applications, Gil's group focuses heavily on case studies, tasks that spotlight diverse tax areas and show through fact-based exercises how changes in tax laws will affect clients. Much of the writing from Gil's group is used online and for hard-copy textbook, hands-on activities, and face-to-face classroom instruction.

Accuracy: Gil and his staff make every effort to produce training materials that are as free from error as possible. With a staff of several writers, consistency is always a challenge. For example, the writers must ensure that H&R Block's dozens of courses consistently use periods at the end of bulleted lists (or not); write out, abbreviate, or use a symbol for section codes in legal documents ("Section," "section," "Sec.," "sec.," or §); capitalize or not capitalize "Medicare" and "Social Security"; spell out or not spell out ages ("34" or "thirty-four"). To supply up-to-date tax information that is correct and consistent, Gil's writers follow rigorous review sequences (writers review the work of other writers; proofreaders review written work; editors read all text before publication).

Gil and his staff are talented people whose jobs demand multitasking. They must

- Know tax laws
- Be able to write clearly, concisely, and correctly
- Be expert in desktop publishing

Most importantly, Gil and his writers must be empathetic. They must ask themselves about their target audience: "What will our tax preparers need to know when they are preparing their client's tax return?" After all, the 80,000 end users of H&R Block tax training materials will impact millions of people just like you and me. Before our tax preparers can help us, Gil and his staff must anticipate the needs of these tax preparers. Knowing tax law is important, but audience recognition is essential to the successful conveyance of complex tax information.

THE WRITING PROCESS AT WORK

Now that you know what should be included in an instruction, it is time to write. The best way to accomplish your technical communication challenge is by following a process: Prewrite to gather data and determine objectives, write a rough draft of your instruction, and rewrite to revise.

The Writing Process

Prewriting	Writing	Rewriting

• Use the reporter's questions to clarify what content you will need to include.
• Determine whether your audience is high tech, low tech, or lay. This will help you decide which abbreviations and acronyms need to be defined.
• Visualize the steps by flowcharting.
	• Organize your instructions chronologically.
	• Include visual aids and callouts.
	• Use headings and subheadings for easy navigation.
		• Revise your draft by
		  • adding details
		  • deleting wordiness
		  • simplifying words
		  • enhancing the tone
		  • reformatting your text
		  • proofreading and correcting errors

Prewriting

Check Online Resources

www.prenhall.com/gerson
For more information about the writing process, visit our companion website.

For writing an instruction, reporter's questions would be a good place to start gathering data and determining objectives. Ask yourself:

- *Who* is my reader?
  - High tech?
  - Low tech?
  - Lay?
- *What* detailed information is needed for my audience? For example, a low-tech or lay reader might be content with "inflate sufficiently" whereas an instruction geared toward high-tech readers will require "inflate to 25 psi."
- *How* must this instruction be organized?
  - Chronologically
- *When* should the procedure occur?
  - Daily, weekly, monthly, quarterly, biannually, or annually?
  - As needed?
- *Why* should the instruction be carried out?
  - To repair, maintain, install, operate, and so on. (See the list of reasons for using an instruction at the beginning of this chapter.)

Once you have gathered this information, you can use brainstorming to sketch out a rough list of the steps required in the instruction and the detail needed for clarity. After that, your major responsibility is to sequence your steps chronologically. To accomplish this goal, use a prewriting technique called *flowcharting*.

Flowcharting. Flowcharts chronologically trace the stages of an instruction, visually revealing the flow of action, decision, authority, responsibility, input/output, preparation, and termination of process. Flowcharting is not just a graphic way to help you gather data and sequence your instruction, however. It is two-dimensional writing. It provides your reader content as well as a panoramic view of an entire sequence. Rather than just reading an instruction step by step, having little idea what's around the corner, with a flowchart your reader can figuratively stand above the instruction and see where it's going.

FIGURE 12.6 Flowcharting Symbols

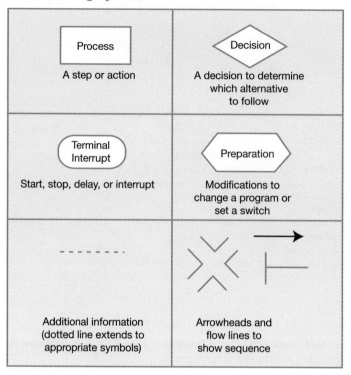

With a flowchart, readers can anticipate cautions, dangers, and warnings, since they can see at a glance where the steps lead.

To create a flowchart, use the International Organization for Standardization (ISO) flowchart symbols shown in Figure 12.6.

Figure 12.7 shows an excellent example of a flowchart, visually depicting an instruction's sequence.

Writing

Once you've graphically depicted your instruction sequence using a flowchart, the next step is to write a rough draft of the instruction.

Review Your Prewriting. In reviewing your prewriting, you will accomplish the following:

- **Clarify your audience.** Now is a good time to review your reporter's questions and to determine whether your initial assumptions regarding audience are still correct. If your attitude toward your audience has changed, redefine your readers and their needs.

- **Determine your focus.** In your prewriting, you answered the question *why*. Are you still certain of your objectives for writing the instruction? If so, write a one-sentence thesis or statement of objectives to summarize this purpose. For example, you could write, "This instruction should be followed so that glitches in the system can be discovered and corrected." This thesis or topic sentence will help you maintain your focus throughout the instruction. If you are uncertain about your purpose for writing the instruction, however, rethink the situation. You will not be able to write an effective instruction until your objectives are clearly stated.

- **Organize your steps.** Review your flowchart to determine whether any steps in the instruction have been omitted or misplaced. If the chronological sequence is incorrect, reorganize your detail. If the chronology is correct, you're ready to move on to the next step in drafting.

FIGURE 12.7 Flowchart of Handling Service Problems

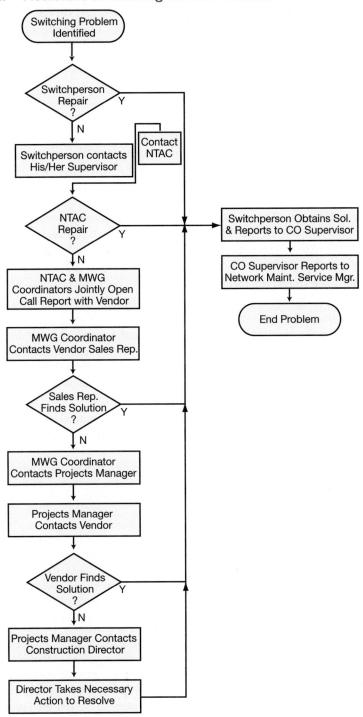

Review the Instruction Criteria. Before you write the rough draft, remind yourself what the instruction will consist of. You'll need an introduction, a discussion with itemized steps, graphics to help your reader perform the requested procedures, and a conclusion.

Write the Draft.   You've gathered your data and determined your objectives. You've correctly sequenced the steps and reminded yourself of the instruction criteria. At this point, you should feel comfortable with the prospect of writing. Follow the sufficing technique again, roughly drafting your instruction. Don't worry about perfect grammar or graphics. Perfecting the text comes in the next step, rewriting.

Rewriting

Rewriting, the last stage in the process, is the time to perfect your instruction. For manuals, this stage calls for *usability testing*. You've written the instruction, but is it usable? If the instruction doesn't help your audience complete the task, what have you accomplished? To determine the success of your manual, test the instruction.

Test for Usability.   To test the usability of an instruction, follow this procedure.

1. **Select a test audience**—The best audience for usability testing would include a representative sampling of individuals with differing levels of expertise. A review team composed only of high-tech readers would skew your findings. In contrast, a review team consisting of high-tech, low-tech, and lay readers would give you more reliable feedback.

2. **Ask the audience to test the instructions**—The audience members would attempt to complete the instructions, following the procedure step by step.

3. **Monitor the audience**—What challenges do the instructions seem to present? For example, while observing the audience, has the audience completed all of the steps easily? Are the correct tools and equipment listed? Are terms defined as needed and presented where they are necessary? Has the test audience abided by the hazard notations? Are any steps overloaded and, thus, too complicated? Has each step been explained specifically?

4. **Time the team members**—How long does it take each member to complete the procedure? More important, why has it taken some team members longer to complete the task?

5. **Quantify the audience's responses**—Once your test audience has completed the procedure, you must debrief these individuals to determine what problems they encountered. Use the Effective Instruction Checklist to help gather quantifiable information about the instruction's usability.

Readers who assess your instructions do not have to suggest ways to fix any problems encountered. Their goal is just to give you feedback. If they say, "I don't understand this word," "Step 3 seems too long," "I could use a graphic here to clarify your meaning," or "More white space would keep me from feeling overwhelmed," you can revise the text accordingly. To do so, reread the instruction, consider the suggestions made during usability testing, and improve the draft by using the rewriting procedures explained in this chapter.

Add Detail for Clarity.   Don't hope you've said enough to clarify your steps. Tell your readers *exactly* how to do it or why to do it. Part of detail involves warnings and cautions, for example. Make sure that you've added all necessary dangers or concerns. In addition, assess each step. Have you added sufficient detail to clarify your intent? For instance, if you are writing to a lay reader, you don't want to say merely, "Insert the guide posts." You must say, "Before inserting the guide post, dig a 2-ft-deep by 2-ft-wide hole. Pour the ready-mixed concrete into the hole. Insert the post into the hole and hold it steady until the concrete starts to set (approximately one minute)."

Delete.   Omit unnecessary words and phrases to achieve conciseness. In addition, delete extraneous information that a high-tech reader might not find useful. This, however, could be dangerous. What you might see as extraneous, your reader might see as necessary. If you delete any information, do so carefully.

Simplify.   Don't overload your steps. If you have too much information in any one stage, divide this detail into smaller units, adding new steps. Doing so will help your readers follow your instructions. Simplify your word usage; refer to the fog index.

Move Information.   Make sure that the correct chronological sequence has been followed.

Reformat Using Highlighting Techniques.   If you include dangers, cautions, warnings, or additional information for clarity, set these off (make them emphatic) through formatting. Use color, text boxes, font size and style, emphatic words, graphics, and so on.

Document Design

See Chapter 8 for more discussion of document design.

Enhance Your Text.   Use the imperative mood for each step in your instruction. Whenever an action is required, begin with a verb to emphasize the action. In addition, you can enhance your text through personalization. Remember, people write to people. To achieve this person-to-person touch, add pronouns to your instruction.

Correct Your Instruction for Accuracy.   Proofread your text for grammatical correctness as well as for contextual accuracy.

Avoid Biased Language.   Your instructions could be read by an international audience. Avoid cultural biases, and write for readers diverse in age and gender.

EFFECTIVE INSTRUCTION CHECKLIST

_____ 1. Does the instruction have an effective title?
  • Is the topic mentioned?
  • Is the reason for performing the instruction mentioned?

_____ 2. Does the instruction have an effective introduction?
  • Is the topic mentioned?
  • Is the reason for performing the instruction mentioned?
  • Is the ease of use or capabilities mentioned?
  • Have the number of steps been listed?
  • Has a list of required tools or equipment been provided?

_____ 3. Are hazard alert messages used effectively?
  • Are the hazard alert messages placed correctly?
  • Is the correct term used (_Danger_, _Warning_, _Caution_, or _Note_)?
  • Is the correct color used with the appropriate term?
  • Does the hazard alert text identify the hazard, provide the consequences, and suggest avoidance steps?
  • Is an effective icon used to depict the hazard?

_____ 4. Is the instruction's discussion effective?
  • Is the instruction organized chronologically?
  • Does each step avoid overloading by presenting one clearly defined action?
  • Does each step begin with a verb?
  • Does the instruction's discussion contain the same number of steps mentioned in the introduction?

_____ 5. Is the instruction well developed?
  • Does the instruction avoid assuming that readers will understand and, instead, clearly explain each point?

_____ 6. Does the instruction recognize audience effectively?
  • Is the instruction written for a high-tech, low-tech, lay audience, or multiple readers?
  • Has sexist language been avoided?
  • Does the text include language appropriate for a multicultural audience?
  • Is the text personalized using pronouns?

_____ 7. Is the conclusion effective?
  • Have warranties been mentioned?
  • Is a sales pitch provided showing ease of use?
  • Has the conclusion reiterated a reason for performing these steps?
  • Are disclaimers provided?

_____ 8. Is the instruction's document design effective?
  • Are highlighting techniques effectively used? These include graphics (such as comic book–look, photographs, or line drawings), varied typefaces and type sizes, color, and white space.

_____ 9. Is correct technical communication style used, abiding by the fog index?

_____ 10. Have grammatical and textual errors been avoided?

PROCESS EXAMPLE

Following is an example of an instruction collaboratively written by a group of students. To write this instruction, the students gathered data, determined objectives and sequenced their instruction through prewriting, drafted an instruction in writing, and finally revised their draft in rewriting.

Prewriting

Figure 12.8 illustrates the students' flowcharting to gather data and to arrange it chronologically.

Writing

After completing the prewriting, the students composed a rough draft and received feedback from other students in the class (Figure 12.9).

Rewriting

The students revised the rough draft, considering all the rewriting techniques discussed in this text. The revision is shown in Figure 12.10.

FIGURE 12.8 Flowchart for Microsoft Project

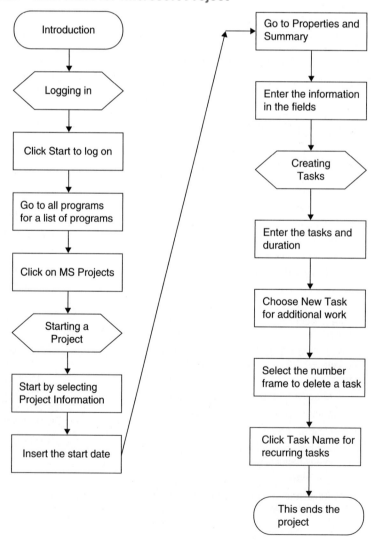

FIGURE 12.9 Instruction Rough Draft with Peer Evaluation Comments

Microsoft Project 2003

TABLE OF CONTENTS

Use a separate page for the title, and include the purpose of this instruction— "Steps for. . ." for example.

For easier navigation, use the IEEE numbering system (1,0, 2.0, etc.), and include the Conclusion in the TOC. Don't forget to add page numbers. And shouldn't you also provide a definition of terms?

Your titles aren't parallel—"logging" and "how to" should read "logging" and "starting."

INTRODUCTION TO MICROSOFT PROJECT 2003

Microsoft Project can help plan and monitor the progress of any project because it can help keep your project team and customers informed by producing reports and other helpful information. The main components of Microsoft Project are the following:

Your first sentence is too long. Cut it in half. Also avoid the repetition of "help."

- **Reports**—Reports help to view and analyze the progress of a project. They communicate project information to many individuals.
- **Charts and Tables**—Visual aides help readers understand data produced in reports.
- **Task Management**—Managing individual steps in a project efficiently helps avoid mistakes by helping readers to keep track of each critical task, summary task, and subtask.

The word "aides" is spelled wrong.

Logging into Project

Consider adding graphics to this introduction, to clarify what each part of Microsoft Project looks like.

Be sure to add 1.0 to the title, and then change the steps to 1.1, 1.2, etc.

> **Warning: Save your progress throughout all these steps to ensure no information is lost or omitted.**

1. Click start button to display Start menu.

You have omitted the article "the" before "start" and "Start" above. And why capitalize one and not the other?

2. Click on the all programs tab to display list.

"all programs tab" sounds odd in the sentence. Maybe it should be capitalized, typed in a different front, put in quotes, or use an icon instead of just listing the words. You need to emphasize this more clearly.

Continued

FIGURE 12.9 Continued

Step number 3 could be separated into two steps. And "scroll over and down" sounds strange—odd style.

3. Click the Microsoft Office tab and scroll over and down to Office.

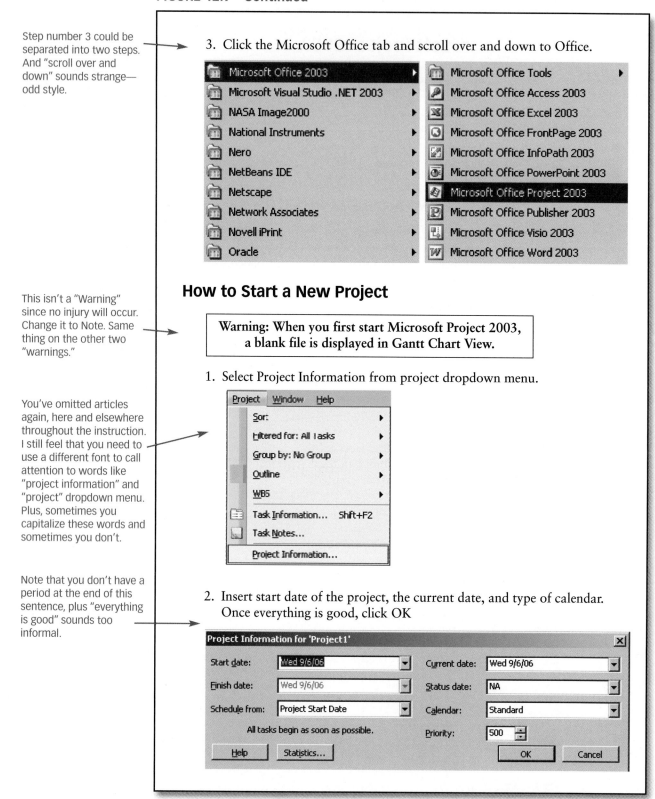

How to Start a New Project

This isn't a "Warning" since no injury will occur. Change it to Note. Same thing on the other two "warnings."

> **Warning:** When you first start Microsoft Project 2003, a blank file is displayed in Gantt Chart View.

1. Select Project Information from project dropdown menu.

You've omitted articles again, here and elsewhere throughout the instruction. I still feel that you need to use a different font to call attention to words like "project information" and "project" dropdown menu. Plus, sometimes you capitalize these words and sometimes you don't.

Note that you don't have a period at the end of this sentence, plus "everything is good" sounds too informal.

2. Insert start date of the project, the current date, and type of calendar. Once everything is good, click OK

FIGURE 12.9 Continued

3. Click Properties from the file menu. This will bring up the Project Properties box. You should select the Summary tab next.

For a step, you need to start with a verb. Rather than "you should select," write, "Select . . ." And make this two separate steps.

4. Next its time to enter the correct information in each field and click OK.

"Its" is spelled wrong—should be "it's"—and start the step with a verb vs. "next"

You need visuals for steps 3 and 4.

CONCLUSION

Microsoft Project 2003 is a great way to manage any project you might encounter.

You need more conclusion, especially a helpline contact number or email address.

USABILITY CHECKLIST

Audience Recognition
_____ 1. Are technical terms defined?
_____ 2. Are graphics used to explain difficult steps at the reader's level of understanding?
_____ 3. Do the graphics depict correct completion of difficult steps at the reader's level of understanding?
_____ 4. Are the tone and word usage appropriate for the intended audience?
_____ 5. Does the introduction involve the audience and clarify how the reader will benefit?

Development
_____ 1. Are steps precisely developed?
_____ 2. Is all required information provided, including hazards, technical descriptions, warranties, accessories, and required equipment or tools?
_____ 3. Is irrelevant or rarely needed information omitted?

Conciseness
_____ 1. Are words, sentences, and paragraphs concise and to the point?
_____ 2. Are the steps self-contained so the reader doesn't have to remember important information from the previous step?

Consistency
_____ 1. Is a consistent and parallel hierarchy of headings used?

_____ 2. Are graphics presented consistently (same location, same use of figure titles and numbers, similar sizes, etc.)?
_____ 3. Does wording mean the same throughout (technical terms, cautions, warnings, notes, etc.)?
_____ 4. Is the same system of numbering used throughout?

Ease of Use
_____ 1. Can readers easily find what they want because instructions include
  • Table of contents
  • Glossary
  • Hierarchical headings
  • Headers and footers
  • Index
  • Cross-referencing or hypertext links
  • Frequently asked questions (FAQs)

Document Design
_____ 1. Do graphics depict how to perform steps?
_____ 2. Is white space used to make information accessible?
_____ 3. Does color emphasize hazards, key terms, or important parts of a step?
_____ 4. Do numbers or bullets divide steps into manageable chunks?
_____ 5. Are hazards (dangers, warnings, cautions, and notes) boldfaced, typed in appropriate colors, and/or placed in textboxes to emphasize importance?

FIGURE 12.10   Completed Instruction

MICROSOFT PROJECT 2003

Instructional Steps
for
Planning and Monitoring Projects

TABLE OF CONTENTS

HAZARDS

This manual defines hazard notations as follows:

Caution

Information needed to help you avoid damaging software and/or losing saved data.

Note

Information needed to help you complete steps correctly.

FIGURE 12.10 Continued

INTRODUCTION TO MICROSOFT PROJECT 2003

Microsoft Project can help plan and monitor the progress of any project. It can keep your project team and customers informed by producing reports. The main components of Microsoft Project are the following:

- **Reports**—Reports help to view and analyze the progress of a project. They communicate project information to many individuals by keeping track of cost, labor, and many other key elements.

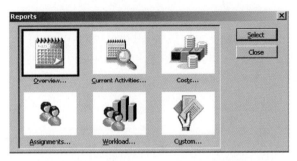

- **Charts and Tables**—Visual aids help readers understand data produced in reports. Charts and tables convey specific information regarding tasks, resources, and assignments.

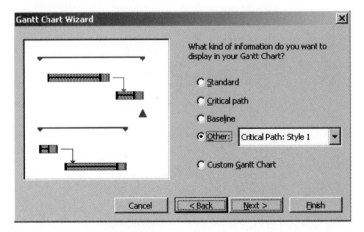

- **Task Management**—Managing individual steps in a project efficiently helps avoid mistakes. Task management includes keeping track of each critical task, summary task, and subtask.

This manual will give the user a basic knowledge of Microsoft Project and its components. After reading this manual, you will be able to manage projects, set tasks, and create tables, charts, and reports.

3

FIGURE 12.10 Continued

1.0 Logging into Project

> **Caution** Save your progress throughout all these steps to ensure no information is lost or omitted.

1.1 Click the **start** button to display the Start menu.

1.2 Click on the **All Programs** tab to display the list.

4

FIGURE 12.10 Continued

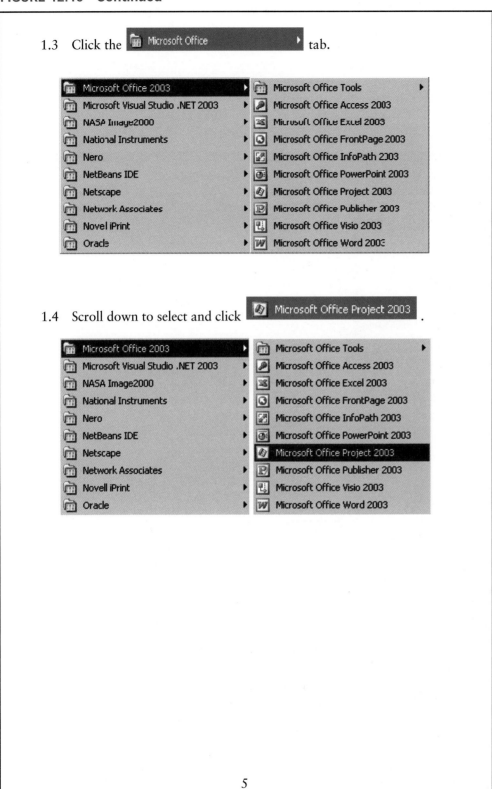

FIGURE 12.10   Continued

2.0   Starting a New Project

> **Note** : When you first start Microsoft Project 2003, a blank file is displayed in Gantt Chart View. Let's take a look at the main window.

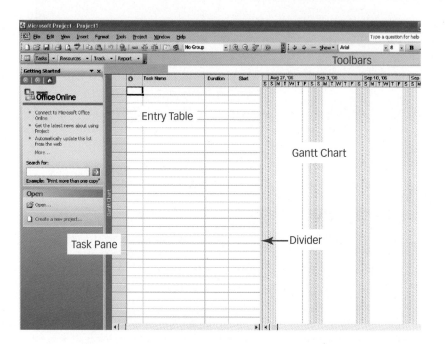

2.1   Select the **Project Information** from the **Project** dropdown menu.

6

FIGURE 12.10   Continued

2.2   Insert the start date of the project, the current date, and type of calendar. Once everything is correct, click [OK].

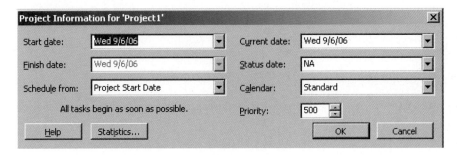

2.3   Click **Properties** from the file menu. This will bring up the **Project Properties** box. Then select the **Summary** tab.

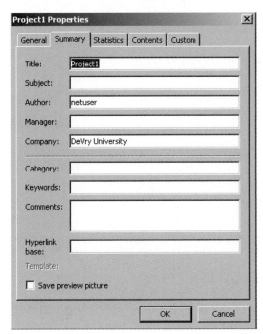

2.4   Enter the correct information in each field and click [OK].

CONCLUSION

Microsoft Project 2003 is a great way to manage any project you might encounter. The options for project management are limitless with this software. Once you become familiar with the program, it can be a great help for your project.

If you have questions or need assistance with your Microsoft Project 2003, contact us on the Web at microsoft_help.com.

7

FIGURE 12.10   Continued

GLOSSARY

Duration	The time it takes to complete a task or project.
Project	A temporary endeavor undertaken to create a unique product, service, or result.
Recurring task	A task that occurs throughout the project.
Report	A breakdown of progress communicating information.

8

CHAPTER HIGHLIGHTS

1. To organize your instruction effectively, include an introduction, a discussion of sequenced steps, and a conclusion.
2. Be sure to number your steps and start each step with a verb.
3. Hazard alerts are important to protect your reader and your company. Place these alerts early in your instruction or before the appropriate step.
4. In the instruction, follow a chronological sequence.
5. To help readers understand the steps, provide precise information.
6. Graphics will help your reader see how to perform each step.
7. Highlighting techniques emphasize important points, thus minimizing damage to equipment or injury to its users.
8. Don't compress several steps into one excessively long and demanding step.
9. The writing process, complete with usability testing, will help you construct an effective instruction.
10. In the introduction to a user manual, you can promote good customer–company relations by using personal pronouns and positive words.

CASE STUDIES

1. PhlebotomyDR needs to provide instructions for its staff (nurses, doctors, and technicians) regarding the correct procedures to ensure cleanliness and employee protection. These include hand washing procedures and the correct use of sterile equipment, such as gloves, masks, aprons, and shoe coverings. Following is a rough draft of one set of instructions for personnel safety. Revise the rough draft according to the criteria provided in this chapter. Correct the order of information, the grammar, and the instruction's content. In addition, improve the instruction by including appropriate graphics.

Hand Washing

- Lather hands to cover all surfaces of hands and wrists.
- Wet hands with water.
- Rub hands together to cover all surfaces of hands and fingers. Pay special attention to areas around nails and fingers. Lather for at least 15 seconds.
- Dry thoroughly.
- Rinse well with running warm water.
- Avoid using hot water. Repeated exposure to hot water can lead to dermatitis.
- Use paper towels to turn off faucet.

Gloves

- Replace damaged gloves as soon as patient safety permits
- Don gloves immediately prior to task.
- Remove and discard gloves after each use involving any bodily fluids.

Masks

Wear masks and eye protection devices (goggles or eye shields) to avoid droplets, spray, or splashes and to prevent exposure to mucous substances. Masks are also worn to protect nurses, doctors, and technicians from infectious elements during close contact with patients.

Aprons and Other Protective Clothing

- Wear aprons or gowns to avoid contact with body substances during patient care procedures.
- Remove and dispatch aprons and other protective clothing before leaving work area.
- Some work areas might require additional protective clothing such as surgical caps and shoe covers or boots.

2. BurgerNet Online Order System is a software product in development. As a technical communicator, you need to write the instruction to accompany this online order system.

BurgerNet Online Order System

1. Description

The BurgerNet application will be used to order burgers for delivery to a customer's home or work.

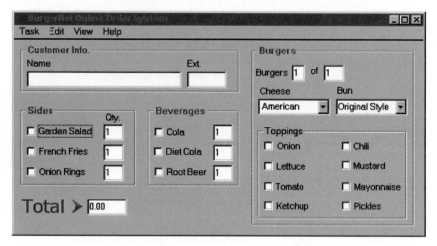

2. Dropdown Menu Bar.
The dropdown menu bar includes a Task, Edit, View, and Help menu.

- Task menu. On the Task menu customers will find a Send Order command, used to electronically send the completed order to the restaurant. The Print Order command on the Task menu lets the customer print a copy of the order. The Exit command terminates the order session.
- Edit menu. On the Edit menu, an Add Favorites command allows customers to add a frequently ordered combination of food items to make future online ordering quick and easy. The Edit Favorites command lets customers modify an existing favorite order.
- View menu. On the View menu, the Preferences command is used to modify default customer. In addition, when favorites have been defined, they are listed under View. Customers can select a favorite for display before sending the order.
- Help menu. On the Help menu, customers can access FAQs.

3. Key Components of BurgerNet.
- Customer Info. frame: Customers type in their name and telephone number in the Name and Extension fields.
- Burgers frame. In the Burgers/Of fields customers specify the number of burgers desired. The Cheese dropdown list lets the customer select either American, Pepperjack, Mozzarella, or Swiss cheese to top their burgers. Bun lets customers choose either original or onion buns from the dropdown list.
- Toppings. Here, customers check what to put on their burgers.
- Sides: Select which side dish or dishes and specify the quantity for each.
- Beverages: Select drink(s) and specify the quantity for each.
- Total field. The total cost of the customer's order is automatically displayed.

Write a step-by-step instruction for completing and sending an online burger order.

INDIVIDUAL AND TEAM PROJECTS

1. Write an instruction. To do so, first select a topic. You can write an instruction telling how to monitor, repair, test, package, plant, clean, operate, manage, open, shut, set up, maintain, troubleshoot, install, use software, and so on. Choose a topic from within your field of expertise or one that interests you. Follow the writing process techniques to complete your instruction. Prewrite (using a flowchart), write a draft (abiding by the criteria for instructions presented in this chapter), and rewrite to perfect your text.

2. Find examples of instructions or user manuals written in your work environment. Bring these to class. Using the criteria for instructions presented in this chapter, decide whether the instructions are successful or unsuccessful. If the instructions are good, show how and why. If they are flawed, explain the problem(s). Then rewrite the instructions to improve them.

3. Good writing demands revision. Following is a flawed instruction. To improve it, rewrite the text, abiding by the criteria for instructions and the rewriting techniques included in this chapter.

DATE: November 30, 2008

TO: Maintenance Technicians

FROM: Second Shift supervisor

SUBJECT: OVEN CLEANING

The convection ovens in kiln room 33 need extensive cleaning. This would consist of vacuuming and wiping all walls, door, roofs, and floors. All vents and dampers need to be removed and a tack cloth used to remove loose dust and dirt. Also, all filters need replacing. I am requesting this because when wet parts are placed in the ovens to cure the paint, loose particles of dust and dirt are blown onto the parts, which causes extensive rework. I would like this done twice a week to ensure cleanliness of product.

---

4. One of the keys to success in writing instructions is audience recognition. One reason flawed instructions fail is lack of audience recognition. For example, an instruction uses high-tech terms but is intended for low-tech readers. Find an example of an instruction geared to a high-tech audience that correctly uses high-tech terms. Then find an example of an instruction geared to low-tech readers that incorrectly uses high-tech words. Next, find an instruction geared to low-tech or lay readers that has correctly defined its terms and explained its technology. Finally, find an example of an instruction that talks down to the reader, using low-tech or lay terms when high-tech words or phrases would have been better.

5. Find examples of instructions that do not provide effective hazard alerts. Using the criteria for effective hazard alerts provided in this chapter, decide where the alerts are flawed. Then improve the instructions or user manuals by creating effective hazard alerts.

PROBLEM-SOLVING THINK PIECE

Read the following instructional steps. Are they in the correct chronological order? How would you reorder these steps to make the instruction more effective?

Changing Oil in Your Car

Run the car's engine for approximately 10 minutes and then drain the old oil.

Park the car on a level surface, set the parking brake, and turn off the car's engine.

Gather all of the necessary tools and materials you might need.

Open the hood.

Jack up and support the car securely.

Place the funnel in the opening and pour in the new oil.

Replace the cap when you have finished pouring in new oil.

Locate the oil filler cap on top of the engine and remove the cap.

Tighten the plug or oil filter if you find leakage.

Run the engine for a minute, then check the dipstick. Add more oil if necessary.

Pour the used oil into a plastic container and dispose of it safely and legally.

WEB WORKSHOP

1. Review any of the following Web site's online instructions. Based on the criteria provided in this chapter, are the instructions successful or not?
   - If the answer is yes, explain why and how the instructions succeed.
   - If the answer is no, explain why the instructions fail.
   - Rewrite any of the flawed instructions to improve them.

Web Sites	Topics
http://www.hometips.com/diy.html	Electrical systems, plumbing, kitchen appliances, walls, windows, roofing, and siding
http://www.hammerzone.com	Kitchen projects, tubs, sinks, toilets, showers, and water heaters
http://dmoz.org/Home/Home_Improvement/	Links to step-by-step procedures for painting, welding, soldering, plumbing, walls, windows, and door repair and installation
http://directory.google.com/Top/Home/Home_Improvement/	Links to sites for instructions on decorating, electrical, flooring, furniture, lighting, painting, plumbing, welding, windows, and doors
http://www.quakerstate.com/pages/carcare/oilchange.asp	Instructions for changing oil
http://www.csaa.com/yourcar/takingcare ofyourcar	AAA instructions for car care, including general maintenance and checking fluids, hoses, drive belts, electrical systems, and tires
http://www.gateway.com/index.shtml	Access the FAQs for upgrading systems or correcting problems with printers, drivers, monitors, memory, and more

2. The U.S. Department of Labor article "Hazard Communication: A Review of the Science Underpinning the Art of Communication for Health and Safety" can be found at http://www.osha.gov/SLTC/hazardcommunications/hc2inf2.html#*.1.4. This article focuses on many important aspects of writing instructions, including the use of icons, readability, and audience variables. Access the article and report your findings either orally or in a memo, letter, or e-mail message.

QUIZ QUESTIONS

1. What are five reasons you would include instructions for your audience?
2. What should you include in the introduction?
3. How can you create an accessible hazard alert?
4. What are the differences among warning, danger, and caution hazard alerts?
5. When is a note used?
6. What colors highlight a warning, a danger, and a caution?
7. What elements are included in a hazard alert?
8. What do graphics achieve in instructions?
9. In what way should you begin each step in an instruction?
10. What are some features of usability testing?

CHAPTER 13

Web Sites, Blogging, and Online Help

COMMUNICATION
at work

In the following scenario, Future Promise is creating a cross-functional team to develop a Web site.

Future Promise is a not-for-profit organization geared toward helping at-risk high school students. This agency realizes that to reach its target audience (teens aged 15–18), it needs an Internet presence.

Future Promise's CEO, Brent Searing, has decided to form a cross-functional team to create the agency's Web site. Brent will encourage the team to work collaboratively to determine the Web site's content, its level of interactivity, and its design features. Brent wants the Web site to include

- College scholarship opportunities
- After-school intramural sports programs
- Job-training skills (resume building and interviewing)
- Service learning programs to encourage civic responsibility
- An FAQ page
- Future Promise's 800-hotline (for suicide prevention, STD information, depression, substance abuse, and peer counseling)
- Additional links (for donors, sponsors, educational options, job opportunities, etc.)

To accommodate these Web components, the Future Promise Web team will consist of the agency's accountant, sports and recreation director, public relations manager, counselor, training

Objectives

When you complete this chapter, you will be able to

1. Recognize the importance of online technical communication.

2. Know the characteristics of e-readers.

3. Understand the characteristics of online communication.

4. Follow criteria to create a successful Web site.

5. Understand the purpose of blogging.

6. Follow criteria to create effective corporate blogging.

7. Understand online help.

8. Follow criteria for effective online help.

9. Apply the usability checklist to create effective Web sites.

---

facilitator, graphic artist, and computer and information systems director. In addition to these Future Promise employees, Brent also has asked two local high school principals, two local high school students, and a representative from the mayor's office to serve on the committee. Future Promise's public relations manager Jeannie Kort will chair the committee.

Jeannie has a big job ahead of her. First, she must coordinate everyone's schedules. The two principals and high school students, for example, can attend meetings only after school hours. Next, she must meet Brent's deadline; he wants the Web site up and running within three months. Jeannie also must manage this diverse team (with varying ages, levels of responsibility, and levels of knowledge).

Another challenge involves hardware and software. Jeannie realizes that not all of her web readers will have state-of-the-art computers. Therefore, she must ask her Web design team to create a Web site that is accessible on many platforms, even low-end hardware. Thus, though Jeannie wants the site to be colorful, interactive, and entertaining, the site also must load quickly, avoid frames, accommodate many different monitor resolutions, and be both PC- and Mac-friendly.

The task is daunting, but the end product will be invaluable for the city and the city's youth. Jeannie and Brent know that by conveying information about jobs, training, scholarships, and counseling to their end users (at-risk teens), Future Promise can improve the quality of many people's lives. Jeannie's Web design team has an exciting project on its hands.

Check Online Resources

www.prenhall.com/gerson
For more information about instructions, visit our companion Web site.

Check out our quarterly newsletters TechCom E-Notes at www.prenhall.com/gerson for dot.com updates, new case studies, insights from business professionals, grammar exercises, and facts about technical communication.

OVERVIEW—THE IMPORTANCE OF ELECTRONIC COMMUNICATION

The written word has undergone significant changes, leaping from the printed page into cyberspace. Correspondence, once limited to hard-copy letters, reports, and memos, is now often online as Web sites, web logs (blogs), and online help. Corporate brochures and newsletters, once paperbound, now are online. Product and service manuals, once paperbound, now are online. Technical communication in the twenty-first century is increasingly electronic.

This trend toward electronic communication is increasing with the growing presence of "Wi/Fi" (wireless fidelity) hotspots. At Wi/Fi hotspots, people can connect to the Internet by way of laptops, tablet PCs, cell phones, and handheld computers. This allows people to work offsite and to communicate with coworkers, clients, vendors, and corporations anywhere and anytime.

THE CHARACTERISTICS OF E-READERS

Electronic communication is an entirely different mode of communication than hard-copy text. Not only do Web sites, blogs, and online help screens differ from paper text, but also the "e-reader" approaches online communication differently than he or she will tackle hard-copy text. What are the differences between paper text and online communication?

E-readers are topic specific.   In libraries or book stores, we wander up and down aisles, looking for any book that interests us. In contrast, e-readers tend to access the Web, blogs, and online help with specific goals in mind. You go online to search for specific information found in specific Web sites: CD prices, automobile loan rates, hotel room availability, restaurant menus, technical specifications for laser printers, the start date for your college's spring semester, and so forth (Moore 2003, 16–17). You access blogs to learn about breaking news at a company. You access online help to find solutions to specific work-related problems.

E-readers want information quickly.   E-readers often scan, skim, and skip over text, looking just for the information they want and ignoring the rest of the text. In fact, readers want to find information in "ten seconds. Your web site visitors love skim-readable pages. They're not lazy. They're just in a hurry. They don't want to waste time and money reading the wrong web page, when there are millions of other pages to choose from" (McAlpine 2002).

E-readers access electronic communication using diverse platforms.   Another difference between an e-reader and one who reads a hard-copy novel is the way in which the reader accesses the document. Most hard-copy text is printed on either book-sized pages or on $8\frac{1}{2} \times 11$ inch pieces of paper. In contrast, e-readers can access electronic communication "using cell phones, PDAs, and other wireless devices" (Moore 2003, 17). Thus, screen resolution and the size of type font are key elements for online readability. Furthermore, e-readers will be using Netscape, Internet Explorer, AOL, Macs and PCs, and a host of other electronic platforms to view your text. Each of these electronic platforms differs in subtle ways.

THE CHARACTERISTICS OF ONLINE COMMUNICATION

Because e-readers are unique, you must alter the way you write online communication. This means changing your mindset as a writer. When you write a blog, a Web site, or online help, you must reconsider the following:

- **The Screen versus the Page**—Forget $8\frac{1}{2} \times 11$. Online help screens are smaller, and good Web screens rarely require scrolling.

- **Skimming versus Linear Reading**—We read books "linearly," line by line. In contrast, online help screens, Web screens, and blogs are skimmed and scanned.
- **Hypertext Links versus Chronological Reading**—We read books from beginning to end, sequentially. Web sites, however, allow us, even encourage us, to leap randomly from screen to screen.

Page Layout

Most hard-copy text is $8\frac{1}{2} \times 11$ inches, with 1-inch margins (top, sides, and bottom). When you get to the bottom of the page, you turn to the next page. In contrast, the size of online help screens, blogs, and Web sites varies.

Page Length. On-screen, less is best. When writing online, "you must be aware that computer screens display smaller amounts of information than a printed page" (Hemmi 2002, 11). A successful Web page should limit pertinent information to one screen, approximately 6 to 7 inches long, without requiring the reader to scroll. This equates to approximately 20 to 22 lines of text, versus the 50 to 55 lines that can fit on $8\frac{1}{2} \times 11$ inch paper. Online help screens are even smaller, rarely filling an entire page. They are more likely the size of 3×5 index cards. Thus, the "material that fits on one printed page might require three to six screens online" (Hemmi, 2002, 11).

Margins. Hard-copy text, when read from margin to margin, is approximately 80 characters long (a character is every letter, punctuation mark, and space). The size and type of your font, of course, affects the number of characters you will type per line. Online, horizontal margins vary from monitor to monitor (21 inches, 19 inches, 17 inches, 15 inches, etc.). On a 17-inch monitor, for example, text runs approximately 12 inches, left margin to right margin. On a 15-inch monitor, text runs approximately 10 inches.

On a Web page, your audience is forced to read more words per line if you use the entire screen. That's difficult. Readers have trouble maintaining their focus (visually and mentally) when expected to read long lines of horizontal text. Vertically, the problem increases. Regardless of the monitor size, cybertext can run forever. The reader can scroll, and scroll, and scroll. Whereas paper sizes are controlled by convention and printing companies, you must control the size of a Web page.

A successful Web page should limit lines of text to perhaps two-thirds of the screen, with a graphic, white space, or hypertext links placed in the remaining third. This breaks up the monotony of long lines into manageable chunks. Thus, the reader can maintain focus more easily (Eddings et al. 1998).

Font. Your computer's word-processing software probably defaults to Times New Roman, 12 point. That is considered the best type and size font for readable hard-copy text. You can enhance your hard-copy text with designer fonts (stylized and cursive), boldface, or underlining for visual effect. In contrast, sans serif fonts, like Arial or **Verdana**, seem to work best for online reading. "Stylized, cursive, italicized, and decorative typefaces are difficult to read online" (Hemmi 2002, 11).

Underlining is especially troublesome online, because online, hypertext links are shown as underlined text. To avoid confusing your readers online, you must avoid using underlining as a highlighting technique.

Structure

Paper text is read sequentially. We read from page 1 through the end of the text. Think of a book: You would never consider reading page 102 before you have read page 18 or Chapter 7 before you have read Chapter 4. In fact, you are expected to read sequentially—page after page after page. After all, the book (or newsletter or manual) is bound; each page is physically stitched, glued, or stapled to the next. In contrast, online help screens and Web sites are composed of randomly "stacked" screens. For online help screens, you access a search mechanism,

The Importance of Audience Recognition and Involvement in Web Design

Shannon Conner is Documentation Developer Lead at Cerner Corporation, one of the world's largest medical software developers. As part of his job, Shannon writes Web-based content for Cerner. When asked what his most important job is as a documentation developer, Shannon says, "Capturing the user—giving the user a reason to come back to our site."

One type of Web site that Cerner develops is marketed to health facilities for use by doctors, nurses, hospital administrators, and patients. This explains the communication challenge that Shannon faces. Cerner's end user is diverse, consisting of lay readers and high-tech specialists.

Lay Readers. Cerner's general audience includes patients with varied reading levels. One end user, for example, could be a 10-year-old boy, who uses the Cerner software to chart his asthma progress and medical treatment. In this instance, the child's reading abilities and knowledge of medical terminology might be limited. In contrast, another Cerner end user might be a senior citizen, using the software to communicate with a primary care physician. The senior citizen might be more familiar with medical terminology but could have limited computer capabilities. Shannon needs to create a Web site that communicates to all of these reading levels and is intuitive enough for all Web users.

In addition, Shannon says that the Web site for the lay readers has "a touch of marketing." Because Cerner wants "the customer to come back to the site," Shannon uses "a lighter touch, jazzier language, and a friendly tone."

Specialized Audiences. "On the flip side," as Shannon says, the Web site must communicate with doctors, nurses, and healthcare administrators. Their Web portal uses less marketing language, assumes a greater knowledge of medical terminology, and provides a more factual tone. The emphasis is on HIPPA laws, reminding the specialized audience to verify their patients' identities before giving them access to secure medical records.

International Audiences. A final challenge for Shannon involves Cerner's global audience. Patients and medical professionals from Australia, Germany, Austria, the United Kingdom, Singapore, and Malaysia use Cerner's products and services. Shannon, like all good technical communicators, knows the value of graphics. A screen capture of a Web application will be more effective than hundreds of words in helping an end user perform a task. In the United States, a screen capture of a Window's dialog box will read "Save," "Exit," or "Next." This will not communicate to all members of Shannon's multilingual audience. The dialog box has to be translated into many languages, and then new visuals must be made for each application.

If a Web help screen has many visuals, suddenly Shannon's task becomes more challenging and more costly for Cerner. Usability for the Web audience is paramount for Shannon. The audience must be able to say, "I can use this Web site, and I can benefit from it." To accomplish these goals, Shannon and his team must consider Cerner's global customers, their varied knowledge of the subject matter, and their diverse computer skills.

To accommodate Cerner's varied end users, Shannon constantly must recognize his audience's unique characteristics and then find ways to involve this reader in the text.

find the topic you want, and click on a link to access it. For a Web site, you scan the home page, and then you decide what you want to read and when you want to read it. Once you make this decision, you click on hypertext buttons, access any screen, and read content in any order. Thus, you receive "just in time" or "as needed" information. That is, if you want to access the linked text, you can. If you don't want to read this screen, you don't have to. Again, on-screen, less is best for the random reader who plans to skim and scan (Goldenbaum and Calvert 1998). Hard-copy documentation follows a page-by-page format, whereas online text is nonlinear. The structure of online text is modular and flexible (Hemmi 2002, 11).

Noise

Though hard-copy text is rigidly linear, it can be easier to read than online text. Paper is dull and absorbs light. Most hard-copy text is colorless and motionless; a book is composed of black ink on white paper. For online text, however, the screen is composed of glass, which reflects light, creating visual glare. Many of our screens are smudged where we have used our fingertips to point out interesting images.

To help e-readers, you need to consider a key challenge of reading online: computer "noise"—sound and visual distractions. Extended viewing of a computer screen is more demanding than continued reading of paper text. Web sites often contain lots of color, blinking text, animated graphics, frames of layered text, and sound and video. Noise—multiple distractions—inundates the screen. In such a busy communication vehicle as a Web site, the less we give the reader on one screen, the better.

WEB SITES

Web sites can convey any amount of information about any topic. Web sites are created by companies, organizations, schools, and government agencies, not to mention individuals. That's why a Web site's URL (Uniform Resource Locator—the Web site's address) reads *.com* (commercial), *.org* (organization), *.edu* (education), *.gov* (government), and so on.

The Democratic National Party has a Web site. So do McDonald's, Ford Motor Company, the city of Las Vegas, the University of Texas, and Greenpeace. There are hundreds of Web sites dedicated to Elvis Presley. Whatever topic you are interested in researching, you can find it online in someone's Web site.

Following are criteria for creating a successful Web site.

Home Page

The home page on a Web site is your welcome mat (Figure 13.1). It is the first thing your reader sees; thus, the home page sets the tone for your site. A successful home page should consist of the following components.

FIGURE 13.1 Web Site Home Page with Lead-in Information

Identification Information. Who are you? What is the name of your company, product, or service? How can viewers get in touch with you if they want to purchase your product? A good home page should clearly name the company, service, or product. In addition, a successful home page should provide the reader access to contact information such as a corporate phone number, an e-mail address, a fax number, and an address (Berst 1998).

Graphic. Don't just tell the reader who or what you are. Show the reader. An informative, attractive, and appealing graphic depicting your product or service will convey more about your company than mere words will. Check out various Web sites to prove this point. At the writing of this textbook, McDonald's Web site (www.mcdonalds.com) creates a family-friendly environment, complete with a sparkling golden arch. iPod's home page (http://www.apple.com/itunes/) shows multiple versions of its iPod Shuffle, with the tag line "Put some color on." Starbuck's Web site (www.starbucks.com) depicts an animated coffee cup to highlight the customer's ability to customize a drink. Your home page's graphic, like a corporate logo, should represent your company's product, service, and ideals.

Lead-in Introduction. In addition to a graphic, provide a phrase or sentence that tells your reader who and what you do. What's your company's focus? To clarify its goals to prospective clients, Creative Courseware adds the following lead-in introduction on its Web site home page.

example	Creative Courseware helps employers **hire, train, develop,** and **manage** their employees within the values and goals of the organization.

Navigation Bar. By providing the reader with hypertext links, the home page's navigation bar acts as an interactive table of contents or index. The reader selects a topic, clicks on that link, and jumps to a new screen. There is no one way to present these hypertext links on the home page. You could provide a vertical or horizontal navigation bar listing your hypertext links.

Linked Pages

Once your readers click on the hypertext links from the home page, they will jump to the designated linked pages. These linked pages should contain the following information.

Headings and Subheadings. Readers click on the home page link and arrive at a new screen. Will readers know where they are now? Or will they be lost in cyberspace? To ensure that readers know where they are in the context of the Web site, you need to use headings. These give the readers visual reminders of their location. Successful headings on linked pages are in the same location, font size, and font type.

Development. As in all correspondence, you should develop your ideas thoroughly. In Chapter 3, we suggest responding to reporter's questions and specifying. The same applies online. Each linked page will develop a new idea. Prove your points precisely.

Navigation

Reporter's Questions

See Chapter 3 for more discussion of reporter's questions.

If readers are viewing one screen but want to return to a screen they looked at before, how do they get there? There are no pages to turn. You need to help the readers navigate through cyberspace. You can do this in two ways.

- **Home buttons**—Readers must be able to return to the home page easily from any page of a Web site. Remember, the home page acts as a table of contents or index for all the pages within the site. By returning to the home page, readers can access any of the other pages. To ensure this easy navigation, you need to provide a hypertext-linked Home button on each page.

- **Links between Web pages**—Why make readers return to the home page each time they want to access other pages within a site? If each page has a navigation bar with a hypertext link to all pages within the site, then readers can access any page, in any order of discovery.

Document Design

You can do amazing things on a Web site. You can add distinctive backgrounds, colored fonts, different font faces and sizes, animated graphics, frames, and highlighting techniques (such as lines, icons, and bullets). However, just because you can, doesn't mean you should. Document design should enhance your text and promote your product or service, not distract from your message. As mentioned before, on a Web site, less is best.

Background.   On Web sites, you can add backgrounds, running the gamut from plain white, to various colors, to an array of patterns: fabrics, marbled textures, simulated paper, wood grains, grass and stone, psychedelic patterns, water and cloud images, and waves. They all look exciting but are not necessarily effective. When choosing a background, you want to consider your corporate image. If you are an engineering company, do you want a pink background? Perhaps a subtle stone or brick background would be more effective in portraying your corporate identity. If you are a child care center, do you want a dark background? Maybe this will be too depressing. In contrast, a white background with toys as a watermark might more successfully convey your center's mission statement.

In addition, remember that someone will attempt to read your Web site's text. Very few font colors or styles are legible against a psychedelic background—or against most patterned backgrounds, for that matter. Instead, to achieve readability, you want the best contrast between text and background. Despite the vast selection of backgrounds at your disposal, the best contrast is still black text on a white background.

Font.   How hard can selecting a font be? You just use any font type and size you want, right? No. As already noted, it's hard to read text online. Text tends to "flicker," glare impedes, and we get eye strain. Thus, font selection is very important. Consider these suggestions (Williams 2000, 389):

- Use sans serif fonts, like Arial, or one specifically designed for the Web—Verdana. Conventional wisdom has always suggested that serif fonts, like Times New Roman, work best for correspondence. However, online, a serif (the "foot" at the bottom of text) displays irregularly on low-resolution screens.
- Use 12- to 14-point type, the size that most readers with normal vision find easiest to read. A smaller font will cause problems for readers. Then use a larger font for headings.
- Avoid overusing bold and italics. First, these should only be used for emphasis, and then used sparingly. Boldface is often poorly formed on-screen because it must add pixels to a letterform. Italics, due to the oblique orientation, do not always work well with the horizontal and vertical orientation of the pixel grid.
- Avoid using underlining, which readers will confuse with a hypertext link.
- Avoid using all caps. As with e-mail, text typed in all caps makes you look as if you are shouting. In addition, studies show that we read lowercase text about 13 percent faster than we read text in all caps. Uppercasing, thus, will add additional reading problems for your audience, already stressed by the challenges of reading online.

Color. When you create a Web site, you can use any color in the spectrum, but do you want to? What corporate image do you have in mind? The font color should be suitable for your Web site, as well as readable. A yellow font on a blue background will cause headaches for your reader. A red font on a black background is nightmarish. The issue, however, is not just esthetics. A primary concern is contrast. Red, blue, and black font colors on a white background are very legible because their contrast is optimum. Other combinations of color don't offer this contrast; thus, the reader will have trouble deciphering your words (Goldenbaum and Calvert 1998).

In addition, when you use color on your Web site, the color you see on your monitor will *not* necessarily be the color your reader sees. As a webmaster, you cannot solve this problem because you cannot control other people's monitors. All you can do is try to reduce the problem. First, use restraint with color. Stick with black, red, and blue font color for text. Variations such as salmon, lime green, cornflower blue, medium aquamarine, and goldenrod won't be true. Next, test your use of color on several different monitors. You may well be surprised at what you see, but now you can change your font color if needed for clarity (Eddings et al. 1998).

Graphics. Graphics affect download time. For example, a "small" graphic (approximately 90×80 pixels, $1\frac{1}{2} \times 1\frac{1}{2}$ inches) consumes about 2.5 kilobytes. A "medium" graphic (approximately 300×200 pixels, 5×4 inches) consumes about 25 kilobytes. A "large" graphic (approximately 640×480 pixels, 10×8 inches) consumes about 900 kilobytes. When you use graphics with varying amounts of color depth, your file size increases. The larger your file, the longer it will take for the images to load. The longer it takes for your file to load, the less interested your reader will be in your Web site. As mentioned already, less is best. Limit the size and number of your graphics.

Graphics

See Chapter 9 for more discussion of graphics.

Furthermore, use suitable graphics. If you work for an accounting firm, do you want graphics depicting a smiling calculator operated by an oversized flamingo? Create and insert your graphics carefully.

Highlighting Techniques. Font sizes and styles, lines, icons, bullets, frames, java applets, animation, video, audio—you can do it all on your Web page, but should you? Nothing is more distracting than a Web site full of highlighting techniques. Avoid overwhelming your Web site with an excessive number of design elements.

Effective Online Highlighting	Ineffective Online Highlighting
1. *Lines* (horizontal rules) can separate headings and subheadings from the text.	1. *Frames* are considered to be one of the worst highlighting techniques. Jesse Berst, editorial director of ZDNet, considers frames to be one of the "Seven Deadly Web Site Sins." He says, "Too many frames . . . produce a miserable patchwork effect." If you want to achieve the frame look, but without the hassle, use tables instead. They are easier to create and revise, they load faster, and they are less distracting to your readers.
2. *Bullets and icons* enliven your text and break up the monotony of wall-to-wall words.	2. *Italics* and *underlining*. Both are hard to read online. Plus underlining looks like a hypertext link.
3. *First- and second-level headings,* achieved by changing your font size and style, separate key ideas.	3. *Java applets* take a long time to load.
4. *Boldface* also emphasizes important points.	4. *Video* requires an add-on users must download before they can enjoy your creation.
5. *White space*, created by indenting, makes text more readable (Spyridakis 2000, 367; Yeo 1996, 12–14).	5. *Animation* can be sophomoric and distracting.

Style

Conciseness is important in all technical correspondence. Conciseness is even more important in a Web site. As noted earlier in this chapter, a successful Web page should be limited to one viewable screen, and a line of text should rarely exceed two-thirds of the screen.

The Internet is a friendly medium of communication, in contrast to many books and journals. On a Web site, you are allowed, and in fact encouraged, to use pronouns, contractions, and positive words. Create a personalized tone that engages your readers.

Grammar

Perhaps because of the friendly, casual nature of the Internet, many Web sites are poorly proofread. That's a problem. Casualness is one thing; lack of professionalism is another. If your Web site represents your company, you don't want prospective clients to perceive your company as lacking in quality control. In fact, that's what poor grammar online denotes—poor attention to detail.

It's easy to see why many sites are grammatically flawed. First, anyone can go online almost instantaneously. All you do is code in your HTML and transfer the file to your Web server, and you are online. That's great for business, but it's bad for proofreading. Good proofreading takes time, a concept almost antithetical to the Internet. Next, though most word-processing packages now have spell checks and grammar checks that tell you immediately when you have made an error, many Web conversion programs do not contain these amenities. Thus, when constructing your Web site, you must be vigilant. The integrity of your Web site demands correct grammar.

Samples

Figures 13.2 through 13.6 are examples of well-designed Web pages. Notice how the Missouri Department of Transportation's (MoDOT) home page (Figure 13.2) is divided approximately into thirds. The left third contains the navigation bar. The middle third contains a visual to catch the reader's attention. The final third provides a pull-down menu for additional information about individual counties.

DOT-COM UPDATES

For Web design tips, check out these sites.

- Website Tips (http://www.websitetips.com/), complete with articles, tutorials, and resources.
- Grantastic Design (http://www.grantasticdesigns.com/tips.html) complete with design tips, rules for effective web layout, and readability and usability ideas.
- Tips and Tricks (http://www.tips-tricks.com/) complete with HTML guides for tables, frames, and sound.
- Web Design Features (http://www.ratz.com/features.html), complete with discussion of good Web design ideas and examples of problems associated with bad Web designs.
- HTML Goodies (http://www.htmlgoodies.com/), one of the most complete sites for all kinds of Web design hints.

FIGURE 13.2 Missouri Department of Transportation (MoDOT) Web Site Home Page

Reprinted with the permission of the Missouri Department of Transportation.

In Figure 13.3, MoDOT uses color to aid accessibility. When the cursor hovers over the "Business" link, the color change helps orient the e-reader to that Web page's content. In Figure 13.4, the same color is used for the Web page's heading, ensuring that the e-reader does not get lost in cyberspace. Also note how the "Business" page's text is placed in the right third

FIGURE 13.3 MoDOT Home Page with Pull-down Screen and Cursor Pointing at Business Hyperlink

Pull-down screen with color change →

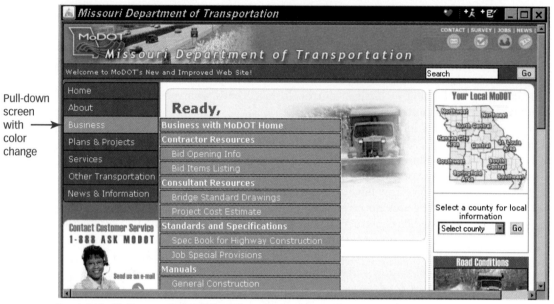

Reprinted with the permission of the Missouri Department of Transportation.

FIGURE 13.4 MoDOT Business Page

MoDOT logo used as a "Home" button

Banner color coded to match heading

Heading for screen

Graphic for visual appeal

Reprinted with the permission of the Missouri Department of Transportation.

of the screen. This helps the reader access the text without having to read from margin to margin. MoDOT's state icon in the top left screen location acts as a "Home" button.

Figure 13.5 shows how color is used again on the navigation bar to differentiate one Web page from the others. Finally, Figure 13.6 highlights the value of graphics for visual appeal.

FIGURE 13.5   MoDOT Home Page with Pull-down Screen and Cursor Pointing at Other Transportation Hyperlink

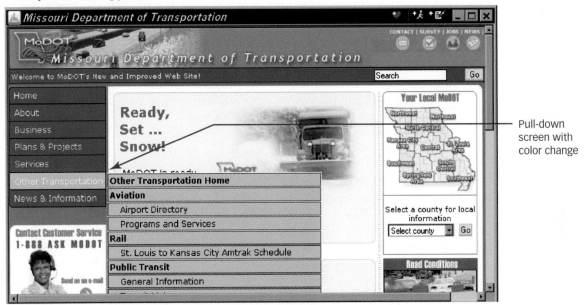

Reprinted with the permission of the Missouri Department of Transportation.

FIGURE 13.6   MoDOT Other Transportation Page

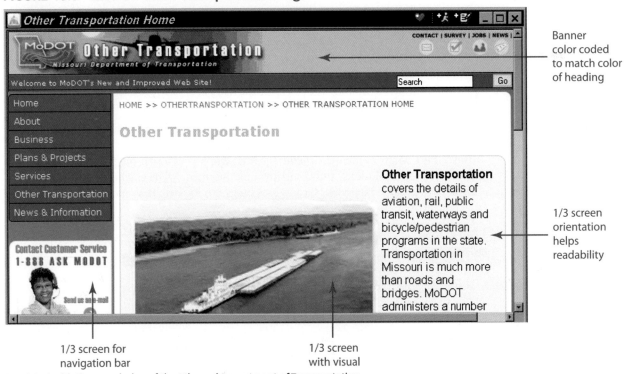

Reprinted with the permission of the Missouri Department of Transportation.

BLOGGING FOR BUSINESS

Jonathan Schwartz, president and chief operating officer of Sun Microsystems, says that blogging is a "must-have tool for every executive. . . . It'll be no more mandatory that they have blogs than that they have a phone and an e-mail account" (Kharif 2004). Bill Gates, Microsoft's CEO, says that blogs could be a better way for firms to communicate with customers, staff, and partners than e-mail and Web sites. "More than 700 Microsoft employees are already using blogs to keep people up to date with their projects" ("Gates backs blogs for businesses" 2004). Marc Cuban, owner of the Dallas Mavericks basketball team, has a blog (Ray 2004). IBM is planning "the largest corporate blogging initiative" by encouraging its 320,000 employees to become active bloggers with the goal of achieving "thought leadership" in the global information technology market (Foremski 2005). Nike has launched an "adverblog" to market its products.

What is "blogging," and why are so many influential companies and business leaders becoming involved in this new communication channel?

Blogging—A Definition

Web logs, also knows as "blogs," are Web-based journals posted online. Blogs allow individuals to create diary entries, consisting of observations, thoughts, insights, "news, opinions, ideas and brainstorms" (Wuorio "5 Ways" 2005).

Blogs are a unique type of communication channel and are more informal than reports and letters. Thus, in tone, blogs are similar to e-mail messages or instant messaging. Like Web sites, blogs appear online. However, blogs are different from Web sites, e-mail messages, and instant messaging because blogs are even more casual in tone and style.

Blogging is a growing phenomenon. Blog readership grew by 58 percent in 2004 (Blumberg 2005). As of late 2004, "an estimated 4.8 million blogs" existed, with up to 6 million Americans reading blogs, "up from just 100,000" in 2002 (Gard 2004). Blogging is growing at a faster pace than the Internet did in its early years (Cross 2003). To put that number in perspective, 4 million bloggers equals more people than readers of the popular newspaper *USA Today* per day (Bruner 2004).

Ten Reasons for Blogging

The rapid growth of blogging might be reason enough for businesses to consider using this online communication channel. However, following are 10 more reasons why companies are beginning to blog.

1. **Communicate with colleagues.** Chris Winfield, president of an online marketing company, says that he uses blogging to improve "communication flow to his employees" (Ray 2004). Many companies encourage their employees to use blogging for project updates, issue resolutions, and company announcements. The engineering department at Disney ABC Cable Networks Group uses blogs "to log help desk inquiries" (Li 2004).

FAQs Why is blogging considered the next big technological wave?

A: Many pundits predict that blogging will make the next big technology splash for the following reasons.

- Young adults, the future in business and technology, read blogs three times more frequently than older adults.
- Bloggers, mostly male, affluent, and online connected, are an attractive demographic for advertisers.

2. **Communicate with customers.** In contrast to private, intranet-based blogs used for internal corporate communication, a company also can have public blogs. Through public blogs, a company can initiate question/answer forums, respond to customer concerns, allow customers to communicate with each other, create interactive newsletters, and build rapport with customers, vendors, and stakeholders.

3. **Introduce new product information.** Because blogs are quick and current, they allow companies to provide up-to-date information about new products and services.

4. **Reach an influential audience segment.** One reason why companies are turning to blogging is that blogs appeal to an influential and emerging audience—youth. Young adults "between the ages of 18 and 24" read blogs "three times more frequently than older adults" (Li 2004). This age segment of online consumers provides an attractive target for companies.

5. **Improve search engine rankings.** Marketing is a key attraction of corporate blogs. A blog post using keywords, allowing for comments and responses, and providing references and links to other sites, tends to rank in the top 10 to 20 listings in Internet search engines (Ray 2004).

6. **Network through "syndication."** To access a Web site, a reader must know the URL or use a search engine, such as Google. Blogs can be distributed directly to the end users through a blog "feed." By using feed programs, such as RSS ("Really Simple Syndication" or "Rich Site Summary"), Atom, or Current Awareness, bloggers can syndicate their blogs or be notified when topics of interest are published. Thus, blogs can be very personalized, essentially delivered to your door on a moment's notice.

7. **Facilitate online publishing.** A key value to blogging is its ease and affordability. Blogging services and software, such as Blogger, Movable Type, TypePad, Xanga, Live Journal, and Weblogger, give bloggers access to easy-to-use Web-based forms.

8. **Encourage "bubble-up" communication.** Blogging promotes brainstorming and public forums. A corporate blog builds networks, allows people to dialogue, and encourages conversation.

9. **Track public opinion.** "Trackback" features, available from many blog services, let companies track blog usage. Tracking lets companies monitor their brand impact and learn what customers are saying (good news or bad news) about their products or services.

10. **Personalize your company.** Personal responses to customer comments help personalize a company. Blogging offers a refreshing option that reaches out to customers, offering a human dimension.

Ten Guidelines for Effective Corporate Blogging

If you and your company decide to enter the "blogosphere," the world of blogging, follow these 10 guidelines.

1. **Determine if you need to blog.** Not every company needs a blog, any more than every company needs a Web site. If blogging does not fit into your corporation's culture, you should avoid the blogosphere.

2. **Identify your audience.** As with all technical communication, audience recognition and involvement are crucial. Before blogging, decide what topics you want to focus on, what your unique spin will be, what your goals are in using a blog, and who your blog might appeal to.

DOT-COM UPDATES

To see examples of corporate blogs and blogs related to the job search, check out these sites.

- http://garmin.blogs.com (**Garmin**)
- http://businessblog.sprint.com (**Sprint**)
- http://www.ibm.com/ibm/syndication/ (**IBM**)
- http://blogs.msdn.com/ (**Microsoft**)
- http://stonyfieldfarm.com/weblog/ (**Stonyfield Farm**)

DOT-COM UPDATES

For more information about blogging, check out the following links.

- "Gates backs blogs for business." *BBCNews*. 21 May 2005. http://news.bbc.co.uk/2/hi/technology/3734981.stm.
- Foremski, Tom. "IBM is preparing. *SiliconValleyWatcher*". 13 May 2005. http://www.siliconvalleywatcher.com/mt/archives/2005/05/can_blogging_bo.php.
- Jackson, Lorrie. "Blogging basics: Creating student journals on the web." 18 July 2006. http://www.education-world.com/a_tech/techtorial/techtorial037print.shtml.
- Kharif, Olga. "Blogging for business." *BusinessWeek Online*. 9 Aug. 2004. http://www.businessweek.com/technology/content/aug2004/tc2004089_3601_tc024.htm.
- Li, Charlene. "Blogging: Bubble or big deal?" *Forrester*. 5 Nov. 2004. http://www.forrester.com/Research/Print/Document/0,7211,35000,00.html.

3. **Achieve customer contact.**   Blogs are innately personal. Take advantage of this feature. Make your blogs fun and informal. You can give your blog personality and encourage customer outreach by including "interesting news of the day, jokes," personnel biographies, question/answer forms, updates, an opportunity to add comments, as well as information about products and services (Ray 2004). In addition, make sure the blog is interactive, allowing for readers to comment, check out new links, or add links.

4. **Determine where to locate the blog.**   You can locate and market blogs in many different places: Add a blog link to your existing Web site, create a totally independent blog site, distribute your blog through RSS aggregators, add a blog link to your e-mail signature, or mention your blog in marketing collateral (Wuorio "Blogging" 2005; Li 2004).

5. **Start "blogrolling."**   Once you have determined audience and blog location, it's time to start blogrolling. Start talking. Not only do you need to start the dialogue by adding content to your blog, but also you want to link your blog to other sites (Wuorio "Blogging" 2005).

6. **Emphasize keywords.**   By mentioning keywords in your blog titles and text, you can increase the number of hits your blog receives from Internet search engines.

7. **Keep it fresh.**   Avoid a stale blog. By posting frequently (daily or weekly), you encourage bloggers to access your site and return to it often.

8. **Respond quickly to criticism.**   In a communication channel like the blogosphere where dialogue is the desired end, companies inevitably will receive criticism about products or services. Blogs allow companies to respond quickly to and manage bad news.

9. **Build trust.**   Don't use the blog only to promote new corporate products. Transparent marketing tends to backfire on bloggers who want sincere content, an opportunity to learn more about a company's culture, and a chance to engage in dialogue.

10. **Develop guidelines for corporate blog usage.**   Because blogs encourage openness from customers as well as employees, a company must establish guidelines for corporate blog usage. One potential problem associated with blogging, for example, is divulging company information, such as financial information or trademark secrets. Figure 13.7 offers a sample code of ethics created by Forrester Research (Li 2004).

DOT-COM UPDATES

For more information about products used to create blogsites, check out these links.

- Blogger.com
- Livejournal.com
- Xanga.com
- Weblogg-ed.com
- Edublogs.org
- FunnyMonkey.com
- Rollerweblogger.org
- WordPress.com
- MovableType.com

ONLINE HELP

Imagine this possibility. Your boss calls you into a meeting and says, "As of today, we are no longer producing paper documents" (McGowan 2000, 23). It's not as farfetched as it sounds. Online Web help is a major part of the technical communicator's everyday job for several reasons, including

- The increased use of computers in business, industry, education, and the home
- The reduced dependence on hard-copy manuals by consumers
- The need for readily available online assistance
- Proof that people learn more effectively from online tutorials than from printed manuals (Pratt 1998, 33)

In addition, online help screens provide the practical value of portability. Rather than take up computer archival space with thousands of pages of PDF files or print out hundreds of pages of instructional help, online help screens allow companies to save space and provide readers ease of use.

FIGURE 13.7   Blogging Code of Ethics

Blogging Code of Ethics

1. I will tell the truth.
2. I will write deliberately and with accuracy.
3. I will acknowledge and correct mistakes promptly.
4. I will preserve the original post, using notations to show where I have made changes so as to maintain the integrity of my publishing.
5. I will never delete a post.
6. I will not delete comments unless they are spam or off-topic.
7. I will reply to e-mails and comments when appropriate and will do so promptly.
8. I will strive for high quality with every post—including basic spell-checking.
9. I will stay on topic.
10. I will disagree with other opinions respectfully.
11. I will link to online references and original source materials directly.
12. I will disclose conflicts of interest.
13. I will keep private issues and topics private because discussing private issues would jeopardize my personal and work relationships.

What is an online help system? Online help systems, which employ computer software to help users complete a task, include procedures, reference information, wizards, and indexes (see Figure 13.8). Typical online help navigation provides readers hypertext links, tables of contents, and full-text search mechanisms.

Help menus on your computer are excellent examples of online help systems. As the computer user, you pull down a help menu, search a help list, and click on the topic of your choice, revealed as hypertext links. When you click on a link, you could get a *pop-up* (a small window superimposed on your text), or your computer might *link* to another full-sized screen layered over your text. In either instance, the pop-up or hyperlink gives the following information about your topic.

- **Overviews**—explanations of why a procedure is required and what outcomes are expected
- **Processes**—discussions of how something works
- **Definitions**—an online glossary of terms
- **Procedures**—step-by-step instructions for completing a task
- **Examples**—feedback verifying the completion of a task or graphic depictions of a completed task
- **Cross-references**—hypertext links to additional information
- **Tutorials**—opportunities to practice online

Online help systems allow the technical communicator to create interactive training tools and informational booths within a document. These online systems can be created using a wide variety of authoring tools. Popular ones include *RoboHELP, HelpBreeze,*

FIGURE 13.8   Microsoft Word 2007 Help Screen with a Search for "Margins" and the Resulting Pop-up Help Screen

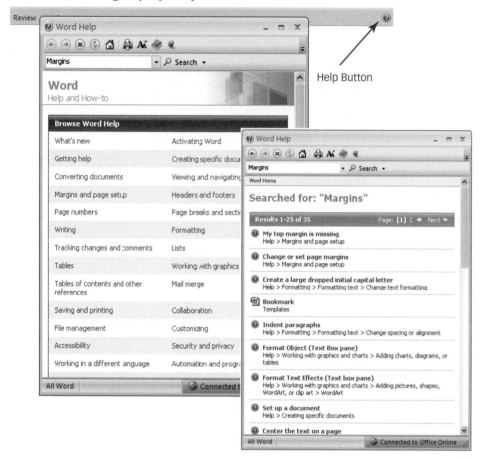

ForeHelp, Help Magician, Visual Help, WYSI–Help, AuthorIT, and *Flare.* The value of online help is immense. Because online help offers "just in time" learning or "as needed" information, readers can progress at their own pace while learning a program or performing a task.

Techniques for Writing Effective Online Help

To create effective online help screens, consider the following suggestions.

Organize Your Information for Easy Navigation. Poorly organized screens lead to readers who are "lost in cyberspace." Either they cannot find the information they need, or they have accessed so many hypertext links that they are five or six screens deep into text.

You can avoid such problems and help your readers access information in various ways.

- Allow users to record a history of the screens they've accessed through a bookmark or a Help Topics pull-down menu.
- Provide an online Contents menu, allowing users to access other, cross-referenced help screens within the system.

- Provide a Back button or a Home button to allow the readers to return to a previous screen.
- Provide links such as "In this section" and "Related topics" on each screen for immediate access to topic-specific content.

A good test is the "three clicks rule." Readers should not have to access more than three screens to find the answer they need. Similarly, readers should not have to backtrack more than three screens to return to their place in the original text.

Recognize Your Audience.   Online help must be user oriented. After all, the only goal of online help is to help the user complete a task. This means that technical communicators have to determine a user's level of knowledge.

If the system is transmitted by way of an intranet or extranet, you might be tempted to write at a high- or low-tech level. Your readers, you assume, will work within a defined industry and possess a certain level of knowledge. That, of course, is probably a false assumption. Even within a specific industry or within a specific company, you will have coworkers with widely diverse backgrounds: accountants, engineers, data processors, salespeople, human resource employees, management, technicians, and so on.

Don't assume. Find out what information your readers need. You can accomplish this goal through usability testing, focus groups, brainstorming sessions, surveys, and your company's hotline help desk logs. Then don't scrimp on the information you provide. Don't just provide the basic or the obvious. Provide more detailed information and numerous pop-ups or links. Pop-up definitions are an especially effective tool for helping a diverse audience. Remember, with online help screens, readers who don't need the information can skip it. Those readers who do need the additional information will appreciate your efforts. They will more successfully complete the tasks, and your help desk will receive fewer calls.

Achieve a Positive, Personalized Tone.   Users want to be encouraged, especially if they are trying to accomplish a difficult task. Thus, your help screens should be constructive, not critical. The messages also should be personalized, including pronouns to involve the reader.

Audience Definition and Tones

See Chapter 4 for more discussion of audience, definitions, and tone.

Excellent examples of this positive, personalized tone can be seen in Microsoft's Office Assistants, found in Microsoft Word, Excel, and Outlook. This online help system allows you to choose your help wizard from several options. For example, you could choose "Links," an animated cat, who says, "If you're on the prowl for answers in Office, Links can chase them down for you." Another wizard, Mother Nature, "provides gentle help and guidance." A third option, the Dot, introduces itself by stating, "When you need help of any kind, just give me a click."

Design Your Document.   How will your help screens look? Document design is important because it helps your readers access the information they need. (See Figure 13.9.) To achieve an effective document design, consider the following points.

Document Design

See Chapter 8 for more discussion of document design.

- **Use color sparingly**—Color causes several problems in online documents. Bright colors and too many colors strain your reader's eyes. Furthermore, a color that looks good on a high-resolution monitor might be difficult to read on a monitor with poor resolution. Your primary goal is contrast. To help your audience read your text, you want to maximize the contrast between the text color and the background color. Black text on a white background offers the optimum contrast. Many help screens, such as those provided by Microsoft Office, use blue text on white background. This combination provides contrast and differentiates the help screens from the running text.

FIGURE 13.9 Help Screen

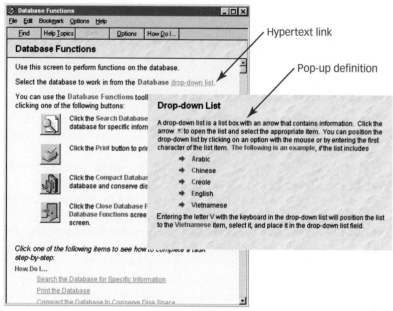

Courtesy of Earl Eddings, technical communications specialist.

- **Be consistent**—Pick a color scheme and stick with it. Your headings should be consistent, along with your word usage, tone, placement of help screen links and pop-ups, graphics, wizards, and icons. Readers expect to find things in the same place each time they look. If your help screens are inconsistent, readers will be confused.

- **Use a 10-point sans serif font**—A 12-point type size is standard for most printed documents, but 10-point type will save you valuable space online. Serif fonts are the standard for most technical writing. A serif font, with small, horizontal "feet" at the bottom of each letter, helps guide the eye while reading printed text. However, on lower resolution monitors, serif text is more difficult to read.

- **Use white space**—Don't clutter your help screens. Avoid excessive emphasis techniques. For example, typical PowerPoint "fly-in" effects not only are distracting, but also they negatively impact the online screen's performance, slowing down load time dramatically. Minimize your reader's overload by adding ample horizontal and vertical white space. Online, less is best.

Conciseness

See Chapter 3 for more discussion of clarity and conciseness.

Be Concise. "It takes 20 to 30 percent longer to read on-screen than in print, so you must minimize text" (Timpone 1997). Furthermore, online space is limited. Chapter 3 provides you many techniques for achieving conciseness. In addition to limiting word and sentence length, a help screen, just like any Web page, should limit horizontal and vertical scrolling. Each screen should include one self-contained message. If readers need additional information, your "In this section" and "Related topics" links allow them to access the follow-up text.

Be Clear.   Your audience reads the help screen only to learn how to perform a task. Thus, your only job is to meet the reader's needs—clearly. To accomplish this goal, you need to be specific. In addition, clarity online could include the following:

- Tutorials to guide the reader through a task. Microsoft gives its readers a "How?" button to meet this need. WordPerfect's help system offers "How Do I?" links to show step-by-step procedures.
- Graphics that depict the end result. Microsoft gives its readers a "Show me" button followed by screen captures. WordPerfect's help system provides an "Examples" link with pictures of what you can do and how to do it.
- Cross-references. WordPerfect provides an "Additional Help" link, complete with a glossary, an index, customer support, a Contents button, and a speech recognition system (available on CD-ROM). Microsoft programs also provide a help index and a table of contents.
- Pop-up definitions with expanding blocks for longer definitions.

Provide Access on Multiple Platforms.   You will want every reader to be able to access your online help, not just those using a PC or a Mac or UNIX. After all, if readers can't access your online help, how have you helped them? (The only exception would be if you were creating online help for an intranet or extranet over which your company had complete control.) To ensure that all readers can access your system,

- Do not use platform-specific technologies, such as ActiveX, which only runs on 32-bit Windows.
- Create content that can be viewed on multiple browsers, including Internet-Exchange and Navigator.
- Create your online help for the oldest version that you will support—the lowest common denominator. Remember, all of your users won't have the latest version browser, software, or hardware.
- Test your online help on at least three different browsers, versions, and platforms. Though you might be creating your online help screens on a 30-inch widescreen, your audience might be accessing the text on a 15-inch panel. Make sure they can see what you see, even if they are viewing text on smaller screens.

Correct Your Grammar.   As in all technical communication, incorrect grammar online leads to two negative results: a lack of clarity and a lack of professionalism. Don't embarrass your company or confuse your reader with grammatical errors. Proofread.

THE WRITING PROCESS AT WORK

Effective writing follows a process of prewriting, writing, and rewriting. Each of these steps is sequential and yet continuous. The writing process is dynamic, with the three steps frequently overlapping. Follow the writing process to create an effective Web site.

Prewriting

Before you begin to construct your Web site, gather data to determine your Web site's goal, audience, focus, and content. Answer these **reporter's questions.**

1. *Why* are you creating a Web site? Are you writing to inform, instruct, persuade, or build rapport? Are you creating the Web site to
   - Inform the public about products and services?
   - Sell products via online order forms?
   - Encourage new franchisees to begin their own businesses?

Check Online Resources

www.prenhall.com/gerson
For more information about the writing process, visit our companion website.

The Writing Process

Prewriting	Writing	Rewriting

• Use the reporter's questions to clarify what content you will need to include.
• Determine whether your audience is high tech, low tech, or lay. This will help you decide which abbreviations and acronyms need to be defined.
• Visualize the Web site's layout by storyboarding.
	• Create a navigation bar and home buttons.
	• Use headings and subheadings for easy navigation in cyberspace.
	• Consider enhancing the Web site with elements of document design.
		• Revise your draft by
		  • adding details
		  • deleting wordiness
		  • simplifying words
		  • enhancing the tone
		  • reformatting your text
		  • proofreading and correcting errors

Communication Process

See Chapter 2 for more discussion of prewriting, writing, and rewriting.

- Advertise new job openings within the company?
- Provide a company profile, including annual reports and employee resumes?
- List satisfied clients?
- Provide contact information?
- All of the above?

2. *Who* is your audience? Are you writing to high- or low-tech users, prospective employees, the general lay public, or stockholders? Will your audience be
   - Customers looking for product information?
   - Current users of a product or service looking for upgrades, enhancements, or changes in pricing?
   - Customers looking for online help with instructions or procedures?
   - Vendors hoping to contact customer service?
   - Prospective new hires looking for information about your company?
   - People looking for franchise information?

3. Answering these questions will help you decide *what* to include in your site. You could include
   - Order forms
   - Forms for customer comments
   - Frequently asked questions (FAQs)
   - Online resumes of your personnel
   - Product or service descriptions
   - Product or service support
   - Job openings
   - Customer contacts
   - Company history, qualifications, affiliations
   - Costs
   - Storefront locations
   - Blog sites

4. *How* will you construct the site? Will you use HTML coding or HTML converters and editors? Will you download existing graphics, modified to meet legal and ethical standards, or will you create new graphics?

FIGURE 13.10   Web Site Storyboard

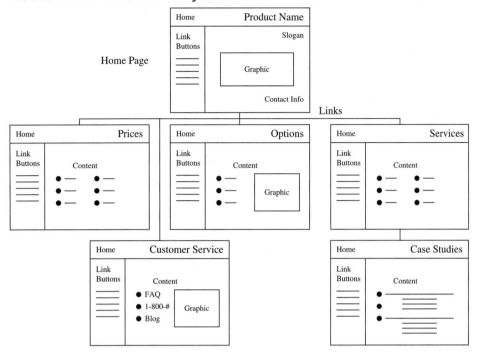

5. *Where* will your Web site reside? Will it be an Internet site open to the World Wide Web, or will you focus on an intranet or extranet site?

Once you have gathered this information, consider using **storyboarding** for a preliminary sketch of your Web site's layout.

- **Sketch the home page.**   How do you want the home page to look? What color scheme do you have in mind? How many graphics do you envision? Where will you place your graphics and text—margin left, centered, margin right?

- **Sketch the linked page(s).**   How many linked pages do you envision? Will they have graphics, online forms, e-mail links, audio or video, headings and subheadings, or links to other Internet, intranet, or extranet sites? Will you use frames or tables? See Figure 13.10 for a sample Web site storyboard.

Writing

To create a rough draft of your Web site, follow this procedure.

Study the Web site usability checklist.   This will remind you what to include in your rough draft and how best to structure the Web site.

Review your prewriting.   Have you answered all the reporter's questions? How does your Web site storyboarding sketch look? Do you see any glaring omissions or unnecessary items? If so, add what's missing and omit what's unneeded.

Draft the Web site.   You can create your rough draft in several ways.

- Write a rough draft in any word-processing program. Then, after revision, you can save your text as a Web page.

- Draft your text directly in an HTML editor, such as FrontPage, Dreamweaver, Ant HTML, HotDog Web Editor, Netscape Navigator 4.0 and above, Microsoft Internet Assistant for Word 6.0 for Windows and above, Corel WordPerfect 8.0, and others.

- Write your draft using Hypertext Markup Language (HTML) coding.

Rewriting

Once you have completed drafting your Web site, is it what you had in mind? If not, edit and revise the site to ensure its usability. Usability is a way to determine whether your reader can "use" the Web site effectively, whether your site meets your user's needs and expectations. Not only do the readers want to find information that helps them better understand the topic, but also the audience wants the Web site to be readable, accurate, up to date, and easy to access. Thus, usability focuses on three key factors (Dorazio 2000).

- **Retrievability**—the user wants to find specific information quickly and be able to navigate easily between the screens.
- **Readability**—the user wants to be able to read and comprehend information quickly and easily.
- **Accuracy**—the user wants complete, correct, and up-to-date information.

The goal of usability testing is to solve problems that might make a Web site hard for people to use. By successfully testing a site's usability, a company can reduce customer and colleague complaints; increase employee productivity (reducing the time it takes to get work done); increase sales volume because customers can purchase products or services online; and decrease help desk calls and their costs. To achieve user satisfaction, revise your Web site to meet the goals listed in our Web Site Usability Checklist. In revising a Web site, consider the following options.

1. **Add new detail for clarity**—Have you said all you need to say? What additional information should you include (contact information, forms to help your customers order goods online, job openings, an online catalog, etc.)?
2. **Delete anything unnecessary for conciseness**—Do your text and graphics all fit on one viewable screen? Does your Web site take too long to load due to excessively large graphics, java scripting, animation, or audio and video plug-ins? Delete the unnecessary add-ons, dead words and phrases, and excessively large graphics.
3. **Simplify words and phrases**—A Web site should be a "friendly" type of correspondence. Revise your text, striving for easy-to-understand words, short sentences, and short paragraphs.
4. **Move information for emphasis**—As in real estate, location, location, location sells. Decide what is most important on your Web site, and then place this information where it can be seen clearly.
5. **Reformat for access**—No communication medium depends more on layout than a Web site. Your Web site's design will make or break it. How does your Web site look? Is there ample white space? Do the graphics integrate well with the text? Have you used headings and subheadings for easy navigation? Have you used color effectively, making sure to achieve optimum contrast between text and background?
6. **Enhance the tone and style of your Web site**—Web sites tend to be informal. Don't write a Web site using the same tone that you would use when writing a formal report. Enhance the tone of your site by using contractions, positive words, and pronouns.
7. **Correct errors**—On the World Wide Web, millions of people might read your Web site. Grammatical errors, therefore, are magnified. You must proofread so your company or organization looks professional.
8. **Make sure it works**—Test your Web site. How do your graphics and colors look on multiple browsers, platforms, and monitors? Do your hypertext links work? Test links to external sites frequently. No matter how beautiful your Web site looks on your monitor, the site will not be successful unless all the components work.

Using Microsoft Word 2007 to Create a Web Site

Formerly, Microsoft's FrontPage was a popular tool for creating Web sites. However, Microsoft's "Next Generation" of Web authoring tools replace FrontPage (discontinued in 2006) with two new products, SharePoint and Expression Web.

SharePoint—The Next Generation Web Designing Tool

As of this textbook's writing, many companies have yet to decide on whether to adopt SharePoint. Instead of using either FrontPage, SharePoint, or Expression Web, which might not be loaded on your computers, you can use Microsoft Word 2007 to create your Web site. Follow these steps to do so.

1. From the **Insert** tab, insert a 3×2 table.

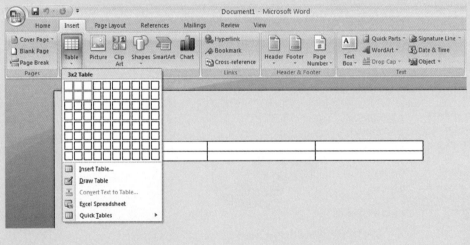

2. Merge the top right two columns.

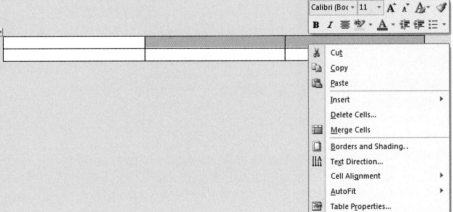

3. Place your cursor in the right rows and hit the Enter key to add space. At this point, you have the beginning design for a Web page.
4. Add graphics, text, color, and font types just as you would with any Word document.

5. To create hypertext links, highlight selected text, right click, and scroll to **Hyperlink**.

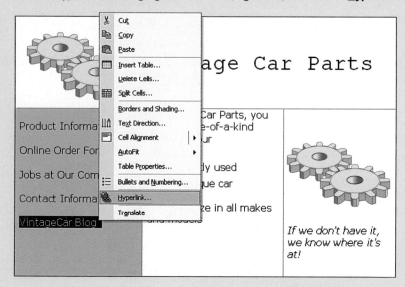

6. To create your Web site, click on the Word 2007 **Office Button**, scroll to **Save As,** and then to **Other Formats**.

7. In the pop-up **Save As** window, click on the **Save as type** down arrow and click on **Web Page.** This saves your work as an HTML (Hypertext Markup Language) file which you can open in a Web browser.

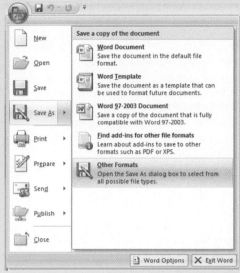

WEB SITE USABILITY CHECKLIST

Audience Recognition and Involvement

_____ 1. Does the Web site meet your reader's needs?

_____ 2. Does the Web site give your audience a reason to return (tutorials, tips, comics, links to other interesting sites, regular updates, etc.)?

_____ 3. Does the Web site solicit user feedback for interactivity and display customer comments?

_____ 4. Does the Web site make it easy for your audience to purchase online through online forms?

_____ 5. Does the Web site use pronouns to engage the reader?

Home Page

_____ 1. Does the home page provide identification information (name of service or product, company name, e-mail, fax, city, state, street address, etc.)?

_____ 2. Does the home page provide an informative and appealing graphic that represents the product, service, or company?

_____ 3. Does the home page provide a welcoming and informative introductory phrase, sentence, or paragraph?

_____ 4. Does the home page provide hypertext links connecting the reader to subsequent screens?

Linked Pages

_____ 1. Do the linked pages provide headings clearly indicating to the reader which screen he or she is viewing?

_____ 2. Do the linked pages develop ideas thoroughly (appropriate amount of detail, specificity, valuable information)?

_____ 3. Are the linked pages limited to one or two primary topics?

Navigation

_____ 1. Does the Web site allow for easy return from linked pages to the home page?

_____ 2. Does the Web site allow for easy movement between linked pages?

_____ 3. Does the Web site provide access to other Web sites for additional information?

Document Design

_____ 1. Does the Web site provide an effective background, suitable to the content and creating effective contrast for readability?

_____ 2. Does the Web site use color effectively, in a way suitable to the content and creating effective contrast for readability?

_____ 3. Does the Web site use a consistent document design (colors, background, graphics, font), carried throughout the entire site?

_____ 4. Does the Web site use headings and subheadings for easy navigation?

_____ 5. Does the Web site vary font size and type to create a hierarchy of headings?

_____ 6. Does the Web site use graphics effectively—suitable to the content, not distracting, and load quickly?

_____ 7. Does the Web site use highlighting techniques effectively (white space, lines, bullets, icons, audio, video, frames, font size and type, etc.) in a way that is suitable to the content and not distracting?

Style

_____ 1. Is the Web site concise? (Remember that reading online is a challenge to end users.)
- Short words (1–2 syllables)
- Short sentences (10–12 words per sentence)

- Short paragraphs (4 typed lines maximum)
- Line length limited to approximately two-thirds of the screen (40–60 characters per line)
- Text per Web page limited to one viewable screen, minimizing the need to scroll

Accuracy

_____ 1. Does the Web site avoid grammatical errors?

_____ 2. Does the Web site ensure that information (phone numbers, e-mail, fax numbers, addresses, content) is current and correct?

_____ 3. Do the internal and external links work?

PROCESS SAMPLES

The Future Promise Web design team created a rough draft home page and second page mockup for their CEO Brent Searing to review. Once he read the Web pages shown in Figures 13.11 and 13.12, he made revision suggestions.

FIGURE 13.11   Rough Draft Home Page with Revision Suggestions and Comments from Future Promise's CEO

Why have we listed the links in the order shown? Did someone on the Web team do a survey or create a questionnaire to find out what our target audience really wants the most? For example, should jobs be listed first?

Future Promise

Scholarships

Intramurals

Volunteerism

Job Links

FAQ

"Preparing Today's Youth for Tomorrow's Challenges"

1894 West King's Highway   San Antonio, TX 78532
Phone: 814-236-5482   E-mail: bsearing@fp.org   Fax: 814-236-5480

Where did the photo come from? Since I don't recognize any of the kids, I'm guessing we downloaded the graphic from the Internet. Do we need permission for its use? Talk to Future Promise lawyers about that.

Should we have a separate link for contact information? If we did, we could use the bottom of the home page for scrolling announcements.

FIGURE 13.12   Rough Draft Intramurals Page with Suggestions and Comments from Future Promise's CEO

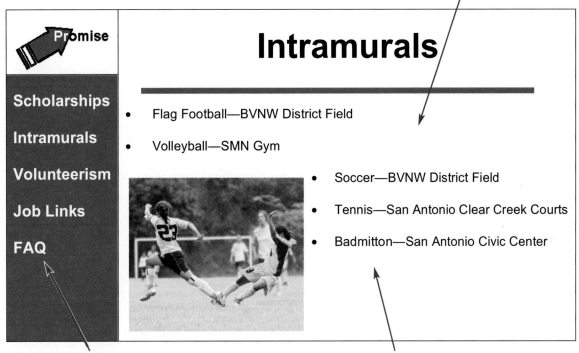

Does our target audience know how to find the various sites? And do the readers understand the abbreviations "BVNW" and "SMN"? Maybe we need to create Mapquest links, for example, for each venue.

We need a link to our contact information on this screen. That way, our audience can call or e-mail us to find out if the sports are co-ed, the dates and times for the intramural events, and whether equipment is provided (tennis racquets, for example).

Notice that "Badminton" is spelled incorrectly.

CHAPTER HIGHLIGHTS

1. To be an effective communicator in business and industry today, you have to communicate electronically.
2. Weblogs (blogs) are Web-based journals posted online and allow users to share opinions or observations.
3. Hard-copy documentation follows page-by-page layout, whereas online text is nonlinear.
4. E-readers are topic specific and want information quickly.
5. Blogging can allow your company to reach an important audience, 18 to 24 year olds who blog frequently.
6. Avoid a stale blog by posting frequently to encourage your audience to return to the blog.
7. Interactive online help systems provide "just in time" and "as needed" information to help your audience complete a task.
8. Developing guidelines for corporate blogging can ensure ethical and appropriate use.
9. The usability checklist can ensure that you create an effective Web site.
10. The process approach to writing applies to all technical communication including Web site construction.

CASE STUDIES

1. **Future Promise** is a not-for-profit organization geared toward helping at-risk high school students. This agency realizes that to reach its target audience (teens aged 15–18), it needs an Internet presence.

   To do so, it has formed a 12-person team, consisting of the agency's accountant, sports and recreation director, public relations manager, counselor, technical writer, graphic artist, computer and information systems director, two local high school principals, two local high school students, and a representative from the mayor's office. Jeannie Kort, the PR manager, is acting as team leader.

   The team needs to determine the Web site's content, design, and levels of interactivity. Jeannie's boss, Brent Searing, has given the team a deadline and a few components that must be included in the site.

   - College scholarship opportunities
   - After-school intramural sports programs
   - Job-training skills (resume building and interviewing)
   - Service learning programs to encourage civic responsibility
   - FAQ page
   - Future Promise's 800-hotline (for suicide prevention, STD information, depression, substance abuse, and peer counseling)
   - Additional links (for donors, sponsors, educational options, job opportunities, etc.)

   Jeannie has a big job ahead of her.

Assignment

Form a team (or become a member of a team as assigned by your instructor) and design Future Promise's Web site. To do so, follow the criteria for Web design provided in this chapter. Then, prewrite, write, and rewrite as follows:

Prewrite

- Research the topics listed above (either in a library, online, or through interviews) to gather details for your Web site.
- Consider your various audiences and their respective needs and interests.
- Focus on your Web site's purpose—are you writing to inform, persuade, instruct, or build rapport?
- Draw a storyboard of how you would like the Web site to look or use an organizational chart to lay out the Web design.
- Divide your labors among the team members (who will research, write, create graphics, etc.?).

Write

- Draft your text either in a word-processing program (you can save as an HTML file) or directly in FrontPage (or any other Web design program of your choice).

Rewrite

- Review the Web site criteria in this chapter.
- Add, delete, simplify, move, enhance, and correct your Web site.
- Test the site to make sure all links work and that it meets your audience's needs.

2. The following short proposal recommends that upper management approve the construction of a corporate Web site. The company's management has agreed with the proposal. You and your team have received a memo directing you to build the Web site. Using the proposal, build your company's Web site.

DATE: November 11, 2008
TO: Distribution
FROM: Shannon Conner
SUBJECT: RECOMMENDATION REPORT FOR NEW CORPORATE
            WEB SITE

Introduction

In response to your request, I have researched the impact of the Internet on corporate earnings. Companies are still striving to position themselves on the World Wide Web and maximize their earnings potential: I concluded that from 2000 to 2002 only 16 percent of companies with Internet sites were making a profit from their online services. This fact, however, has changed. From 2005 to 2008 many more corporate Web sites earned a profit for their companies. The time is right for our company, Java Lava, to go online.

Discussion

Why should we go online? The answer is simple: We can maximize our profits by making our local product a global product. How can we accomplish this goal?

1. *International bean sales.* Currently, coffee bean sales account for only 27 percent of our company's overall profit. These coffee bean sales depend solely on walk-in trade. The remaining 73 percent stems from over-the-counter beverage sales. If we go global via the Internet, we can expand our coffee bean sales dramatically. Potential clients from every continent will be able to order our coffee beans online.

2. *International promotional product sales.* Mugs, T-shirts, boxer shorts, jean jackets, leather jackets, key chains, paperweights, and calendars could be marketed. Currently we give away items imprinted with our company's logo. By selling these items online, we could make money while marketing our company.

3. *International franchises.* We now have three coffeehouses, located at 1200 San Jacinto, 3897 Pecan Street, and 1801 West Paloma Avenue. Let's franchise. On the Internet, we could offer franchise options internationally.

4. *Online employment opportunities.* Once we begin to franchise, we'll want to control hiring practices to ensure that Java Lava standards are met. Through a Web site, we could post job openings internationally and list the job requirements. Then potential employees could submit their resumes online.

In addition to this information, used to increase our income, we could provide the following.

- A map showing our three current sites.
- Our company's history—founded in 1948 by Hiram and Miriam Coenenburg, with money earned from their import/export bean business. "Hi" and "Mam," as they were affectionately called, handed their first coffeehouse over to their sons (Robert, John, and William) who expanded to our three stores. Now a third generation, consisting of Rob, John, and Bill's six children (Henry, Susan, Andrew, Marty, Blake, and Stefani) could take us into the next millennium.
- Sources of our coffee bean—Guatemala, Costa Rica, Columbia, Brazil, Sumatra, France, and the Ivory Coast.
- Freshness guarantees—posted ship dates and ground-on dates; 100 percent money-back guarantee. Corporate contacts (addresses, phone numbers, e-mail, fax numbers, etc.).

Conclusion

Coffee is a "hot" commodity now. The international market is ours for the taking. We can maximize our profits and open new venues or expansion. The Web is our tarmac for takeoff. I'm awaiting your approval.

INDIVIDUAL OR TEAM PROJECTS

Online Help

1. Study the help menus in your word-processing package. Distinguish between the pop-up windows and the hyperlinks. What are the differences? What are the similarities?

2. In a small group, have each individual ask a software-related question regarding a word-processing application. These could include questions such as "How do I print?" and "How do I set margins?" and so on. Then, using a help menu, find the answers to these questions. Are the answers provided by way of pop-up windows or hyperlinks? Rewrite any of these examples to improve their conciseness or visual layout.

3. Take an existing document (one you have already written in your technical writing class or writing from your work environment) and rewrite it as online help. To do so, find "hot" topics in the text and expand on them as either pop-up windows or hyperlinks. Then highlight new hot buttons within the expanded pop-ups or hyperlinks for additional hypertext documents. Create four or five such layers.

   If you have the computer capability to do so, use your computer system. If you don't have the computer capabilities, use 3 × 5 index cards for the layered pop-ups or hyperlinks.

4. Using our suggestions in this chapter, rewrite the following example as an effective online help screen. To do so, reformat the information for access on a smaller screen and create hot-button words or phrases. Determine which of the hot buttons can be expanded with a short pop-up window and which require a longer hyperlink. Research any data with which you are unfamiliar to provide additional information for your hypertext surfer.

Because your company is located in the East Side Commercial Park, the outside power is subject to surges and brownouts. These are caused when industrial motors and environmental equipment frequently start and stop. The large copiers in your office also cause power surges that can damage electronic equipment.

As a result of these power surges and outages, your three network servers have become damaged. Power supplies, hard drives, memory, and monitors have all been damaged. The repair costs for 2008 damages totaled $55,000.

Power outages at night also have caused lost data and failed backups and file transfers. Not being able to shut down the servers during power failures is leading to system crashes. Data loss requires time-consuming tape restorations. If the failures happen before a scheduled backup, the data are lost, which requires new data entry time. Ten percent of your data loss can not be restored. This will lead to the loss of customer confidence in the reliability of your services.

Finally, staff time losses of $40,000 are directly related to network downtime. Data-entry staff are not able to work while the servers are being repaired. These staff also must spend extra time reentering lost data. Fifteen percent of the data management staff time is now spent recovering from power-related problems.

To solve your problems, you need to install a BACK-UPS 9000 system. It has the configuration and cost-effectiveness appropriate for your current needs.

Web Sites

1. Create a corporate Web site. To do so, make up your own company and its product or service. Your company's service could focus on dog training, computer repair, basement refinishing, vent cleaning, Web site construction, child care, auto repair, personalized aerobic training, or online haute cuisine. Your company's product could be paint removers, diet pills, interactive computer games, graphics software packages, custom-built engines, flooring tiles, or duck decoys. The choice is yours. To create this Web site, follow our writing process. Prewrite by listing answers to reporter's questions. Then sketch your site through storyboarding. Next, draft your Web site, using HTML coding. Finally, rewrite by adding, deleting, simplifying, moving, reformatting, enhancing, correcting, and making sure your links work.

2. Create a Web site for your technical communication, business writing, or professional writing class. To create this Web site, follow our writing process. Prewrite by listing answers to reporter's questions. Then sketch your site through storyboarding. Next, draft your Web site, using HTML coding. Finally, rewrite by adding, deleting, simplifying, moving, reformatting, enhancing, correcting, and making sure your links work.

3. Create a Web site for your high school, college, fraternity or sorority, church or synagogue, or professional organization. To create this Web site, follow our writing process. Prewrite by listing answers to reporter's questions. Then sketch your site through storyboarding. Next, draft your Web site using HTML coding. Finally, rewrite by adding, deleting, simplifying, moving, reformatting, enhancing, correcting, and making sure your links work.

4. Create a personal Web site for yourself or for your family. To create this Web site, follow our writing process. Prewrite by listing answers to reporter's questions. Then sketch your site through storyboarding. Next, draft your Web site using HTML coding. Finally, rewrite by adding, deleting, simplifying, moving, reformatting, enhancing, correcting, and making sure your links work.

5. Research several Web sites, either corporate or personal. Use our Web Site Usability Checklist to determine which sites excel and which sites need improvement. Then write a report justifying your assessment. In this report, clarify exactly what makes the sites successful. To do so, you could use a table, listing effective traits and giving examples from the Web sites to prove your point. Next, explain why the unsuccessful sites fail. To do so, again use a table to list effective traits and give examples from the inferior sites to show why they are unsuccessful. Finally, suggest ways in which the unsuccessful sites could be improved.

PROBLEM-SOLVING THINK PIECES

1. FlyHigh Travel does not book trips to Washington, DC, for tours of national monuments, to Las Vegas to see shows, or to Boston to visit historical sites. Instead, FlyHigh specializes in "adventure trips . . . in the air, sea, and land." If it's shark watching off the Barrier Reef, paragliding from an Acapulcan cliff, or mountain climbing in Nepal, FlyHigh is the traveler's answer. FlyHigh books adventures like feeding stingrays in the Caribbean, animal photo shoots in Kenya, and kayaking down the Colorado River. FlyHigh also makes travel arrangements for scuba diving in Hawaii, spelunking in French caves, and skydiving anywhere in the world.

   To reach as wide an audience as possible, FlyHigh is building its Web site (www.WeFlyHigh.com). FlyHigh's CEO wants this Web site to include trip information, testimonials from satisfied travelers, pricing, and photos that highlight each trip's excitement.

Assignment

What content would you include in this Web site?
- What specific trip information would all travelers need to plan their trips?
- What unique information will FlyHigh's international customers need?
- What unique information will FlyHigh's clients need who travel with families?
- What unique information will FlyHigh's "fit and fearless" travelers need whose travels will involve extreme sports?
- What other audience types can you envision, and how would FlyHigh appeal to their needs?

2. Web site design is challenging. Some Web sites are outstanding; others do not succeed. Check out *http://www.webpagesthatsuck.com/* and *http://www.webpractices.com/samplesites.htm*, two URLs that assess flawed Web sites. Do you agree with their assessments? Write a memo to your instructor or give an oral presentation explaining your decisions.

3. Check out Stonyfield Farm's blog site at http://www.stonyfield.com/weblog/index.cfm. This site says, "Welcome to the Stonyfield Farm blog Cow'munities" where they provide "a chance for you to look inside Stonyfield and get to know us, and us to know you." Read entries from their two blog sites: "Baby Babble" and "The Bovine Bugle." Then answer the following questions.
- What are Stonyfield's goals in creating these two blogs, seemingly unrelated to their product?
- How do these blogs relate to corporate business?
- If you were hired by a company of your choice, what storylines would your blogs feature and why?

WEB WORKSHOP

1. Access any company's Web site and study the site's content, layout (color, graphics, headings, use of varying font sizes and types, etc.), links (internal and external), ease of navigation, tone, and any other considerations you think are important. Then, determine how the Web site could be improved if you were the site's webmaster. Once you have made this

determination, write a memo recommending the changes that you believe will improve the site. In this memo,

- Analyze the Web site's current content and design, focusing on what is successful and what could be improved.
- Provide feasible alternatives to improve the site.
- Recommend changes.

2. Audience recognition involves a reader's level of knowledge as well as his or her gender, religion, age, culture, and language. No communication channel allows for a more diverse audience than the Internet, where anyone can click on a link and enter a site.

   Still, effective Web sites recognize and appeal to specific audiences. To better understand how successful workplace communication recognizes and involves the audience, click on the following sites.

   - http://www.mcdonalds.com/ (food)
   - http://www.gap.com/browse/home.do (clothing)
   - http://www.harley-davidson.com/ (motorcycles)

   Who are the intended audiences within each Web site, and what techniques do the Web sites use to involve and recognize their unique audience types.

3. Bloggers provide up-to-date information on newsbreaking events and ideas.

   - For e-commerce news, visit http://blog.clickz.com/ (ClickZ Network—Solution for Marketers).
   - For business blogging, visit http://www.businessweek.com/the_thread/blogspotting/ (Business Week Online) to learn where the "worlds of business, media and blogs collide."
   - For technology news (information technology, computer information systems, biomedical informatics, and more), visit http://blogs.zdnet.com/. Once in this site, use the search engine to find a technology topic that interests you.
   - For information about criminal justice, visit http://lawprofessors.typepad.com/ crimprof_blog/criminal_justice_policy/ (Law Professor Blog Network).
   - For information about science, visit http://www.scienceblog.com/cms/index.php (ScienceBlog), which has links to blogs about aerospace, medicine, anthropology, geoscience, and computers.
   - The Car Blog (http://thecarblog.com/) has news items about automotives.
   - Every news agency, such as ABC, NBC, CBS, CNBC, has a news blog.

To see what's new in your field of interest, check out a blog. Then, report key findings to your instructor in an e-mail message.

QUIZ QUESTIONS

1. What are the characteristics of e-readers?
2. Why should a corporation consider blogging?
3. What is blogging?
4. How does online communication differ from hard-copy documents?
5. Why is the use of online help growing?
6. How can you achieve effective online help screens?
7. How can you achieve a pleasant tone in your help screen?
8. In what three ways can you design an effective screen?
9. Why should you be concerned with length in your help screen?
10. What are characteristics of successful Web sites?
11. Why should you avoid "noise" on your Web site?
12. What information is usually included on a home page in a Web site?
13. What information should be included in a linked page, and how is navigation important?
14. How can you achieve effective document design on a Web page?
15. Which highlighting techniques should you avoid on a Web site?

CHAPTER 14

Summaries

COMMUNICATION *at work*

The Deer Creek Construction scenario shows how poor communication skills, both oral and written, can negatively impact a company's success and how a consulting firm used summaries to share its researched findings.

Deer Creek Construction had a wonderful reputation in the city, county, and region for providing construction services—primarily roads, bridges, on- and off-ramps, and cloverleafs—that were built on time and within budget. Both civic leaders and community end users praised Deer Creek's quality. However, lately, Deer Creek has not been getting selected for new projects. Their proposals continually were coming in second and third to competitors.

To determine why this was occurring, Deer Creek's management hired a consulting firm. This firm studied Deer Creek's written proposals and oral presentations. The consultants also performed primary research by interviewing clients who had worked with the construction company. Finally, the consulting firm conducted secondary research on best practices for proposal writing.

Objectives

When you complete this chapter, you will be able to

1. Understand why you will write a summary.

2. Apply the criteria for effective summaries.

3. Use an appropriate style for writing summaries.

4. Write an objective summary.

5. Follow the writing process to write a summary.

6. Use the checklist to evaluate your summary.

After doing so, the consulting firm found that Deer Creek's written proposals were as follows:

- Stiff and dry, instead of creating a more personalized tone that achieved audience involvement
- Gender insensitive, using phrases like "when an engineer visits your site, *he* will . . ."
- Visually unappealing, leaning too heavily on long, dense paragraphs instead of graphics and other highlighting techniques

Furthermore, the consulting firm found that Deer Creek's oral presentations had problems.

- The presenters used handwritten flip charts instead of more slick PowerPoint slides.
- The presenters fidgeted and slumped against the lecterns while they spoke.

The consultants also discovered from their research that Deer Creek was not adhering to several best practices for proposal writing.

Despite Deer Creek's quality work, the firm was losing business due to poor communication skills: oral, written, and nonverbal. Something had to change—*fast*. To share their findings with Deer Creek, the consultants wrote a summary of the problems the company was experiencing, based on their research.

WHY WRITE SUMMARIES?

You might be required to write a summary if, for example, your boss is planning to give a presentation at a civic meeting (Better Business Bureau, Rotary Club, or Businesswomen's Association) and needs some up-to-date information for the presentation. Bosses often are too busy to perform research, wade through massive amounts of researched data, or attend a conference at which new information might be presented. Therefore, your boss asks you to research the topic and provide the information in a shortened version. He or she wants you to provide a summary of your findings.

Similarly, your boss might need to give a briefing to upper-level management. Again, time is a problem. How can your boss get the appropriate information and digest it rapidly? The answer is for *you* to read the articles, attend the conference or meeting, and then summarize your research. At other times, you will be asked to summarize your own writing. For example, if you are writing a long report, you might include an **executive summary,** which is an overview of the report's key points.

Perhaps your boss is trying to decide on a course of action regarding new facilities, new software, new hiring practices, new client possibilities, or changes to existing procedures. Before he or she makes the decision, however, the boss has asked you to research the topic and then summarize your findings. Finally, as a class assignment, your teacher might want you to write a summary of a formal research report. The summary, though shorter than a research report, would still require that you practice research skills, such as reading, note taking, and paraphrasing.

To write a summary, you'll need to study the research material. Then, in condensed form (a summary is no more than 5 to 15 percent of the length of the original source), you'll report on the author's main points. Preparing a summary puts to good use your research, analyzing, and writing skills.

The following will help you write a summary.

- Criteria for writing summaries
- The writing process at work

Executive Summaries

See Chapter 16 for more discussion of executive summaries.

CRITERIA FOR WRITING SUMMARIES

A well-constructed summary, though much shorter than the original material being summarized, highlights the author's important points. A summary is like a *Reader's Digest* approach to writing. Although the summary will not cover every fact in the original, after reading the summary you should have a clear overview of the original's main ideas. There are several criteria for accomplishing this.

Overall Organization

As in any well-written document, a summary contains an introduction, a discussion, and a conclusion.

Introduction. Begin with a topic sentence. This sentence will present the primary focus of the original source and list the two or three major points to be discussed. You must also tell your reader what source you are summarizing. You can accomplish this in one of three ways. You can either list the author's name and article title in the topic sentence; preface your summary with a works cited notation providing the author's name, title, publication date, and page numbers; or follow your topic sentence with a footnote and give the works cited information at the bottom of the page.

Name and Title in Topic Sentence

In "Organizational Implications of the Future Development of Technical Communication: Fostering Communities of Practice in the Workplace" (*Technical Communication*, August 2005: 277–287), Lori Fisher and Lindsay Bennion state that over the last 10 to 15 years, the technical communication profession has undergone changes.

Works Cited Prefacing Topic Sentence

Fisher, Lori, and Lindsay Bennion. "Organizational Implications of the Future Development of Technical Communication: Fostering Communities of Practice in the Workplace." *Technical Communication*. August 2005: 277–287.

Lori Fisher and Lindsay Bennion state that over the last 10 to 15 years, the technical communication profession has undergone changes.

Topic Sentence Followed by Explanatory Footnote

Lori Fisher and Lindsay Bennion state that over the last 10 to 15 years, the technical communication profession has undergone changes.[*]

_____

_____

_____

_____

_____

_____

[*]This paragraph summarizes Lori Fisher and Lindsay Bennion's article "Organizational Implications of the Future Development of Technical Communication: Fostering Communities of Practice in the Workplace." *Technical Communication*. August 2005: 277–287.

Discussion. In this section, briefly summarize the main points covered in the original material. To convey the author's ideas, you can paraphrase, using your own words to restate the author's point of view.

Paraphrasing
See Chapter 5 for more discussion of paraphrasing.

Conclusion. To conclude your summary, you can either reiterate the focus statement, reminding the reader of the author's key ideas; highlight the author's conclusions regarding his or her topic; or state the author's recommendations for future activity.

Internal Organization

How will you organize the body of your discussion? Because a summary is meant to be objective, you should present not only what the author says but also how he or she organizes the information. For example, if the author has developed his or her ideas according to a problem/solution format, your summary's discussion should also be organized as a problem/solution. This would give your audience the author's content and method of presentation. Similarly, if the author's article is organized according to cause/effect, comparison/contrast, or analysis (classification/division), this would determine how you would organize your summary.

Development

To develop your summary, you'll need to focus on the following:

- **Most important points**—Because a summary is a shortened version of the original, you can't include all that the author says. Thus, you should include only the two or three key ideas within the article. Omit irrelevant details, examples, explanations, or descriptions.
- **Major conclusions reached**—Once you've summarized the author's key ideas, then state how these points are significant. Show their value or impact.
- **Recommendations**—Finally, after summarizing the author's major points and conclusions, you'll want to tell your audience if the author recommends a future course of action to solve a problem or to avoid potential problems.

Style

The summary, like all technical communication, must be clear, concise, accurate, and accessible. Watch out for long words and sentences. Avoid technical jargon. Most important, be sure that your summary truly reflects the author's content. Your summary must be an unbiased presentation of what the author states and include none of your opinions.

Length

As mentioned earlier, the summary will be approximately 5 to 15 percent of the length of the original material. To achieve this desired length, omit references to the author (after the initial reference in the works cited or topic sentence). You also may need to omit some types of material, such as the following:

- Past histories
- Definitions
- Complex technical concepts
- Statistics
- Tables and figures
- Tangential information (such as anecdotes and minor refutations)
- Lengthy examples
- Biographical information

Audience Recognition

Audience

See Chapter 4 for more discussion of audience.

You must consider your audience in deciding whether to omit or include information. Although you usually would omit definitions from a summary, this depends on your audience. Technical communication is useless if your reader does not understand your content. Therefore, determine whether your audience is high tech, low tech, or lay.

Grammar and Mechanics

As always, flawed grammar and mechanics will destroy your credibility. Not only will your reader think less of your writing and research skills, but also errors in grammar and mechanics might threaten the integrity of your summary. Your summary will be inaccurate and, therefore, invalid.

THE WRITING PROCESS AT WORK

Writing your summary will be easiest if you follow the same step-by-step process detailed throughout this text. Remember that the writing process is dynamic, with the steps frequently overlapping.

The Writing Process

Prewriting	Writing	Rewriting

• Decide whether you are writing to inform, instruct, or persuade.
• Determine whether your audience is internal, external, or both. This will help you decide which abbreviations and acronyms need to be defined.
• Conduct your research and determine which information to summarize.
	• Organize your content using modes such as problem/solution, cause/effect, comparison, argument/persuasion, analysis, chronology, etc. Your summary's organization will reflect the original document's organization.
	• Abide by rules for correct paraphrasing.
	• Document your sources correctly.
		• Revise your draft by
		  o adding details
		  o deleting wordiness
		  o deleting biased comments and technical terms
		  o simplifying words
		  o enhancing the tone
		  o reformatting your text
		  o proofreading and correcting errors

Prewriting

To gather your data and determine your objectives, do the following:

Locate Your Periodical Article, Book Chapter, or Report. If you are summarizing a meeting or seminar, take notes. If your boss or instructor gives you your data, one major obstacle has been overcome. If, however, you need to visit a library or go online to research your topic, refer to Chapter 5 for helpful hints on research skills.

Write Down Your Works Cited or References Information. Once you've located your research material, *immediately* write down the author's name, the article's title, the name of the periodical or book in which this source was found, the date of publication, and the page numbers. Doing so will save you frustration later. For example, if you lose your copy of the article and have not documented the source of your information, you will have to begin your research all over again. Instead, spend a few minutes at the beginning of your assignment documenting your sources. This will ensure that you do not spend several harried hours the night before the summary is due frantically searching for the missing article.

Read the Article to Acquire a General Understanding of Its Content. Determine exactly what the author's thesis is and what main points are discussed. You can take notes in one of two ways. Either you can underline important points, or you can make marginal or interlinear notes on a photocopy. For a summary of an article or chapter in a book, we suggest that you write notes directly on the photocopy. To do so, read and reread a paragraph. After you've read and understood the paragraph, write a one- or two-word notation in the margin that sums up that paragraph's main idea.

> **Research and Documentation**
>
> See Chapter 5 for more discussion of research and documentation.

Writing

Once you've gathered your data and determined your author's main ideas, you're ready to write a rough draft of your summary.

Review Your Prewriting.   Review your marginal notes and your underlining. Have you omitted any significant points? If you have, now is the time to include them. Have you included any ideas that are insignificant or tangential? If so, delete these to ensure that your summary is the appropriate length.

Write a Rough Draft, Using the Sufficing Technique.   Once you've reviewed the data you have gathered, quickly draft the text of your summary. Follow the criteria for writing an effective summary discussed earlier in this chapter, including

- A topic sentence clearly stating the author's main idea. Also provide the works cited information here, in a title, or in an explanatory footnote. Providing accurate documentation of the source is an ethical consideration and an important reflection of your professional skills.
- Organization *paralleling* the author's method of organization (comparison/contrast, argument/persuasion, cause/effect, analysis, chronology, spatial, importance, etc.).
- Clear transitional words and phrases.
- Development through paraphrases. Restate the author's ideas in your own words. Avoid using quoted material.
- A conclusion in which you reiterate the main points discussed, state the significance of the author's findings, or recommend a future course of action.

Rewriting

Once you've written a rough draft of your summary, it's time to revise. Follow the six-step revision techniques to ensure that your finished product will be acceptable to your readers.

Add Detail for Accuracy.   When reviewing your draft, be sure that you've covered all of the author's major points. If you haven't, now is the time to add any omissions. Be sure that your summary accurately covers the author's primary assertions.

Delete Unnecessary Information, Biased Comments, and Dead Words and Phrases. Deleting is especially important in a summary. Because a summary must be no more than 5 to 15 percent of the length of the original source, your major challenge is brevity. Therefore, review your draft to see if you have included any of the author's points that are *not* essential. Such nonessential information may include side issues that are interesting but not mandatory for your reader's understanding.

Another type of information you'll want to delete will be complex technical theories. Although such data might be valuable, a summary is not the vehicle for conveying this kind of information. If you have included such theories, delete them. These deletions serve two purposes. First, your summary will be stronger since it will focus only on key ideas and not on tangential arguments. Second, the summary will be shorter due to your deletions.

Delete biased comments. If you inadvertently have included any of your attitudes toward the topic, remove these biases. A summary should not present your ideas or your subjective comments and opinions; it should reflect only what the author says.

Finally, as with all good technical communication, conciseness is a virtue. Long words and sentences are not appreciated. Reread your draft and delete any unnecessary words and phrases. Strive for an average of 15-word sentences and one- or two-syllable words. We're not suggesting that every multisyllabic word is incorrect, but keep them to a minimum. If an author has written about telecommunications or microbiology, for example, you shouldn't simplify these words. You should, however, avoid long words that aren't needed and long sentences caused by wordy phrases.

Simplify "Impressive" Words and Complex Technical Terms. Good technical communication doesn't force the reader to look up words in a dictionary. Your goal in a summary is to present complex data in a brief and easily understandable package. This requires that you avoid difficult words. In a summary, *difficult* means two things. First, as with all good technical communication, you want to use words that readers understand immediately. Don't write *supersede* when you mean *replace*. Don't write *remit* when you mean *pay*. Second, you'll also need to simplify technical terms. To do so, you can define them parenthetically or merely replace the technical term with its definition.

Move Information within the Summary to Parallel the Author's Organization. Does your summary reflect the order in which the author has presented his or her ideas? Your summary must tell the reader not only what the author has said but also how the author has presented these ideas. Make sure that your summary does this. If your summary fails to adhere to the author's organization, move information around. Cut and paste. Maintain the appropriate comparison/contrast, argument/persuasion, or cause/effect sequence.

Correct Errors. Reread your draft and look for grammatical and mechanical errors. Don't undermine your credibility by misspelling words or incorrectly punctuating sentences. In addition, accurate information is mandatory. You might be writing this summary for your boss, who is giving a speech at a civic meeting or briefing upper-level management. Your boss is trusting you to provide accurate and ethical information. Therefore, to ensure that your boss is not embarrassed by errors, review your text against the original source. Make sure that all the information is correct.

Avoid Biased Language. Biased language is always inappropriate and potentially offensive to readers.

As one of our regular rewriting tips, we ask you to *enhance* the text by adding pronouns and positive words. For summaries, however, such enhancements would be incorrect. Because the summary must only restate what the author has written and be devoid of your attitude, total objectivity is necessary.

Use the following Summary Checklist to help you write an effective summary.

SUMMARY CHECKLIST

_____ 1. Does your summary provide the works cited information (documentation) for the article that you're summarizing?

_____ 2. Is the works cited information correct?

_____ 3. Does your summary begin with an introduction clearly stating the author's primary focus?

_____ 4. Does your summary's discussion section explain the author's primary contentions and omit secondary side issues?

_____ 5. In the discussion section, do you explain the author's contentions through pertinent facts and figures while avoiding lengthy technicalities?

_____ 6. Is your content accurate? That is, are the facts that you've provided in the summary exactly the same as those the author provided to substantiate his or her point of view?

_____ 7. Have you organized your discussion section according to the author's method of organization?

_____ 8. Did you use transitional words and phrases?

_____ 9. Have you omitted direct quotations in the summary, depending instead on paraphrases?

_____ 10. Does your conclusion either reiterate the author's primary contentions, reveal the author's value judgment, or state the author's recommendations for future action?

_____ 11. Is your summary completely objective, avoiding any of your own attitudes?

_____ 12. Have you used an effective style, avoiding long sentences and long words?

_____ 13. Are your grammar and mechanics correct?

_____ 14. Have you avoided biased comments and omitted personal opinions?

PROCESS EXAMPLE

Following is one student's rough draft summary of an article (Figure 14.1), the student's rough draft with editing suggestions (Figure 14.2), and the revised copy with revisions (Figure 14.3).

FIGURE 14.1   Rough Draft

The student wrote a rough draft summary of a journal article

In Organizational Implications of Future Development of Technical Communication: Fostering Communities of Practice in the Workplace (Technical Communication, 2005), Lori Fisher and Lindsay Bennion states that over the last 10 to 15 years, the technical communication profession has undergone changes. In addition to many practitioners having formal training in the field of technical communication, technical communicators often work as members of multidisciplinary teams. These teams shares goals for the organization and common business objectives. The team of specialists ensures the success of the business.

Because technological advances are constantly evolving, the technical communication environment is both complex and dynamic, technical communicators need to be a part of a "community" of professionals. In this community, technical communicators are relying on both formal and informal organizational structures for support. Within these organizational structures, methods for best practices are evolving to apply to the field of technical communication and support the business communicators.

"Communities of practice" in business are comprised of people with shared issues, a way to communicate, and methods for teamwork. It is very important for these communities to be supported formally by corporate sponsored and discipline specific report structures or by informal means such as e-mail messages, listservs, or by collaborative work spaces. "The visual designers in our organization made clear that they feel physical proximity of designers is key to their relationship, citing the need for frequent 'over the shoulder' critiques of each other's work" (281). The reason that this is so important is because within these communities based on the skill set of the members, people can learn by sharing examples of projects, ideas for new work techniques, best practices in the workplace, or simply through Q and A. Through such formal and informal interaction, these skill based communities can successfully ensure the transfer and understanding of knowledge crucial to business success.

Leaders and mentors in the business can help by supporting these skill based communities. To be fully supportive of communities of practice, organizations have to commit both time and resources. In the future, technical communicators will have to exhibit skills at operating in several diverse communities of practice at the same time. The most successful technical communities, supported through organizations, will adhere to best practices, rely on mentoring, interact with teams, and share skills and knowledge. With the success of technical communicators communities of practice companies will achieve a higher standard of success.

FIGURE 14.2   Rough Draft with Tracking Changes Showing Editing Suggestions

In Organizational Implications of Future Development of Technical Communication: Fostering Communities of Practice in the Workplace (Technical Communication, 2005), Lori Fisher and Lindsay Bennion states that over the last 10 to 15 years, the technical communication profession has undergone changes. In addition to many practitioners having formal training in the field of technical communication, technical communicators often work as members of multidisciplinary teams. These teams shares goals for the organization and common business objectives. The team of specialists ensures the success of the business.

Because technological advances are constantly evolving, the technical communication environment is both complex and dynamic, technical communicators need to be a part of a "community" of professionals. In this community, technical communicators are relying on both formal and informal organizational structures for support. Within these organizational structures, methods for best practices are evolving to apply to the field of technical communication and support the business communicators.

"Communities of practice" in business are comprised of people with shared issues, a way to communicate, and methods for teamwork. It is very important for these communities to be supported formally by corporate sponsored and discipline specific report structures or by informal means such as e-mail messages, listservs, or by collaborative work spaces. "The visual designers in our organization made clear that they feel physical proximity of designers is key to their relationship, citing the need for frequent 'over the shoulder' critiques of each other's work" (281). The reason that this is so important is because within these communities based on the skill set of the members, people can learn by sharing examples of projects, ideas for new work techniques, best practices in the workplace, or simply through Q and A. Through such formal and informal interaction, these skill based communities can successfully ensure the transfer and understanding of knowledge crucial to business success.

Leaders and mentors in the business can help by supporting these skill based communities. To be fully supportive of communities of practice, organizations have to commit both time and resources. In the future, technical communicators will have to exhibit skills at operating in several diverse communities of practice at the same time. The most successful technical communities, supported through organizations, will adhere to best practices, rely on mentoring, interact with teams, and share skills and knowledge. With the success of technical communicators communities of practice companies will achieve a higher standard of success.

Steve Gerson 1/7/07 1:16 PM
Comment: add quotation marks around article title

Steve Gerson 1/7/07 1:16 PM
Comment: note agreement error: use "state" vs. "states"

Steve Gerson 1/7/07 1:17 PM
Comment: agreement error again: write "teams share" vs. "shares"

Steve Gerson 1/7/07 1:17 PM
Comment: run-on sentence, needs a period or semicolon

Steve Gerson 1/7/07 1:19 PM
Comment: avoid biased opinions in a summary

Steve Gerson 1/7/07 1:23 PM
Comment: add hyphens to "corporate-sponsored" and "discipline-specific" for compound adjectives

Steve Gerson 1/7/07 1:18 PM
Comment: avoid using quotes in a summary

Steve Gerson 1/7/07 1:19 PM
Comment: another biased comment

Steve Gerson 1/7/07 1:20 PM
Comment: "skill-based" needs to be hyphenated

Steve Gerson 1/7/07 1:20 PM
Comment: add hyphen

Steve Gerson 1/7/07 1:21 PM
Comment: add apostrophe = communicators'

Steve Gerson 1/7/07 1:21 PM
Comment: add comma

FIGURE 14.3   Summary with Revisions

> In "Organizational Implications of the Future Development of Technical Communication: Fostering Communities of Practice in the Workplace" (*Technical Communication*, August 2005), Lori Fisher and Lindsay Bennion state that over the last 10 to 15 years, the technical communication profession has undergone changes. In addition to many practitioners having formal training in the field of technical communication, technical communicators often work as members of multidisciplinary teams. These teams share goals for the organization and common business objectives. The team of specialists ensures the success of the business.
>
> Because technological advances are constantly evolving, and the technical communication environment is both complex and dynamic, technical communicators need to be a part of a "community" of professionals. In this community, technical communicators are relying on both formal and informal organizational structures for support. Within these organizational structures, methods for best practices are evolving to apply to the field of technical communication and support the business communicators.
>
> "Communities of practice" in business are comprised of people with shared issues, a way to communicate, and methods for teamwork. These communities can be supported formally by corporate-sponsored and discipline-specific report structures or by informal means such as e-mail messages, listservs, or by collaborative work spaces. Within these communities based on the skill set of the members, people can learn by sharing examples of projects, ideas for new work techniques, best practices in the workplace, or simply through Q and A. Through such formal and informal interaction, these skill-based communities can successfully ensure the transfer and understanding of knowledge crucial to business success.
>
> Leaders and mentors in the business can help by supporting these skill-based communities. To be fully supportive of communities of practice, organizations have to commit both time and resources. In the future, technical communicators will have to exhibit skills at operating in several diverse communities of practice at the same time. The most successful technical communities, supported through organizations, will adhere to best practices, rely on mentoring, interact with teams, and share skills and knowledge. With the success of technical communicators' communities of practice, companies will achieve a higher standard of success.

CHAPTER HIGHLIGHTS

1. A summary is a compressed version of a much longer document or speech.
2. You can summarize chapters, books, speeches, reports, material from the Internet, and more.
3. A well-written summary is about 5 to 15 percent of the length of the original.
4. Include an introduction, a discussion, and a conclusion to give your summary a coherent design.
5. Include source citations when appropriate.
6. The writing process helps you create an effective summary.
7. Delete biased comments from a summary because a summary should not reflect your ideas.
8. Avoid using long words and sentences in a summary.
9. Organize your summary in the same way the author organized the original source.
10. In a summary, focus only on key ideas and not on tangential arguments.

CASE STUDY

Due to Deer Creek's problems getting new clients (refer to the scenario at the beginning of this chapter), Deer Creek management has decided to hire a manager of corporate communication to

- Act as a liaison between Deer Creek's engineers, city/county/state government agencies, and end-user clients.
- Prepare proposals (gather information through meetings and write the reports).
- Make oral presentations to potential clients.

You have been asked to write the job description for this position, to be advertised in the local newspaper, in Deer Creek's employment office, and online in the company's Web site: DeerCreekConstruction.com.

Assignment

Research the position of manager of corporate communication (either online, in the *Occupational Outlook Handout,* or at a local company's work site). Find out the salary range for this job and educational requirements. What additional job responsibilities will this position entail? What skills should this manager of corporate communication have? Based on your research, write a summary of your findings.

INDIVIDUAL AND TEAM PROJECTS

1. Locate an article that interests you (one within your field of expertise, your degree program, or an area that you would like to pursue). Study this article and summarize it according to the criteria provided in this chapter.
2. Locate an article that interests you. After reading it, take marginal notes (one- to three-word notations per paragraph) highlighting the article's key points. Then, write either a topic or a sentence outline.
3. Read three to five articles. Then determine what method of organization the authors have used. Have the authors used analysis? Others might use division, focusing on parts of a whole, comparison/contrast, argument/persuasion, cause/effect, and so on.
4. After attending a lecture, meeting, or conference, summarize its content. Provide the speaker's name, the location of the presentation, and the date of presentation for the source citation.

Outlining

See Chapter 5 for more information about outlining.

WEB WORKSHOP

1. Using the Internet, research 10 companies in your major field. Write a summary of each home page and connecting links.
2. Many companies provide annual stockholder's reports online. Access a Web browser and find a company's online stockholder's report. Using the techniques discussed in this chapter, summarize the report.
3. Many companies issue press releases or news releases online to update their employees or stakeholders of important corporate information. Access a Web browser and find a company's press release. Using the techniques discussed in this chapter, summarize your findings.
4. Access a Web browser to research articles on current issues, such as *outsourcing, telecommuting, globalization, multiculturalism, flextime, microrobotics, job sharing, virtual teams, best practices,* and *retooling.* Summarize your findings by following the techniques presented in this chapter.

QUIZ QUESTIONS

1. What constitutes a well-constructed summary?
2. Explain the function of a topic sentence in a summary.
3. In what ways can you organize a summary?
4. What types of material are omitted from a summary?
5. What key aspects of writing style are evident in a summary?
6. Approximately how long is a summary?
7. What is an executive summary?
8. Why should you avoid biased comments in a summary?
9. What two primary goals do you achieve when you delete unnecessary information or biased comments from a summary?
10. Why should you avoid complex technical terms in a summary?

CHAPTER 15

....

Short, Informal Reports

COMMUNICATION *at work*

In this scenario, a director of information technology must write a variety of reports, including meeting minutes, trip reports, feasibility/recommendation reports, and progress reports.

Cindy Katz is director of information technology at **WIFINATION** WIFINation, an electronics communication company with branches in 35 cities and a home office in Austin, TX.

Cindy has been traveling to Austin from her office site in Miami, Florida, one week each month for a year. She is part of a team being trained on WIFINation's new corporate software. Her team consists of staff members from human resources, information technology, accounting, payroll, and administrative services. The software they are learning will be used to manage WIFINation's electronic communication systemwide. It will allow for

- Electronic payroll reports
- Employee benefits
- Corporate blogging
- WIFINation's intranet
- WIFINation's Web site
- WIFINation's e-mail system

Cindy must document her activities monthly and biannually. First, as the project team's recording secretary, Cindy keeps her team's **meeting minutes**. Next, she must record her travel expenses and the team's achievements, necessitating monthly **trip reports** submitted to her manager. As a member of her team's technology impact task force subcommittee, Cindy also

Objectives

When you complete this chapter, you will be able to

1. Understand the purposes of a report.
2. Distinguish the differences between short, informal reports and long, formal, researched reports.
3. Follow guidelines for writing short reports.
4. Use headings and talking headings to organize your short reports.
5. Write different types of short reports including incident reports, investigative reports, trip reports, progress reports, lab reports, feasibility/recommendation reports, and meeting minutes.
6. Evaluate your reports using the report checklist.

has been asked to study technology options and emerging technology concerns. This means that she will write a **feasibility/recommendation report** following her study to justify the implementation of the new technology systems the team decides on.

Finally, when her project is completed, Cindy will collaborate with her team members to write a **progress report** for WIFINation's board of directors. Though Cindy's area of expertise is computer information systems, her job requires much more than programming or overseeing the networking of her corporation's computer systems. Cindy's primary job has become communication with colleagues and administrators. Writing reports is a major component of this job requirement.

WHAT IS A REPORT?

Report is a simple word that is hard to define. Reports come in different lengths and levels of formality, serve different and often overlapping purposes, and can be conveyed to an audience using different communication channels. In fact, report types and characteristics can overlap. You could write a short, informal e-mail progress report that conveys information about job-related projects. However, you might write a longer, more formal progress report, in a letter format for an external audience, which provides facts, analyzes these findings, and recommends follow-up action.

Your reports will satisfy one or all of the following needs.

- Supply a record of work accomplished
- Record and clarify complex information for future reference
- Present information to a large number of people
- Record problems encountered
- Document schedules, timetables, and milestones
- Recommend future action
- Document current status
- Record procedures

FAQs: Reports

Q: What's a report? I've seen reports that look like memos and reports that look like letters? What's the difference?

A: Reports come in all shapes and sizes, and different companies define the word *report* differently. For one company, a report could consist of an informal, half-page e-mail message, a page-long memo, or a three-page letter. Other companies might define a report as a formal, long document, such as a 30-page report on acquiring new clients or a 100-page annual report to stockholders.

Few companies use the same terms for their reports or even write the same kinds of reports. For example, note how the following industries use different terminology for their reports.

- Banking supervisors might write a **Job in Jeopardy** report to document employee problems.
- If a customer's loan request is denied, the bank will report this by writing an **Adverse Action Letter** to the client, documenting this unfavorable finding.
- Accountants write **Financial Statements,** reports that document if a client's finances are normal, if some aspect of the client's finances does not meet industry standards, or if rules have not been followed and violations have occurred.
- Accounting **Valuation Reports** are sent to the IRS for estate planning and gift tax reporting.
- Software developers write **Flash Notification** reports informing clients of vital flaws discovered in software that could lead to financial loss.
- Software **Release Notes,** sent electronically via a company's Web site or CD accompanying software, are reports that explain corrections or enhancements and include how-to instructions for testing software functionality.

Each industry sector writes different kinds of reports using different communication channels. However, all reports have certain characteristics in common, as this chapter will illustrate.

Long? Short? Formal? Informal? To clarify unique aspects of different kinds of reports, look at Table 15.1.

TABLE 15.1 Unique Aspects of Reports

Report Features	Distinctions	Definition of Unique Characteristics
Length and Scope	Short	A typical short report is limited to one to five pages. Short reports focus on topics with limited scope. This could include a limited timeframe covered in the report, limited financial impact, limited personnel, and limited impact on the company.
	Long	Long reports are more than five pages long. If a topic's scope is large, including a long timeframe, significant amounts of money, research, many employees, and a momentous impact on the company, a long report might be needed.
Formality (Tone)	Informal	Most short reports are informal, routine messages, written as letters, memos, or e-mail.
	Formal	Formal reports are usually long and contain standardized components, such as a title page, table of contents, lists of illustrations, abstracts, appendices, and works cited/references.
Audience	Internal (high-tech, low-tech)	Colleagues, supervisors, or subordinates within your company are an internal audience. Usually you would write an e-mail or memo report.
	External (multiple audience levels)	An external audience is composed of vendors, clients, customers, or companies with whom you are working. Usually you would write a report in letter format.
	Internal and external	If a report is being sent to both an internal and external audience, you would write either an e-mail, memo, or letter report.
Purpose	Informational	Informational reports focus on factual data. They are often limited in scope to findings: "Here's what happened."
	Analytical	Analytical reports provide information but analyze the causes behind occurrences. Then, these analytical reports draw conclusions, based on an interpretation of the data: "Here's what happened and why this occurred."
	Persuasive	Persuasive reports convey information and draw conclusions. Then, these reports use persuasion to justify recommended follow-up action: "Here's what happened, why this occurred, and what we should do next."
Communication Channels	E-mail	E-mail reports, written to internal and external audiences, are short and informal.
	Memo	Memo reports are written to internal audiences and are usually short and informal.
	Letter	Letter reports are sent to external audiences. These reports can be either long or short, formal or informal, depending on the topic, scope, purpose, and audience.
	Hard-copy PowerPoint printouts	PowerPoint slides can be printed and then bound for presentation. These can be formal or informal, depending on the topic, purpose, scope, and audience.
	Electronic (online)	Many reports can be accessed via a company's Web site. These reports can be downloaded and printed out and often are "boilerplate"—text that can be used repetitively. Electronic reports also provide interactivity, allowing end users to fill out the report online and submit it to the intended audience.

TYPES OF REPORTS

Many reports fall into the following categories.

- Incident reports
- Investigative reports
- Trip reports
- Progress reports
- Lab reports
- Meeting minutes
- Feasibility/recommendation reports
- Research reports
- Proposals

How much time do you spend writing reports and why are the reports important in the accounting field?

Linda M. Freeman, CPA, says she and her colleagues throughout the office of Marks, Nelson, Vohland & Campbell spend approximately "20 to 25 percent of our day on report writing, including background research time, drafting, and proofreading."

On a daily basis, Linda and her colleagues write Compilation Reports, Review Reports, and Audit Reports. Because these reports are "subject to guidelines of the American Institute of Certified Public Accountants," the reports, in large part, are boilerplate with prescribed wording.

Linda also writes Valuation Reports on a daily basis. Some Valuation Reports are sent to the IRS for estate planning and gift tax reporting; others "go to the courts for settling disputes between business owners or divorcing couples." Because Valuation Reports are judgment calls on the part of the CPAs, these reports are not boilerplate and entail less prescribed language.

Private Letter Ruling Requests are less-frequently written reports. They are "high-level tax-planning" documents that require intensive research. In these formal requests sent to the IRS, a client proposes a new business or activity structured in a unique way. The accountant projects that the venture will be taxed according to one standard and asks the IRS if they agree with this assumption.

Times are changing. Communication is "going more electronic," Linda says. Reports to federal agencies, for example, are "template driven." Accountants are encouraged to use online forms and software from federal agencies, to help the government cut back on paperwork.

Writing takes up much of Linda's time. Her reports must address many precise accounting standards. Linda's ultimate goal in report writing is quality assurance: adherence to legalities.

Proposals

See Chapter 17 for additional information.

This chapter will focus on short, informal reports (incident reports, investigative reports, trip, progress, lab reports, feasibility/recommendation reports, and meeting minutes). Chapter 16 will concentrate on long, formal reports requiring research. Chapter 17 will discuss long, formal proposals.

CRITERIA FOR WRITING REPORTS

Although there are many different types of reports and individual companies have unique demands and requirements, certain traits, including format, development, and style, are basic to all report writing.

Organization

Every short report should contain five basic units: identification lines, headings and talking headings, introduction, discussion, and conclusion/ recommendations.

Identification Lines.   You must identify the date on which your report is written, the names of the people to whom the report is written, the names of the people from whom the report is sent, and the subject of the report (as discussed in Chapter 6, the subject line should contain a *topic* and a *focus*). In a short internal report written using a memo format, the identification lines will look like this:

Subject Lines

See Chapter 6 for additional information.

Date: March 15, 2008
To: Rob Harken
From: Stacy Helgoe
Subject: Report on Usenet Conference

Headings and Talking Headings. To improve your page layout and make content accessible, use headings and talking headings. Headings—words or phrases such as "Introduction," "Discussion," "Conclusion," "Problems with Employees," or "Background Information"—highlight the content in a particular section of a document. Talking headings, in contrast, are more informative than headings. Talking headings, such as "Human Resources Committee Reviews 2008 Benefits Packages," informatively clarify the content that follows.

Headings

See Chapter 8 for additional information and for discussion of how to create a hierarchy of headings.

Introduction. The introduction supplies an overview of the report. It can include three or more optional subdivisions, such as the following:

- **Purpose**—a topic sentence(s) explaining why you are submitting the report (rationale, justification, objectives) and the exact subject matter of the report
- **Personnel**—names of others involved in the reporting activity
- **Dates**—what period of time the report covers

In this introductory section, use headings or talking headings to summarize the content. These can include headings for organization, such as "Overview" or "Purpose," or more informative talking headings, such as "Purpose of Report," "Committee Members," "Conference Dates," "Needs Assessment," and "Problems with Machinery." The headings will depend on the type of report and the topic of discussion.

---

example

Introduction

Report Objectives: I attended the Southwest Regional Conference on Workplace Communication in Fort Worth, TX, to learn more about how our company can communicate effectively. This report addresses the workshops I attended, consultants I met with, and pricing for training seminars.

Conference Dates: August 5–8, 2008

Committee Members: Susan Lisk and Larry Rochelle

Some businesspeople omit the introductory comments in reports and begin with the discussion. They believe that introductions are unnecessary because the readers know why the reports are written and who is involved.

These assumptions are false for several reasons. First, it is false to assume that readers will know why you're writing the report, when the activities occurred, and who was involved. Perhaps if you are writing only to your immediate supervisor, there's no reason for introductory overviews. Even in this situation, however, you might have an unanticipated reader for the following reasons.

- Immediate supervisors change—they are promoted, fired, retire, or go to work for another company.
- Immediate supervisors aren't always available—they're sick for the day, on vacation, or off-site for some reason.

Second, avoiding introductory overviews assumes that your readers will remember the report's subject matter. This is false because reports are written not only for the present, when the topic is current, but also for the future, when the topic is past history. Reports go on file—and return at a later date. At that later date, the following may occur.

- You won't remember the particulars of the reported subject matter.
- Your colleagues, many of whom weren't present when the report was originally written, won't be familiar with the subject.
- You may have outside, lay readers who need additional detail to understand the report.

An introduction—which seemingly states the obvious—is needed to satisfy multiple readers, readers other than those initially familiar with the subject matter, and future readers who are unaware of the original report.

Discussion.   The discussion section of the report can summarize many topics, including your activities, the problems you encountered, costs of equipment, and warranty information. This is the largest section of the report requiring detailed development (discussed below). Use headings or talking headings to organize your content.

Conclusion/Recommendations.   The conclusion section of the report allows you to sum up, to relate what you have learned, or to state what decisions you have made regarding the activities reported. The recommendation section allows you to suggest future action, such as what the company should do next. Not all reports require recommendations. You can use headings or talking headings to organize the content.

Talking headings, such as "Benefits of the Conference" and "Proposed Next Course of Action" provide more focus than simple headings, such as "Conclusion" and "Recommendation."

The conclusion shows how the writer benefited.

The recommendation explains what the company should do next and why.

Conclusion/Recommendation

Benefits of the Conference: The conference was beneficial. Not only did it teach me how to improve my technical communication but also provided me contacts for workplace communication training consultants.

Proposed Next Course of Action: To ensure that all employees benefit from the knowledge I acquired, I recommend hiring a consultant to provide technical communication training.

Development

Now that you know what subdivisions are traditional in short reports, the next questions might be, "What do I say in each section? How do I develop my ideas?"

First, answer the reporter's questions.

1. *Whom* did you meet or contact, who was your liaison, who was involved in the accident, who was on your technical team, and so on?
2. *When* did the documented activities occur (dates of travel, milestones, incidents, etc.)?
3. *Why* are you writing the report and why were you involved in the activity (rationale, justification, objectives)? Or, for a lab report, for example, why did the electrode, compound, equipment, or material act as it did?

4. *Where* did the activity take place?
5. *What* were the steps in the procedure, what conclusions have you reached, or what are your recommendations?

Second, when providing information, *quantify.* Do not be vague or imprecise. Specify to the best of your abilities with photographic detail.

The following justification is an example of vague, imprecise writing.

BEFORE

Installation of the machinery is needed to replace a piece of equipment deemed unsatisfactory by an Equipment Engineering review.

Which machine are we purchasing? Which piece of equipment will it replace? Why is the equipment unsatisfactory (too old, too expensive, too slow)? When does it need to be replaced? Where does it need to be installed? Why is the installation important? A department supervisor will not be happy with the preceding report. Instead, supervisors need information *quantified,* as follows:

AFTER

The *exposure table* needs to be installed by *9/08* so that we can *manufacture printed wiring products with fine line paths and spacing (down to 0.0005 inch).* The table will replace the *outdated printer* in *Dept. 76.* Failure to install the table *will slow the production schedule by 27%.*

Note that the italicized words and phrases provide detail by quantifying.

Audience

Since reports can be sent both internally and externally, your audience can be high-tech, low-tech, lay, or multiple. Before you write your report, determine who will read your text. This will help you decide if terminology needs to be defined and what tone you should use. In a memo report to an in-house audience, you might be writing simultaneously to your immediate supervisor (high-tech), to his or her boss (low-tech), to your colleagues (high-tech), and to a CEO (low-tech). In a letter report to an external audience, your readers could be high-tech, low-tech, or lay. To accommodate multiple audiences, use parenthetical definitions, such as Cash in Advance (CIA) or Continuing Property Records (CPR).

In reports, audience determines tone. For example, you cannot write directive reports to supervisors mandating action on their part. It might seem obvious that you can write directives to subordinates or a lay audience, but you should not use a dictatorial tone. You will determine the tone of your report by deciding if you are writing vertically (up to management or down to subordinates), laterally (to coworkers), or to multiple readers.

Style

Style includes conciseness, simplicity, and highlighting techniques. You achieve conciseness by eliminating wordy phrases. Use *consider* rather than *take into consideration;* use *now* rather than *at this present time.* You achieve simplicity by avoiding old-fashioned words: *utilize* becomes *use, initiate* becomes *begin, supersedes* becomes *replaces.*

The value of highlighting has already been shown in this chapter. The parts of reports reviewed earlier use headings. Graphics can also be used to help communicate content, as

Effective Technical Communication Style

See Chapter 3 for additional information.

Highlighting Techniques

See Chapter 8 for additional information.

evident in the following example. A recent demographic study of Kansas City predicted growth patterns for Johnson County (a large county south of Kansas City), as follows:

BEFORE

Johnson County is expected to add 157,605 persons to its 1980 population of 270,269 by the year 2010. That population jump would be accompanied by a near doubling of the 96,925 households the county had in 1980. The addition of 131,026 jobs also is forecast for Johnson County by 2010, more than doubling its employment opportunity.

The information is difficult to access readily. We are overloaded with too much data. Luckily, the report provided a table (Table 15.2) for easier access to the data. Through highlighting techniques (tables, white space, headings), the demographic forecast is made accessible at a glance.

AFTER

Table 15.2   Johnson County Predicted Growth by 2010

	Population	Households	Employment
1980	270,269	96,925	127,836
2010	427,874	192,123	258,862
% change	+58.3%	+98.2%	+102%

SHORT, INFORMAL REPORTS

All reports include identification information, an introduction, a discussion, and conclusion/recommendations. However, different types of short reports customize these generic components to meet specific needs. Let's look at the criteria for seven common types of reports: incident reports, investigative reports, trip reports, progress reports, lab reports, feasibility/recommendation reports, and meeting minutes.

Incident Reports

Purpose and Examples.   An *incident report* documents an unexpected problem that has occurred. This could be an automobile accident, equipment malfunction, fire, robbery, injury, or even problems with employee behavior. In this report, you will document what happened. If a problem occurs within your work environment that requires analysis (fact finding, review, study, etc.) and suggested solutions, you might be asked to prepare an incident report (also called a trouble report or accident report), as follows:

- **Sales**—One of your sales representatives has been involved in a car accident while on job-related travel. The sales representative or his or her supervisor must document this problem.
- **Biomedical technology**—A CAT scan in the radiology department is not functioning correctly. This has led to the department's inability to read x-rays. To avoid similar problems, you need to report this incident.
- **Hospitality management**—An oven in your restaurant caught fire. This not only injured one of your cooks but also damaged the oven, requiring that it be replaced with more fire-resistant equipment.
- **Retail**—A customer was hurt while in your showroom. Incidents also could include employees who are not abiding by company policy. Maybe one of your retail locations has experienced a burglary. The police have been contacted, but as site manager, you believe the problem could have been avoided with better in-store security. Your incident report will document the event and show how to avoid future problems.

Criteria. Companies requiring maintenance reports rarely provide employees with easy-to-complete forms. To write an incident report when you have not been given a printed form, include the following components.

1. Introduction

   Purpose: In this section, document when, where, and why you were called to perform maintenance. What motivated your visit to the scene of the problem?

   Personnel: *Who* was involved, and *what* role do you play in the report? That could entail listing all of the people involved in the accident or event. These might be people injured, as well as police or medical personnel answering an emergency call.

   In addition, *why* are you involved in the activity? Are you a supervisor in charge of the department or an employee? Are you a police officer or medical personnel writing the report? Are you a maintenance employee responsible for repairing the malfunctioning equipment?

2. Discussion (body, findings, agenda, work accomplished)

   Using subheadings or itemization, quantify what you saw (the problems motivating the report). List your findings in chronological order and be specific. Include the following information:

   - Make or model of the equipment involved
   - Police departments or hospitals contacted
   - Names of witnesses
   - Witness testimonies (if applicable)
   - Extent of damage—financial and physical
   - Graphics (sketches, schematics, diagrams, layouts, etc.) depicting the incident visually
   - Follow-up action taken to solve the problem

3. Conclusion/recommendations

   Conclusion: Explain what caused the problem.

   Recommendations: Relate what could be done in the future to avoid similar problems.

Often, on the job you will write reports to colleagues and supervisors with whom you work on a day-to-day basis. Using e-mail allows the audience to quickly and conveniently access this short, informal report and respond to any issues. Figure 15.1 presents an example of an incident report, written as an e-mail.

Investigative Reports

Purpose and Examples. As the word *investigate* implies, an *investigative report* asks you to examine the causes behind an incident. Something has happened. The report does not just document the incident. It focuses more on why the event occurred. You might be asked to investigate causes leading up to a problem in the following instances.

- **Security**—You work in a bank's security department. You are responsible for investigating theft, burglary, fraud, vandalism, check forging, and other banking illegalities. One of your clients, a college student at the local university, reports losing her purse at a campus party. Within hours of the theft, checks bearing her name are showing up across the city. Your job now is to investigate the incident and report your findings.

- **Engineering**—A historic, 100-year-old bridge crossing your city's river is buckling. The left lane is now two inches higher than the right lane, and expansion joints are separating beyond acceptable specifications. You must visit the bridge site, inspect the damage, and report on the causes for this construction flaw.

- **Medicine**—As radiographic technologist, you have administered a bone scan to a patient. This scan detected a shadowy area in the right medial humerus. This

FIGURE 15.1 E-mail Incident Report for an Internal Audience

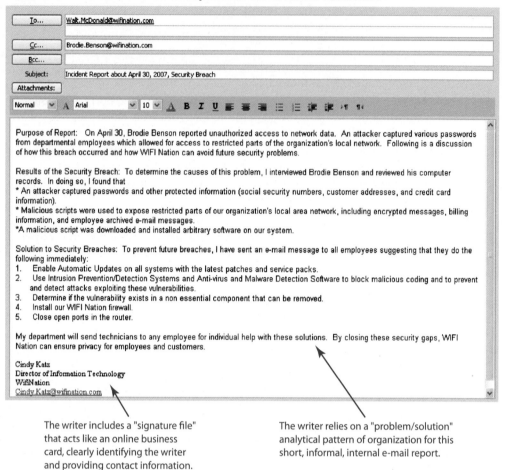

The writer includes a "signature file" that acts like an online business card, clearly identifying the writer and providing contact information.

The writer relies on a "problem/solution" analytical pattern of organization for this short, informal, internal e-mail report.

E-Mail Criteria

See Chapter 6 for additional information.

medical problem could be arthritis or inflammation. However, the shadow also could indicate a metastatic condition. With the help of a pathologist and radiologist, you must submit an investigative report explaining the causes for this aberration and suggesting a follow-up procedure.

- **Computer technology**—You work in a college's technology department. Primarily, your job is to help faculty members with their technology applications. The college requires that all student grades be kept on a newly installed, campus-wide database and then be submitted electronically when the semester ends. For some reason, faculty cannot access their students' records for grade inputting. You must investigate the causes behind this technology glitch and solve the problems—*now!* The semester grades are due within 24 hours.

Criteria.   Following is an overview of what you might include in an effective investigative report.

1. Introduction (overview, background)

    Purpose. In the purpose section, document the date(s) of the incident. Then comment on your objectives or rationale. What incident are you reporting on and what do you hope to achieve in this investigation?

You might also want to include the following optional subheadings.

    Location. Where did the incident occur?

Personnel. Who was involved in the incident? This could include those with whom you worked on the project or those involved in the situation.

Authorization. Who recommended or suggested that you investigate the problem?

2. Discussion (body, findings, agenda)

This is the major part of the investigation. Using subheadings, document your findings. This can include the following:

- A review of your observations. This includes physical evidence, descriptions, lab reports, testimony, and interview responses. Answer the reporter's questions: who, what, when, where, why, and how.
- Contacts—people interviewed
- Difficulties encountered
- Techniques, equipment, or tools used in the course of the investigation
- Test procedures followed, organized chronologically

3. Conclusion/recommendations

Conclusion. What did you accomplish? What did you learn? What discoveries have you made regarding the causes behind the incident? Who or what is at fault?

Recommendations. What do you suggest next? Should changes be made in personnel or in the approach to a particular situation? What training is required for use with the current technology, or should technology be changed? What is the preferred follow-up for the patient or client? How can the problem be fixed?

Use letter formats to write short, informal reports to an external audience. Figure 15.2 illustrates an investigative report, written in letter format. This example is written at a

FIGURE 15.2 Investigative Report in Letter Format to an External, High-Tech Audience

Frog Creek Wastewater Treatment Plant
9276 Waveland Blvd.
Bowstring City, UT 86554

September 15, 2008

Bowstring City Council
Arrowhead School District 234
Bowstring City, UT 86721

Subject: INVESTIGATIVE REPORT ON FROG CREEK WASTEWATER
           POLLUTION

On September 7, 2008, teachers at Arrowhead Elementary School reported that over a five-day period (September 2–6), approximately 20 students complained of nausea, lightheadedness, and skin rashes. On the fifth day, the Arrowhead administration called 911 and the Arrowhead School District (ASD 234) in response to this incident.

Bowstring City paramedics treated the children's illnesses, suggesting that the problems might be due to airborne pollutants. The Bowstring City Council contacted the Frog Creek Wastewater Treatment Plant (FCWTP) to investigate the causes of this problem. This report is submitted by Mike Moore (Director of Public Relations), Sue Cottrell (Wastewater Engineer), and Fred Mittleman (Wastewater Engineer), in response to Bowstring City's request.

Follow letter format when you write a short, informal report for an external audience.

The subject line provides a topic (Frog Creek Wastewater Pollution) and a focus (Investigative Report).

In the letter's introduction, reporter's questions clarify when and where the event occurred, who was involved, and why the report is being written.

Criteria for Successful Letters and Letter Format

See Chapter 6 for additional information.

FIGURE 15.2 (Continued)

Committee Findings

Impact on School Children: Arrowhead Elementary School administrators reported the following:

- Monday, September 2—two children reported experiencing nausea.
- Tuesday, September 3—two children reported experiencing nausea, and one child experienced lightheadedness.
- Wednesday, September 4—three children experienced skin rashes.
- Thursday, September 5—two children complained of nausea, one child was lightheaded, and two children showed evidence of skin rashes.
- Friday, September 6—two children reported nausea, three reported skin rashes, and two lightheadedness.

After Friday's occurrences, Arrowhead Elementary School administrators called 911. Bowstring paramedics reported that (with parental approval) the children were treated with antacids for nausea, antihistamines for skin rashes, and oxygen for their lightheadedness. No other incidents were reported in the neighborhood surrounding the school.

Report on Frog Creek Pollutants: Wastewater Engineers Sue Cottrell and Fred Mittleman took samples of Frog Creek on September 7–10 and found the following:

Typical Frog Creek Readings	Readings from September 7–10
Low alkalinity: generally <30 mg/l (milligrams per liter)	High alkalinity readings: <45 mg/l
Low inorganic fertilizer nutrients (phosphorous and nitrogen): <20 mg/l	High phosphorous and nitrogen readings: <25 mg/l
Limited algae growth: 1 picometers	High algae readings: 4 picometers

Explanation of Findings:
Despite normally low readings, in late summer, with heat and rain, these readings can escalate. Higher algae-related odors above the 3–6 picometer thresholds, along with increased alkalinity (<50 mg/l) can create health problems for youth, elderly, or anyone with respiratory illnesses.

These studies showed that algae, alkalinity, and fertilizer nutrients were higher than usual.

The above elevated readings were caused by three factors (heat, rain, and northeasterly winds).

- Heat—On the days of the Arrowhead Elementary School incident, the temperature ranges were 92 degree F–95 degrees, unusually high for early September. Algae and chemical growth increases in temperatures above 84 degree F.

Include a new-page notation on every page, after page 1. In this notation, provide the writer's name, page number, and date. You can use a header to create this notation.

The findings use chronological organization to document the incidents.

The findings not only investigate the causes of the incident, but also document with specific details. To achieve a readable format, the text is made accessible through boldfaced headings, italicized subheadings, bulleted details, and a comparison/contrast table.

FIGURE 15.2 (Continued)

Mike Moore
Page 3
September 15, 2008

- Rain—In addition, on September 4–6, Bowstring City received 2″ of rain, swelling Frog Creek to 3′ above its normal levels. Studies show that rain-swollen creeks and rivers lead to increased pollutants, as creek bottom silt rises.
- Wind—On September 4–5, a prevailing northeasterly wind blew from Frog Creek toward Arrowhead Elementary School's playground.

Follow-up Studies: On September 11–14, our engineers rechecked Frog Creek, finding that the chemical levels had returned to a normal, acceptable range.

Conclusion about Incident

Frog Creek normally has acceptable levels of algae, alkalinity, and fertilizer nutrient levels. The heat and higher water levels temporarily led to elevated pollutant readings. These levels subsequently returned to normal. Wind directions during the school incidents also had an impact on the children's illnesses. On follow-up questionnaires, FCWTP employees found that the school children's ailments had subsided.

The Arrowhead Elementary School situation appears to have been an isolated incident due to atmospheric changes.

Recommendations for Future Course of Action

Though we constantly monitor Frog Creek for safety, FCWTP HAZMAT employees would be happy to work with parents and teachers to provide additional health information. In a one-hour workshop, presented during the school day or at a Parent-Teacher Organization meeting, FCWTP could offer the following information:

- Scientific data about stream and creek pollutants
- The effects of rain, wind, and heat on creek chemicals
- Useful preventive medical emergency techniques

This information would explain real-world applications for science classes and provide valuable health tips for parents and teachers. We have enclosed for your review information about our proposed training sessions.

The letter's conclusion provides options for the readers and a positive tone appropriate for the intended audience.

Please let us know if you would like to benefit from this free-to-the-public workshop. We would be happy to schedule one at your convenience.

Mike Moore
Frog Creek Wastewater Treatment Plant Director of Public Relations

Enclosure

high-tech level because the city council needs to understand the scientific information about the incident. Moreover, this report will be kept on file not only for documentation, but also for potential litigious reasons.

In contrast to Figure 15.2, the high-tech investigative report, Figure 15.3 is written at a lay level for an external audience consisting of the elementary school administration, the teachers, and the parents of the school children affected by the incident in Frog Creek. The highly technical and scientific information is omitted, the material is condensed, and the writer includes sufficient details to alleviate the concerns of the audience. In addition, the writer emphasizes actions taken and assurances of how and why this incident will not recur.

FIGURE 15.3 Investigative Report in Letter Format (Lay Audience)

Because this letter report is sent to multiple audiences, no reader address is given. Instead, an "attention line" is used.

Frog Creek Wastewater Treatment Plant
"Safety is Our Number 1 Concern."
9276 Waveland Blvd.
Bowstring City, UT 86554

September 15, 2008

Attention: Parents and Arrowhead Elementary School Teachers and Administrators

Subject: Report on September 2-6 Frog Creek Incident

To communicate effectively with the public, the writer used pronouns and a pleasant tone. Words like "unusual" and "rare" are used to lessen the parents' concerns.

Last week, your children at Arrowhead Elementary School experienced nausea, lightheadedness, and skin rashes. This was due to an unusual environmental situation at Frog Creek, a rare case of airborne pollutants caused by high winds and rain.

For headings, the writer used questions that an audience of concerned parents and teachers might have.

How Did This Happen?

Algae is a good thing. The crayfish, snails, and minnows that your children love seeing in Frog Creek thrive on algae and lichen (the small, green plants that are the food base for most marine life). However, when temperatures rise above 84 degrees, algae can grow to an unhealthy level.

The same thing applies to the acid level in water. When acid is regulated by alkalinity, algae growth is controlled. When acid levels rise, however, algae can bloom or marine creatures can die. Both of these problems lead to unusual odors.

That's what happened last week. Our studies showed that algae and alkalinity were higher than usual. The causes were increased heat, rain, and wind:

Because the audience is low-tech, the writer uses simple sentence structure ("Algae is a good thing"; "The same applies to acid levels in water.") and avoids complex scientific discussion, focusing on temperature, rain levels, and wind direction.

- <u>Heat</u>—On the days of the Arrowhead Elementary School incident, the temperature ranges were 92 degree F to 95 degrees F, unusually high for early September.
- <u>Rain</u>—In addition, on September 4-6, Bowstring City received 2" of rain, swelling Frog Creek to 3' above its normal levels. The rain forced the creek bottom silt to rise, which led to increased pollutants.
- <u>Wind</u>—On September 4-5, a northeasterly wind blew from Frog Creek toward Arrowhead Elementary School's playground.

FIGURE 15.3 Continued

page 2
September 15, 2008

What Can We Do to Help?

Could heat, rain, and wind lead to similar situations in the future? Yes. But . . . the incidents from last week were very rare. Please do not expect a repeat occurrence any time soon. Tell your children to enjoy Frog Creek for its beauty and natural resources.

We would be happy to meet with you and your children to explain the science of this environmental event. Plus, we'd like to provide techniques for managing simple ailments like nausea, skin rashes, and lightheadedness.

In a one-hour workshop, presented during the school day or at a Parent-Teacher Organization meeting, FCWTP could offer the following information:

- Scientific data about stream and creek pollutants—with hands-on tutorials for your students.
- The effects of rain, wind, and heat on creek chemicals—complete with graphics and an age-appropriate PowerPoint presentation.
- Useful preventive medical emergency techniques—which every parent and child should know.

This information would explain real-world applications for science classes, as well as provide valuable health tips for parents and teachers. We have enclosed for your review information about our proposed training sessions.

Please let us know if you would like to benefit from this free-to-the-public workshop. We would be happy to schedule one at your convenience. Frog Creek Wastewater Treatment Plant (FCWTP) wants to assure you that your child's "safety is our number 1 concern."

Sincerely,

Mike Moore
Frog Creek Wastewater Treatment Plant Director of Public Relations

Enclosure

The writer emphasizes the positive by using words such as "please," "enjoy Frog Creek for its beauty," and "We would be happy to meet with you."

The writer focuses on the audience's future concerns and suggests age-appropriate material to explain the incident to the elementary school children. This attempt at community outreach connects the wastewater treatment plant to concerned constituents.

Creating Headers and Footers in Microsoft Word 2007

To create headers and footers (useful for new-page notations in reports), follow these steps.

1. Click on the Insert tab on your toolbar. You will see the following ribbon.

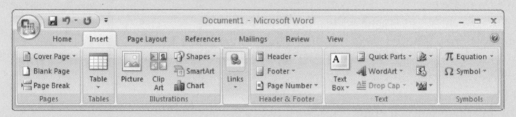

2. Click on either Header or Footer. When you click on your choice of either Header or Footer, you will see a dropdown menu, such as shown below.

3. Choose the type of Header or Footer you want to use in your document and type your content (date, name, page number, etc.), as shown in the following example.

Mary Subicas	Page 2	October 12, 2008

Header

Trip Reports

Purpose and Examples. A *trip report* allows you to report on job-related travel. When you leave your worksite and travel for job-related purposes, your supervisors not only require that you document your expenses and time while off-site, but also want you to keep up to date on your work activities. For example, you might be engaged in work-related travel as follows:

- **Information technology**—You go to a conference to learn about the latest hardware and software technologies for the workplace. There, you meet with vendors, participate in hands-on technology workshops, and learn what other companies are doing to manage their technology needs. When you return, you write a trip report documenting your activities.

- **Heating, ventilating, and air conditioning**—One of your clients is building an office site. Your company has been hired to install their heating, ventilating, and air conditioning (HVAC) system. You travel to your client's home office to meet with other contractors (engineering and architectural) so all team members can agree on construction plans. At the conclusion of your job-related travel, you will write a trip report about your meeting.

- **Engineering**—To be a responsible member of your community, your company has partnered with the local school district. One of your jobs is to visit the city's high schools and discuss engineering job preparedness. To do so, you give an oral presentation about the engineering job market, desired job skills, and the importance of technical communication in the workplace. Upon returning to your office, your supervisor wants a trip report not only to document how the company is working with the community but also to learn how else the company can meet community needs.

- **Biomedical equipment sales**—Four days a week, every week, you are on the road making sales calls. Each month, you must document your job-related travel to show that you are making your quota and to receive recompense for travel expenditures.

Criteria. Following is an overview of what you will include in an effective trip report.

1. Introduction (overview, background)

   Purpose. In the purpose section, document the date(s) and destination of your travel. Then comment on your objectives or rationale. What motivated the trip, what did you plan to achieve, what were your goals, why were you involved in job-related travel?

   You might also want to include the following optional subheadings.

   Personnel. With whom did you travel?

   Authorization. Who recommended or suggested that you leave your worksite for job-related travel?

2. Discussion (body, findings, agenda)

   Using subheadings, document your activities. This can include a review of your observations, contacts, seminars attended, or difficulties encountered.

3. Conclusion/recommendations

   Conclusion. What did you accomplish—what did you learn, whom did you meet, what sales did you make, what benefit to yourself, colleagues, or your company occurred?

   Recommendations. What do you suggest next? Should the company continue on the present course (status quo) or should changes be made in personnel or in the approach to a particular situation? Would you suggest that other colleagues attend this conference in the future, or was the job-related travel not effective? In your opinion, what action should the company take?

Figure 15.4 presents an example of an informal trip report written in memo format. The memo format, providing identification lines (Date, To, From, Subject), is written to an internal audience and creates a hard-copy document.

DOT-COM UPDATES

For more information about trip reports, check out the following links.

- Disney Trip Report Archive lets you read reports for all trips taken during an indicated period, plus you can add your own trip report. http:// www. mouseplanet.com/dtp/ trip.rpt/

- Vanguard, an independent consulting company specializing in customer contact and convergence, provides their employees a sample trip report template. http:// www. vanguard.net/documents/ CCT%20Trip%20Report%2 06-02.doc

FIGURE 15.4 Trip Report in Memo Format

DATE: February 26, 2008   The "topic"   The "focus"
TO: Debbie Rulo
FROM: Oscar Holloway
SUBJECT: Trip Report—Unicon West Conference on Electronic Training

Introduction

Purpose of the Meeting: On Tuesday, February 23, 2008, I attended the Unicon West Conference on Electronic Training, held in Ruidoso, NM. My goal was to acquire hands-on instruction and learn new techniques for electronic training, including the following:

- Online discussion groups
- E-based tutorials
- Intranet instruction
- Videoconference lecture formats

Conference Participants: My coworkers Bill Cole and Gena Sebree also attended the conference.

Discussion

Presentations at the Conference:

Gena, Bill, and I attended the following sessions:

- *Online Discussion Groups*
  This two-hour workshop was presented by Dr. Peter Tsui, a noted instructional expert from Texas State University, San Marcos, TX. During Dr. Tsui's presentation, we reviewed how to develop online questions for discussion, post responses, interact with colleagues from distant locales, and add to streaming chats. Dr. Tsui worked individually with each seminar participant.

- *E-based Tutorials*
  This hour-long presentation was facilitated by Debbie Gorse, an employee of Xenadon E-Learning, Inc. (Colorado Spring, CO). Ms. Gorse used video and overhead screen shots to give examples of successful E-based, computer-assisted instructional options.

- *Intranet Instruction*
  Dr. Randy Towner and Dr. Karen Pecis led this hour-long presentation. Both are professors at the University of Nevada, Las Vegas. Their workshop focused on course development, online instructional methodologies, customizable company-based example, and firewall-protected assessment. The professors provided workbooks and hands-on learning opportunities.

- *Videoconferencing*
  Denise Pakula, Canyon E-Learning, Tempe, AZ, spoke about her company's media tools for teleconferenced instruction. These included lapel microphones, multi-directional pan/tilt video cameras, plasma display touch screens, wideband conference phones, recessed lighting ports with dimming preferences, and multimedia terminals.

The introduction section answers the "reporter's questions"—who, what, when, where, and why.

Different "heading levels" and highlighting techniques are used to make the information more accessible, to help the readers navigate your text.

FIGURE 15.4 (Continued)

Oscar Holloway
Page 2
February 26, 2008

A "new-page notation" helps your readers avoid losing or misplacing pages.

Conclusion

The conclusion focuses on the primary findings to give the audience direction.

Presentation Benefits:

Every presentation we attended was beneficial. However, the following information will clarify which workshop(s) would benefit our company the most:

1. Dr. Tsui's program was the most useful and informative. His interactive presentation skills were outstanding and included hands-on activities, small-group discussions, and individual instruction. In addition, his online discussion techniques offer the greatest employee involvement at the most cost-effective pricing. We met with Dr. Tsui after the session, asking about his fees for onsite instruction. He would charge only $90 per person (other people we researched charged at least $150 per person). Dr. Tsui's fees should fit our training budget.

2. E-based tutorials will not be a valid option for us for two reasons. First, Xenadon's products are prepackaged and allow for no company-specific examples. Second, Ms. Gorse's training techniques are outdated. Videos and overhead projections will not create the interactivity our employees have requested in their annual training evaluations.

3. The Towner/Pecis workshop was excellent. Intranet instruction would be ideal for our needs. We will be able to customize the instruction, provide participants individualized feedback, and ensure confidentiality through firewall-protection. Furthermore, Drs. Towner and Pecis used informative workbooks and hands-on learning opportunities in their presentation.

4. Ms. Pakula's presentation on videoconferencing focused more on state-of-the-art equipment than on instruction. We believe that the price of the equipment exceeds both our budget and our needs. Our current videoconference equipment is satisfactory. If the Purchasing Department is looking for new vendors, they might want to contact Canyon.

Recommendations

The recommendation suggests the next course of action.

Gena, Bill, and I suggest that you invite Dr. Tsui to our site for further consultation. We also think you might want to contact Drs. Towner and Pecis for more information on their training.

Progress Reports

Purpose and Examples.   A *progress report* lets you document the status of an activity, and explain what work has been accomplished and what work remains. Supervisors and customers want to know what progress you are making on a project, whether you are on schedule, what difficulties you might have encountered, and what your plans are for the next reporting period. Because of this, your audience might ask you to write progress (or activity or status) reports—daily, weekly, monthly, quarterly, or annually.

- **Biomedical technology**—You and your team are developing a new heart monitor. This entails researching, patenting, building, testing, and marketing. You have been working on this project for months. What is your status? A progress report will tell your investors and supervisors where you stand, if you are on schedule, and when the project will conclude.

- **Hospitality management**—The city's convention center is considering new catering options. Your job has been to compare and contrast catering companies to see which ones would best be suited for the convention center's needs. The deadline is arriving for a decision. What is the status? Whom have you considered, what are their prices and food choices, what additional services do they offer, and so forth? You need to submit a progress report so management can determine what the next steps should be.

- **Project management**—Your company is renovating its home office. Many changes have occurred. These include new carpeting, walls moved to create larger cubicles, the construction of larger conference rooms, a new cafeteria and fitness center, and improved lighting. Other changes are still in progress, such as increased parking spaces, exterior landscaping, a child care center, and handicapped accessibility. The supervisor wants to know when these renovations will be concluded. You need to write a progress report to quantify what has occurred, what work remains, and when work will be finished.

- **Automotive technology**—Your company recently suffered negative publicity due to product failures. As manufacturing supervisor, you have initiated new procedures for automotive manufacturing to improve your product quality. How are these procedural changes going? Your company CEO needs an update. To provide this information, you must write a progress report.

Criteria.   Following is an overview of what you will include in an effective progress report.

1. Introduction (overview, background)

    Objectives: These can include the following:
    - Why are you working on this project (what's the rationale)?
    - What problems motivated the project?
    - What do you hope to achieve?
    - Who initiated the activity?

    Personnel: With whom are you working on this project (i.e., work team, liaison, contacts)?

    Previous activity: If this is the second, third, or fourth report in a series, remind your readers what work has already been accomplished. Bring them up to date with background data or a reference to previous reports.

2. Discussion (findings, body, agenda)

    Work accomplished: Using subheadings, itemize your work accomplished either through a chronological list or a discussion organized by importance.

Work remaining: Tell your reader what work you plan to accomplish next. List these activities, if possible, for easy access. A visual aid, such as a Gantt or pie chart, fits well after these two sections. The chart will graphically depict both work accomplished and work remaining.

Problems encountered: Inform your reader(s) of any difficulties encountered (late shipments, delays, poor weather, labor shortages) not only to justify your possibly being behind schedule but also to show the readers where you'll need help to complete the project.

3. Conclusion/recommendations

Conclusion: Sum up what you've achieved during this reporting period and provide your target completion date.

Recommendations: If problems were presented in the discussion, you can recommend changes in scheduling, personnel, budget, or materials that will help you meet your deadlines.

Figure 15.5 presents an example of a progress report.

Create Successful Graphics

See Chapter 9 additional information and discussion of how to create Gantt and pie charts.

Check Online Resources

www.prenhall.com/gerson

For more information about creating successful graphics, like Gantt and pie charts, visit our companion Web site.

Lab Reports

Purpose and Examples. A *lab report* lets you document the status of and findings from a laboratory experiment, procedure, or study. Professionals in electronics, engineering, medical fields, the computer industry, and other technologies often rank the ability to communicate as highly as they do their technical skills. Conclusions drawn from a technical procedure are worthless if they reside in a vacuum. The knowledge acquired from a laboratory activity *must* be communicated to colleagues and supervisors so they can benefit from your discoveries. You write a lab report after you have performed the lab to share your findings.

- **Biomedical technology**—You have performed a pathology study on tissue, reviewed a radiological scan, or drawn blood. What have you found? To help nurses and doctors provide the best patient care, you must write a lab report documenting your findings.

- **Electronics**—Your company manufactures global positioning systems (GPSs) to correctly inform a user of an exact location. The GPS receiver must compare the time a signal is transmitted by a satellite with the time it is received. Your company's receptors are malfunctioning as are the units' electronic maps. Why? Your job is to study the electronic systems on randomly selected GPS units and write a lab report documenting your findings.

- **Information technology**—Customers are calling your company's 1-800 hotline almost daily, complaining about hard drive error readings. This is bad for business and profitability. To solve these hard drive malfunctions, you must study units to find the problem. Then, you will write a lab report to document your discoveries.

You write a lab report after you've performed a laboratory test to share with your readers

- Why the test was performed
- How the test was performed
- What the test results were
- What follow-up action (if any) is required

FIGURE 15.5 Progress Report in Memo Format

TO: Buddy Ramos
FROM: Pat Smith
DATE: April 2, 2008
SUBJECT: First Quarterly Report—Project 80 Construction

Purpose of Report

The introduction explains *why* the report has been written and *what* topic will be discussed.

In response to your December 20, 2007, request, following is our first quarterly report on Project 80 Construction (Downtown Airport). Department 93 is in the start-up phase of our company's 2007 build plans for the downtown airport and surrounding site enhancements. These construction plans include the following:

1. *Airport construction*—terminals, runways, feeder roads, observation tower, parking lots, maintenance facilities.
2. *Site enhancements*—northwest and southeast collecting ponds, landscaping, berms, and signage.

Work Accomplished

In this first quarter, we have completed the following:

1. *Subcontractors*: Toby Summers (Project Management) and Karen Kuykendahl (Finance) worked with our primary subcontractors (Apex Engineering and Knoblauch and Sons Architects). Toby and Karen arranged site visitations and confirmed construction schedules. This work was completed January 12, 2008.
2. *Permits*: Once site visitations were held and work schedules agreed upon, Toby Summers and Wilkes Berry (Public Relations) acquired building permits from the city. They accomplished this task on January 20, 2008.

The discussion provides quantified data and dates for clarity, such as "76.4% pass" and "January 30, 2008." The discussion also clarifies who worked on the project and lists other primary contacts.

3. *Core Samples*: Core sample screening has been completed by Department 86 with a pass/fail ratio of 76.4% pass to 23.6% fail. This meets our goal of 75%. Sample screening was completed January 30, 2008.
4. *Shipments*: Timely concrete, asphalt, and steel beam shipments this quarter have provided us a 30-day lead on scheduled parts provisions. Materials arrived February 8, 2008.
5. *EPA Approval*: Environmental Protection Agency (EPA) agents have approved our construction plans. We are within guidelines for emission controls, pollution, and habitat endangerment concerns. Sand cranes and pelicans nest near our building site. We have agreed to leave the north plat (40 acres) untouched as a wildlife sanctuary. This will cut into our parking plans. However, since the community will profit, we are pleased to make this concession. Our legal department also informs us that we will receive a tax break for creating this sanctuary. EPA approval occurred on February 15, 2008.

Pat Smith
Page 2
April 2, 2008

Problems Encountered

Core samples are acceptable throughout most of our construction site. However, the area set aside for the northwest pond had a heavy rock concentration. We believed this would cause no problem. Unfortunately, when Anderson Brothers began dredging, they hit rock, which had to be removed with explosives.

Since this northwest pond is near the sand crane and pelican nesting sites, EPA told us to wait until the birds were resettled. The extensive rock removal and wait for wildlife resettlement have slowed our progress. We are behind schedule on this phase.

This schedule delay and increased rock removal will affect our budget.

Work Remaining

To complete our project, we need to accomplish the following:

1. *Advertising*: Our advertising department is working on brochures, radio and television spots, and highway signs. Advertising's goal is to make the construction of a downtown airport a community point of pride and civic celebration.

2. *Signage*: With new roads being constructed for entrance and exit, our transportation department is working on street signage to help the public navigate our new roads. In addition, transportation is working with advertising on signage designs for the downtown airport's two entrances. These signs will juxtapose the city's symbol (a flying pelican) with an airplane taking off. The goal is to create a logo that simultaneously promotes the preservation of wildlife and suggests progress and community growth.

3. *Landscaping*: We are working with Anderson Brothers Turf and Surf to landscape the airport, roads, and two ponds. Our architectural design team, led by Fredelle Schneider, is selecting and ordering plants, as well as directing a planting schedule. Anderson Brothers also is in charge of the berms and pond dredging. Fredelle will be our contact person for this project.

4. *Construction*: The entire airport must be built. Thus, construction comprises the largest remaining task.

Project Completion/Recommendations

Though we have just begun this project, we have completed approximately 15% of the work. We anticipate a successful completion, especially since deliveries have been timely.

Only the delays at the northwest pond site present a problem. We are two weeks behind schedule and $3,575.00 over cost. With approximately 10 additional personnel to speed the rock removal and with an additional $2,500, we can meet our target dates. Darlene Laughlin, our city council liaison, is the person to see about corporate investors, city funds, and big-ticket endowments. With your help and Darlene's cooperation, we should meet our schedules.

A "Problems Encountered" section helps justify delays and explain why more time, personnel, or funding might be needed to complete a project.

Numbered points with italicized subheadings help readers access the information more readily. This is especially valuable when an audience needs to refer to documents at a later date.

The conclusion sums up the overall status of the project: "15%" complete.

The recommendation explains how the problems discussed in the "Problems Encountered" section can be solved and what is needed to complete the job: "additional personnel" and "increased funds." It also states who to contact for help, "Darlene Laughlin."

FIGURE 15.5 (Continued)

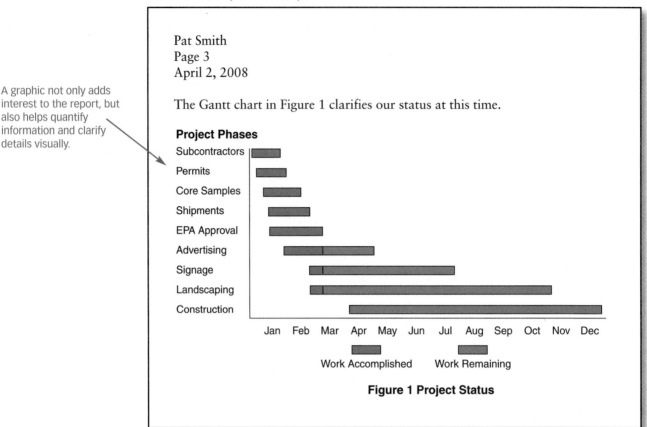

Pat Smith
Page 3
April 2, 2008

A graphic not only adds interest to the report, but also helps quantify information and clarify details visually.

The Gantt chart in Figure 1 clarifies our status at this time.

Project Phases

Subcontractors
Permits
Core Samples
Shipments
EPA Approval
Advertising
Signage
Landscaping
Construction

Jan Feb Mar Apr May Jun Jul Aug Sep Oct Nov Dec

Work Accomplished   Work Remaining

Figure 1 Project Status

Criteria.   The following are components of a successful lab report.

1. Introduction (overview, background)

   Purpose: Why is this report being written? To answer this question, provide any or all of the following:
   - Rationale (What problem motivated this report?)
   - Objectives (What does this report hope to prove?)
   - Authorization (Under whose authority is this report being written?)

2. Discussion (body, methodology)

   How was the test performed? To answer this question, provide the following:
   - Apparatus (What equipment, approach, or theory have you used to perform your test?)
   - Procedure (What steps—chronologically organized—did you follow in performing the test?)

3. Conclusion/recommendations

   Conclusion. The conclusion of a lab report presents your findings. Now that you've performed the laboratory experiment, what have you learned or discovered or uncovered? How do you interpret your findings? What are the implications?

Recommendations. What follow-up action (if any) should be taken?

You might want to use graphics to supplement your lab report. Schematics and wiring diagrams are important in a lab report to clarify your activities, as shown in Figure 15.6.

FIGURE 15.6 Lab Report in Memo Format

DATE: July 18, 2008
TO: Dr. Jones
FROM: Sam Ascendio, Lab Technician
SUBJECT: LAB REPORT ON THE ACCURACY OF DECIBEL
            VOLTAGE GAIN (A) MEASUREMENTS

INTRODUCTION

Purpose

Technical Services has noted inaccuracies in recent measurements. In response to its request, this report will present results of tested A (gain in decibels) of our ABC voltage divider circuit. Measured A will be compared to calculated A. This will determine the accuracy of the measuring device.

DISCUSSION

Apparatus

- Audio generator
- Decade resistance box
- 1/2 W resistor: four 470 ohm, two 1 kilohm, 100 kilohm
- AC millivoltmeter

The discussion includes the equipment used and a step-by-step procedure.

Procedure

1. Figure A shows a voltage divider. For each value of R (resistances) in Table 1, voltage gain was calculated (table attached).
2. An audio generator was adjusted to give a reading of 0 dB for input voltage.
3. Output voltage was measured on the dB scale. This reading is the measured A and is recorded in Table 1.
4. Step 3 was repeated for each value of R listed in Table 1.
5. Figure B shows three cascaded voltage dividers. $A1 = v2/v1$, $A2 = v3/v2$, and $A3 = v4/v3$. These voltage gains added together give the total voltage, as recorded in Table 2.
6. The circuit in Figure B was connected.
7. Input voltage was set at 0 dB on the 1-V range of the AC millivoltmeter.
8. Values V2, V3, and V4 were read and recorded in Table 3.

FIGURE 15.6 (Continued)

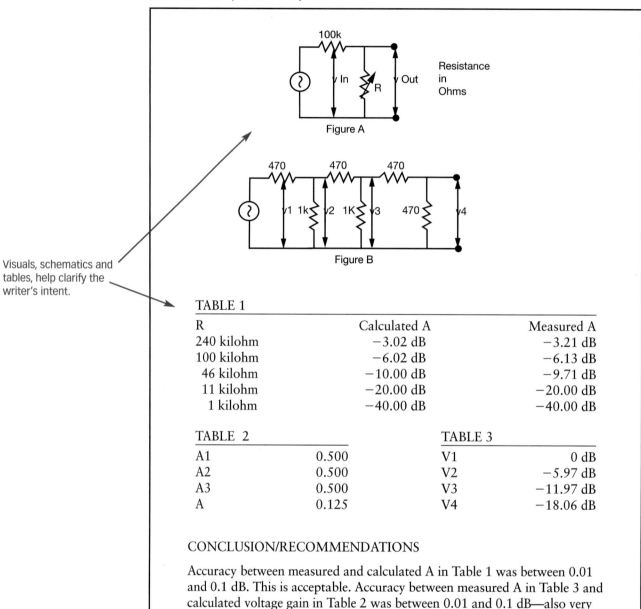

Visuals, schematics and tables, help clarify the writer's intent.

TABLE 1

R	Calculated A	Measured A
240 kilohm	−3.02 dB	−3.21 dB
100 kilohm	−6.02 dB	−6.13 dB
46 kilohm	−10.00 dB	−9.71 dB
11 kilohm	−20.00 dB	−20.00 dB
1 kilohm	−40.00 dB	−40.00 dB

TABLE 2			TABLE 3	
A1	0.500		V1	0 dB
A2	0.500		V2	−5.97 dB
A3	0.500		V3	−11.97 dB
A	0.125		V4	−18.06 dB

CONCLUSION/RECOMMENDATIONS

Accuracy between measured and calculated A in Table 1 was between 0.01 and 0.1 dB. This is acceptable. Accuracy between measured A in Table 3 and calculated voltage gain in Table 2 was between 0.01 and 0.1 dB—also very accurate. These tests show minimal error. No further action should be taken.

Feasibility/Recommendation Reports

Purpose and Examples. A *feasibility/recommendation report* accomplishes two goals. First, it studies the practicality of a proposed plan. Then, it recommends action. Occasionally, your company plans a project but is uncertain whether the project is feasible. Will the plan work, does the company have the correct technology, will the idea solve the problem, or is there enough money? One way a company determines the viability of a project is to perform a feasibility study, to document the findings, and then to recommend the next course of action.

- **Manufacturing**—Your company is considering the purchase of new equipment but is concerned that the machinery will be too expensive, the wrong size for your facilities, or incapable of performing the desired tasks. You need to research and analyze the options, determining which equipment best suits your company's needs. Then, you will recommend purchase.

- **Accounting**—Your company wants to expand and is considering new locations. The decision makers, however, are uncertain whether the market is right for expansion. Are interest rates good? Are local property taxes and sales taxes too high? Will the city provide tax rebate incentives for your company's growth? You need to study the feasibility of expansion at this time and report your recommendations.

- **Web design**—Your company wants to create a Web site to market your products and services globally. The company CEO wants this Web site to be unique—different from the competitors' sites. The CEO wants to be sure that online checkout is easy, that pricing is cost effective, that products are depicted in a visually appealing way, and that the site loads quickly. How will you make your Web site stand out from the competition? You must write a feasibility report to present the options as well as to offer your recommendations.

- **Health management**—It is time to update your health information system. With increasingly complex insurance and regulatory challenges, your current system is outdated. What are your options? You could install software to help code and classify patient records. You could hire consultants to help comply with inpatient and outpatient regulations. You could outsource your patient load to home healthcare agencies. You could upgrade your intranet system to provide decision makers with more accurate information. A feasibility report is needed to study the options before you recommend changes.

Criteria.   One way a company determines the viability of a project is to perform a feasibility study and then write a feasibility report documenting the findings. The following are components of an effective feasibility report.

1. Introduction (overview, background)

    Objectives: Under this subheading, you can answer any of the following questions.

    - What is the purpose of this feasibility report? Until you answer this question, your reader doesn't know. As mentioned earlier in this chapter, it's false to assume prior knowledge on the part of your audience. One of your responsibilities is to provide background data. To answer the question regarding the report's purpose, you should provide a clear and concise statement of intent.

    - What problems motivated this study? To clarify for your readers the purposes behind the study, *briefly* explain either

        —what problems cause doubt about the feasibility of the project (i.e., is there a market, is there a piece of equipment available that would meet the company's needs, is land available for expansion?).

        —what problems led to the proposed project (i.e., current equipment is too costly or time consuming, current facilities are too limited for expansion, current net income is limited by an insufficient market).

    - Who initiated the feasibility study? List the name(s) of the manager(s) or supervisor(s) who requested this report.

Personnel: Document the names of your project team members, your liaison between your company and other companies involved, and your contacts at these other companies.

2. Discussion (body, findings)

Under this heading, provide accessible and objective documentation.

Criteria: State the criteria upon which your recommendation will be based. Criteria are established so you have a logical foundation for comparison of personnel, products, vendors, costs, options, schedules, and so on.

Analysis: In this section, compare your findings against the criteria. In objectively written paragraphs, develop the points being considered. You might want to use a visual such as a table to organize the criteria and to provide easy access.

3. Conclusion/recommendations

Conclusion: In this section, you go beyond the mere facts as evident in the discussion section: You state the significance of your findings. Draw a conclusion from what you have found in your study. For example, state that "Tim is the best candidate for director of personnel" or "Site 3 is the superior choice for our new location."

Recommendations: Once you have drawn your conclusions, the next step is to recommend a course of action. What do you suggest that your company do next? Which piece of equipment should be purchased, where should the company locate its expansion, or is there a sufficient market for the product?

Figure 15.7 presents an example of a feasibility/recommendation report.

Check Online Resources

www.prenhall.com/gerson

For more information about feasibility/recommendation report samples, visit our companion Web site.

FIGURE 15.7 Feasibility Report in Memo Format

FROM: Cindy Katz, Director of Information Technology
TO: Shamir Rammalah, Accounts Payable
DATE: August 13, 2008
SUBJECT: Feasibility Study for Technology Purchases

The "Purpose" reminds the reader why this report has been written and what the report's goal is.

Purpose of the Report

The purpose of this report is to study which technology will best meet your communication needs and budget. After analyzing the feasibility of various technologies, we will recommend the most cost-effective technology options.

The "Problem" details what issues have led to this report.

Technology Problems

According to your memo dated August 1, 2008, your department needs new communication technologies for the following reasons:

- Your department has hired three new employees, increasing your headcount to ten.
- Currently, your department has only five laptops. This slows down the productivity of the department's ten employees.
- We also must update training because WIFINation has purchased new software.
- Accounts Payable's current printer is also insufficient due to your increased headcount.

FIGURE 15.7 (Continued)

Cindy Katz
Page 2
August 13, 2008

Vendor Contacts

Our vendor contacts for the laptops, printers and software are as follows:

Electek	**Tech On the Go**	**Mobile Communications**
Steve Ross	Jay Rochlin	Karen Allen
stever1@electek.com	jrochlin@tog.com	karen.allen@mobcom.net

The "Vendor" section provides contact information (names and e-mail addresses).

Criteria for Vendors

The following criteria were considered to determine which communication technology would best meet your department's needs:

1. *Trainers*—Because of the unique aspects of the software, we need trainers who are familiar with password creation, privacy laws, malware, spyware, and corporate e-mail policy.
2. *Maintenance*—We need to purchase equipment and software complete with either quarterly or biannual service agreement (at no extra charge).
3. *Service Personnel*—The service technicians should be certified to repair and maintain whatever hardware we purchase. In addition, the vendors must also be able to train our personnel in hardware usage.
4. *Warranties*—The warranties should be for at least one year with options for renewal.
5. *Cost*—The total budgeted for your department is $15,000.

The "Criteria" states the topics used to research the report and includes precise details. In this way, the audience can understand the rationale for later decisions.

Needs Assessment

Purchasing agrees that the Accounts Payable hardware and software needs exceed their current technology. Not only are the department's laptops and printer insufficient in number, but also they do not allow the personnel to access corporate e-mail, the Internet, or word-processing packages. Updated equipment is necessary.

Vendor Evaluation

- **Electek**—Having been in business for 10 years, this company is staffed by highly trained technicians and sales staff. All Electek employees are certified for software training. The company promises a biannual maintenance package and subcontracted personnel if employees cannot repair hardware problems. They offer manufacturers' guarantees with extended service warranties costing only $100 a year for up to 5 years. Electek offers 20% customer incentives for purchases of over $2,000.
- **Tech On the Go (TOG)**—This company has been in business for two years. TOG provides only subcontracted service technicians for hardware repair. TOG's employees are certified in software training. The owners do not offer extended warranty options beyond manufacturers' guarantees. No special customer pricing incentives are offered though TOG sells retail at a wholesale price.

FIGURE 15.7 (Continued)

Cindy Katz
Page 3
August 13, 2008

The organizational mode comparison/contrast is used to analyze the strengths and weaknesses of each vendor.

- **Mobile Communications**—Having been in business for 5 years, Mobile has certified technicians and sales representatives. All repairs are provided in house. The company offers quarterly maintenance at a fee of $50 ($200 per year). Mobile offers a customer incentive of 10% discounts on purchases over $5000.

Cost Analysis

The discussion provides specific details to prove the feasibility of the plan or project.

- *Laptops*—Accounts Payable requests one laptop per departmental employee. Our analysis has determined that the most affordable laptops we can purchase (with the requested software and wireless Internet connections) would cost $1,500 per unit. Thus, 10 laptops would cost $15,000. Even with discounts, this exceeds your department's budget.
- *Printers*—Accounts Payable requests three additional printers, with the capability to print double-side pages, staple, collate, print in color, and print three different sizes of envelopes and three different sizes of paper. Our analysis has found that the most affordable printers meeting your specifications would cost $2,500 each. Again, when combining this cost with that of laptops, you exceed your budget.
- *Training*—The manufacturer can provide training on our new software. Training can be offered Monday through Friday, starting in September. The manufacturer says that effective training must entail at least 20 hours of hands-on practice. The cost for this would be $5,000.

The following table compares the three vendors we researched on a scale of 1–3, 3 representing the highest score.

Graphics depict the findings more clearly and more concisely than a paragraph of text.

Table 1: Criteria Comparison

Criteria	Electek	TOG	Mobile
Maintenance	3	2	3
Personnel	3	3	3
Warranties	3	2	2
Cost	3	2	2
Total	**12**	**9**	**10**

Summary of Findings

We can not purchase the number of laptops and printers you have requested. Doing so exceeds the budget. Training is essential. Your department must adjust its budget accordingly to accommodate this need.

The conclusion sums up the findings, explaining the feasibility of a course of action—why a plan should or should not be pursued.

All three vendors have the technology you require. However, TOG and Mobile do not meet the criteria. In particular, these companies do not provide either the maintenance packages, warranties, or pricing required.

FIGURE 15.7 (Continued)

Cindy Katz
Page 4
August 13, 2008

Recommended Action

Given the combination of cost, maintenance packages, warranties, and service personnel, Electek is our best choice.

We suggest the following options for printers and laptops: purchasing five laptops instead of ten; purchasing one additional printer instead of three; and/or sharing printers with nearby departments.

The recommendation explains what should happen next and provides the rationale for this decision.

Conveying Information, Analyzing, and Recommending

See Chapter 16 for additional information.

Meeting Minutes

Purpose and Examples.   *Meeting minutes* document the results of a meeting—what was discussed, proposed, voted on, and planned for future meetings. As the recording secretary for your project team, a community organization, your city council, or a departmental meeting, your job is to record the meeting's minutes. In meeting minutes, you record who attended the meeting, when it began, when it ended, and where it took place. You report on the topics discussed, decisions arrived at, and plans for the future.

Criteria.   Meeting minutes should include the following key components.

1. Introduction
   Include the following:
   - **Date, time, and place.** At the beginning of the minutes, list the date on which the meeting is held, what time the meeting began, and where the meeting was located.
     September 7, 2008
     7:00 P.M.
     Conference Room C
   - **Attendees.** List the names of those who attended the meeting.
   - **Approval of last meeting's minutes.** After asking participants to read the last meeting's minutes, vote to accept them.

2. Discussion (findings, agenda)
   This is the most important part of your minutes. In this long section, you report on the agenda items. Recording secretaries have shared with us a major concern. They believe that their job requires them to report *every* word that has been spoken. This is not true. In fact, *Robert's Rules of Order* says exactly the opposite.
   - A reporting secretary should keep a record of the proceedings, stating what was done and *not* what was said. Your job is not to report every committee members' comments. Instead, focus on decisions made, conclusions arrived at, issues confronted, opposing points of view, and votes taken.
   - If resolutions are agreed upon, the reporting secretary must document the wording of this resolution exactly.
   - Finally, all content within this discussion section must be reported objectively, without "criticism, favorable or otherwise." *(Robert's Rules of Order Revised)*

3. Conclusion
   - **Old/new business.** The last topic of a meeting's agenda should be a review of any "old" topics still unresolved and needing further discussion. Similarly, toward the end of a meeting, you will need to report on new topics, perhaps ones that will need to be covered in future meetings.
   - **Next meeting.** Report when the committee will meet next, providing the date, time, and location.
   - **Time of adjournment.** Report when the meeting ended.
   - **Signature.** A typical lead-in statement reads, "Respectfully submitted by _____." You can sign your name beneath the typed signature (unless the minutes will be submitted electronically).

Organization. Typically, meeting minutes are organized chronologically. When you write your minutes, you can organize the content according to the sequential order of the topics discussed. This sequential organization allows readers to follow a meeting's agenda. However, minutes also can be organized according to importance. Though chronology is an easy approach to documenting a meeting, your audience might better understand the meeting's focus if you document which topics received the greatest discussion or which topics will have the largest impact.

Layout. Reporting secretaries often contend that they must write essay-like reports, following a paragraph format. This is neither true nor advisable. In fact, an essay-like report, including wall-to-wall words, will turn off your readers. Instead, use headings, subheadings, and bulleted lists to make the minutes more accessible.

Figure 15.8 shows an example of meeting minutes.

DOT-COM UPDATES

For more information about meeting minutes, check out the following links:

- Free Management Library offers a sample agenda with typical format and content of board meeting minutes. http://www.managementhelp.org/boards/minutes.htm
- About.com provides information on "Taking Meeting Minutes." http://careerplanning.about.com/cs/communication/a/minutes.htm

FIGURE 15.8 Meeting Minutes

**Employee Benefits Project Team
Meeting Minutes**

Date: April 12, 2008

Time: 7:00 P.M.–9:30 P.M.

Objectives: In this second project team meeting, the goal was to discuss employee benefits options that could be added to the company's cafeteria plan.

Attendees: Stacy Helgoe, Christie Pieburn, Darren Rus, Andrew McWard, Bill Lamb, Jessica Studin

Team Leader: Phyllis Goldberg, Director of Employee Benefits

Agenda:
1. Flex Time—Stacy and Christie reported on flextime options. They stated that in a survey dated March 15, 76% of employees favored flex time benefits. These included beginning our workdays at 7:00 A.M. and extending work hours until 6:30 P.M. Doing so would allow employees to either arrive for work earlier than standard and/or work later than usual. This work-schedule flexibility would allow employees to manage child care needs, appointments, and other concerns that might fall within the traditional workday.

FIGURE 15.8 (Continued)

Page 2
Meeting Minutes
April 12, 2008

The downside of flextime, according to management, includes increased security coverage, increased utilities (heating, lighting, etc.), and inconsistent coverage of office hours. The benefits of flextime include increased employee morale, less congestion in the parking lots at traditionally prime times, and employee empowerment.

Our project team voted 6-1 to promote flex time to management at next month's meeting.

2. Stock Options—At last month's meeting, Darren was asked to study the possibility of adding stock options to our cafeteria plan. He met with Accounting, our external Auditor, and three proprietary stock brokerage companies (Bull and Bear, Market Trend, and BuyItNow). Please see the attached brokerage firm proposals. The proposals offered exciting information.

However, both Accounting and our Auditor convinced Darren that this year would not be the best time to offer our employees stock options. Our profit margin ratio is down, we cannot meet our quarterly forecasts, and we cannot distribute dividends to stockholders.

Future European expansion plans should increase our revenues. Therefore, our project team voted 5-2 to table this issue until the end of next quarter.

3. Personal Leave Days—Currently, our employees are allotted five sick leave days a year. Our proposal asks for an additional two personal leave days. These days would have to be used within the calendar year. The personal leave days would not roll over, nor could they be banked. Andrew McWard surveyed the employees regarding this topic. Ninety-eight percent of those surveyed favored the personal leave proposal.

Our project team voted 6-1 to promote personal leave days as a cafeteria plan option.

4. Sick Bank—Last year, a survey suggested that employees might be willing to share their unused sick leave days with other employees facing catastrophic illnesses or family emergencies. This is a common practice at other corporations of our size. This not only gives employees a feeling of ownership, but also it is a method of teambuilding.

Bill revisited this issue with employees. He attended departmental meetings across the company, asking for input. Bill reported that employees do not seem to favor this proposal at the moment. His informal survey shows only a 30% interest.

Unless these numbers change, our committee sees no reason to pursue a sick bank. We voted 4-3 against proposing a sick bank option.

5. Short and Long-term Disability—Jessica visited three insurance carriers (In-need Insurance, Evergreen Insurance Inc., and Employee Associates General). Her goal was to get quotes on short- and long-term disability insurance. Our employee survey ranked this as a number one priority.

FIGURE 15.8 (Continued)

Page 3
Meeting Minutes
April 12, 2008

Our committee agreed with the survey and highly recommended this as an add-on to our employee cafeteria plan (7-0 vote). In fact, without this benefit, employees with illnesses longer than five days or any banked hours have no salary or job insurance. This fact will negatively impact both employee retention and recruitment. Most major firms offer this benefit. To stay competitive, we must do so also.

Old Business—Andrew reminded us that the company holiday party and family picnic needed to be planned. We will contact the hospitality committee to determine their status.

New Business—In next month's meeting, we will focus on ways to implement our recommendations.

Respectfully submitted by Jessica Studin, Recording Secretary

Attachment:
Brokerage Firm Proposals
Insurance Proposals

THE WRITING PROCESS AT WORK

the writing process

Prewriting		Writing		Rewriting
• Write to inform, analyze, persuade, and/or recommend. • Determine whether your audience is internal or external high-tech, low-tech, lay or multiple. • Choose the correct communication channel (e-mail, memo format, letter, etc.) for your audience. • Research your topic to gather information.		• Organize your content using modes such as problem/solution, cause/effect, comparison, argument/persuasion, analysis, chronology. • Use figures and tables to clarify content. • Use headings and talking headings for access.		• Revise your draft by • Adding details • Deleting wordiness • Simplifying words • Enhancing the tone • Reformatting your text • Proofreading and correcting errors

Writing Process

See Chapter 2 for additional information.

Now that you know the criteria for reports in general and for specific types of reports (incident, investigative, trip, progress, lab, feasibility/recommendation, and meeting minutes), the next step is to construct these documents. How do you begin? As always, *prewrite, write,* and *rewrite*. Remember that the process is dynamic and the steps frequently overlap.

Prewriting

We have presented several techniques for prewriting—reporter's questions, clustering/mind mapping, flowcharting, and brainstorming/listing. An additional technique is *outlining*.

Outlining is ideally suited for short reports. You can focus on whether the primary subject is a trip report, progress report, lab report, feasibility/recommendation report, incident report, investigative report, or meeting minutes in the *main idea*. In addition, the *subordinate points* easily correspond to the introduction, discussion, and conclusion/recommendations. Finally, you can develop your ideas more specifically in the *subheadings*.

Writing

Once you've outlined your report, the next step is to write the text. To write your report, do the following:

Reread Your Prewriting. Review your outline. Determine whether you've covered all important information. If you believe you've omitted any significant points, add them for clarity. If you've included any irrelevant ideas, omit them for conciseness.

Assess Your Audience. Your decisions regarding points to omit or include will depend on your audience. If your audience is familiar with your subject, you might be able to omit background data. However, if you have multiple audiences or an audience new to the situation, you'll have to include more data than you might have assumed necessary in your prewriting.

Audience

See Chapter 4, for additional information.

Draft the Text. Focusing on your major headings (introduction, discussion, conclusion/recommendations), use the sufficing technique mentioned in earlier chapters. Just get your ideas down on paper in a rough draft without worrying about grammar.

Organize Your Content. You can organize the discussion portion of your report using any one of these four modes: *chronology, importance, comparison/contrast,* and *problem/solution*. Which of these modes you use depends on your subject matter.

For example, if the subject of your trip report is a price check of sales items at two stores, comparison/contrast would be appropriate, as in Figure 15.9.

If the subject of your incident report is a site evaluation, chronology might work in the report's discussion, as in the following example.

Example

Chronology

DISCUSSION

8:00 A.M.	I arrived at the site and met with the supervisor to discuss procedures.
9:00 A.M.	We checked the water tower for possible storm damage. Only 10 shingles were missing.
10:00 A.M.	We checked the irrigation channel. It was severely damaged, the wall cracked in six places and water seeping through its barriers. Surrounding orchards will be flooded.
11:00 A.M.	We checked fruit bins. No water had entered.
1:00 P.M.	We checked the freezer units. The storm had disrupted electricity for four hours. All contents were destroyed.
2:00 P.M.	The supervisor and I returned to his office to evaluate our findings.

FIGURE 15.9 Trip Report Organized by Comparison/Contrast

TO: Meagan Clem
FROM: Mary Jane Post
DATE: January 12, 2008
SUBJECT: PRICE CHECK REPORT—HANDY SANDY HARDWARE

INTRODUCTION

On Thursday, January 8, 2008, I compared our sale prices on plumbing items with those of Handy Sandy Hardware, 1000 W. 29th St., Newtown, Wisconsin.

DISCUSSION

Item	Hughes's Sale Price	Handy Sandy's Sale Price
1/2″ copper tubing	$ 4.89 (10′)	$ 4.99 (10′)
3/4″ copper tubing	10.50 (10′)	9.99 (10′)
1/2″ CPVC pipe	2.39 (10′)	2.99 (10′)
3/4″ CPVC pipe	3.49 (10′)	5.99 (10′)
1 1/2″ PVC pipe	3.99 (10′)	4.99 (10′)
Acme 800 faucet	40.95	39.95
Acme 700 faucet	22.95	25.95

The table uses comparison/contrast for organization.

CONCLUSION

On five of the seven items (71.4 percent), Hughes's had the lower price.

RECOMMENDATIONS

We should continue to compare our prices to Handy Sandy's. We should also try to lower any prices that exceed theirs.

As discussed in earlier chapters, chronology is an easy method of organization to use and to follow, but it is not always your most successful choice. Chronology inadvertently buries key data. For instance, in the report findings on page 469, the most important discoveries occur at 10:00 A.M. and at 1:00 P.M., hidden in the middle of the list. To avoid making your readers guess where the important information is, use the third method of organization—importance—in which you list the most important point first, lesser points later, as in the following example.

Example

Importance

DISCUSSION

1. Freezer units: Electricity was out for four hours. All contents were destroyed.
2. Irrigation channel: The wall was cracked in six places. Seepage was significant and sure to affect the orchards.
3. Water tower: Minor damage to shingles.
4. Fruit bins: No damage.

Organizational Modes

See Chapter 3 for additional information.

If you're writing a progress report to document a problem and suggest a solution, use problem/solution to organize your data, as follows:

Example

Problem/Solution

DISCUSSION

Problem—Production schedules on our M23 and B19 are behind three weeks due to machinery failures. Our numerical control device no longer maintains tolerance. This is forcing us to rework equipment, which is costing us $200 per reworked piece.

Solution—We must purchase a new numerical control device. The best option is an Xrox 1234. This machine is guaranteed for five years. Any problems with tolerance during this period are covered. Xrox either will correct the errors on site or provide us a loaner until the equipment is fixed. To avoid further production delays, we must purchase this machine by 1/18/08.

Rewriting

After a gestation period in which you let the report sit so you can become more objective about your writing, retrieve the report and rewrite. Perfect your text by using the following rewriting techniques.

Add Detail for Clarity.   Have you answered the reporter's questions as thoroughly as needed? You often have multiple readers, many of whom do not know your motivations or objectives. They need clarity. For instance, avoid "several new PCs," writing "26 new PCs."

Delete Dead Words and Phrases for Conciseness.   For example, in your recommendations, don't say, "Acme's opinion is that apparently the proposed ideas will not successfully supersede those already implemented." Instead, simply write, "Recommendations: No further action is required."

Simplify Old-Fashioned Words and Phrases. What does *pursuant* mean? How about *issuance of this report?* The ultimate goal of technical writing is to communicate, not to confuse.

Move Information Within Your Discussion for Emphasis. Make sure that you've either maintained chronology (if you're documenting an agenda) or used importance to focus on your main point. If you've confused the two, cut and paste—move around information to achieve your desired goals.

Proofreading Tips

See Chapter 3 for additional information.

Reformat Your Text for Accessibility. Highlight your key points with underlining or boldface. Use graphics to assist your readers. Don't overload your readers with massive blocks of impenetrable text.

Enhance Your Text for Style. Be sure to quantify when needed, personalize with pronouns for audience involvement, and be positive. Stress words like *benefit, successful, achieved,* and *value* rather than using negative words like *cannot, failed, confused,* or *mistaken.*

Proofread and Correct the Report for Grammatical and Contextual Accuracy. Refer to Chapter 3 for proofreading tips.

Audience and Biased Language

See Chapter 4 for additional information.

Avoid Biased Language *Foreman* should be *supervisor, chairman* should be *chairperson* or *chair,* and *Mr. Swarth, Mr. James,* and *Sue* must be *Mr. Swarth, Mr. James,* and *Ms. Aarons.*

Use the following checklist to help you in writing your short report.

SHORT REPORT CHECKLIST

_____ 1. Have you chosen the correct communication channel (e-mail, letter, or memo) for your short, informal report?

_____ 2. Does your subject line contain a topic and a focus? If you write only "Subject: Trip Report" or "Subject: Feasibility/Recommendation Report," you have not communicated thoroughly to your reader. Such a subject line merely presents the focus of your correspondence. But what's the topic? To provide both topic and focus, you need to write "Subject: Trip Report on Solvent Training Course, ARCO Corporation—3/15/08" or "Subject: Feasibility Report on Company Expansion to Bolker Blvd."

_____ 3. Does the introduction explain the purpose of the report, document the personnel involved, or state when and where the activities occurred?

_____ 4. When you write the discussion section of the report, do you quantify what occurred? In this section, you must clarify precisely. Supply accurate dates, times, calculations, and problems encountered.

_____ 5. Is the discussion accessible? To create reader-friendly ease of access, use highlighting techniques, such as headings, boldface, underlining, and itemization. You also might want to use graphics, such as pie charts, bar charts, or tables.

_____ 6. Have you selected an appropriate method of organization in your discussion? You can use chronology, importance, comparison/contrast, or problem/solution to document your findings.

_____ 7. Does your conclusion present a value judgment regarding the findings presented in the discussion? The discussion states the facts; the conclusion decides what these facts mean.

_____ 8. In your recommendations, do you tell your reader what to do next or what you consider to be the appropriate course of action?

_____ 9. Have you effectively recognized your audience's level of understanding (high tech, low tech, lay, management, subordinate, colleague), internal, or external and written accordingly?

_____10. Is your report accurate? Correct grammar and calculations make a difference. If you've made errors in spelling, punctuation, grammar, or mathematics, you will look unprofessional.

PROCESS EXAMPLE

Let's look at how one student used the writing process (prewriting, writing, and rewriting) to construct her progress report.

Prewriting. The student used a simple topic outline to gather data and determine her objectives (Figure 15.10).

Writing. The student drafted her progress report, focusing on the information she discovered in prewriting (Figure 15.11).

FIGURE 15.10 Topic Outline for Prewriting

Sales Proposal to Round Rock Hospital

I. Introduction
    A. When—Oct. 10
    B. What—Sales Proposal to hospital
    C. Why—in response to your memo

II. Discussion
    A. What points to cover?
        1. Containers
        2. Sterilization
        3. Documentation
        4. Cost
        5. Disposal
    B. What's next?
        1. Future work requirements

III. Conclusion/Recommendations
    A. Conclusion—When? On schedule, 50% completed
    B. Recommendations—Comply with regulations

Prewriting Techniques

See Chapter 2 for additional information.

FIGURE 15.11 Student Rough Draft

November 6, 2008
To: Carolyn Jensen
From: Shuan Wang.
Subject: Progress Report.

Purpose: This is a progress report on the status of a sales proposal you requested in your memo of October 10, 2008. The objective of this proposal is the sale of a total program for infectious waste control and disposal to a Round Rock Hospital. The proposal will cover the following five things.

1. Containers
2. Steam sterilization
3. Cost savings
4. Landfill operating
5. Computerized documentation

Work Completed: The following is a list of items that are finished on the project.

FIGURE 15.11 (Continued)

Containers
I have finished a description of the specially designed containers with disposable biohazard bag.

Steam sterilization
I have set a instructions for using the Biological Detector, model BD 12130 as a reliable indicator of sterilization of infectious wastes.

Cost savings
Central Hospital sent me some information on installing a pathological incinerator as well as the constant personal and maintenance costs to operate them. I ran this information through our computer program "Save-Save". The results show that a substantial savings will be realized by the use of our services. I have data to support this, since more and more of our customer base is online with the system proving that expenses to other companies has been reduced.

Future Work: I have an appointment to visit Round Rock Hospital on November 7, 2008. At that time I will make on-site evaluations of present waste handling practices at Round Rock Hospital. After carefully studing this evaluation, I will report to you on needed changes to comply with governmental guidelines.

Conclusion: The project is proceeding on schedule. Approximately 50% of work is done (see graph). I don't see any problem at this time and should be able to meet the target date. I will update you on our continuing progress in the near future.

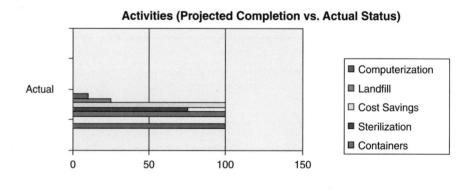

Activities (Projected Completion vs. Actual Status)

Rewriting. After drafting this report, the student submitted her copy to a peer review group, which helped her edit and revise her text (Figure 15.12).

FIGURE 15.12 Peer Group Revision Suggestions

November 6, 2008
To: Carolyn Jensen
From: Shuan Wang.
Subject: Progress Report.

Omit the period and clarify the topic of this report.

Purpose: This is a progress report on the status of a sales proposal you requested in your memo of October 10, 2008. The objective of this proposal is the sale of a total program for infectious waste control and disposal to a Round Rock Hospital. The proposal will cover the following five <u>things</u>.

1. Containers
2. Steam sterilization
3. Cost savings
4. Landfill operating
5. Computerized documentation

The second sentence above is long. Remember, our teacher told us to limit sentences to 10–15 words.

"Things" = weak word!

Work Completed: The following is a list of items that are finished on the project.

Boldface the headings for emphasis.

Containers
I have finished a description of the specially designed containers with disposable biohazard bag.

Steam sterilization
I have <u>set a</u> instructions for using the Biological Detector, model BD 12130 as a reliable indicator of sterilization of infectious wastes.

Correct the typo = "a set of instructions"

Cost savings
Central Hospital sent me some information on installing a pathological incinerator as well as the constant personal and maintenance costs to operate them. I ran this information through our computer program "Save-Save". The results show that a substantial savings will be realized by the use of our services. I have data to support this, since more and more of our customer base is online with the system proving that expenses to other companies has been reduced.

Correct the spelling error

Future Work: I have an appointment to <u>visit Round Rock</u> Hospital on November 7, 2008. At that time I will make on-site <u>evaluations</u> of present waste handling practices at Round Rock Hospital. After carefully <u>studing</u> this evaluation, I will report to you on needed changes to comply with governmental guidelines.

Conclusion: The project is proceeding on schedule. Approximately 50% of work is done (see graph). I don't see any problem at this time and should be able to meet the target date. I will update you on our continuing progress in the near future.

Why is there an empty cell in the graphic?

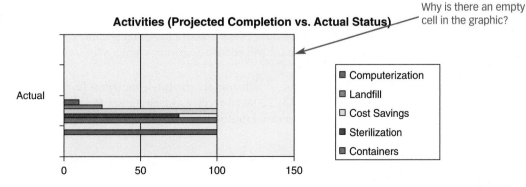

Activities (Projected Completion vs. Actual Status)

- Computerization
- Landfill
- Cost Savings
- Sterilization
- Containers

After discussing the suggested changes with her peer review group, the student revised her draft and submitted her finished copy (Figure 15.13).

FIGURE 15.13 Finished Progress Report in Memo Format

Date: November 6, 2008
To: Carolyn Jensen
From: Shuan Wang
Subject: Progress Report on Round Rock Hospital

Purpose of Report:

In response to your October 10, 2008, request, following is a progress report on our Round Rock Hospital sales proposal. This proposal will present Round Rock Hospital our total program for infectious waste control and disposal. The proposal will cover the following five topics:

- Containers
- Steam sterilization
- Cost savings
- On-site evaluations
- Landfill disposal

Work Completed:

1. *Containers*—On October 5, I finished a description of the specially designed containers with biohazard bags. These were approved by Margaret Chase, Round Rock's Quality Control Administrator.
2. *Steam sterilization*—On November 4, I wrote instructions for using our Biological Detector, model BD 12130. This mechanism measures infectious waste sterilization levels.
3. *Cost savings*—Round Rock Hospital's accounting manager, Lenny Goodman, sent me their cost charts for pathological incinerator expenses and maintenance costs. I used our computer program "Save-Save" to evaluate these figures. The results show that we can save the hospital $15,000 per year on annual incinerator costs, after a two year break-even period. I have data to support this from other customers (see the attachment).

Work Remaining:

1. *On-site evaluation*—When I visit Round Rock Hospital on November 7, I will evaluate their present waste handling practices. After studying these evaluations, I will report necessary changes for governmental compliance.
2. *Landfill disposal*—I will contact the Environmental Protection Agency on November 13, 2008, to receive authorization for disposing sterilized waste at our sanitary landfill.

FIGURE 15.13 (Continued)

Wang
Page 2
November 6, 2008

Completion of Project:
The project is proceeding on schedule. Approximately 50% of work is done (see Figure 1).

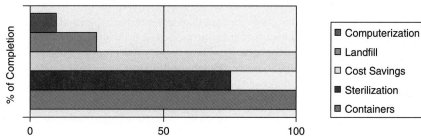

Figure 1: Completion Status of Planned Activities

I see no problems at this time. We should meet our December 31, 2008, target date. I will update you on our continuing progress in two weeks.

CHAPTER HIGHLIGHTS

1. Reports are used to document many different occurrences on the job and written to both internal and external audiences.
2. Use headings and talking headings, such as "Introduction," "Discussion," "Conclusion/recommendations," "Projected Cost of the Project," "Information about Laptops," and "Needs Assessment for Travel-Related Expenses" when designing your report.
3. E-mail, letters, and memos are effective communication channels for short, informal reports.
4. Progress reports recount work accomplished and work remaining on a project.
5. Lab reports document the findings from a lab analysis.
6. Feasibility/recommendation reports are used to determine the viability of a proposed project.
7. Outlining is a prewriting technique that will help you write effective short, informal reports.
8. An incident report documents an unexpected problem that has occurred.
9. An investigative report asks you to examine the causes behind an incident.
10. Meeting minutes document the results of a meeting.

CASE STUDIES

In small groups, read the following case studies and write the appropriate report.

1. You manage an engineering department at WIFINATION. Your current department supervisor is retiring. Thus, you must recommend the promotion of a new supervisor to the company's executive officer, Kelly Adams. You know that WIFI seeks to promote individuals who have the following traits.

   - Familiarity with modern management techniques and concerns, such as total quality management (TQM), teamwork, global economics, and the management of hazardous materials
   - An ability to work well with colleagues (subordinates, lateral peers, and management)
   - Thorough knowledge of one's areas of expertise

   You have the following candidates for promotion. Using the information provided about each and the criteria for feasibility/recommendation reports discussed in this chapter, write your report recommending your choice for a new supervisor.

   a. **Pat Jefferson.** Pat has worked for WIFI for 12 years. In fact, Pat, has worked up to a position as a lead engineer by having started as an assembler, then working in test equipment, quality control, and environmental safety and health (ESH). As an engineer in ESH, Pat was primarily in charge of hazardous waste disposal. Pat's experience is lengthy, although Pat has only taken two years of college coursework and one class in management techniques. Pat is well liked by all colleagues and is considered to be a team player.

   b. **Kim Kennedy.** Kim is a relatively new employee at WIFI, having worked for the company for two years. Kim was hired directly out of college after earning an MBA degree from the Mountaintop College School of Management. As such, Kim is extremely familiar with today's management climate and modern management techniques. Kim's undergraduate degree was a BS in business with a minor in engineering. Currently, Kim works in the engineering department as a departmental liaison, communicating the engineering department's concerns to WIFI's other departments. Kim has developed a reputation as an excellent coworker who is well liked by all levels of employees.

   c. **Chris Clinton.** Chris has a BS degree in engineering from Poloma College and an MBA degree from Weatherford University. Prior to working for WIFI, Chris served on the IEEE (Institute of Electrical and Electronic Engineering) Commission for Management Innovation, specializing in global concerns and total quality management. In 1991, Chris was hired by WIFI and since then has worked in various capacities. Chris is now lead engineer in the engineering department. Chris has earned high scores on every yearly evaluation, especially regarding knowledge of engineering. Whenever you have needed assistance with new management techniques, Chris has been a valued resource. Chris's only negative points on evaluations have resulted from difficulties with colleagues, some of whom regard Chris as haughty.

   Whom will you recommend for supervisor? Write your feasibility/recommendation report stating your decision.

2. You are the accountant at AAA Computing, a retail store specializing in computer hardware. Your boss states that all new computers sold should be accompanied by an optional service contract. This service will be held by an outside vendor. Your job is to research several vendors and write a feasibility/recommendation report recommending the best choice for your company. To do so, you know your boss will emphasize the following criteria.

   - **Years in business/expertise**—to be sure that customers receive quality service, the vendor should have a good track record and be familiar with computer hardware innovations.

- **Quick response/turnaround**—because many of your customers depend on their computers for daily business operations, the vendor should provide on-site service for minor repairs and 24-hour turnaround service on major repairs.
- **Cost-effective pricing**—the less the vendor charges, the more you can mark up the service cost. Thus, AAA can receive more profit.

The following vendors have proposed their services. Using the information provided about each company and the criteria for feasibility reports provided in this chapter, write your report recommending a vendor for AAA's service contracts.

a. **QuickBit.** This company has 16 months' experience in computer hardware service. QuickBit, although without a lengthy track record, is co-owned by three graduates of the Silicon Valley Institute of Technology's renowned BA program in computer information services. The three individuals are very knowledgeable about computers, having received superior instruction and hands-on training using today's most up-to-date technology. Because QuickBit is small, it can provide immediate, 24-hour on-call service. Furthermore, QuickBit owns a 12-wheeler truck fully stocked with parts. Therefore, all service can be handled on-site. QuickBit charges $50 per hour plus parts.

b. **ROM on the Run.** This company has been in business for 10 years. It employs 50 servicepeople, has a lengthy track record, and has provided service for ARC Telecommunications, Capital Bank, Helping Hand Hospital, and the State Penitentiary at Round Rock. Because of its years in business and the fact that all employees have at least two-year certificates from vocational colleges, it charges $85.50 per hour plus parts. ROM on the Run promises on-site service within 24 hours on all service requests.

c. **You Bet Your Bytes (YBYB).** YBYB has been in business for four years, employs 10 servicepeople (all of whom have at least a BS degree in computer systems), and specializes in retail outlets. YBYB charges $70 per hour plus parts. It advertises that it responds within 30 minutes for on-site service calls. Furthermore, it advertises that most parts are carried on its service trucks. Any repairs requiring unavailable parts can be made within 24 hours if the service call is received by 3:30 P.M. Calls after that time require two working days.

Which company do you recommend? Write the feasibility/recommendation report stating your choice.

3. Your company, Telecommunications R Us (TRU), has experienced a 45 percent increase in business, a 37 percent increase in warehoused stock, and a 23 percent increase in employees. You need more room. Your executive officer, Polina Gertsberg, has asked you to research existing options. To do so, you know you must consider the following criteria.

- **Ample space for further expansion.** Gertsberg suggests that TRU could experience further growth upward of 150 percent. You need to consider room for parking, warehouse space, additional offices, and a cafeteria—approximately 20,000 square feet total.
- **Cost.** Twenty million dollars should be the top figure, with a preferred payback of five years at 10 percent.
- **Location.** Most of your employees and customers live within 15 miles of your current location. This has worked well for deliveries and employee satisfaction. A new location within this 15-mile radius is preferred.
- **Aesthetics.** Ergonomics suggest that a beautiful site improves employee morale and increases productivity.

After research, you've found three possible sites. Based on the following information and on the criteria for feasibility/recommendation reports discussed in this chapter, write your report recommending a new office site.

a. **Site 1 (11717 Grandview).** This four-story site, located 12 miles from your current site, offers three floors of finished space equaling 18,000 square feet. The fourth floor is an unfinished shell equaling an additional 3,000 square feet. As is, the building will sell for $19 million. If the current owner finishes the fourth floor, the addition would cost $4 million more. For the building as is, the owner asks for payment in five years at 12 percent interest. If the fourth floor is finished by the owner, payment is requested in seven years at 10 percent. The building has ample parking space but no cafeteria, although a building next door has available food services. Site 1 is nestled in a beautifully wooded area with hiking trails and picnic facilities.

b. **Site 2 (808 W. Blue Valley).** This one-story building offers 21,000 square feet that includes 100 existing offices, a warehouse capable of holding 80 storage bins that measure 20 feet tall × 60 yards long × 8 feet wide, and a full-service cafeteria. Because the complex is one story, it takes up 90 percent of the lot, leaving only 10 percent for parking. Additional parking is located across an eight-lane highway that can be crossed via a footbridge. The building, located 18 miles from your current site, has an asking price of $22 million at 8.75 percent interest for five years. Site 2 has a cornfield to its east, the highway to its west, a small lake to its north where flocks of geese nest, and a strip mall to its south.

c. **Site 3 (1202 Red Bridge Avenue).** This site is 27 miles from your current location. It has three stories offering 23,000 square feet, a large warehouse with four-bay loading dock, and a cafeteria with ample seating and vending machines for food and drink. Because this site is located near a heavily industrialized area, the asking price is $15 million at 7.5 percent interest for five years.

Which site do you recommend? Write your feasibility/recommendation report stating your choice.

INDIVIDUAL AND TEAM PROJECTS

1. Write a progress report. The subject of this report can involve a project or activity at work. Or, if you haven't been involved in job-related projects, write about the progress you're making in this class or another course you're taking. Write about the progress you're making on a home improvement project (refinishing a basement, constructing a deck, painting and papering a room). Write about the progress you're making on a hobby (rebuilding an antique car, constructing a computer, or making model trains, etc.). Whatever your topic, first prewrite, then write a draft, and finally rewrite, revising the text. Follow the criteria presented in this chapter regarding progress reports.

2. Write a lab report. The subject of this report can involve a test you're running at work or in one of your classes. Whatever topic you select, follow the three stages in the writing process to construct your report: prewrite, write a draft, and then revise the draft in rewriting. Use the criteria regarding lab reports presented in this chapter to help you write the report.

3. Write a feasibility/recommendation report. You can draw your topic either from your work environment or home. For example, if you and your colleagues were considering the purchase of new equipment, the implementation of a new procedure, expansion to a new location, or the marketing of a new product, you could study this idea and then write a report on your findings. If nothing at work lends itself to this topic, then consider plans at home. For example, are you and your family planning a vacation, the purchase of a new home or car, the renovation of your basement, or a new business venture? If so, study this situation. Research car and home options, study the market for a new business, and get bids for the renovation. Then write a feasibility/recommendation report to your family documenting your findings. Whether your topic comes from business or home, gather your data in prewriting, draft your text in writing, and then revise in rewriting. Follow the criteria for feasibility/recommendation reports provided in this chapter to help you write the report.

4. Write an incident report. You can select a topic either from work or home. If you have encountered a problem at work, write an incident report documenting the problem and providing your solutions to the incident. If nothing has happened at work lending itself to this topic, then look at home. Has your car broken down, did the water heater break, did you or any members of your family have an accident of any sort, did your dog or cat knock over the vase your mother-in-law gave you for Christmas? Consider such possibilities, and then write an incident report documenting the incident. Follow the criteria for incident reports provided in this chapter, and use writing process techniques (prewriting, writing, and rewriting).

5. Revision is the key to good writing. An example of a seriously flawed progress report that needs revising is shown below. To improve this report, form small groups and first decide what's missing according to this chapter's criteria for good progress reports. Then use the rewriting techniques presented in this chapter to rewrite and revise the report.

Site Visit—Alamo Manufacturing
November 1

Sam, I visited our Alamo site and checked on the following:

1. Our plant facilities suffered some severe problems due to the recent wind and hail storms. The west roof lost dozens of shingles, leading to water damage in the manufacturing room below. The HVAC unit was submerged by several feet of water, shorting out systems elsewhere in the plant. In addition, our north entryway awning was torn off its foundation due to heavy winds. This not only caused broken glass in our front entrance door and a few windows bordering the entrance, but also the entryway driveway now is blocked for customer access.

2. Maintenance and security failed to handle the problems effectively. Security did not contact local police to secure the facilities. Maintenance responded to the problems far too many hours late. This led to additional water and wind damage.

3. Luckily, the storm hit early in the day, before many of our employees had arrived at work. Still, a few cars were in the parking lot, and they suffered hail damage. Are we responsible?

4. The storm, though hurting manufacturing, will not affect sales. Our 800-lines and e-mail system were unaffected. But, we might have problems with delivery if we can't fix the entryway impediment. I think I have some solutions. We could reroute delivery to one of our new plants, or maybe we should consider direct delivery to our sales staff (short term at least). Any thoughts?

5. Meanwhile, I have gotten on Maintenance's case, and it is working on repairs. Brownfield HVAC service has given us a quote on a new sump pump system that has failsafe programs (not too bad an extra cost, given what we can save in the long run). Plainview Windows & Doors is already on the front entrance problem. It promises replacement soon.

6. As for future plans, I think we need to reevaluate both Maintenance and Security. Training is an option. We could also add new personnel, reconsider our current management in those departments, or maybe just a few, good, hard, strongly stated comments from you would do the trick.

Anyway, we're up and running again. Upfront repair costs are covered by insurance, and long-term costs are minimal because sales weren't hurt. Our only remaining challenges are preventive, and that depends on what we do with our Maintenance and Security staff. Let me know what you think.

PROBLEM-SOLVING THINK PIECES

1. Angel Guerrero, computer information technologist at HeartHome Insurance, has traveled from his home office to a branch location out of town. While on his job-related travel, he encountered a problem with his company's remanufactured laptop computer. He realized that the problem had been ongoing not only for this laptop but also for six other remanufactured laptops that the company had recently purchased.

   Angel thinks he knows why the laptops are malfunctioning and plans to research the issue. When he returns to home office from his travel, he needs to write a report. What type of report should he write? Explain your answer, based on the information provided in this chapter.

2. Minh Tran is a special events planner in the Marketing Department at Thrill-a-Minute Entertainment Theme Park. Minh and her project team are in the middle of a long-term project. For the last eight months, they have been planning the grand opening of the theme park's newest sensation ride—*The Horror*—a wooden roller coaster that boasts a 10 g (gravity) drop.

   During one of its weekly project meetings, the team has hit a roadblock. The rap group "Bite R/B Bit" originally slated to play at the midnight unveiling has cancelled at the last minute. The team needs to get a replacement band. One of Minh's teammates has researched the problem and presented six alternatives bands (at varying prices and levels of talent) for consideration. Minh needs to write a report to her supervisor.

   What type of report should she write? Explain your decision, based on the criteria for reports provided in this chapter.

3. Toby Hebert is human resource manager at Crab Bayou Industries (Crab Bayou, LA), the world's largest wholesaler of frozen Cajun food. She and her staff have traveled to New Orleans to attend meetings held by five insurance companies that planned to explain and promote their employee benefits packages.

   During the trip, Toby's company van is sideswiped by an uninsured driver. When Toby returns to Crab Bayou, she has to write a report. What type of report should she write? Base your decision on the criteria for reports provided in this chapter.

4. Bill Baker, claims adjuster for CasualtyU Insurance Company (**CUIC**), traveled for a site visit. Six houses insured by his company were in a neighborhood struck by a hailstorm and 70 mile per hour winds. The houses suffered roof and siding damages. Trees were uprooted, with one falling on a neighbor's car. Though no injuries were sustained, over $50,000 worth of property damage occurred.

   When he returns to CUIC, what type of report should he submit to his boss? Defend your decision based on criteria for reports provided in this chapter.

WEB WORKSHOP

1. More and more, companies and organizations are putting report forms online. The reason for doing this is simple—ease of use.

   Go online and access online report forms. All you need to do is use any Internet search mechanism, and type in "online _____ report form." (In the space provided, type in "trip," "progress," "incident," "investigative," or "feasibility/recommendation.") Once you find examples, evaluate how they are similar to and different from the written reports shown in this chapter. Share your findings with others in your class, either through oral presentations or written reports.

2. Every company writes reports—and you can find examples of them online. Use the Internet to access Report Gallery, found at *http://www.reportgallery.com/*. Report Gallery bills itself as "the largest Internet publisher of annual reports." From this Web site, you can find the annual reports from 2,200 companies, including many Fortune 500 companies.

   Study the annual reports from 5 to 10 different companies. What do these reports have in common? How do they differ from each other? Discuss the reports' page layout, readability, audience involvement, content, tone, and development. How are the reports similar to and different from those discussed in this chapter?

   Share your findings with others in your class, either through oral presentations or written reports.

QUIZ QUESTIONS

1. What are some purposes of a report?
2. What are some common types of short, informal reports?
3. List the four basic units of a report.
4. What types of material are included in the discussion section?
5. Explain how the reporter's questions can help you prepare your report.
6. Why would you use e-mail, memo, or letter formats for short, informal reports?
7. What organizational modes are useful for writing short, informal reports?
8. When do you write a trip report?
9. When do you write a progress report?
10. When do you write a lab report?
11. When do you write a feasibility/recommendation report?
12. When do you write an incident report?
13. When do you write an investigative report?
14. What content do you include in meeting minutes?
15. How do visual aids assist the reader in a report?
16. What do you include in the identification lines?
17. What do you include in the conclusion?
18. What do you include in the recommendation?
19. What is the writing style appropriate for a report?
20. How do you satisfy multiple readers in the introduction?

CHAPTER 16

Long, Formal Reports

COMMUNICATION *at work*

In the following scenario, an information technology consulting company needs to write a long, formal report recommending a backup data storage facility for a client.

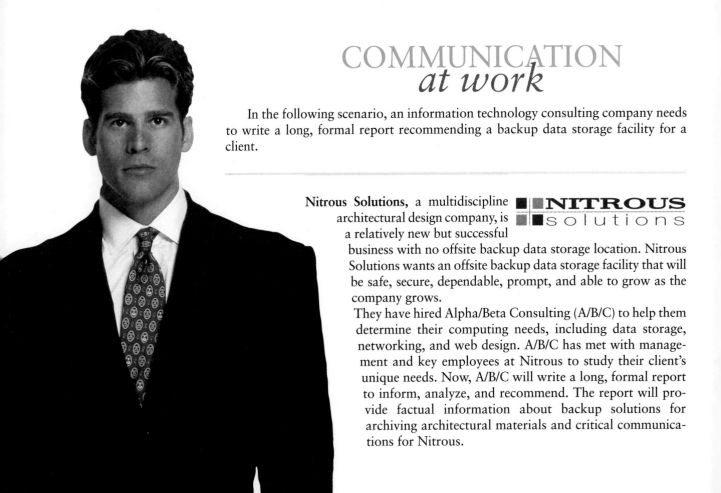

Nitrous Solutions, a multidiscipline architectural design company, is a relatively new but successful business with no offsite backup data storage location. Nitrous Solutions wants an offsite backup data storage facility that will be safe, secure, dependable, prompt, and able to grow as the company grows.

They have hired Alpha/Beta Consulting (A/B/C) to help them determine their computing needs, including data storage, networking, and web design. A/B/C has met with management and key employees at Nitrous to study their client's unique needs. Now, A/B/C will write a long, formal report to inform, analyze, and recommend. The report will provide factual information about backup solutions for archiving architectural materials and critical communications for Nitrous.

Objectives

When you complete this chapter, you will be able to

1. Understand the purposes of writing long, formal reports.

2. Distinguish the differences among informational, analytical, and recommendation reports.

3. Follow guidelines for writing long, formal reports.

4. Include the major components of long, formal reports: front matter, text, and back matter.

5. Distinguish between primary and secondary research to develop your content.

6. Evaluate your long, formal reports using the report checklist.

The long, formal report will then analyze options, focusing on pros and cons of various solutions. In this analysis, A/B/C will draw conclusions about pricing, support, scalability, reliability, ease of use, and archiving choices. Based on their findings and analysis of the options, A/B/C will draw a conclusion. Then, the consulting company will recommend the best options to help Nitrous fulfill their company's mission statement.

Check out our quarterly newsletters TechCom E-Notes at www.prenhall.com/gerson for dot.com updates, new case studies, insights from business professionals, grammar exercises, and facts about technical communication.

WHY WRITE A LONG, FORMAL REPORT?

Short, Informal Reports

See Chapter 15 for more discussion of short, informal reports.

In some instances, your subject matter might be so complex that a short report will not thoroughly cover the topic. For example, your company asks you to write a report about the possibility of an impending merger. This merger will require significant commitments regarding employees, schedules, equipment, training, facilities, and finances. Only a long report, complete with research, will convey your content sufficiently and successfully.

You could have to write a long, formal report requiring research for a variety of reasons such as the following:

1. When a subject is important
2. When documentation will reveal the importance of the topic
3. When large sums of money are involved
4. When large numbers of people are affected
5. When time and resources are devoted to development of the report
6. When research will explain and support the topic

Topics for Long, Formal Reports

When you devote considerable time and energy to long reports, you will be dealing with topics that often have serious and complex issues to be considered. The following are titles of long, formal research reports.

- The Effect of Light Rail on Wyandotte County: An Economic Impact Study on Infrastructure
- The Increasing Importance of Mobile Communications: Security, Procurement, Deployment, and Support
- Managing Change in a Technologically Advancing Marketplace at EBA Corp.
- Information Stewardship at The Colony, Inc.: Legal Requirements for Protecting Information
- Remote Access Implementation and Management for New Hampshire State University

The above examples share several common denominators. These include the significant costs required for implementation of large-scale changes, the impact on large numbers of people, the time and resources required to research the topics, and the challenges of writing the long, formal reports expected by the intended audiences. Short reports would not suffice.

TYPES OF LONG, FORMAL REPORTS: INFORMATIVE, ANALYTICAL, AND RECOMMENDATION

Reports can provide information, be analytical, and/or recommend a course of action. Occasionally, you might write a report that only informs. You might write a report that just analyzes a situation. You could write a report to recommend action persuasively. Usually, however, these three goals will overlap in your long, formal reports.

For example, let's say that your company is expanding globally and will need a WAN (a wide area network that spans a large geographical area). In a long, formal report to management, you might first write to inform the audience that this WAN will meet the strategic goals of storing and transmitting information to your global coworkers. The report also will analyze the ways in which this network will be safe, reliable, fast, and efficient. Finally, the report will recommend the best designs for providing secure communications to clients, partners, vendors, and coworkers. Ultimately, you must persuade the audience, through thorough research, that your envisioned network's design, hardware specifications, software, and estimated budget will satisfy both internal and external needs. Such a report would have to be informative, analytical, and persuasive to convince the audience to act on the recommended suggestion.

Information

When you provide information to your audience, focus on the facts. These facts will help your readers better understand the situation, the context, or the status of the topic. For example, Nitrous Solutions asked Alpha/Beta Consulting to report on backup data storage options for software archives (see the opening scenario on page 484). To do so, A/B/C needed to present factual information about the importance of backup storage. In a long, formal report, A/B/C informed its client about the following: the threats to data, the cost of lost data, how often data should be backed up, and what data should be archived (see Figure 16.1).

Analysis

When you analyze for your audience, you begin with factual information. However, you expand on this information by interpreting it and then drawing conclusions. Once Alpha/Beta Consulting presented the informational findings about backup data storage facilities, they followed this information with a more in-depth analysis of backup options and drew a conclusion for the client (see Figure 16.2).

Recommendation

After providing information and analysis, you can recommend action as a follow-up to your findings. The recommendation allows you to tell the audience why they should purchase a product, use a service, choose a vendor, select a software package, or follow a course of action. Alpha/Beta Consulting presented findings and analyzed information for its client Nitrous Solutions. Based on their analysis, the consulting company then made a recommendation (see Figure 16.3).

MAJOR COMPONENTS OF LONG, FORMAL REPORTS

Because short reports run only a few pages, you can assume that your readers will be able to follow your train of thought easily. Thus, short reports merely require that you use headings such as "Introduction," "Discussion," and Conclusion/Recommendation" to guide your readers through the document or talking headings to summarize the content more thoroughly. Long reports, however, place a greater demand on readers. Your readers could be overwhelmed with many pages of information and research. A few headings won't be enough to help your readers wade through the data.

Check Online Resources

www.prenhall.com/gerson
See our Companion Web site for more long, formal report samples.

Tables and Figures
See Chapter 9 for more discussion of tables and figures.

Check Online Resources

www.prenhall.com/gerson
For more information about short, informal reports, visit our companion Web site.

Headings and Talking Headings
See Chapter 15 for more discussion of headings and talking headings.

FIGURE 16.1 Information about Backup Data Storage

Information is presented in a variety of ways. The consulting company details data in bulleted lists, asks questions and provides answers, and explains facts and findings in paragraphs.

Description of Backup Data Storage:
The following information provides facts about permanent storage to help Nitrous Solutions maintain a high degree of security in an offsite backup data storage location.

Facts:
Due to events such as 9/11 and Hurricane Katrina, the importance of maintaining mission-critical electronic data and backups has become evident for companies that wish to avoid a catastrophic outcome. Moving data to a separate offsite location dilutes risk and protects against data loss when, for example, the company's IT infrastructure or its critical electronic information is damaged by the following:

Threats to Your Data	Dangers and Costs of Data Loss
➤ Fire	➤ Loss of mission-critical files
➤ Flood	➤ Corrupted database
➤ Hurricane or tornado	➤ Corrupted operating system
➤ Lightning strike	➤ Loss of all files on hard drive
➤ Earthquake	➤ Mac laptop damage, theft or loss
➤ Heat, sunlight	➤ Damaged computer hardware
➤ Humidity, moisture, spilled liquids	➤ Backup media loss or theft
➤ Smoke, dust and dirt	➤ Total loss of all data at your site
➤ Electrical surge or power failure	➤ Compromised data security
➤ Media failure	➤ Competitor access to your data
➤ Hard drive failure	➤ Lost business records

When and What Data Should be Backed Up?
Some basic questions that are commonly asked when determining when and what data should be backed up are as follows:

How often should desktop PCs or Macs be backed up?
- Data on desktop PCs and Macs should be backed up at a minimum of at least once a week. This will prevent anyone losing more than one week's worth of information.

Should everything on my PC or Mac be backed up once a week?
- This is not necessary. Only data needs to be backed up once a week. Software, operating systems, and even static data do not need to be backed up that often. Backing up everything on your PC or Mac adds time to the process and requires substantially more backup tapes, backup disks, or other types of media.

How long should I keep computer backups?
- Some companies keep original backups up to seven years and have a set schedule that includes five different increments of backups: daily, weekly, monthly, quarterly, and yearly.

This is a topic for discussion with your key management, legal aids, information systems technicians, records management, and key personnel. Balancing the needs of a company's computer system protection, and good records management, takes informed decisions on the parts of all these departments.

FIGURE 16.2 Analysis of Backup Data Storage Options

Analysis of Backup Data Storage Options

Industry analysts claim that two out of five businesses that experience a disaster will go out of business within five years of the event due to information and service loss as a consequence of disasters.

This is an alarming statistic considering the high cost of not "expecting the unexpected." In the event of a disaster, a corporation could lose hundreds, thousands, or even millions of dollars through lost productivity. At worst the corporation could go out of business without the possibility of a second chance.

The two main types of offsite data storage include the following:

Offsite Data Storage Facility	Internet Data Storage Facility
Tapes, CDs, DVDs, or hard drives are sent to a predetermined offsite location for security and disaster recovery purposes, thus allowing a business to be safe in the knowledge that they will be able to recover quickly and seamlessly from any disaster. With simply a walk across the street or a drive downtown to the storage facility, the business can quickly begin to rebuild from the data they have stored.	Offsite Internet data storage, although relatively new, has quickly become one of the fastest growing arenas in the records and information management industry. A business can simply select the files and the time that they wish to have their files backed up with the offsite backup software. Their files are compressed and encrypted and sent to a backup server over their existing Internet connection. The business can upload or download your files as often as needed, with usually no additional charges.

Offsite Location for Data Storage Options	
Offsite Residential Location	**Offsite Facility Location**
Costs	
• Completely Free	• $5–$650 per month depending on space and features
Reliability	
• Data are accessible when needed. • Save and archive as much as you need with no restrictions or extra fees.	• Insurance on data is automatic with contract. • Data are fully accessible during normal business hours, after hours by emergency only.

FIGURE 16.2 Continued

Security	
• More secure than leaving your data in your office building	• Vault and/or building monitored by 24-hour surveillance and alarm systems • Security patrolled

Scalability	
• Can store as little or as much data as needed, with no restrictions, guidelines, rules, or regulations free of charge	• Just like the Internet storage, as your company grows, so can data storage. Of course, an increase of storage means an increase of monthly payment.

Ease of Use	
• Your house, your rules • Maintenance of data management can be completed in your spare time.	• Easy to set up and manage • Step-by-step instructions available • Some companies offer free technical help.

Conclusion

Regardless of how much or how little a business uses a computer, your company will create important and unique data. The unique data can include financial and project budget records, digital images, client profiles, and marketing sales. The data are priceless and constantly at risk.

When analyzing information, A/B/C presented facts and figures, pros and cons, and specific details. Then, the consulting company interpreted this information in a conclusion.

Data loss is likely to occur for many reasons:

- Hardware failure = 42%
- Human error = 35%
- Software corruption = 13%
- PC viruses = 7%
- Hardware destruction = 3%

For these reasons, Nitrous Solutions should choose backup storage for its valuable data.

FIGURE 16.3 Analysis and Recommendation

Offsite Location for Permanent Storage:
Analysis and Recommendation

Weighted Analysis:

Type		Offsite Residential		Offsite Facility	
Criteria	Weight	*Rating	Score	*Rating	Score
Cost	15%	5	.75	3	.45
Performance	25%	-	-	-	-
Reliability	30%	4	1.20	4	1.20
Security	15%	3	.45	4	.60
Scalability	10%	5	.50	5	.50
Ease of use	5%	4	.20	4	.20
Total	100%		3.1		2.95

*Rating: The weighted analysis is based on a 1 to 5 scale; 1 is poor and 5 is excellent.

Recommendation:

Because Nitrous Solutions does not currently have a network infrastructure in place, the data storage decision should be considered a high priority since all new data that the company creates will be significant to the company's overall success. Alpha/Beta Consulting recommends that Nitrous Solutions use the residential offsite data storage option due to the slight advantage it has over a professional storage facility. The one and only real advantage between the two is cost. Whereas residential storage is free with no contracts or monthly charges, a storage facility will charge anywhere from $5 to $650 per month, depending on the size of data and equipment that needs to be accompanied with such data.

If Nitrous Solutions determines the need to have multiple backup solutions, rather than a single reinforcement to the backup data, another highly recommended means of storage would be that of an Internet storage facility. Due to the popular demand and increasing technology, many different companies are now offering free trials of their Internet storage facilities and software. Due to the unknown intentions of Nitrous Solutions, only the features for Internet storage will be researched due to the fact that almost all options for Internet companies are either compatible or identical.

Based on information and an analysis of the findings, provided in a weighted table, A/B/C recommended a backup solution for the client.

In addition to the basic components of a short report, a long, formal report includes the following (see Figure 16.4):

- **Front matter**—(title page, cover letter, table of contents, list of illustrations, and an abstract or executive summary)
- **Text**—(introduction including purpose, issues, background, and problems; discussion; and conclusion/recommendation)
- **Back matter**—(glossary, works cited or references page, and an optional appendix)

Title Page

The title page serves several purposes. On the simplest level, a title page acts as a dust cover or jacket keeping the report clean and neat. More important, the title page tells your reader the following:

Routine Correspondence and Cover Letters

See Chapter 6 for more discussion of routine correspondence and cover letters.

- Title of the long report (thereby providing clarity of intent)
- Name of the company, writer, or writers submitting the long report
- Date on which the long report was completed

If the long report is being mailed outside your company to a client, you also might include on the title page the audience to whom the report is addressed. If the long report is being submitted within your company to peers, subordinates, supervisors, or owners, you might want to include a routing list of individuals who must sign off or approve the report. Following are two sample title pages. Figure 16.5 is for a long report with routing information; Figure 16.6 is for a long report without routing information.

Cover Letter

Check Online Resources

www.prenhall.com/gerson

For more information about letters, visit our companion Web site.

Your cover letter prefaces the long report and provides the reader an overview of what is to follow. It tells the reader

- Why you are writing
- What you are writing about (the subject of this long report)
- What exactly of importance is within the report

FIGURE 16.4 Components of a Long, Formal Report

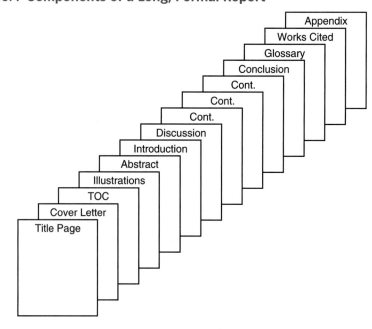

- What you plan to do next as a follow-up
- When the action should occur
- Why that date is important

Table of Contents

Long reports are read by many different readers, each of whom will have a special area of interest. For example, the managers who read your reports will be interested in cost concerns, timeframes, and personnel requirements. Technicians, in contrast, will be interested in technical descriptions and instructions. Not every reader will read each section of your long report.

Technical Descriptions and Instructions

See Chapter 11 for more discussion of technical descriptions and instructions.

FIGURE 16.5 Title Page for Long Report (with Routing Information)

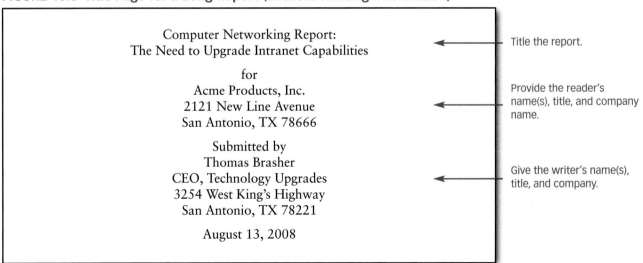

Report on Multicultural Workforce at StartCo Insurance

Prepared by: _____ Date: _____
    Pete Niosi
    Assistant Director, Human Resources

Reviewed by: _____ Date: _____
    Leah Workman
    Manager, Accounting

Recommended by: _____ Date: _____
    Greg Foss
    Department Supervisor, Customer Service

Recommended by: _____ Date: _____
    Shirley Chandley
    Director, Human Resources

Approved by: _____ Date: _____
    Ralph Houston
    Vice President

Provide routing information for the long report.

FIGURE 16.6 Title Page for a Long Report (without Routing Information)

Computer Networking Report:
The Need to Upgrade Intranet Capabilities

for
Acme Products, Inc.
2121 New Line Avenue
San Antonio, TX 78666

Submitted by
Thomas Brasher
CEO, Technology Upgrades
3254 West King's Highway
San Antonio, TX 78221

August 13, 2008

Title the report.

Provide the reader's name(s), title, and company name.

Give the writer's name(s), title, and company.

Your responsibility is to help these different readers find the sections of the report that interest them. One way to accomplish this is through a table of contents. The table of contents should be a complete and accurate listing of the main *and* minor topics covered in the report. In other words, you don't want just a brief and sketchy outline of major headings. This could lead to page gaps; your readers would be unable to find key ideas of interest. In the table of contents in Figure 16.7, we can see that the discussion section contains approximately 16 pages of data. What is covered in those 16 pages? Is anything of value discussed? We don't know.

BEFORE

FIGURE 16.7   Flawed Table of Contents

This poorly written table of contents omits second-level headings and has too many page gaps.

Table of Contents

In contrast, an effective table of contents fleshes out this detail so your readers know exactly what is covered in each section. By providing a thorough table of contents, you will save your readers time and help them find the information they want and need. Figure 16.8 is an example of a successful table of contents.

In the example, note that the actual pagination (page 1) begins with the introductory section. Page 1 begins with your main text, not the front matter. Instead, information prior to the introduction is numbered with lowercase Roman numerals (i, ii, iii, etc.). Thus, the title page is page i, and the cover letter is page ii. However, you don't need to print the numbers on these two pages. Therefore, the first page with a printed number is the table of contents. This is page iii, with the lowercase Roman numeral printed at the foot of the page and centered.

List of Illustrations

If your long report contains several tables or figures, you will need to provide a list of illustrations. This list can be included below your table of contents, if there is room on the page, or on a separate page. As with the table of contents, your list of illustrations must be clear and informative. Don't waste your time and your reader's time by providing a poor list of illustrations like Figure 16.9. Instead, provide a list like that shown in Figure 16.10.

AFTER

FIGURE 16.8 Effective Table of Contents

Table of Contents

iii

Second-level and third-level headings help the audience access all parts of the report in any order.

FIGURE 16.9 Flawed List of Illustrations

List of Illustrations

Fig. 1	2
Fig. 2	4
Fig. 3	5
Fig. 4	5
Fig. 5	9
Table 1	3
Table 2	6

AFTER

FIGURE 16.10 Effective List of Illustrations

List of Illustrations

Figure 1.	Revenues Compared to Expenses	2
Figure 2.	Average Diesel Fuel Prices Since 2005	4
Figure 3.	Mainshaft Gear Outside Face	5
Figure 4.	Mainshaft Gear Inside Face	5
Figure 5.	Acme Personnel Organization Chart	9
Table 1.	Mechanism Specifications	3
Table 2.	Costs: Expenditures, Savings, Profits	6

Check Online Resources

www.prenhall.com/gerson

For more information about graphics, visit our companion Web site.

Abstract

As mentioned earlier, a number of different readers will be interested in your long report. Because your readers are busy with many different concerns and might have little technical knowledge, they need your help in two ways: They need information quickly, and they need it presented in low-tech terminology. You can achieve both these objectives through an abstract or executive summary.

The abstract is a brief overview of the proposal or long report's key points geared toward a low-tech reader. To accomplish the required brevity, you should limit your abstract to approximately one to two pages.

Each long report you write will focus on unique ideas. Therefore, the content of your abstracts will differ. Nonetheless, abstracts might focus on the following: (a) the *problems* necessitating your report, (b) your suggested *solutions*, and (c) the *benefits* derived when your suggestions are implemented.

For example, let's say you are asked to write a formal report suggesting a course of action (limiting excessive personnel, increasing your company's workforce, improving your

Audience

See Chapter 4 for more discussion of audience.

Creating a Hierarchy of Headings Using Microsoft Word 2007

When writing your long, formal reports, you will want to include headings and subheadings. To help your readers navigate the text, create a clear hierarchy of headings (also called "cascading headings") that distinguish among "first-level headings," "second-level headings," "third-level headings," and "fourth-level headings."

Microsoft Word 2007 will help you create this hierarchy of headings. You can apply a "style" to your entire document in Word's "formatting text by using styles" tool.

For example, let's say that you want your headings and subheadings to look as follows:

Hierarchy of Headings	Description
HEADING ONE	First-level headings: Arial, all cap, boldface, 16 pt. font.
Heading two	Second-level headings: Times New Roman, italics, boldface, 14 pt. font.
Heading three	Third-level headings: Times New Roman, boldface, 12 pt. font.
Heading four	Fourth-level headings: Times New Roman, italics, 12 pt. font.

To create a hierarchy of headings, follow these steps.

1. Click on the **Home** tab and then click on the down arrow to the right of **Heading 2**.

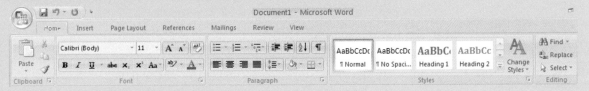

The following styles and formatting box will pop up.

2. Right click on the **Heading 1** button and scroll to **Modify**. Here, you can change font type (2.1), size (2.2), and color (2.3).

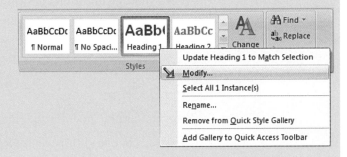

corporation's physical facilities, etc.). First, your abstract should specify the problem requiring your planned action. Next, you should mention the action you are planning to implement. This leads to a brief overview of how your plan would solve the problem, thus benefiting your company.

Another approach to an abstract for a long report is to summarize the content being analyzed in the report, setting the stage for the major issues to be discussed. In this abstract, you will highlight all the main points in the long report, omitting any supporting facts and documentation (to be developed in the Discussion section of the report). You not only want to be brief, focusing on the most important issues, but also you should avoid high-tech terminology and concepts. The purpose of the abstract is to provide your readers with an easy-to-understand summary of the entire report's focus. Your readers want the bottom line, and they want it quickly. Therefore, either avoid all high-tech terminology completely, define your terms parenthetically, or refer readers to a glossary.

Executive Summary

An executive summary, similar to an abstract but generally longer, is found at the beginning of either a formal report or a proposal and summarizes the major topics covered in the document. By reading the executive summary, your audience (the decision makers) gets an overview of the much longer report or proposal. Executive summaries can even take the place of the much longer formal report or proposal. In this way, you save your reader's time and communicate only the most relevant parts of the report or proposal. To create an effective executive summary, follow these key points.

Proposals

See Chapter 17 for more discussion of proposals.

1. Begin with a purpose statement for the report or proposal.
2. Give an overview of the key ideas discussed in the report or proposal.
3. Inform the reader if any problems will affect the outcome of the report or proposal.
4. Suggest solutions to the problems, or encourage the reader to suggest solutions.
5. Conclude the executive summary and recommend a course of action.

Key Parts of an Executive Summary. To write an effective and concise executive summary, use only the most important details and supporting statistics or information. Omit details or technical content which your reader either does not need to know or would unnecessarily confuse the reader. Details are used in the discussion section of the report or the proposal. To limit the length of your executive summary to one or two pages, choose the most important elements, and omit those that are secondary. An effective summary can include the following:

- Purpose and scope of the report or proposal, stating the problem or need and your ability to improve the situation
- Research or methods used to develop your content
- Conclusions about your analyses of the topic
- Your qualifications showing that you can resolve the issue
- A project management plan and timetable
- The total project budget
- Recommendations based on your findings

Introduction

Your introduction should include the following: (1) purpose, (2) background, and (3) problem.

Purpose. In one to three sentences or a short paragraph, tell your readers the purpose of your long report. This purpose statement informs your readers *why* you are writing or *what* you hope to achieve. This statement repeats your abstract to a certain extent. However, it's not redundant; it's a reiteration. Although numerous people read your report, not all of them read each line or section of it. They skip and skim.

The purpose statement, in addition to the abstract, is another way to ensure that your readers understand your intent. It either reminds them of what they have just read in the abstract or informs them for the first time if they skipped over the abstract. Your purpose statement is synonymous with a paragraph's topic sentence, an essay's thesis, the first sentence in a letter, or the introductory paragraph in a shorter report.

Background and Problem.   Whereas the purpose statement should be limited to one to three sentences or a short paragraph for clarity and conciseness, your discussion of the problem is more detailed. For example, if you are writing a report about the need for a new facility, your company's current work and storage space must be too limited. Your company has a problem that must be solved.

Your introduction's focus on the problem and background, which could average one to two pages, is important for two reasons. First, it highlights the importance of your report and establishes a context for the reader. The introduction emphasizes for your readers the report's priority. In this section, you persuade your readers that a problem truly exists and needs immediate attention.

Second, by clearly stating the problem and background information, you also reveal your knowledge of the situation. This section reveals your expertise. Thus, after reading this section of the introduction, your audience should recognize the importance of the issue and trust you to solve it or understand the complexity of the topic being discussed.

Discussion

The discussion section of your long report constitutes its body. In this section, you develop the detailed content of the long report. As such, the discussion section represents the major portion of the long report, perhaps 85 percent of the text. What will you focus on in this section? Because every report will differ, we can't tell you exactly what to include. However, your discussion can contain any or all of the following:

- Analyses
  - Existing situation
  - Solutions
  - Benefits
- Product specifications of mechanisms, facilities, or products
- Comparison/contrast of options
- Assessment of needs
- Features of the systems or products
- Optional approaches or methodologies for solving the problems
- Managerial chains of command (organizational charts)
- Biographical sketches of personnel
- Corporate and employee credentials
  - Years in business
  - Satisfied clients
  - Certifications
  - Previous accomplishments
- Schedules
  - Implementation schedules
  - Reporting intervals
  - Maintenance schedules
  - Delivery schedules

- Completion dates
- Payment schedules
- Projected milestones (forecasts)
- Cost analyses
- Profit and loss potential
- Documentation and researched material
- Survey results
- Lab report results
- Warranties
- Maintenance agreements
- Online help
- Training options
- Impact on the organization (time, personnel, finances, customers)

Check Online Resources

www.prenhall.com/gerson
For more information about audience, visit our companion Web site.

You will have to decide which of these sections will be geared toward high-tech readers, low-tech readers, or a lay audience. Once this decision is made, you will write accordingly, defining terms as needed. In addition to audience recognition, you should also enhance your discussion with figures and tables for clarity, conciseness, and cosmetic appeal.

Conclusion/Recommendation

You must sum up your long report in a page or so, providing your readers with a sense of closure. The conclusion can restate the problem, the important implications of your analysis, your solutions, and the benefits to be derived. In doing so, remember to quantify. Be specific—state percentages and amounts.

Your recommendation will suggest the next course of action. Specify when this action will or should occur and why that date is important. The conclusion/recommendation section can be made accessible through highlighting techniques, including headings, subheadings, underlining, boldface, itemization, and white space.

Glossary

Because you will have numerous readers with multiple levels of expertise, you must be concerned about your use of high-tech language (abbreviations, acronyms, and terms). Although some of your readers will understand your terminology, others won't. However, if you define your terms each time you use them, two problems will occur: You might insult high-tech readers, or you will delay your audience as they read your text. To avoid these pitfalls, use a glossary. A glossary is an alphabetized list of high-tech terminology placed after your conclusion/recommendation.

A glossary is invaluable. Readers who are unfamiliar with your terminology can turn to the glossary and read your definitions. Those readers who understand your word usage can continue to read without stopping for unneeded information.

Research and Documentation

See Chapter 5 for more discussion of research and documentation.

Works Cited (or References)

If you use research to write your long report, you will need to include a works cited or references page. This page(s) documents the sources (books, periodicals, interviews, computer software, Internet sites, etc.) you have researched and from which you have quoted or paraphrased.

Appendix

A final, optional component is an appendix. Appendices allow you to include any additional information (survey results, tables, figures, previous report findings, relevant letters or memos, etc.) that you have not built into your long report's main text. The contents of your appendix should not be of primary importance. Any truly important information should be incorporated within the report's main text. Valuable data (proof, substantiation, or information that clarifies a point) should appear in the text where it is easily accessible. Information provided within an appendix is buried, simply because of its placement at the end of the report. You don't want to bury key ideas. An appendix is a perfect place to file nonessential data that provides either documentation for future reference or further explanation or support of content in the report.

USING RESEARCH IN LONG, FORMAL REPORTS

You can use researched material to support and develop content in your long, formal report. Use quotes and paraphrases to develop your content. Technical communicators often ask how much of a long, formal report should be *their* writing, as opposed to researched information. A general rule is to lead into and out of every quotation or paraphrase with your own writing. In other words,

- Make a statement (your sentence).
- Support this generalization with a quotation or paraphrase (referenced material from another source).
- Provide a follow-up explanation of the referenced material's significance (your sentences).

Check Online Resources

www.prenhall.com/gerson
For more information about research, visit our companion Web site.

Research Includes Primary and Secondary Sources

Primary research is research performed or generated by you. You do not rely on books or periodicals. Instead you create original research by preparing a survey or a questionnaire targeting a group of respondents, by networking to discover information from other individuals, by visiting job sites, or by performing lab experiments. You may perform this research to determine for your company the direction a new marketing campaign should take, the importance of diversity in the workplace, the economic impact of relocating the

FAQs

Q: Why do long, formal reports include research? Aren't research papers just assignments we have in freshman composition classes?

A: Research, a major component of long, formal reports, helps you develop your report's content. Often, your own comments, drawn from personal experience, will lack sufficient detail, development, and authority to be sufficiently persuasive. You need research for the following reasons:

- Create content
- Support commentary and content with details
- Prove points
- Emphasize the importance of an idea
- Enhance the reliability of an opinion
- Show the importance of a subject to the larger business community
- Address the audience's need for documentation and substantiation

Who Writes Long, Formal Reports?

City, state, and the federal governments constitute the largest employer in the United States. The government hires engineers, financial analysts, accountants, scientists, administrative assistants, fire fighters, police officers, employees in public works and utilities, and parks and recreation workers.

Do these employees have to write long, formal reports? The answer is "yes." **Vicki Charlesworth**, Assistant City Manager/City Clerk of Shawnee, Kansas, says that her city's engineers, police officers, fire chiefs, financial advisors, and parks and recreation employees often write formal reports to city council members and the city's mayor. These long reports help the city council members "make informed decisions" that impact the city's residents and business owners.

Topics for the reports bubble up from citizens. A citizen contacts one of the city council members, for example, to share a concern about streets, sidewalks, animal control, traffic, or taxes. Maybe a citizen wants to expand a business, improve safety guard training for school crosswalks, or build a skateboard park for resident recreation. City council members ask the city's staff to research the issues, write a report that provides information on relevant statutes and/or other cities' best practices for handling issues, analyze options, and then recommend follow-up actions.

In one recent instance, the city needed to expand an existing road. The two-lane, winding road, initially built in the early 1940s, no longer met the city's growing population needs. To accommodate city growth, the road had to be widened to four lanes and straightened to meet increased traffic and speed limits. This change would necessitate significant research and resources. City engineers needed to study the extent to which a road change would impact

- An adjacent animal refuge
- Current homeowners' land
- Sewage lines
- Easements
- Electrical and water lines
- New signage
- Traffic signals

The roadway project would be time consuming and costly. Only through a long, formal report could city engineers thoroughly provide the city council information about this project, analyze the options, and recommend appropriate construction steps.

Vicki says that poor communication damages the city employees' credibility and diminishes their professionalism. Flawed reports—ones that aren't well written or clearly developed—suggest that the writers "lack interest in the city's residents and business owners," are unresponsive to their constituents, or "just don't want to help" the city's taxpayers. That's bad PR, potentially leading to citizen complaints and inefficient government. It's not surprising that Vicki says, "One of the most important assets you can have in a government is writing skills."

company to a new office site, the usefulness of a new product, or the status of a project. You also may need to interview people for their input about a particular topic. For example, your company might be considering a new approach to take for increased security on employee computers. You could ask employees for a record of their logs which would highlight the problems they have encountered with their computer security. With primary research, you will be generating the information based on data or information from a variety of sources that might include observations, tests of equipment, interviews, networking, surveys, and questionnaires.

When you conduct *secondary research*, you rely on already printed and published information taken from sources including books, periodicals, newspapers, encyclopedias, reports, proposals, or other business documents. You might also rely on information taken from a Web site or blog. All of this secondary research requires parenthetical source citations.

the writing process

Prewriting	→ ←	Writing	→ ←	Rewriting

Prewriting	Writing	Rewriting

• Write to inform, analyze, and/or recommend.
• Determine whether your audience is internal or external.
• Research your topic to gather information.
	• Organize your content using modes such as problem/solution, cause/effect, comparison/contrast, argument/persuasion, chronology, etc.
	• Develop your ideas by informing, analyzing, and recommending.
	• Use figures and tables to clarify content.
	• Use headings and talking headings for access.
		• Revise your draft by
		  • adding details
		  • deleting wordiness
		  • simplifying words
		  • enhancing the tone
		  • reformatting your text
		  • proofreading and correcting errors

Now that you know the criteria for long, formal reports and the importance of research, the next step is to write your report. To do so, follow the recursive, step-by-step process of *prewriting*, *writing*, and *rewriting*. Remember that the process is dynamic and the steps frequently overlap.

Check Online Resources

www.prenhall.com/gerson

For more information about process, visit our companion Web site.

Prewriting

Throughout this textbook, we have presented several techniques for prewriting—reporter's questions, clustering/mind mapping, flowcharting, brainstorming/listing, and outlining. Another excellent way to gather data and to determine your objectives is to conduct a survey.

Surveys, like interviews, questionnaires, or conversations, are primary sources of research. In contrast to secondary research from books, periodicals, databases, or online searches, surveys require help from other people. To ensure that you get other people's assistance during your survey, consider doing the following:

- Ask politely for their assistance.
- Explain why you need the interview and information.
- Explain how you will use the information.
- Tell the survey participants approximately how long the survey should take.
- Make a convenient appointment for the interview or to fill in the survey.
- Come prepared. Research the subject matter so you will be prepared to ask appropriate questions. Write your interview questions or the survey before you meet with the person. Take the necessary paper, forms, pencils, pens, laptop, electronic notepads, handheld PDAs, or recording devices you will need for the meeting.
- Create survey questions that allow for quantifiable answers. In other words, try to avoid simple "yes/no" questions. Give survey participants a choice from a value-based scale, such as "4 = absolutely critical; 3 = important; 2 = somewhat important; 1 = not important."

- When the interview ends or the individuals complete the survey, thank them for their assistance.
- Ask the survey participant(s) for permission to use their responses in your research.

Figure 16.11 is a sample survey. An information technology manager at Design International, Inc. used this primary research to substantiate his long, formal report recommending that the company switch from a Microsoft Windows operating system to a Linux operating system.

FIGURE 16.11   Survey Used to Conduct Primary Research

User Survey on Computer Operating Systems

Our company is considering migrating from a Windows-based computing system to a Linux-based system. To help us make this decision, please answer the following questions, using a 4 pt. scale.

1. Would you like to customize your desktop, choosing your own GUI links to computer operations (graphical images that you choose as iconic buttons to click on)?

___ 4 = absolutely critical
___ 3 = important
___ 2 = somewhat important
___ 1 = not important

2. Your DOS commands in Windows are a "text mode interface." Are you comfortable with our current system that relies on limited scripting/commands, or would you prefer a text mode interface that allows for multiple commands and is customizable?

___ 4 = absolutely critical
___ 3 = important
___ 2 = somewhat important
___ 1 = not important

3. Windows impacts your departmental budget because Microsoft allows only a single copy of Windows to be used on only one computer. In contrast, if we purchased a Linux system, you can run it on any number of computers for no additional charge. Such a purchase would require training to help computer users learn the new system. Realizing that a change in systems would necessitate training, how important is this budgetary issue to your department?

___ 4 = absolutely critical
___ 3 = important
___ 2 = somewhat important
___ 1 = not important

4. Windows comes with an operating system. Linux-based machines require that consumers install the operating system. Our company would provide you the software and hardware; an individual or individuals within your department would be required to perform this installation. Would this requirement make you vote *against* a purchase of a Linux system?

___ 4 = absolutely critical in a vote of "no"
___ 3 = important in a vote of "no"
___ 2 = somewhat important in a vote of "no"
___ 1 = not important—our department would not vote "no" due to this requirement

---

FIGURE 16.11 Continued

5. Has spyware or malware become a problem in your work environment?

___ 4 = absolutely critical
___ 3 = important
___ 2 = somewhat important
___ 1 = not important

6. Has the misuse of user IDs and passwords become a problem in your department, leading to security issues?

___ 4 = absolutely critical
___ 3 = important
___ 2 = somewhat important
___ 1 = not important

7. If your department has different types of and ages of computers, this could cause problems with operating systems with different hardware platforms. Windows, for example, does not run on different platforms. Linux systems can run on many different platforms. How important is it in your department to allow for different hardware platforms?

___ 4 = absolutely critical
___ 3 = important
___ 2 = somewhat important
___ 1 = not important

8. Does your department need computers to be clustered for multiple use and networking?

___ 4 = absolutely critical
___ 3 = important
___ 2 = somewhat important
___ 1 = not important

9. How important is storing data, backing it up, and moving stored data from one computer to another?

___ 4 = absolutely critical
___ 3 = important
___ 2 = somewhat important
___ 1 = not important

10. Windows supports many different kinds of printers, while Linux does not. If your department has different makes and models of printers, the operating system is an important choice. If your department does not have different makes and models of printers, the choice of an operating system is not important. How important is the choice of an operating system in your department, regarding printers?

___ 4 = absolutely critical
___ 3 = important
___ 2 = somewhat important
___ 1 = not important

Writing

Once you've gathered data for your report through primary and secondary research, the next step is to write the text. To write your report, do the following:

Reread Your Prewriting. Review your survey, other research, and the criteria for a successful long, formal report. Determine whether you've covered all important information. If you believe you've omitted any significant points, add them for clarity. If you've included any irrelevant ideas, omit them for conciseness.

Assess Your Audience. Your decisions regarding points to omit or include will depend on your audience. If your audience is familiar with your subject, you might be able to omit background data. However, if you have multiple audiences or an audience new to the situation, you'll have to include more data than you might have assumed necessary in your prewriting. Determine which acronyms, abbreviations, and high-tech terms you will need to define for your audience. You can define these terms within the text; however, you might prefer to use a glossary for your definitions.

Draft the Text. Focusing on your major headings (introduction, discussion, conclusion/recommendations), use the sufficing technique mentioned in earlier chapters. Just get your ideas down on paper in a rough draft without worrying about grammar.

Organize Your Content. You can organize the various portions of your report using any one of these four modes: *chronology, importance, comparison/contrast,* and *problem/solution*. For example, the introduction of your report lends itself to a problem/solution organization. The discussion unit in your long, formal report will benefit from comparison/contrast as a means of analyzing content. When concluding your report, you can use importance to highlight the benefits of your recommendation.

Develop Your Ideas. To prove your points, inform by providing information, findings, and results. Use analysis to draw conclusions. Recommend action based on information and analysis.

Rewriting

After a gestation period in which you let the report sit so you can become more objective about your writing or have the report reviewed by a coworker, retrieve the report and rewrite. Perfect your text by using the following rewriting techniques.

Add Detail for Clarity. Have you answered the reporter's questions (*who, what, when, where, why,* and *how*) as thoroughly as needed? You often have multiple readers of long, formal reports. Many of them do not know your motivations or objectives. They need clarity. Much of this clarity can be accomplished by writing an effective table of contents (TOC) and list of illustrations. Successful TOCs and lists of illustrations include headings, subheadings, and titles. Add these details for clarity.

Delete Dead Words and Phrases for Conciseness. A long, formal report, by definition, will be expansive. You don't want to make reading even harder for your audience by cluttering the text with excessively long words, phrases, sentences, and paragraphs. Try to limit word length to two syllables (except for terms which cannot be shortened, like *telecommunications, technology, engineer,* and more). Limit sentences to 10 to 15 words when you can. Limit paragraphs to four to six lines of text.

LONG, FORMAL REPORT CHECKLIST

_____ 1. Have you included the major components for your long, formal report (front matter, text, and back matter)?

_____ 2. In your long, formal report, have you written to inform, analyze, and/or recommend?

_____ 3. Have you used research effectively, incorporating quotes and paraphrases successfully?

_____ 4. Have you used primary and/or secondary research?

_____ 5. Did you correctly give credit for the source of your research?

_____ 6. Have you correctly used a hierarchy of headings to help your audience navigate the text?

_____ 7. Did you use tables and/or figures to develop and enhance content?

_____ 8. Have you met your audience's need for definitions of acronyms, abbreviations, and high-tech terms by providing a glossary?

_____ 9. Have you achieved clarity and conciseness?

_____10. Is your text grammatically correct?

DOT-COM UPDATES

For samples of and more information about long, formal reports, check out the following links.

- World Wide Web Consortium (W3C) Reports http://www.w3.org/TR/
- The EPA Acid Rain Report http://www.epa.gov/airmarket/cmprpt/arp05/
- The National Institute on Aging's 2005 Report on Alzheimer's Disease http://www.nia.nih.gov/Alzheimers/Publications/ADProgress2004_2005/
- The June 2004 Progress Report on the Federal Building and Fire Safety Investigation of the World Trade Center http://wtc.nist.gov/progress_report_june04/progress_report_june04.htm
- The World Health Organization's Report on HIV/AIDS http://www.who.int/hiv/fullreport_en_highres. pdf

Simplify Old-Fashioned Words and Phrases. What does _pursuant_ mean? How about _issuance of this report?_ The ultimate goal of technical communication is to communicate, not to confuse. Use words that are easy to understand.

Move Information within Your Discussion for Emphasis. Make sure that you've used organizational modes, such as problem/solution, comparison/contrast, cause/effect, argument/persuasion, and chronology. These help your audience follow your thoughts, make connections between ideas, and see the rationale for your recommendations.

Reformat Your Text for Accessibility. Highlight your key points with underlining or boldface. Use graphics to assist your readers. Include headings, subheadings, and talking headings to help your audience navigate the text.

Enhance Your Text for Style. Be sure to quantify when needed, personalize with pronouns for audience involvement, and be positive. Stress words like _benefit, successful, achieved,_ and _value_ rather than bowing to negative words like _cannot, failed, confused,_ or _mistaken._

Proofread and Correct the Report for Grammatical and Contextual Accuracy. Careful proofreading will enhance your credibility.

Avoid Biased Language. _Foreman_ should be _supervisor, chairman_ should be _chairperson_ or _chair,_ and _Mr. Swarth, Mr. James,_ and _Sue_ must be _Mr. Swarth, Mr. James,_ and _Ms. Aarons._

See Figure 16.12 for a sample long, formal report.

FIGURE 16.12 Long, Formal Report Recommending a Linux Operating System

Linux Desktop in Enterprise Settings
at Design International, Inc.

Submitted by
John Staples
Manager of Information Technology

March 2, 2008

FIGURE 16.12 Continued

Date:	March 2, 2008
To:	Tiffany Steward, CEO
From:	John Staples, Manager Information Technology
Subject:	Recommendation for Migration to a Linux Operating System

Ms. Steward, thank you for allowing me to report on a possible solution to Design International's operating system challenges. This report is in response to a survey my team conducted and the Information Technology Department's focus on improving internal and external communication.

Currently, we use Microsoft Windows as our computing system. However, Windows has a few inherent challenges. These include limited security, limited customization, and increased cost. Based on substantial research, my team and I would like to present an option—Linux. This report will provide you the following information:

- A definition of Linux   page 2
- A comparison/contrast of Linux vs. Windows   page 2
- An explanation of how Linux works   page 2–3
- The benefits of migrating to Linux   page 8

Information Technology believes migrating to a Linux Operating System can benefit Design International. I appreciate your thoughts on this report and look forward to answering any questions you might have. If you could assess this report by March 20, we would be able to initiate a Linux migration before the next quarter begins.

FIGURE 16.12 Continued

Table of Contents

List of Illustrations

iii

FIGURE 16.12 Continued

Abstract

The Microsoft Windows desktop has established a dominant presence in the enterprise desktop marketplace. Currently, the Windows desktop has nearly total control over the enterprise desktop market primarily due to

- An aggressive marketing strategy
- Its intuitive ease of use
- Its apparent ease of installation and administration
- Readily available support

Windows, however, is not the only viable desktop solution in the enterprise marketplace. Because of potential risks and benefits associated with any technology product, as IT manager I must be alert for alternatives that are secure, reliable, and cost effective.

This report provides technical information regarding the issue of migration to a Linux enterprise desktop solution, analyzes the pros and cons of migrating to a Linux solution, and recommends appropriate action for our company, Design International, Inc.

Introduction

Purpose

We need to consider changing from a Windows-based desktop to a Linux-based desktop for Design International.

Background of Enterprise Desktop Computing

Throughout the history of desktop computing solutions, Microsoft has clearly prevailed in the marketplace. The near monopoly enjoyed by Microsoft over users in all aspects of desktop computing has been established and maintained for many reasons including the user-friendly graphic environment which has been at the center of virtually all Microsoft products.

Despite the dominance of the Microsoft desktop in the desktop marketplace, other alternatives are emerging. One promising alternative is the family of Linux desktop operating systems.

Challenges Created by Relying Only on Windows

Though Microsoft Windows is the standard in the marketplace, it presents certain challenges for end users. At Design International, we have encountered the following challenges:

- Security—In Windows, the operating system and key applications are combined, which can lead to security issues for e-mail clients. In addition, users of Windows are dependent upon the vendor for recognition and correction of security issues, as well as the implementation of security patches.
- Customization—Prepackaged IT systems, such as Windows, disallow Design International to customize our computerization.
- Cost—Windows requires that Design International pay costly software licensure fees. This cost is compounded by the fact that Windows is platform dependent, limiting server selection.

1

Summarizing the report in the Abstract allows the reader to get a "snapshot" of the content. Since most people think that Microsoft Windows is the only option for desktop computing, this report discusses an alternative—Linux.

The IT manager lets the reader see that a recommendation will be made based on the facts in the report.

Providing a background for the report establishes and explains context for the audience.

Highlighting problems in the introduction shows the importance of your topic and the need to consider other options.

FIGURE 16.12 Continued

Discussion

What the Linux Desktop Environment Is

The current industry standard for the computing desktop is the Microsoft XP operating system. This system, like all other Microsoft products, is distributed to consumers, including enterprise consumers, in machine code only. The source code is kept as a tightly guarded secret. The Microsoft operating system comes to the consumer as an "end-to-end solution" that includes operating system, desktop, common applications, and utilities all bundled into one package.

In contrast, Linux is not necessarily distributed as machine code, nor is it necessarily an end-to-end solution. "Strictly speaking, Linux refers to the kernel maintained by Linus Torvalds and distributed under the same name through the main repository and various mirror sites." (Yaghmour 2003).

Embedding secondary research helps to develop the analysis and gives authority to your argument.

The Linux kernel is open source. As such, a custom operating system may be built to suit a particular need based on the kernel. However, most distributions of Linux include common user applications and are marketed by one of several vendors in the open source community such as RedHat, SuSe, Mandrake, or Caldera. Linux is available either by way of machine code or in its native source code format. In this manner, and by virtue of the general public license, the user can modify the operating system, so long as the source code remains open.

Linux vs. Windows Design Issues

The Microsoft Windows operating system follows the monolithic design. Most of the features of the system, including the graphics user interface (GUI) by way of the Internet Explorer, are built into one single unit, the kernel (Petreley 2004, 20).

Comparing and contrasting facts about the options helps to explain the topic to the reader. By focusing on pros and cons, you analyze the topic, clarifying the issue and supporting the conclusions you will draw later.

While the Linux kernel is also characterized in the literature as being "monolithic" in relationship to traditional Unix microkernel design, fewer of the utilities and drivers are part of the kernel. What further differentiates the design of the Windows operating system and the design of Linux is that Linux is fully modular in nature (Petreley 2004, 21). Microsoft developed the Internet Explorer browser as an integrated part of the operating system, ostensibly to provide for more convenience for the user.

Unfortunately, the highly integrated graphic nature of Windows tends to cause the most problems. Because of this underlying architecture, a flaw in the graphics renderings can be detected and exploited to the extent that the entire computer can be taken over or destroyed by malicious code. According to Petreley, a common cause of the "blue screen of death" is something as innocuous as a bug in a graphics card driver (26).

In contrast, the Linux operating system does not for the most part allow graphics drivers to run inside the kernel. Again, according to Petreley, Linux runs almost all graphics drivers outside of the kernel. "A bug in a graphics driver may cause the graphical desktop to fail, but not cause the entire system to fail. If this happens, one simply restarts the graphical desktop" (Petreley 2004, 25).

How Linux Works

Proprietary software primarily exists for the purpose of generating revenue for the owner of the copyright to the software. Proprietary software is not usually

2

FIGURE 16.12 Continued

sold to the user. Rather, its use is permitted in accordance with a licensing agreement. According to Kenston (2005), software licenses serve three basic functions:

1. They create profit for the developer.
2. They provide a means for controlling the distribution of the product.
3. They protect the underlying source code from modification. (6)

Vendors of proprietary software have placed a premium on protecting the source code from modification so that product development remains under the tight control of the vendor's development group. Raymond characterized software developed under the cathedral model as "carefully crafted by individual wizards or small bands of magi working in splendid isolation, with no beta released before its time" (29).

These concepts of business and development have two primary advantages: protecting the vendor's profit motive and providing for centralized control of changes and development.

Hardware Support for a Linux-based Platform

Support for hardware varies from device to device and configuration to configuration. This support is either provided by the Linux kernel or through drivers provided by the device manufacturer. The Linux 2.4 kernel provides full support for the Intel x86 family of processors, the most common platform from which Linux is run.

In terms of I/O (Input/Output) devices, the Linux kernel supports a broad range of serial port and parallel port devices, along with keyboards, pointing devices, printers, scanners, displays, and other common business tools. However, Linux does not support all such devices. In implementing a Linux desktop solution, as IT manager I had to research the transition of current devices and any new device purchases to ensure that drivers are available and that performance is established.

File Systems for Linux

The Second Extended File system (ext2fs) was developed specifically for Linux. According to Dalheimer et al. (2003), "The ext2fs is one of the most efficient and flexible file systems" (35). This file system was designed to provide an implementation of UNIX file commands. This system is currently the most common file system found in Linux distributions. Some of the features that are supported by ext2fs are

- Standard Unix file types, directories, device files, and links
- Large file systems, up to 32TB
- Root user reserved files
- Root user block sizes that can improve I/O performance
- File system state tracking (Dalheimer et al. 2003, 6)

3

Using subheadings organizes information for the reader and makes the content accessible. You will use subheadings to develop your table of contents.

Because long report topics are complex, you want to help your reader navigate the text easily. To create a readable style in your long report, use short sentences, short paragraphs, bulleted lists, and headings and subheadings.

FIGURE 16.12 Continued

Network Support for Linux

The Linux kernel provides for a full implementation of the TCP/IP networking protocols. Some of the features included with the kernel are

Device drivers for a wide range of Ethernet cards	Parallel Line Internet Protocol (PLIP)	IPv6 protocol suite	DHCP
Appletalk	IRDA	FTP	Simple Mail Transfer Protocol (SMTP)
DECnet	AX.25 for packet radio networks	Telnet	NNTP

(Dalheimer et al., 2003, 42)

In addition, Linux comes with support for a network firewall. This means that virtually any Linux computer can be configured as a router and a firewall.

How Linux Addresses the Enterprise Desktop Need at Design International

The User and the Enterprise Desktop

Two examples of desktops that are emerging as the front runners for enterprise use are the KDE and the GNOME desktops. Both provide a colorful graphic user environment that is highly customizable, even to the extent that it can be made to look a lot like the familiar Windows environment. Support for other features including drop and drag, multiple languages, and session management is now available.

What are Design International's Options for Accessing Linux?

We can load Linux systems on our computers at Design International in the following ways (Freedman 2006, 69).

- **Live CD:** The most risk-free approach to accessing Linux is achieved by inserting CDs in all our computers, rebooting, and circumventing Windows once the system restarts. Computers started this way will function as Linux machines. This solution, however, is a quick fix and not advisable for long-term functionality.

- **Parallel installation:** We can download and install Linux software and run it parallel to Windows. This way, our end users can choose between the two operating systems at start-up.

- **Windows replacement:** If we choose to use Linux exclusively, we must remove all Windows files and install Linux. Then, if we would want access to Windows, we would need to perform a complete Windows reinstall.

Persuasion is a key component of recommendations. To communicate persuasively, you might need to refute opposing points of view. Notice how the writer analyzes information in this section but then rebuts the argument by pointing out its drawbacks.

Use of a Linux Environment at Design International

Is Design International ready for a migration to a Linux Environment? Will our employees benefit from such a computing change? To find answers to these questions, I conducted an online survey of our 3,000 nationwide employees. The

4

FIGURE 16.12 Continued

employees were asked to assess 10 topics related to their computer usage. Specifically, the survey focused on issues such as security, printer usage, budget, installation of hardware, networking, and archiving of files. They ranked their responses on the following four-point scale.

___ 4 = absolutely critical
___ 3 = important
___ 2 = somewhat important
___ 1 = not important

I received a 75 percent response rate, equaling 2,250 responses to the survey (see Appendix). The results of this survey are shown in Figure 1.

Conducting primary research, such as through a survey, is an excellent way to add content that is customized for your particular audience and topic.

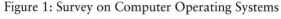
Figure 1: Survey on Computer Operating Systems

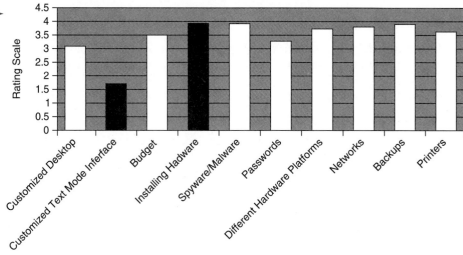

In Figure 1, notice that two bars in the bar chart are dark colored ("Customized Text Mode Interface" and "Installing Hardware"). The bar noting "Customized Text Mode Interface" scores low, suggesting that our employees are not interested in creating customized coding. The "Installing Hardware" scores high, but this is a strong "No" vote, suggesting that our employees *do not* want to be responsible for installing their own hardware. However, all of the other topics scored highly, suggesting that a Linux environment would meet and exceed our employees' computing needs, regarding cost, security, printer options, backup storage, and more.

A key component of analysis is interpretation. Informing allows you to convey data. In analysis, you explain what this data means to provide the audience a better understanding of the consequences of your findings.

Explanation of Survey Results

Based on the 10 questions asked, the results strongly suggest that our employees would profit from a migration to a Linux environment. To clarify why this is true, Table 1 compares and contrasts key facts regarding Windows and Linux, as represented in the survey.

5

FIGURE 16.12 Continued

Table 1 uses comparison/contrast to analyze Windows vs. Linux. This analysis provides information and then draws conclusions.

Table 1 Comparison/Contrast of Key Facts Regarding Windows vs. Linux	
Windows	**Linux**
Graphical User Interface	
Windows offers some customization of the GUI.	Linux allows advanced users to customize GUIs to their liking.
Cost	
Windows is expensive. Microsoft allows a single copy of Windows to be used on only one computer.	Linux is very cheap and/or free. Once you have purchased Linux, you can run it on any number of computers for no additional charge.
Getting the Operating System	
Windows comes with an operating system.	PC vendors sell Linux-based machines only as servers, not to consumers, so you must buy a new computer with the operating system preinstalled or as a consumer you must install the Linux system yourself.
Viruses and Spyware	
Spyware is the worst problem affecting Windows-based computers.	One of the major reasons for using Linux is that viruses and spyware are far less troublesome.
Users and Passwords	
Windows XP allows the omission of the password. Windows software is not designed to be used by a restricted user. Therefore, viruses and malware can access Windows machines easily.	Linux supports root and restricted users, which limits viruses and malware.
Printer Drivers	
Windows supports many printers.	Linux does not support as many printers as Windows.

Migration Planning to the Linux Environment

Migration planning is critical to deploying the Linux desktop. During the initial process of planning the migration, a critical task will be identifying groups of users by way of patterns of IT usage. This accomplishes groupings by way of "role based client segmentation" (Almond et al. 2005, 28). To accomplish this,

6

FIGURE 16.12 Continued

you must assess the use of IT services and then group the users into functional usage patterns. Table 2 depicts a typical example of client segmentation.

A major concern with altering desktop solutions is how such a change will affect the end user. This table highlights key points of the analysis and clarifies pros and cons.

Table 2 Client Segmentation				
Fixed Function	**Technical Workstation**	**Transactional Workstation**	**Basic Office**	**Advanced Office**
Limited Use of Business Apps	Limited Use of Business Apps	Applications Drive Business Process	Applications Drive Business Process	Applications Drive Business Process
Limited Office Productivity	Simple Office Productivity	Simple Office Productivity	Simple Office Productivity	Advanced Office Productivity
No E-mail	Simple E-mail	Simple E-mail	Simple E-mail	Advanced E-mail
No Instant Messaging	No Instant Messaging	No Instant Messaging	No Instant Messaging	Instant Messaging
Simple Browser Access	Simple Browser Access	Simple Browser Access	Simple Browser Access	Advanced Browser Applications
Filepoint Systems Management; Network Access; Host Emulation	Filepoint Systems Management; Network Access; Host Emulation	Filepoint Systems Management; Network Access; Host Emulation	Filepoint Systems Management; Network Access; Host Emulation	Filepoint Systems Management; Network Access; Host Emulation

(Almond et al., 2005, 28)

Once general user roles have been identified along with the applications that they use, you must establish functional continuity by identifying potential technological solutions for the new network environment.

Solutions to Migration Issues

According to Almond et al. (2005), six potential technological solutions for solving migration issues are

- Bridging applications, or applications that work in both the old and new environment.
- Similar applications or applications that work in the new environment that have the look and feel of the application used in the old environment.
- Server-based applications or applications that run solely on the server.
- Independent Software Vendor applications, or custom software purchased for the new environment.

7

FIGURE 16.12 Continued

- Web-based functional equivalents, or applications that run in the browser, regardless of the operating system.
- Custom porting of custom applications or modifying the code of custom applications so that they operate in the new environment as they did in the old. (33–36)

Managing User Satisfaction in Migration Planning

Perhaps the most critical issue in migration planning will be managing user satisfaction. As with any change, the reaction by users may be anything from enthusiastic acceptance, to outright rebellion. Some strategies suggested by Almond et al. (2005) that can be used to help move users forward during the migration process are as follows:

- **Communicate.** Make sure that users understand what changes are upcoming, what the schedule is, and how the changes will have a positive impact on the organization.
- **Use bridging applications.** Some Linux-based software such as OpenOffice runs well on Microsoft systems. Where possible, give users the opportunity to gradually become acclimated to the new environment prior to the migration.
- **Provide a similar look and feel.** The GNOME and KDE desktops can be customized to provide the user with a familiar Windows look-alike work environment.
- **Provide users with the opportunity to try it before the migration.** Many Linux distributions are available in the form of a bootable CD. The user can test the operating system on their own computer without actually installing it by merely booting from the CD (Almond et al. 2005, 34–36).

Benefits of Using Linux at Design International

The Linux desktop operating system is a reliable and viable alternative to Microsoft Windows. According to Dalheimer et al. (2003), key attributes that make Linux a viable operating system for Design International include the following:

- **Linux is free.** Organizations using the Linux desktop can eliminate the need for paying software licensure fees. In fact, an independent study by Enterprise Management Associates concluded that "Linux may, in many cases, be substantially less expensive to own than Windows" (Galli 2006, 16).
- **Linux is powerful.** The Linux operating system is efficient, and makes good use of hardware. Many companies are finding that older hardware can be kept in service for a longer period of time through the use of Linux.
- **Linux is compatible.** Linux can exist on the same computer as Windows, and even access Windows files that are in a different partition on the same disk as the Linux installation. Additionally, Linux can access Windows files across a network.
- **Linux has inherently superior desktop security.** Its modular design separates the operating system from key applications that are at the root of many security problems, including the browser and the E-mail client.
- **Linux provides for efficiency through customization.** The user can make the Linux system into exactly what the organization needs, thus making most efficient use of equipment (Dalheimer et al. 2003, 11–19).

8

Focusing on the key benefits to using Linux helps your readers understand the implications of your topic and substantiates the recommendation.

FIGURE 16.12 Continued

Conclusion

As previously noted, the Microsoft Windows desktop has established a dominant presence in the enterprise desktop marketplace. Nonetheless, as the IT manager at Design International, I must be alert for opportunities to improve service delivery that is safe, secure, and cost effective. Among the options that are currently available in the marketplace is the range of Linux-based desktop solutions. While the Linux desktop solution is a viable candidate for many enterprise environments, as the IT manager I must carefully weigh the strengths and weaknesses of each available system to ensure that the best solution for the individual circumstances is chosen.

Recommendation

We should consider migrating to a Linux-based desktop system at Design International. A further study should be conducted to consider the economic impact of migration on our budget, resources, and personnel.

Glossary

Term or Abbreviation	Definition
Drivers	An electronic component used to control another electronic component. The term refers to a specialized chip that controls high-power transistors.
GNOME	An international project to create an easy-to-use computing system built from free software.
GPL	The General Public License allows anyone to use or distribute source code.
I/O	Input/Output. Input allows a computer system user to manipulate the system; Output allows the system to produce the "what" the user has asked for.
KDE	K Desktop Environment is a free desktop environment that uses GUIs.
Kernel	The central component of most operating systems (OSs). It manages the system's resources and allows for the communication between hardware and software.
Open Source Software	Computer software that permits users to change and to redistribute the software in the revised form.
OS	A computer's Operating System manages the computer's hardware and software.
RPC	Remote Procedure Call is a situation where one computer initiates a call on another computer or network.

9

FIGURE 16.12 Continued

Works Cited

Almond, Chris, et al. "Linux client migration cookbook." *IBM Redbooks*. 2004. 20 Apr. 2005. http://publib-b.boulder.ibm.com/redbooks.nsf/RedbookAbstracts/sg246380.html?Open

Dalheimer, Matthias, et al. *Running Linux*. Sebastapol, CA: O'Reilly Safari Books, 2003.

Freedman, David H. "Tech geeks have long praised open-source software. Now's the time to see what the fuss is about." *Inc.* 28, 6 (June 2006): 69–71.

Galli, Peter. "OSDL answers Microsoft claims; Study: Linux is cost-effective alternative to Windows." *eWeek*. 23, 7 (Feb. 13, 2006): 16.

Kenston, Geoff. "Software licensing issues." *Faulkner Information Services 2004*. 14 Feb. 2005. http://www.faulkner.com/products/faccts/

Petreley, Nicholas. "Security report: Windows vs. Linux." *The Register* (October 22, 2004): 20–28.

Raymond, Eric. *The cathedral and the bazaar*. Sebastapol, CA: O'Reilly Safari Books, 1999.

Yaghmour, Karim. *Building embedded Linux systems*. Sebastapol, CA: O'Reilly Safari Books, 2003.

10

CHAPTER HIGHLIGHTS

1. Long, formal reports require time, resources of people and money, and have far-reaching effects.

2. Long, formal reports often deal with topics that consider serious and complex issues.

3. Long reports can provide information, be analytical, and/or recommend a course of action.

4. Long, formal reports include front matter (title page, cover letter, table of contents, list of illustrations, and abstract); text (introduction, discussion, conclusion/recommendation); and back matter (glossary, works cited, and optional appendix).

5. To help your readers navigate the report, create a clear hierarchy of headings.

6. When you write your long report, consider your audience and enhance your discussion with figures and tables for clarity, conciseness, and cosmetic appeal.

7. Primary research is performed and generated by you.

8. Surveys or questionnaires are primary sources of research.

9. When you conduct secondary research, you rely on already printed and published information.

10. Secondary research is from books, periodicals, databases, and online searches.

CASE STUDY

Alpha/Beta Consulting (A/B/C) solves customer problems related to hardware, software, and Web-based applications. The consulting company helps clients assess their computing needs, provides training, installs required peripherals to expand computer capabilities, and builds e-commerce applications, such as Web sites and corporate blogs. A/B/C's client base includes international companies and academic institutions.

A new client, Nitrous Solutions, has asked A/B/C to build its e-commerce opportunities. Nitrous Solutions' current Web site is outdated. Since the Web site was created, the company has added new services, including land planning, municipal engineering, surveying, transportation studies, residential development, and water resource management. In addition, Nitrous Solutions has expanded globally. It now has a presence in the Far East (Japan, Taiwan, and Singapore), the Mideast (Dubai, Jordan, and Israel), and South Africa. Nitrous Solutions' new services and new service locations are not evident on the current Web site.

An additional challenge for Nitrous Solutions is client contact. Currently, the company depends on a hard-copy newsletter. However, corporate communications at Nitrous Solutions believes that a corporate blog would be a more effective means of connecting with its modern client.

A/B/C can meet Nitrous Solutions' needs by building a new Web site and by creating a corporate blog.

Assignment

1. Write the abstract that will preface A/B/C's long, formal report to Nitrous Solutions.
2. Write the introduction that will preface A/B/C's long, formal report to Nitrous Solutions.
3. Create a survey questionnaire asking Nitrous Solutions' employees what they would like to see on the company's Web site and corporate blog.
4. Outline what you believe Nitrous Solutions' new Web site should include to cover its new international focus. This constitutes the discussion section of the report.
5. Conduct any research you need to find more information about either Web design or the development of a corporate blog.

INDIVIDUAL AND TEAM PROJECTS

1. Read a long, formal report. It might be one your instructor provides you or one you obtain from a business or online (see the links provided in this chapter's Web Workshop and in this chapter's Dot.com Updates). If the report contains an abstract or executive summary, is it successful, based on the criteria provided in this chapter? Explain your decision. If the abstract or executive summary could be improved upon, revise it to make it more successful. If the report does not have an abstract or executive summary, write one.
2. Read the following abstract from Alpha/Beta Consulting, written for the client Nitrous Solutions. It needs revision. How would you improve its layout and content?

Nitrous Solutions, a multidiscipline architectural design, is a start-up business with plans to begin operation in October 2008. The company currently has no information technology network for internal and external communication. Alpha/Beta Consulting will recommend a server to meet Nitrous Solutions' unique needs, a backup solution for archiving architectural materials and critical communications, software for digital asset management, and a proposed network design. After implementing these suggestions, Nitrous Solutions will have an IT network that will be safe, secure, dependable, and prompt. The system that Alpha/Beta Consulting is recommending also can be restructured with simplicity as the company's needs grow. By installing the IT network recommended in this report, Nitrous Solutions will be able to fulfill their company's mission statement.

3. Alpha/Beta Consulting, as agreed upon in a meeting with its client Nitrous Solutions, determined that its research for a contracted long, formal recommendation report would take no less than 350 hours and no more than 400 hours. As detailed in the following time sheet A/B/C's four employees assigned to the project worked 376 hours. Using the information provided in the time sheet (Table 16.1), create a line graph that tracks the employees' hours and a pie chart showing the percentage of time each employee worked.

TABLE 16.1 Time Sheet

Staff	Hours Worked Each Week														Total Hours for Project
	wk 1	wk 2	wk 3	wk 4	wk 5	wk 6	wk 7	wk 8	wk 9	wk 10	wk 11	wk 12	wk 13	wk 14	
Kenyon Patel	6.0	10.5	7.0	7.5	7.0	8.5	6.0	11.0	8.0	8.0	7.0	6.5	12.0	8.0	113.00
Randy Butler	6.5	5.5	5.5	3.0	5.5	8.0	4.0	5.5	6.0	8.0	7.0	9.5	15.0	8.0	97.00
Maria Villatega	0.0	0.0	3.5	4.0	4.5	5.0	3.0	1.5	10.0	17.0	8.0	9.0	8.0	4.0	77.50
James Soto	4.5	3.5	4.0	5.5	5.0	5.0	3.0	2.0	14.0	12.0	8.5	10.5	7.5	3.5	88.50
Total	17.00	19.50	20.00	20.00	22.00	26.50	16.00	20.00	38.00	45.00	30.50	35.50	42.50	23.50	376.00

4. At the conclusion of their long, formal report, the employees at A/B/C wrote a recommendation. Their recommendation, though excellent in terms of its content, needed to be reformatted for easier access and better emphasis of key points. Read the employees' recommendation and improve its layout.

Recommendations to Meet Nitrous Solutions' Network Needs

To ensure a secure and dependable network that can be restructured to meet Nitrous Solutions' developing needs, Alpha/Beta Consulting recommends the following:

You can improve your computing system's "Application Layer." The applications software programs that Alpha/Beta Consulting recommends consist of the Adobe Creative Suite 2, for imaging, editing, illustrations, file sharing, Web designs, and digital processing; Microsoft Office, for spreadsheet and Excel documents, word-processing files, project timelines, and maintenance; and Internet communication utilities, such as Web browsers and FTP clients, to guarantee rapid deployment of company communications and contract sales. All these applications will be run on both the current Macintosh and Microsoft operating systems.

You also need to add a "Data Layer" to your new computing system. The data transferred on the new network will consist of large graphic files and fonts being accessed from a centralized server, postscript printing, voice, and messaging. TCP/IP will be the main communication protocol, but Appletalk may also be used.

To meet Nitrous Solutions' network infrastructure requirements, Alpha/Beta Consulting recommends Category 6 (Cat 6) cables connected to 1Gbps switches. These constitute your "Network Layer." The logical star topology network will enable a higher degree of performance for all users associated with the production of architectural designs, meet all organizational duties and communications requirements, and provide a centralized server as the main point of access for all the client machines.

Finally, once installed, the physical network must have a "Technology Layer." This will enable the company to perform at a competitive level. This communication system will be a vital piece of the overall success of the company. The server will provide quick and reliable data storage, meet Nitrous Solutions' data backup needs, and allow for streamlined work flow.

PROBLEM-SOLVING THINK PIECES

Read the following text from a long, formal report written by Alpha/Beta Consulting. Based on the explanation in this chapter, decide whether the text informs, analyzes, and/or recommends. Explain your decisions.

example

Purpose Statement

This report will recommend the design parameters, hardware specifications, and estimated expenses needed to build and install external and internal computer network technology to meet Nitrous Solutions' communication needs.
This report's recommendation will focus on the following key areas:
- A solution for Nitrous Solutions' server needs
- Recommendations for a backup solution
- Possible software for digital asset management
- A proposed network design

Needs Analysis

1. Company Background

Nitrous Solutions, a multidiscipline architectural design firm located in Raleigh, North Carolina, is a start-up company. Nitrous Solutions will occupy a 2,000-square-foot office space.
The firm plans to begin operations in October 2008. The company will provide industrial, interior, landscape, and green architecture services for an international client base. Communication is a major component of the business plan. Thus, a reliable, efficient, creative communications/IT technology system is mandatory.

2. Basic IT Requirements

Currently, Nitrous Solutions has no network. The company needs an infrastructure plan that allows for the following:
- An IT platform that will be compatible with Macintosh and Windows.
- Eight Windows work stations and four Macintosh platforms.
- A network based on WiMAX impact.
- An internal and external network that is safe, secure, reliable, and fast.
- An internal and external network that can store, execute, and transmit architectural materials.
- A network that will allow for successful communications to clients, partners, and vendors.
- A data backup solution.
- Digital asset management software.

3. Detailed IT Requirements

In addition to the basic network requirements for Nitrous Solutions, the company also has asked for recommendations to meet the following micro network needs.

- Information about the most efficient server operating system for reliable communication in a cross-platform environment
- Requirements for different file and storage server hardware and software
- An analysis of which FTP server software would transfer secure data reliably to external clients
- The need for scripts to scan recently modified files and back them up to a server
- Reliable backup solutions and backup media types
- A comparison of different software applications that would prevent data duplication in the permanent backup archives
- A comparison of software that could catalog backup storage
- A software solution allowing Microsoft Word documents to be converted and catalogued into one PDF file

WEB WORKSHOP

You can find many examples of long, formal reports online. Compare the format of the online reports with the criteria for long, formal, researched reports discussed in this chapter. Explain where the online reports are similar to or different from this chapter's criteria. Do the online reports successfully communicate information and analyze issues and problems? Check out these sites, for example.

- Formal Reports: Carnegie Commission on Science, Technology, and Government (http://www.carnegie.org/sub/pubs/ccstfrep.htm)
- Reports from the American Society of Civil Engineers following Hurricane Katrina (http://www.asce.org/static/hurricane/whitehouse.cfm)
- International Narcotics Control Strategy Report, Released by the Bureau for International Narcotics and Law Enforcement Affairs (http://www.state.gov/p/inl/rls/nrcrpt/2003/vol2/html/29910.htm)
- The Use of Radio Frequency Identification by the Department of Homeland Security to identify and track suspicious individuals (http://www.dhs.gov/xlibrary/assets/privacy/privacy_advcom_rpt_rfid_draft.pdf)
- Report on "GreenDC Week: Earth Day Celebration" (http://dceo.dc.gov/dceo/lib/dceo/Green_DC_Week.pdf)
- The United States Government Accountability Offices' Report on "Women's Participation in the Sciences" (http://www.gao.gov/new.items/d04639.pdf)

QUIZ QUESTIONS

1. What are reasons for writing long, formal reports?
2. Explain informative formal reports.
3. Explain analytical formal reports.
4. Explain recommendation formal reports.
5. What are the components of a long, formal report?
6. What do you include in an abstract?
7. What are the reasons for using research in a long, formal report?
8. Explain primary research.
9. Explain secondary research.
10. What is the purpose for using a hierarchy of headings in a long, formal report?

CHAPTER 17

Proposals

In this scenario, Bellaire Educational Supplies/Technologies uses proposals to sell its products.

BEST (Bellaire Educational Supplies/Technologies) manufactures and markets school supplies for all levels of education (K–12, public or parochial, and college).

These supplies include the following:

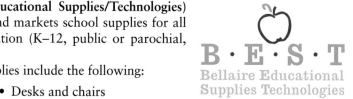

B·E·S·T
Bellaire Educational
Supplies Technologies

- Desks and chairs
- Modular computer workstations
- Blackboards and whiteboards
- Markers, chalk, and erasers
- Overhead projectors (both mobile and fixed computer projection systems)
- Pull-down screens
- Pens and pencils
- Grade books

BEST produces catalogs and mails fliers and brochures, but the majority of its large-ticket sales (to entire school systems) come from external proposals. These proposals, which usually are written in response to letters of inquiry or as follow-ups to person-to-person sales meetings, focus on the following components.

Objectives

When you complete this chapter, you will be able to

1. Format proposals effectively using components such as a title page, cover letter, table of contents, list of illustrations, abstract, introduction, discussion, conclusion/recommendation, and glossary.

2. Write effective internal proposals to persuade corporate decision makers to address issues and provide resources.

3. Write effective external proposals to sell a new service or product to a potential customer.

4. Distinguish among common proposal terms including RFP, T&C, SOW, boilerplate, solicited proposals, and unsolicited proposals.

5. Apply research techniques to gather information for proposals.

6. Evaluate your proposals using the proposal checklist.

- Cover letters prefacing the proposal to personalize the enclosed text as well as direct the audience to key parts of the proposal
- Table of contents itemizing the major headings and subheadings within the text
- A directory of tables and figures
- An executive summary, geared toward low-tech readers with limited time to read the entire proposal
- An introduction highlighting the reader's unique educational supplies problems and explaining how BEST will meet the client's needs
- Text that details costs, services, delivery, maintenance, guarantees, and product quality
- A conclusion summing up the proposal's key points
- A glossary to define unfamiliar words, acronyms, and abbreviations

Many of BEST's proposals are actually hard-copy printouts of PowerPoint slides. BEST's sales force and technical communicators have concluded that most readers want less text and easy-to-read documentation, complete with ample graphics. PowerPoint is the solution to this need. Not only can BEST's sales staff use PowerPoint to make oral presentations about its products, but also they can print out the PowerPoint slides as $11 \times 8\frac{1}{2}$ inch landscape pages and bind them as handouts. The hard-copy pages are easy to create, easy to read, and meet the audience's need for conciseness.

BEST's products, once used, sell themselves. However, the proposals open the door. Through proposals, BEST has found a perfect way to present clear, concise, and thorough documentation of the company's assets.

Check out our quarterly newsletters TechCom E-Notes at www.prenhall.com/gerson for dot.com updates, new case studies, insights from business professionals, grammar exercises, and facts about technical communication.

WHY WRITE A PROPOSAL?

When you write a proposal, your goal is to sell an idea persuasively. Consider this scenario: Your company is growing rapidly. As business increases, you believe that several changes must occur to accommodate this growth. For example, you think that the company needs a larger facility. This new building could be located in your city's vibrant new downtown expansion corridor, in a suburban setting, or entail the expansion of your current site. A new building or expansion should include amenities to improve recruitment of new employees, such as workout facilities, daycare, restaurant options, and even a gaming room. Finally, as part of new employee recruitment, you believe that the company must increase its diversity hiring practices.

Short, Informal Reports, and Long, Formal Reports

See Chapters 15 and 16 for more discussion of short, informal reports and long, formal reports.

Internal Proposals

How will you convey these ideas to upper-level management? The topic is large and will require extensive financial obligations, time for planning, and a commitment to new staffing. A short, informal report will not suffice. In contrast, you will have to write a type of longer, formal report—an *internal proposal* for your company's management.

Additional examples of internal proposals include the following:

- Your company needs to improve its mobile communication abilities for employees who work at diverse locations. To accomplish this goal, you write an internal proposal requesting the purchase of Wi/Fi-compatible laptops, handheld computers, and cell phones with Internet access.
- Your company's insurance coverage is skyrocketing. As a member of the human resources staff, you have researched insurance carriers and now will propose insurance options to upper-level management.
- Your company is migrating to a new software platform. Employees will need training to use the software. In a long, formal report, you propose consulting companies who can offer the training, optional schedules, funding sources, and posttraining certification.

External Proposals

Whereas *internal proposals* are written to management within your company, *external proposals* are written to sell a new service or product to an audience outside your company. Your biotechnology company, for example, has developed new software for running virtual cell cultures. The software simulates cell runs and displays synchronous strip charts for sterile monitoring. Data from the runs are graphed for comparison purposes. Not only will your company sell the software, but also the company provides consulting services to train clients in the software use. Your responsibility is to write an *external proposal* selling the benefits of this new corporate offering to a prospective client.

Requests for Proposals

Many external proposals are written in response to *requests for proposals* (RFPs). Often, companies, city councils, and state or federal agencies need to procure services from other corporations. A city, for example, might need extensive road repairs. A governmental agency needs Internet security systems for its offices. A hospital asks engineering companies to submit proposals about facility improvements. An insurance company needs to buy a fleet of cars for its adjusters. To receive bids and analyses of services, the city will write an RFP, specifying the scope of its needs. Competing companies will respond to this RFP with an external proposal.

FAQs

Q: When I read about proposals, I hear strange terms like *RFP, T&C, SOW, boilerplate,* and *solicited* and *unsolicited*. What do these words mean?

A: Here's a table defining these common proposal terms.

Proposal Terms	Definitions
RFP	**Request for Proposals**—means by which external companies and agencies ask for proposals
T&C	**Terms and Conditions**—the exact parameters of the request and expected responses
SOW	**Scope of Work *or* Statement of Work**—a summary of the costs, dates, deliverables, personnel certifications, and/or company history
Boilerplate	Any content (text or graphics) that can be used in many proposals
Solicited Proposal	A proposal written in response to a request
Unsolicited Proposal	A proposal written on your own initiative

In each of these instances, you ask your readers to make significant commitments regarding employees, schedules, equipment, training, facilities, and finances. Only a long, formal proposal, possibly complete with research, will convey your content sufficiently and successfully.

Research and Documentation

See Chapter 5 for more discussion of research and documentation.

CRITERIA FOR PROPOSALS

To guide your readers through a proposal, you will need to provide the following:

- Title page
- Cover letter
- Table of contents
- List of illustrations
- Abstract
- Introduction
- Discussion (the body of the proposal)
- Conclusion/recommendation
- Glossary
- Works cited (or references) page
- Appendix

Check Online Resources

www.prenhall.com/gerson
For more information about sample proposals, visit our companion Web site.

Check Online Resources

www.prenhall.com/gerson
For more information about long, formal reports, visit our companion Web site.

For information about title pages, cover letters, tables of contents, lists of illustrations, discussion, glossaries, works cited, and appendices, see Chapter 16. Each of these components typical of long, formal reports is thoroughly covered in that chapter. Following is information specifically related to your proposal's abstract, introduction, and conclusion/recommendation.

Abstract

Your audience for the proposal will be diverse. Accountants might read your information about costs and pricing, technicians might read your technical descriptions and process analyses, human resources personnel might read your employee biographies, and shipping/delivery

DOT-COM UPDATES

For information about proposals, check out these Web sites.

- **Capture Planning** (http://www. captureplanning.com/) provides proposal samples, templates, tips, and case studies

- **SERA Learning** (http://www.sera.com/). Go to their "Funding" link for proposal tips and templates.

- **Smith Richardson Foundation** (http://www. srf.org/grants/template. php#) provides a proposal template to help applicants apply for funding. Their template gives a great overview of typical proposal components.

- **"Templates make it easier to write winning proposals."** (http://www. usatoday.com/money/ smallbusiness/columnist/ abrams/2005-03-11-abrams_x.htm). *USA Today* discusses the value of and components for proposal templates.

might read your text devoted to deadlines. One group of readers will be management—supervisors, managers, and highly placed executives. How do these readers' needs differ from others? Because these readers are busy with management concerns and might have little technical knowledge, they need your help in two ways: They need information quickly, and they need it presented in low-tech terminology. You can achieve both these objectives through an abstract or executive summary.

The abstract is a brief overview of the proposal's key points geared toward a low-tech reader. If the intended audience is upper-level management, this unit might be called an executive summary. To accomplish the required brevity, limit your abstract to approximately 3 to 10 sentences. These sentences can be presented as one paragraph or as smaller units of information separated by headings. Each proposal you write will focus on unique ideas. Therefore, the content of your abstracts will differ. Nonetheless, abstracts should focus on the following: (a) the *problem* necessitating your proposal, (b) your suggested *solution*, and (c) the *benefits* derived when your proposed suggestions are implemented. These three points work for external as well as internal proposals.

For example, let's say you are asked to write an internal proposal suggesting a course of action (limiting excessive personnel, increasing your company's workforce, improving your corporation's physical facilities, etc.). First, your abstract should specify the problem requiring your planned action. Next, you should mention the action you are planning to implement. This leads to a brief overview of how your plan would solve the problem, thus benefiting your company.

If you were writing an external proposal to sell a client a new product or service, you would still focus on problem, solution, and benefit. The abstract would remind the readers of their company's problem, state that your company's new product or service could alleviate this problem, and then emphasize the benefits derived.

In each case, you not only want to be brief, focusing on the most important issues, but also you should avoid high-tech terminology and concepts. The purpose of the abstract is to provide your readers with an easy-to-understand summary of the entire proposal's focus. Your executives want the bottom line, and they want it quickly. They don't want to waste time deciphering your high-tech terms. Therefore, either avoid all high-tech terminology completely or define your terms parenthetically.

The following is an example of a brief, low-tech abstract for an internal proposal.

<table>
<tr>
<td>

Abstract

Due to deregulation and the recent economic recession, we must reduce our workforce by 12%.

Our plan for doing so involves

- Freezing new hires
- Promoting early retirement
- Reassigning second-shift supervisors to our Desoto plant
- Temporarily laying off third-shift line technicians

Achieving the above will allow us to maintain production during the current economic difficulties.

</td>
<td>

example

An effective abstract highlights the problem, possible solutions, and benefits in the proposal.

</td>
</tr>
</table>

Introduction

Your introduction should include two primary sections: purpose and problem.

Audience
See Chapter 4 for more discussion of meeting the needs of the audience.

Purpose.   In one to three sentences, tell your readers the purpose of your proposal. This purpose statement informs your readers *why* you are writing or *what* you hope to achieve. This statement repeats your abstract to a certain extent. However, it's not redundant; it's a reiteration. Although numerous people read your report, not all of them read each line or section of it. They skip and skim.

The purpose statement, in addition to the abstract, is another way to ensure that your readers understand your intent. It either reminds them of what they have just read in the abstract or informs them for the first time if they skipped over the abstract. Your purpose statement is synonymous with a paragraph's topic sentence, an essay's thesis, the first sentence in a letter, or the introductory paragraph in a shorter report.

The following is an effective purpose statement.

A purpose statement expresses the goal of the proposal.

<table>
<tr>
<td>

Purpose Statement: The purpose of this report is to propose the immediate installation of the 102473 Numerical Control Optical Scanner. This installation will ensure continued quality checks and allow us to meet agency specifications.

</td>
<td>

example

</td>
</tr>
</table>

Problem (Needs Analysis).   Whereas the purpose statement should be limited to one to three sentences for clarity and conciseness, your discussion of the problem must be more detailed. For example, if you are writing an internal proposal to add a new facility, your company's current work space must be too limited. You have a problem that must be solved. If you are writing an external proposal to sell a new piece of equipment, your prospective client must need better equipment. Your proposal will solve the client's problem.

Your introduction's focus on the problem, which could average one to two pages, is important for two reasons. First, it highlights the importance of your proposal. It emphasizes for your readers the proposal's priority. In this problem section, you persuade your readers that a problem truly exists and needs immediate attention. Second, by clearly stating the problem, you also reveal your knowledge of the situation. The problem section reveals your expertise. Thus, after reading this section of the introduction, your audience should recognize the severity of the problem and trust you to solve it. One way to help your readers understand the problem is through the use of highlighting techniques, especially headings and subheadings. See Figure 17.1 for a sample introduction.

FIGURE 17.1   Introduction with Purpose Statement and Needs Analysis

Provide specific details to explain the problem. Doing so shows that you understand the reader's need and highlights the proposal's importance.

1.0 Introduction

1.1 Purpose Statement

This is a proposal for a storm sewer survey for Yakima, WA. First, the survey will identify storm sewers needing repair and renovation. Then it will recommend public works projects that would control residential basement flooding in Yakima.

1.2 Needs Analysis

1.2.1. Increased Flooding

Residential basement flooding in Yakima has been increasing. Fourteen basements were reported flooded in 2007, whereas 83 residents reported flooded basements in 2008.

1.2.2. Property Damage

Basement flooding in Yakima results in thousands of dollars in property damage. The following are commonly reported as damaged property.

- Washers
- Dryers
- Freezers
- Furniture
- Furnaces

Major appliances cannot be repaired after water damage. Flooding also can result in expensive foundation repairs.

1.2.3. Indirect Costs

Flooding in Yakima is receiving increased publicity. Flood areas, including Yakima, have been identified in newspapers and on local newscasts. Until flooding problems have been corrected, potential residents and businesses may be reluctant to locate in Yakima.

1.2.4. Special-Interest Groups

Citizens over 55 years old represent 40 percent of the Yakima, WA, population. In city council meetings, senior citizens with limited incomes expressed their distress over property damage. Residents are unable to obtain federal flood insurance and must bear the financial burden of replacing flood-damaged personal and real property. Senior citizens (and other Yakima residents) look to city officials to resolve this financial dilemma.

Discussion

When writing the text for your proposal, you need to sell your ideas persuasively, develop your ideas thoroughly through research, observe ethical technical communication standards, organize your content so the audience can follow your thoughts easily, and use graphics.

Persuasive Writing

See Chapter 10 for more discussion of persuasive communication.

Communicating Persuasively.   A successful proposal will make your audience act. Writing persuasively is especially important in an *unsolicited* proposal since your audience has not asked for your report. A *solicited* proposal, perhaps written in response to an RFP, is written to meet an audience's specific request. Your audience wants you to help them meet a need or solve a problem. In contrast, when you write an *unsolicited* proposal, your audience has not asked for your assistance. Therefore, in this type of proposal, you must convincingly persuade the audience that a need exists and that your proposed recommendations will benefit the reader.

To write persuasively, you should accomplish the following:

- Arouse audience involvement—focus on your audience's needs that generated this proposal.
- Refute opposing points of view in the body of your proposal.
- Give proof to develop your content, through research and proper documentation.
- Urge action—motivate your audience to act upon your proposal by either buying the product or service or adopting your suggestions or solutions.

Researching Content for Proposals. As in any long, formal report, consider developing your content through research. This can include primary and secondary sources such as the following:

- Interviewing customers, clients, vendors, and staff members
- Creating a survey and distributing it electronically or as hard-copy text
- Visiting job sites to determine your audience's needs
- Using the Internet (online access to libraries, search engines, corporate Web sites, etc.)
- Reading journals, books, newspapers, and other hard-copy text

Communicating Ethically. When you write a proposal, your audience will make decisions based on your content. They will decide what amounts of money to budget, how to allocate time, what personnel will be needed to complete a task, and if additional equipment or facilities will be required. Therefore, your proposal must be accurate and honest. You cannot provide information in the proposal that dishonestly affects your decision makers. To write an ethical proposal, you must provide accurate information about credentials, pricing, competitors, needs assessment, and sources of information and research. When using research, for example, you must cite sources accurately to avoid plagiarism, as discussed in Chapter 5.

Organizing Your Content. Your proposal will be long and complex. To help your audience understand the content, use modes of organization. These can include the following:

- **Comparison/contrast**—rely on this mode when offering options for vendors, software, equipment, facilities, and more.
- **Cause/effect**—use this method to show what created a problem or caused the need for your proposed solution.
- **Chronology**—show the timeline for implementation of your proposal, reporting deadlines to meet, steps to follow, and payment schedules.
- **Analysis**—subdivide the topic into smaller parts to aid understanding.

The discussion section of your proposal lends itself to many different modes of organization. Consider applying various modes of organization to the key components of your proposal's body. (See Table 17.1.)

Using Graphics. Graphics, including tables and figures, can help you emphasize and clarify key points. For example, note how the following graphics can be used in your proposal's discussion section.

- **Tables.** Your analysis of costs lends itself to tables.
- **Figures.** The proposal's main text sections could profit from the following figures.
  - **Line charts**—excellent for showing upward and downward movement over a period of time. A line chart could be used to show how a company's profits have decreased, for example.
  - **Bar charts**—effective for comparisons. Through a bar or grouped bar chart, you could reveal visually how one product, service, or approach is superior to another.
  - **Pie charts**—excellent for showing percentages. A pie chart could help you show either the amount of time spent or amount of money allocated for an activity.

Check Online Resources

www.prenhall.com/gerson
For more information about persuasive communication, visit our companion Web site.

Research
See Chapter 5 for more discussion of research and documentation.

Ethics
See Chapter 3 for more discussion of ethics.

DOT-COM UPDATES
For more information about ethical considerations, check out the following link.
- The Online Ethics Center for Engineering and Science, http://www.onlineethics.org/

Methods of Organization
See Chapter 3 for more discussion of organization.

TABLE 17.1 Key Components of the Proposal's Discussion Section

Analyses of the existing situation, your suggested solutions, and the benefits your audience will derive	Spatial descriptions of mechanisms, tools, facilities, or products	Process analyses explaining how the product or service works	Chronological instructions explaining how to complete a task
Comparative approaches to solving a problem	Comparing and contrasting purchase options	Analyzing managerial chains of command	Chronological schedules for implementation, reporting, maintenance, delivery, payment, or completion
Corporate and employee credentials	Years in business	Satisfied clients	Certifications
Previous accomplishments	Biographical sketches of personnel	Projected milestones (forecasts)	Comparative cost charts

Visual Aids

See Chapter 9 for more discussion of figures and tables.

- **Line drawings**—effective for technical descriptions and process analyses
- **Photographs**—effective for technical descriptions and process analyses
- **Flowcharts**—a successful way to help readers understand procedures
- **Organizational charts**—excellent for giving an overview of managerial chains of command

Conclusion/Recommendations

Sum up your proposal, providing your readers closure. The conclusion can restate the problem, your solutions, and the benefits to be derived. Your recommendation will suggest the next course of action. Specify when this action will or should occur and why that date is important. Figure 17.2 is a conclusion/recommendation from an internal proposal.

FIGURE 17.2   Conclusion/Recommendation for Internal Proposal

Summarize the key elements of the proposal.

Recommend follow-up action and show the benefits.

3.0 Solutions for Problem

Our line capability between San Marcos and LaGrange is insufficient. Presently, we are 23% under our desired goal. Using the vacated fiber cables will not solve this problem because the current configuration does not meet our standards. Upgrading the current configuration will improve our capacity by only 9% and still present us the risk of service outages.

4.0 Recommended Actions

We suggest laying new fiber cables for the following reasons. They will

- Provide 63% more capacity than the current system
- Reduce the risk of service outages
- Allow for forecasted demands when current capacity is exceeded
- Meet standard configurations

If these new cables are laid by September 1, 2008, we will predate state tariff plans to be implemented by the new fiscal year.

the writing process

Prewriting	→ ←	Writing	→ ←	Rewriting

• Determine whether you are writing an internal or external proposal.
• Determine whether you are writing a solicited or unsolicited proposal.
• Conduct primary and secondary research to gather information.
• Use a prewriting technique, such as brainstorming, outlining, or answering reporter's questions, to organize your thoughts.
		• Organize your content using modes such as problem/solution, cause/effect, comparison, argument/persuasion, chronology, etc.
		• Write persuasively by arousing your audience's interest, refuting opposing points of view, giving proof to support your contentions, and urging action in the proposal's recommendation.
		• Use figures and tables to clarify content.
		• Use headings and talking headings for access.
				• Revise your draft by
				  • adding details
				  • deleting wordiness
				  • simplifying words
				  • enhancing the tone
				  • reformatting your text
				  • proofreading and correcting errors

Proposals, which include descriptions, instructions, cost analyses, scheduling assessments, and personnel considerations, are more demanding than other kinds of technical correspondence. Therefore, writing according to a process approach is important and will help you to write an effective proposal. For your proposal, in order to persuade your audience to act, you will have data to gather, information to organize, and text to revise. To help you accomplish these tasks, prewrite, write, and rewrite. Remember that the writing process is dynamic, and the steps often overlap.

Prewriting

Throughout this textbook, we have provided numerous prewriting techniques geared toward helping you gather data and determine your objectives. Any or all of these can be used while prewriting your proposal. For example, you might want to use the following:

- *Brainstorming/listing* to determine your audience and to outline the key components of your proposal. Is your audience internal or external? This will affect the terms you might need to define in a glossary.
- *Reporter's questions (who, what, when, where, why,* and *how)* to help you gather data for any of the sections in your proposal
- *Flowcharting* to organize procedures and schedules
- *Outlining* and *mind mapping* to organize your managerial sections (organizational charts for chains of command, personnel responsibilities, etc.)

Process

See Chapter 2 for more discussion of the writing process.

Research

See Chapter 5 for more discussion of research.

Check Online Resources

www.prenhall.com/gerson

For more information about research and documentation, visit our companion Web site.

Organization

See Chapter 3 for more discussion.

In addition to these prewriting techniques, you might need to perform primary and secondary research prior to writing. For instance, let's say your engineering firm is submitting a proposal to an out-of-state corporation. Prior to bidding on the job, you will need to find out what kinds of certifications or licenses this state requires. To do so, you will have to conduct secondary research to find the state's requirements and read state laws regarding construction certifications and licenses. You might also want to conduct primary research by interviewing or surveying your client's employees.

Writing

After gathering your data and organizing your thoughts through prewriting, your next step is to draft your proposal.

Review Your Prewriting. Double-check your brainstorming/listing, mind mapping, flowcharting, reporter's questions, interviews, research, and surveys. Is all the necessary information there? Do you have what you need? If not, now is the time to add new detail or research your topic further. In contrast, perhaps you've gathered more information than necessary. Maybe some of your data are irrelevant, contradictory, or misleading. Delete this information. Focus your attention on the most important details.

Organize the Data. Each of your proposal's sections will require a different organizational pattern. Following are several possible approaches.

- **Abstract**—problem/solution/benefit
- **Introduction**—cause/effect (The problem unit is the cause of your writing; the purpose statement represents the effect.)
- **Main text**—This unit will demand many different methods of organization, including
  —Analysis (cost charts, approaches, managerial chains of command, personnel biographies, etc.)
  —Chronology (procedures, scheduling)
  —Spatial (descriptions)
  —Comparison/contrast (options—approaches, personnel, products)
- **Conclusion/recommendation**—analysis and importance (Organize your recommendations by importance to highlight priority or justify need.)

Write Using Sufficing Techniques. When you draft your text, don't worry about correct grammar, highlighting techniques, or graphics. Just get down the information as rapidly as you can. You can revise the draft during rewriting, the third stage of the writing process.

Format Your Writing According to the Criteria for Effective Proposals. To format your proposal, include the following components.

- Title page
- Cover letter
- Table of contents
- List of illustrations
- Abstract

- Introduction
- Discussion (the body of the proposal)
- Conclusion/recommendation
- Glossary
- Works cited (or references) page
- Appendix

Headings

For more discussion of headings and talking headings, see Chapters 8 and 15.

Remember to use headings and talking headings to help your audience navigate the text.

Rewriting

After you have written a rough draft of your proposal, the next step is to revise it—fine-tune, hone, sculpt, and polish your draft.

Add Detail for Clarity.   In addition to rereading your rough draft and adding a missing *who, what, when, where, why*, and *how* where necessary, add your graphics. Another important addition to your proposal in this rewriting stage is a glossary. Now that you've written your rough draft, go back over each page to decide which abbreviations, acronyms, and high-tech terms must be placed in your glossary. After each of these high-tech words or phrases, add an asterisk. After the first abbreviation, acronym, or high-tech word or phrase, add a footnote stating that subsequent words or phrases followed by an asterisk will be defined in the glossary. Then write your glossary and add it to your proposal.

Delete Dead Words and Phrases for Conciseness.   Because proposals are long, you've already made your reader uncomfortable. People don't like wading through massive pages of text. Anything you can do to help your reader through this task will be appreciated. Avoid making the proposal longer than necessary.

Objectives

See Chapter 3 for more discussion of the objectives of technical communication.

Simplify Old-Fashioned Words.   A common adage in technical communication is, "Write the way you speak." In other words, try to be somewhat conversational in your writing. We aren't suggesting that you use slang or colloquialisms. But a word like *supersede* should be simplified: Instead, write *replace*. Check your text to see whether you've used words that are unnecessarily confusing.

Deleting and simplifying, when used together, will help you lower your fog index. This is especially important in your abstract and conclusion/recommendation sections, which are geared toward low-tech management. Review these sections of your proposal to determine if you have written at the appropriate level. If you have used terminology that is too high tech, simplify.

Move Information.   Each section of your proposal will use a different organizational method. Your abstract, for example, should be organized according to a problem/solution/benefit approach. Your introduction will be organized according to cause (the problem) and effect (the purpose). The organization of your main text will vary from section to section. You might analyze according to importance, set up schedules and procedures chronologically, describe spatially, and so forth. In rewriting, revise your proposal to ensure that each section maintains the appropriate organizational pattern. To do so, move information around (cut and paste).

Document Design

See Chapter 8 for more discussion.

Reformat for Reader-Friendly Ease of Access.   If you give your readers wall-to-wall words, they will have difficulty wading through your proposal. To avoid this, revise your proposal by reformatting. Indent to create white space. Add headings and subheadings. Itemize ideas, boldface key points, and underline important words or phrases. Using graphics (tables and figures) will also help you avoid long, overwhelming blocks

PROPOSAL CHECKLIST

Have you included the following in your proposal?

_____ 1. Title page (listing title, audience, author or authors, and date)

_____ 2. Cover letter (stating why you're writing and what you're writing about; what exactly you're providing the readers; what's next—follow-up action)

_____ 3. Table of contents (listing all major headings, subheadings, and page numbers)

_____ 4. List of illustrations (listing all figures and tables, including their numbers and titles, and page numbers)

_____ 5. Abstract (stating in low-tech terms the problem, solution, and benefits)

_____ 6. Introduction (providing a statement of purpose and a lengthy analysis of the problem)

_____ 7. Discussion (solving the readers' problem by discussing topics such as procedures, specifications, timetables, materials/equipment, personnel, credentials, facilities, options, and costs)

_____ 8. Conclusion (restating the benefits and recommendation for action)

_____ 9. Glossary (defining terminology)

_____ 10. Appendix (optional additional information)

SPOTLIGHT

How to Write an Effective Proposal

In an interview, **Mary Woltkamp,** President of Effective Business Communication, Inc., made the following comments about the importance of proposals in her job.

Q: Tell me about proposals.

A: All of the proposals that I write are for an external audience: potential clients. Thus, the tone is very formal and businesslike. One thing I always keep in mind when I'm writing is word choice. It's very easy to slip into the jargon of our industry, but more often than not, those terms are Greek to our clients. When I have to use industry-specific words, I define them. I'm also very conscious of the length of my words, sentences, and paragraphs. In general, the people who read my proposals are very busy. They don't have time to wade through a lot of unnecessary verbiage. I'm a big fan of headings, subheadings, and bulleted lists, and I always use lots of white space.

Q: What are some components in your proposals?

A: I include the following headings in a project proposal.

- Contact Information—for both the client and my company—always the first page of the document.
- Situation—What is the client's need? This assures the client that we understand their dilemma.
- Business Objectives—How will this project positively affect the client's business and return on investment?
- Project Objectives—What behaviors will be changed as a result of the project or training?
- Scope—A very detailed list of all the tasks that we think will need to be done in order to complete the project.
- Deliverables—What the client gets when we're all done, such as paper materials, electronic files, etc.
- Project Timeline—The milestones that have been identified at this point in the process.
- Time and Cost Estimates—A table that outlines our time and cost estimates for each task listed in the Scope.
- Project Team and Company Background—A short bio on all team members, plus our expertise and references (tailored to the prospective client's industry).

Q: Who constitutes the audiences for your proposals?

A: It varies. Sometimes the proposal is addressed to someone pretty high up the food chain, such as the company's president and CEO. Other times, we're dealing with a department manager or project manager who is responsible for the project.

of text. By reformatting your proposal, you will make the text inviting and ease reader access.

Enhance the Tone of Your Proposal.   Although you want to keep your proposal professional, remember that people write to people. Enhance the tone of your text by using pronouns and positive, motivational words. These are important in your statements of benefit and recommended courses of action. *Sell* your ideas.

Correct Errors.   If you're proposing a sale, you must provide accurate figures and information. Proposals, for example, are legally binding. If you state a fee in the proposal or write about schedules, your prospective client will hold you responsible; you must live up to those fees or schedules. Make sure that your figures and data are accurate. Correcting errors will ensure that you abide by ethical standards when writing your proposal.

Reread each of your numbers, recalculate your figures, and double-check your sources of information. Correct errors prior to submitting your report. Failure to do so could be catastrophic for your company or your client. Of course, you must also check for typographical, mechanical, and grammatical errors. If you submit a proposal containing such errors, you will look unprofessional and undermine your company's credibility.

Avoid Biased Language.   As noted throughout this textbook, both men and women read your writing. Don't talk about *foremen, manpower,* or *men and girls.* Change *foremen* to *supervisors, manpower* to *workforce* or *personnel,* and *men and girls* to *men and women.* Don't address your cover letter to *gentlemen.* Either find out the name of your readers or omit the salutation according to the simplified letter style. In addition, remember that your audience will be diverse. Avoid biased language.

Simplified Format

See Chapter 6 for more discussion of simplified letter format.

SAMPLE INTERNAL AND EXTERNAL PROPOSALS

Figure 17.3 illustrates a sample external proposal that you can use as a model for your writing. Figure 17.4 shows a sample internal proposal written in PowerPoint.

FIGURE 17.3 External Proposal

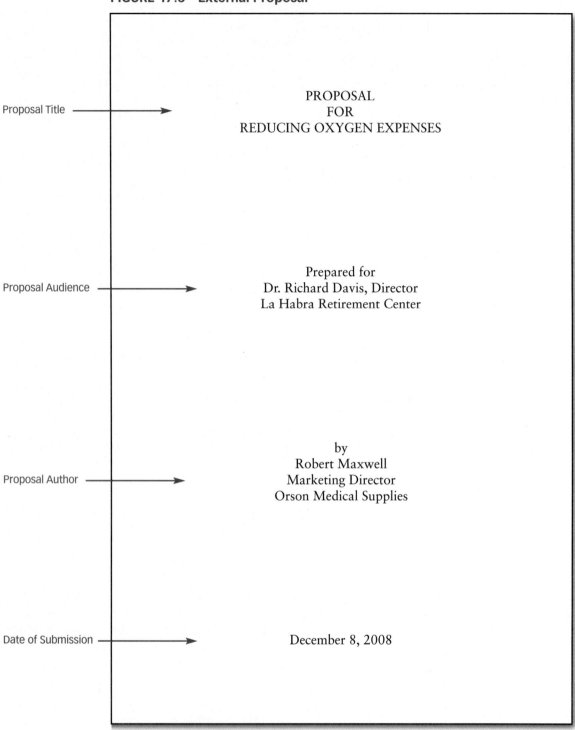

Proposal Title

PROPOSAL
FOR
REDUCING OXYGEN EXPENSES

Proposal Audience

Prepared for
Dr. Richard Davis, Director
La Habra Retirement Center

Proposal Author

by
Robert Maxwell
Marketing Director
Orson Medical Supplies

Date of Submission

December 8, 2008

FIGURE 17.3   Continued

Orson Medical Supplies
"Improving the Quality of Life"
12345 College Blvd.
Overland Park, OR 90091
976-988-2000

December 8, 2008

Dr. Richard Davis, Director
La Habra Retirement Center
220 Cypress
La Habra, CA 90631

Dear Dr. Davis:

Submitted for your review is our proposal regarding the Electronic
Demand Cannula oxygen-saving system. This document is in response
to your September 25, 2008, letter and our subsequent discussions.

Within our report, you will find the following supporting materials
geared toward your requests:

- Product specifications . . . . pages 3–4
- Operating instructions . . . . page 4
- Qualifications and experience . . . . page 5
- Cost . . . . page 6
- Warranty . . . . page 7

Thank you for your interest in our product. We look forward to serving you
and will call within two weeks to finalize arrangements.

Sincerely,

Robert Maxwell

Robert Maxwell
Marketing Director

Enclosure: Proposal

Itemizing the cover letter body focuses on key points within the proposal and provides the appropriate page numbers for reference. Maintaining a pleasant tone and being persuasive will appeal to your reader.

FIGURE 17.3   Continued

TABLE OF CONTENTS

LIST OF ILLUSTRATIONS

iii

Headings, subheadings, and page numbers help the audience find information and navigate the text.

Figure and table numbers plus titles allow the reader to find the visuals quickly.

FIGURE 17.3   Continued

ABSTRACT

Expenses for medical oxygen have increased steadily for several years. Now the federal government is reducing the amount of coverage that Medicare allows for prescription oxygen.

These cost increases can be reduced through the use of our new Electronic Demand Cannula (EDC). The EDC delivers oxygen to the patient only when the patient inhales. Oxygen does not flow during the exhalation phase. Therefore, oxygen is conserved.

This oxygen-saving feature can reduce your oxygen expenses by as much as 50 percent. Patients who use portable oxygen supplies can enjoy prolonged intervals between refilling, thus providing more freedom and mobility.

Emphasize problems generating the proposal.

Show how the proposal can solve the problems for the client.

Highlight the benefits derived from accepting the proposal. Stating the problem, solution, and benefits gives focus to the proposal.

iv

FIGURE 17.3   Continued

1.0 INTRODUCTION

1.1 PURPOSE

This is a proposal to sell the new Electronic Demand Cannula (EDC)* to the La Habra Retirement Center, La Habra, California. This bid to sell offers you a special discount when you purchase our EDCs in the quantities suggested in this proposal.

1.2 PROBLEMS

1.2.1 *High Costs*
Since 2000, the price of medical-grade oxygen has skyrocketed. It cost $10 per 1,000 cubic feet (cu ft) in 2000. Today, medical-grade oxygen costs $26 per 1,000 cu ft. In fact, you can expect next year's oxygen expenses to double the amount you spent this year.

1.2.2 *Governmental/Insurance Involvement*
Many factors have contributed to this soaring cost, including demand, product liability, and inflation. However, two factors contributed the most. First, legislation reduced the amount that Medicare pays for prescription oxygen. Second, few insurance companies offer programs covering long-term prescription oxygen. Therefore, you, or your patients, must pay the additional expenses.

1.2.3 *Decreased Quality of Service*
Because prescription oxygen has risen in cost so dramatically, few medical service companies can produce affordable EDCs and stay competitive. Since 2001, according to *Medical Digest Bulletin,* 80 percent of medical service vendors have gone out of business. Your ability to receive quality service at an affordable price has diminished.

*This and subsequent terms marked by an asterisk (*) are defined in the glossary.

1

The IEEE numbering system (1.2, 1.2.1) aids document design and allows the writer to organize detailed content.

Headings and subheadings are used to help the reader navigate the text. The writer places these headings and subheadings on the Table of Contents thus giving readers quick access to all parts of the proposal.

FIGURE 17.3 Continued

2.0 DISCUSSION

2.1 IMPLEMENTATION OF ELECTRONIC DEMAND CANNULA

Because the price of oxygen will not go down, you must try to use less while obtaining the same clinical benefits.

Orson Medical Supplies, a leader in oxygen-administering technology, proposes the implementation of our new EDC*. Using state-of-the-art electronics, the EDC senses the patient's inspiratory effort.* When a breath is detected, the EDC dispenses oxygen through the patient's cannula.* The patient receives oxygen only when he or she needs it.

Explanations and details address the audience's need to be persuaded by seeing the benefits of the proposal.

Continuously flowing cannulas waste gas during exhalation and rest. Clinical studies have proved that 50 percent of the oxygen used by cannula patients is wasted during that phase. These same tests also revealed that blood oxygen saturation* does not significantly vary between continuous and intermittent flow cannulas. The patient receives the same benefit from less oxygen. Table 1 explains this in greater detail.

Table 1
BLOOD OXYGEN SATURATION
USING THE ELECTRONIC DEMAND CANNULA
VERSUS
CONTINUOUS FLOW CANNULAS

Prescribed Flowrate (L/min)*	Breaths per Minute (bpm)*	Blood Oxygen Saturation %	
		Intermittent	Continuous
0.5	12	96%	98%
1	12	98%	99%
2	12	99%	100%
3	12	100%	100%
4	12	100%	100%

A table adds visual appeal, makes complex information easier to understand, and enhances the readability of numbers.

We have included a technical description of the EDC to help explain how this system will benefit oxygen cannula users.

2

FIGURE 17.3 Continued

2.2 TECHNICAL DESCRIPTION

The EDC* is an oxygen-administering device that is designed to conserve oxygen. The EDC is composed of six main parts: oxygen inlet connector, visual display indicators (LEDs),* power switch, patient connector, AC adapter connector, and high-impact plastic case (see Figures 1 and 2).

Line drawings combined with dimensions help the reader to visualize the equipment. In this way, the writer anticipates questions that the reader might have about the equipment.

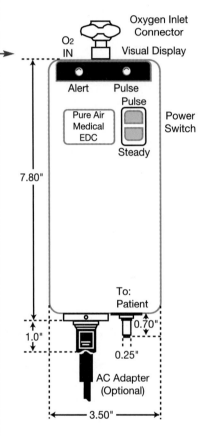

Figure 1 EDC Front View

OXYGEN INLET CONNECTOR: The oxygen inlet connector is a DISS No. 1240 (Diameter index safety system)* and is made of chrome-plated brass.

VISUAL DISPLAY: Two LEDs* provide visual indications of important functions. Alarm functions are monitored by a red LED, Motorola No. R32454. An indication of each delivered breath is given by the pulse display, which is a yellow LED, Motorola No. Y32454.

POWER SWITCH: The power switch is an ALCO No. A72-3 slide switch. The dimensions are $0.5'' \times 0.30''$: button height is $0.20''$. Electrical Specifications: Dry contact rating is 1 amp, contract resistance is 20 milliohms, and the life expectancy is 100,000 actuations.

3

FIGURE 17.3 Continued

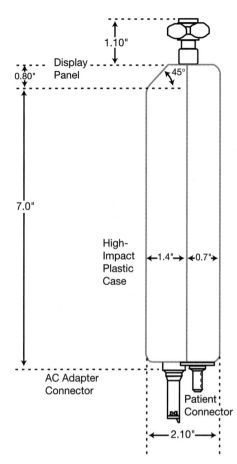

PATIENT CONNECTOR: Attachment of the patient cannula system is made at the patient connector, which is located at the bottom of the case. The white nylon connector, Air Logic No. F-3120-85, is a 10-32, UNF male threaded, straight barbed connector for 1/8″ ID flexible tubing.

AC ADAPTER CONNECTOR: An optional AC adapter* and battery charger assembly, part number PA-32, plugs into the AC adapter connector, which is located at the bottom left-hand side of the case. The connector is a male, D-subminiature, 12-pin flush insert supplied by Dupont Connector Systems. The part number is DCS: 68237009.

HIGH-IMPACT PLASTIC CASE: The case housing is made from an impact-resistant, flame-retardant, oxygen-compatible ABS plastic*.

Using technical descriptions creates a "word picture" of the equipment.

Figure 2 EDC Side View

2.3 OPERATING INSTRUCTIONS

The EDC is an oxygen-saving and administering device (see Figures 1 and 2). By following these five easy steps, you will be able to enjoy the benefits of intermittent demand oxygen.

WARNING: Federal law prohibits the sale or use of this device without the order of a physician.

4

FIGURE 17.3   Continued

1. Attach your oxygen supply to the Oxygen Inlet Connector located at the top of the case.
2. Move the Pulse-Steady Switch to the Pulse position to begin intermittent demand flow.
3. Connect your nasal cannula to the Patient Outlet Connector located at the bottom of the case.
4. Adjust your oxygen supply to the oxygen flow prescribed by your physician.
5. Put on your nasal cannula and breathe normally. The pulse light will turn on when a breath is delivered.

You are now ready to conserve oxygen by as much as 50 percent. Should you have the need to go back to continuous flow, just push the Pulse-Steady Switch to the "steady" position.

2.4 QUALIFICATIONS AND EXPERIENCE

Qualifications and experience help to persuade the reader of your company's expertise, reputation, and ability to do the job.

Orson Medical Supplies has been an international leader in the field of respiratory therapy since 1975. Pure Air introduced the first IPPB* respirator on the market. In 1985, responding to the needs of doctors and therapists, we produced the first life support volume ventilator, the VV-1. The VV-1 became the industry standard by which all other ventilators were measured.

In 1990, Orson Medical Supplies introduced the first computer-controlled life support system, the VV-2. Technology developed for this product has found application in other areas as well. Recently, we introduced one such product, the Electronic Demand Cannula.

Orson Medical Supplies is located in Overland Park, Oregon. The main manufacturing and engineering facility employs 450 people. Regional sales and service branch offices are located throughout the United States.

2.5 PERSONNEL

Each of our engineering facilities is staffed by trained technicians ready to answer your questions. The following individuals have been assigned to La Habra Retirement Center:

5

FIGURE 17.3   Continued

Randy Draper

Randy (BS, Electrical Engineering, South Central Texas University, 1992) has worked at Orson Medical Supplies since 1992. In 16 years at Orson, Randy has been promoted from service technician to supervisor. Randy specializes in developing new medical equipment. He has supervised the development teams that worked on the X29 respirator, the Z284-00 ventilator, and the Omega R-449 sphygmomanometer. Randy was lead development specialist for the Pure Air EDC.

Randy will be in charge of your account. Please contact him directly regarding any questions you might have about the EDC.

Ruth Bressette

Ruth (BS, Mechanical Engineering, Pittsburgh State University, 1997) has worked at Orson Medical Supplies since 1998. She has risen in our company from service technician to manager of troubleshooting/ maintenance. Ruth has received the highest-level certification (Master Technician) offered by the IEEE for service on every piece of equipment developed, manufactured, and sold by Orson.

Ruth will be the manager of your Orson equipment maintenance and troubleshooting crew. Her responsibility is to ensure that your equipment is kept in outstanding working condition. She will schedule maintenance checks and promptly assign technicians to troubleshoot potential malfunctions.

Douglas Loeb

Doug (AA, Electrical Engineering Technology, Plainview Community College, 1992) is one of our most accomplished troubleshooters. Having worked at Orson for 16 years, Doug is commended annually for his speed, accuracy, and skill. Your equipment is in good hands with Doug. He will be your primary troubleshooter and maintenance person.

2.6 COST

Orson Medical Supplies is pleased to offer our Pure Air EDC at cost-effective pricing. Table 2 explains the benefits you'll derive when purchasing in quantity.

6

> Provide detailed credentials and emphasize the qualifications of the team members for the audience. Such details can help persuade the audience to accept the proposal.

FIGURE 17.3 Continued

Table 2 LIST PRICE VS. DISCOUNT PRICE FOR EDC

List Price	Quantity Warranty*	Extended *per EDC*	Total Cost
$350	1–9 units	$75	$425
Discount Price			
$310	10–24 units	$60	$370
$300	25+ units	$50	$350

Visuals, such as tables and figures, break up potentially monotonous text and help to persuade the audience through accessible content.

As you can see, Orson is happy to offer you substantial savings when you purchase our Pure Air EDC in volume. At these prices, and assuming normal use, the oxygen cost savings will exceed your initial investment in less than one year, as shown in Figure 3.

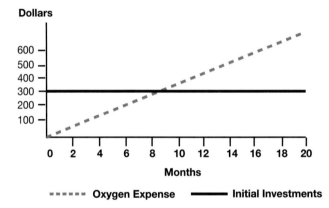

Figure 3 Return on Investment

2.7 WARRANTY

Orson Medical Supplies warrants this product to be free of manufacturing defects for a one-year period after the original date of consumer purchase. This warranty does not include damage done to the product due to accident, misuse, improper installation or operation, or unauthorized repair. This warranty also does not include replacement of parts due to normal wear.

*Orson's extended warranty is discussed in Section 2.8.

7

FIGURE 17.3 Continued

If the product becomes defective within the warranty period, we will replace or repair it free of charge.

This warranty gives you special legal rights. You may also have other rights that vary from state to state. Some states do not allow exclusions or limitations of incidental damage, so the above limitations may not apply to you.

2.8 EXTENDED WARRANTY

In addition to the coverage provided in our unconditional warranty, you might want to take advantage of our extended warranty package. For the prices provided in Table 2, Orson Medical Supplies will extend the warranty to cover a three-year period after the original date of consumer purchase. With this three-year extended warranty, Orson not only covers manufacturing defects but also replaces worn parts free of charge.

Offering the customer an extended warranty could persuade the reader of the benefits of the proposal.

The extended warranty does not cover damage to the product resulting from accident, misuse, improper installation or operation, or unauthorized repair.

For answers to any of your questions regarding repair, replacement, warranty, or extended warranty, please call 1-800-555-ORSN, or write to Manager, Customer Relations, Orson Medical Supplies, 12345 College Blvd., Overland Park, OR 90091.

3.0 CONCLUSION

3.1 MAJOR CONCERN

Prescription oxygen expenses are escalating whereas government support has been reduced. The cost increase to the patient and the health care facility will be enormous.

3.2 RECOMMENDATION

To offset the inevitable rise of oxygen expenses, we recommend the use of the Electronic Demand Cannula.

8

FIGURE 17.3 Continued

4.0 GLOSSARY

ABS plastic	Acrylonitrile butadiene styrene— a durable and long-lasting plastic
AC adapter	A remote power supply used to convert alternating current to direct current
Blood oxygen saturation	The partial pressure of oxygen in alveolar blood recorded in percent
Cannula	A small tube inserted into the nose, specifically for administering oxygen
Electronic Demand Cannula (EDC)	An electronically controlled device that dispenses oxygen only when triggered by an inspiratory effort
Inspiratory effort	The act of inhaling
Light-emitting diode (LED)	A solid-state semiconductor device that produces light when current flows in the forward direction

LIST OF ACRONYMS/ABBREVIATIONS

bpm	Breaths per minute
DISS	Diameter index safety system
EDC	Electronic Demand Cannula
IPPB	Intermittent positive pressure breathing
L/min	Liters per minute

9

Alphabetized glossary defining jargon and technical terms helps both a low-tech and a lay audience to understand the text.

Q: How do businesses use PowerPoint for proposals?

A: Many companies create proposal boilerplates using PowerPoint. Companies have found that PowerPoint is a valuable tool for writing and submitting proposals. Commerce Bank's Trust Department, for example, prepares its proposals using PowerPoint. They say that PowerPoint makes proposals easier to read as well as to write for the following reasons.

- Easy to create
- Easy to transport as a disk versus multiple-paged documents
- Easy to send either electronically or as hard-copy mail
- Easy for an audience to read and access, with only 6 to 10 lines of text
- Allows for color and animation
- Allows for easy-to-create handouts
- Useful for oral presentations

FIGURE 17.4   Proposal Written in PowerPoint

Date: June 12, 2009
To: Leann Towner, Chief Financial Officer
From: Jonathan Bacon, Information Technology Manager
Subject: Proposal for New Corporate Technology Support

Leann, in response to your request, the IT Department is happy to propose a new technology support system.

Among detailed analyses of our current system, this proposal presents the following:
- ✓ BioStaffing's current technology challenges
- ✓ Hardware purchasing suggestions
- ✓ Technology vendor recommendations

We are confident that our suggestions will maintain BioStaffing's competitive edge and increase employee satisfaction. If I can answer any questions, either call me (ext. 3625) or e-mail me at jbacon@biostaff.com.

List of Illustrations

Both figure/table numbers and titles are given for easy reference and clarity.

1.0 Abstract

An abstract often presents the problem necessitating the proposal, provides the proposed solutions, and shows the potential benefits.

1.1 Problem
BioStaffing's current technology policies are not responsive to our evolving needs, our hardware is outdated, and our current vendor is changing ownership.

1.2 Solution
To solve these problems, the Information Technology (IT) Department suggests the following:
- Upgrading our technology needs biannually rather than every five years
- Purchasing new computers, printers, scanners, and digital cameras
- Hiring a new vendor to supply and repair our hardware

1.3 Benefits
A biannual replacement schedule will allow us to stay current with hardware advancements. Purchasing new technologies will help our staff more effectively meet client needs. Hiring a new vendor is key to these goals. Our current vendor agreement ends June 30, 2009. A new vendor will provide better turnaround, pricing, maintenance, and merchandise.

2.0 Introduction

2.1 Purpose

The purpose of this proposal is to improve BioStaffing's current technology. By revising our policies, purchasing new hardware, and hiring a new technology vendor, we can increase both customer and worker satisfaction.

2.2 Problem

BioStaffing faces three key challenges:

2.2.1 *Technology Policies*
2.2.2 *Outdated Hardware*
2.2.3 *Vendor Changes*

Pronouns such as our and we are appropriate for an internal proposal written to coworkers.

2.0 Introduction cont.
Problems

2.2.1 *Technology Policies*
Since 1987, we have replaced hardware and software on a five-year, rotating basis. This was an effective policy initially since new-computer costs were expensive.

However, the current policy is no longer responsive to our needs, for the following reasons:

- <u>*Prices have gone down*</u>. Today's hardware costs are more affordable. In addition, vendors will offer BioStaffing buyer-incentives. We can buy hardware on an as-needed basis by taking advantage of special sales pricing.

  In contrast, our current five-year hardware replacement policy is not responsive. It disregards today's changes in technology pricing, and it disallows us from taking advantage of dealer incentives.

A variety of highlighting techniques, such as boldface headings, underlined subheadings, and bullets, make the text easier to read.

2.0 Introduction cont.
Problems

- <u>*Repair costs have gone up*</u>. By replacing technology only every five years, BioStaffing has resorted to repairing and retrofitting outdated equipment. Costs for these repairs have increased more than five-fold over the last ten years, as noted in the following line graph (Figure 1).

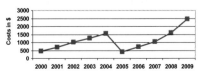

Figure 1: Increase in Costs due to Repairs and Retrofitting Per Workstation

The line graph adds clarity and provides visual appeal.

2.0 Introduction cont.
Problems

2.2.2 Outdated Hardware

We need to purchase new technology items because our current hardware and software are outdated.

- *Our computers have insufficient memory*. Most of our computers have only 2.0 GHz processors with approximately 256 MB of RAM versus the current minimal standard 3.0 GHZ and 512 MB. This negatively impacts speed of document retrieval.*

2.0 Introduction cont.
Problems

The proposal uses specific details to persuade the audience.

- *Our ink jet printers are slow and produce poor quality documents*. Our current black and white printers produce only 7 ppm with a dpi resolution of 1440 x 720.* In contrast, new black and white laser printers will produce up to 20 ppm at a dpi resolution of 5760 x 1440.

> **NOTE**: At 7 ppm (times 60 minutes per hour), BioStaffing personnel can print 420 pages per hour. Current printing requires **9.5 hours**.
>
> In contrast, at 20 ppm (times 60 minutes per hour), improved printers can produce 1200 pages per hour, requiring only **3.3 hours.**
>
> **Laser printers can save our company over 6 hours of lost time each day.**

2.0 Introduction cont.
Problems

To clarify the importance of ppm, look at Figure 2.

Figure 2: Time Spent Printing Based on PPM

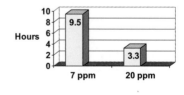

3.0 Discussion

3.1 Revised Technology Replacement Policy
We must revise BioStaffing's technology replacement policy. It is not responsive to our growing technology needs. We suggest this approach

3.1.1 Biannual Rotating Replacements

We propose that BioStaffing divide its 12 departments into two categories. The "Red" team would receive technology upgrades one year (even numbered years), while the "Blue" team would receive technology upgrades the next year (odd numbered years).

IT has divided the departments into teams alphabetically for fairness, as shown in Table 1:

3.0 Discussion cont.

Table 1: Biannual Rotating Technology Replacements	
Red Team (*even* numbered years)	**Blue Team** (*odd* numbered years)
Accounting	Manufacturing
Administrative Services	Personnel
Corporate Communication	Sales
Information Technology	Shipping and Receiving

The table, color, and white space show how formatting aids readability.

3.0 Discussion cont.

3.2. Hardware Purchases
After researching department requirements, IT determined that BioStaffing needs to upgrade business application hardware as follows:

3.2.1 Hardware Allocation Analysis
Based on survey results from each department, IT suggests that the hardware listed in Table 2 be allocated as follows:

- **Accounting**—2 desktops, 1 laptop, 5 handhelds, and 1 laser printer
- **Administrative Services**—3 desktops, 1 laptop, 1 handheld, and 1 laser printer
- **Corporate Communication**—1 laptop, 7 handhelds, 1 scanner, 1 laser printer, and 2 digital cameras
- **Information Technology**—1 laptop, 7 handhelds, 1 scanner, 1 laser printer, and 2 digital cameras
- **Manufacturing**—2 desktops, 1 laptop, 2 handheld, and 1 laser printer
- **Personnel**—2 desktops, 1 laptop, 4 handhelds
- **Sales**—2 desktops, 1 laptop, 10 handhelds, and 1 digital camera
- **Shipping and Receiving**—1 desktop, 1 laptop, 2 handhelds

Providing specific amounts shows how you will meet the reader's needs.

3.0 Discussion cont.

3.2.2 *Hardware Costs*

Table 2: Business Application Hardware Costs		
Hardware	**Individual Costs**	**Total Costs per Item**
12 Desktop Computers	$1,000 each	$12,000
8 Laptop Computers	$700 each	$5,600
38 Handheld computers	$150 each	$5,700
5 Laser printers	$400 each	$2,000
3 Scanners	$170 each	$510
5 digital cameras	$200 each	$1,000
Total Costs for all items		$26,810

3.0 Discussion cont.

Based on this scale, Figure 3 shows our evaluation findings:

Figure 3: Vendor Evaluation Matrix

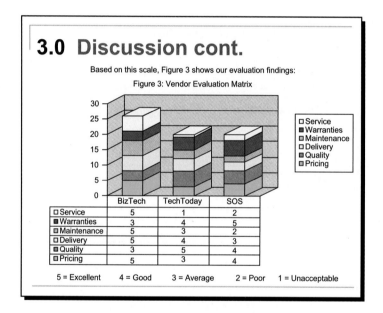

	BizTech	TechToday	SOS
☐ Service	5	1	2
■ Warranties	3	4	5
☐ Maintenance	5	3	2
☐ Delivery	5	4	3
■ Quality	3	5	4
☐ Pricing	5	3	4

5 = Excellent   4 = Good   3 = Average   2 = Poor   1 = Unacceptable

4.0 Conclusion

4.1 Conclusion

BioStaffing's technology needs are critical.

- *Repair costs* due to old hardware have risen five-fold per workstation, from as low as $500 to as high as $2,500.

- *Outdated printers*, producing only 7 pages per minute, require up to 9.5 hours a day to print BioStaffing's 4,000 pages of documentation.

- *Expired technology vendor contracts* need to be renegotiated.

Key to the above challenges is our current 5-year technology replacement policy. Technology advances demand a more responsive policy.

4.0 Conclusion cont.

4.2 Recommendation

The Information Technology Department, based on research and surveys, proposes the following:

- *A biannual technology replacement policy*—this would allow BioStaffing to meet employee technology needs more responsively as well as benefit from vendor pricing incentives.

- *New communication hardware*—this would include desktop, laptop, and handheld computers; laser printers; scanners; and digital cameras.

- *A new technology vendor*—based on our research, we suggest that BizTech Warehouse will best meet our technology needs.

4.0 Conclusion cont.

4.3 Benefits

These changes must occur before June 30, 2009, when our current vendor contract ends. By acting now, BioStaffing will benefit in several ways.

- We can save up to 20% on hardware and maintenance costs.

- We will maintain our competitive edge in the marketplace.

- Most importantly, our employee satisfaction and productivity will increase as they work with the latest technology and software upgrades.

Positive words and phrases such as "benefit," "Save," "competitive edge," "Provide employee satisfaction," and "increase productivity" help sway the audience toward the writer's stance.

5.0 Glossary

Acronym/Abbreviation	Definition
dpi	Dots per inch
GHz	Gigahertz
GB	Gigabyte
HD	Hard drive
MB	Megabyte
MHz	Megahertz
ppm	Pages per minute

Though all readers might not need terms defined, a glossary provides clarity to a less specialized audience.

CHAPTER HIGHLIGHTS

1. You might have multiple readers for a proposal, including internal and external audiences. Consider your audiences' needs. To communicate with different levels of readers, include abstracts, glossaries, and parenthetical definitions.

2. A proposal could include the following:
   - Title page
   - Cover letter
   - List of illustrations
   - Abstract
   - Introduction
   - Discussion
   - Conclusion
   - Recommendation
   - Glossary
   - Works cited (or references)
   - Appendix

3. Subheadings and visual aids will make your proposal more accessible.

4. Use primary and secondary research to develop your content.

5. Write persuasively to convince your audience to act. To accomplish this goal, arouse reader interest, refute opposing points of view, gather details to support your argument, and urge action.

6. Write ethically by documenting sources and making sure your content (prices, timelines, credentials, etc.) is accurate.

APPLY YOUR KNOWLEDGE

CASE STUDIES

1. The technical communication department at Bellaire Educational Supplies/Technologies (BEST) needs new computer equipment. Currently, the department has outdated hardware, outdated word-processing software, an outdated printer, and limited graphics capabilities. Specifically, the department is using computers with 12-inch black-and-white monitors, hard drives with only 256 KB of memory, and one 10 MB hard disk drive. The word-processing package used is WordPro 3.0, a version created in 1996. Since then, WordPro has been updated four times; the latest version is 6.5. The department printer is a black-and-white Amniprint dot matrix machine. The current word-processing package has no clip art. To create art, the department must go off-site to a part-time graphic artist who charges $35 an hour, so the department uses very few graphics.

   Because of these problems, the company's user manuals, reports, and sales brochures are being poorly reviewed by customers. Further, BEST has no Web site for product advertisement or company recognition. The bottom line: BEST is falling behind the curve, and profits are off 27 percent from last year.

   As department manager, you have consulted with your five staff members (Jim Nguyen, Mario Lozano, Mike Thurmand, Amber Badger, and Maya Liu) to correct these problems. As a team, you have decided the company needs to purchase the following new equipment.

- **Six new personal computers**—Each computer must have a 17-inch color monitor, 32 MB of memory, a 2 GB hard drive, a 12-speed CD-ROM drive, and a VGA graphics card.
- **Two laser printers**—These must have a print speed of 24 ppm, resolution of 600×600 dpi, and 4 MB of memory, expandable to 132 MB.
- **Word-processing software**—WordPro 6.5 with these capabilities: voice-activated annotations, typing, and correcting; automatic footnoting and endnoting; envelope labeling; grammar and spell checking; thesaurus; help options; automatic index generating; 50 or more scalable fonts.
- **Graphics software**—For professional-quality newsletters and brochures, BEST needs the capability for quick demonstrations and design tips; a layout checker with at least 10 online views; 20 true type fonts, each scalable; a table creation toolbar; 2,000 clip art images; and a logo creator with 50 border design options.
- **Scanner**—To increase your graphics potential, BEST also needs a flatbed scanner with these specifications: 300 dpi image resolution; 155 ppm gray-scale scanning capability; 8-bit, 256 gray-scale color support; and approximately $8 \frac{1}{2} \times 12$-inch bed size.

Using the criteria provided in this chapter, write an internal proposal to BEST's CEO, Jim McWard. In this proposal, explain the problem, discuss the solution to this problem, and then highlight the benefits derived once the solution has been implemented. These benefits will include increased productivity, better public relations, increased profits, and less employee stress. Develop these points thoroughly, and provide Mr. McWard the names of vendors for the required hardware and software. To find these vendors, you could search the Internet.

2. You own **Buzz Electronics Co.**, 4256 Crawfish Blvd., Rouex, Louisiana 65221. Mr. and Mrs. Allan Thibodeux, 3876 Spanish Moss Drive, Bayside, Louisiana 65223, have asked you to give them a bid on electrical work for a new family room they are adding to their home.

You and Mr. and Mrs. Thibodeux have gone over the couple's electrical needs, including the following: The room, which will measure 18 feet (east to west) by 15 feet (north to south), should have four 110 V outlets for three lamps, a clock, a radio, a CD and DVD player, and a television. The family wants the four 110 V outlets to be placed equidistant throughout the room.

The client wants two 220 V outlets. One 220 V outlet will go by the southwest window on the west wall where the family plans to put a window air-conditioning unit. The window will be located 3 feet in from the south wall. There will be another window on the west wall, located 3 feet in from the north wall. A third picture window, measuring 6 feet wide by 4 feet high, will be centered on the south wall. The other 220 V outlet must be placed on the east wall, where the family plans to put home office equipment (computer, printer, scanner, and fax machine). Their office desk will sit 5 feet from the door leading into the room. The door will be built on the east wall where it comes to a corner meeting the south wall.

Centered in the ceiling, the family wants electrical wiring for a fan with a light package. In addition to this light, the family also wants a light mounted on the east wall above the desk area, so wiring is needed there, approximately 5 to 6 feet up from the floor.

The family wants two light switches in the room: one by the door and one on the north wall, approximately 6 feet in from where the north and west walls meet. The Thibodeuxs plan to have a couch and lamp in that area for reading. They want both to be double switches, one to control the fan and ceiling lights and the other to control additional floor and ceiling lights in the room. All light switches need rheostats for dimming. Finally, the Thibodeuxs plan to have a whole-house vacuum system installed in the walls, and they have asked you if you can provide this service.

Buzz Electronics has been in business since 1995. The company has worked with thousands of satisfied customers, including both residential and business owners. Buzz has long-standing contracts for service with Acme Construction, J&L Builders, Food-to-Go Groceries, the City of Bayou Bend, LA, and Ross and Reed Auto Showroom.

As owner of Buzz Electronics, you have an Associate & Science degree in Electronics from Sandy Shoal Community College, Sandy Shoal, LA. You are ETA-I (Electronics Technicians Association International) Certified; NASTec (National Appliance Service Technician) Certified; and a Certified Industrial Journeyman. You have eight employees, all of whom also are Certified Industrial Journeymen.

Assignment

Write a short (3 to 5 pages) external proposal—bid for contract. To do so, study the Thibodeux's electrical needs, list the parts you will need to complete the job, estimate the time for your labor—including setup, work performed, and cleanup. Then, provide a price quote. You might need to research the wiring and equipment needed for this job, either online, in technical journals, or in parts catalogues. Include *everything* you will need to complete this job. Remember, what you write in your proposal is legally binding. You cannot add on parts and labor after the fact and expect payment or a happy customer. Follow the guidelines provided in the textbook for proposals.

INDIVIDUAL AND TEAM PROJECTS

1. Write an external proposal. To do so, create a product or a service and sell it through a long report. Your product can be an improved radon detection unit, a new fiber optic cable, safety glasses for construction work, bar codes for pricing or inventory control, an improved four-wheel steering system, computer graphics for an advertising agency, and so on. Your service may involve dog grooming, automobile servicing, computer maintenance, home construction (refinishing basements, building decks, room additions, and so on), freelance technical writing, at-home occupational therapy, or telemarketing. The topic is your choice. Draw from your job experience, college coursework, or hobbies. To write this proposal, follow the process provided in this chapter—prewrite, write a draft, and then rewrite to revise.

2. Write an internal proposal. You can select a topic from either work or school. For example, your company or department is considering a new venture. Research the prospect by reading relevant information. Interview involved participants or survey a large group of people. Once you have gathered your data, document your findings and propose to management the next course of action. If you choose a topic from school, you could propose a daycare center, on-campus bus service, improved computer facilities, tutoring services, coed dormitories, pass/fail options, and so on. Write an internal proposal to improve your company's Web site; expand or improve the security of your company's parking lot; create or improve your company's physical site security in light of post 9-11 issues; improve policies for overtime work; improve policies for hiring diversity; or improve your company's policies for promotion. Research your topic by reading relevant information or by interviewing or surveying students, faculty, staff, and administration. Once you have gathered your data, document your findings and recommend a course of action. In each instance, be sure to prewrite, write, and rewrite.

PROBLEM-SOLVING THINK PIECES

1. Stinson, Heinlein, and Brown Accounting, LLC, employs over 2,000 workers, including accountants, computer information specialists, a legal staff, paralegals, and office managers. The company requires a great deal of written and oral communication with customers, vendors, governmental agencies, and coworkers. For example, a sample of their technical communication includes the following:

   - Written reports to judges and lawyers
   - Letters and reports to customers
   - E-mail and memos to coworkers
   - Oral communication in face-to-face meetings, videoconferences, and sales presentations

   Unfortunately, not all employees communicate effectively. The writing companywide is uneven. Discrepancies in style, grammar, content, and format hurt the company's professionalism. The same problems occur with oral communication. George Hunt, a midlevel manager, plans to write an internal, unsolicited proposal to the company's principal owners, highlighting the problems and suggesting solutions. What must Mr. Hunt include in his proposal—beyond the obvious proposal components (a title page, table of contents, abstract, introduction, and so forth)—to persuade the owners to accept his suggestions? Suggest ways in which the problem can be solved.

2. Toby Hebert is sales manager at Crab Bayou Industries (Crab Bayou, Louisiana). In his position, Toby manages a sales staff of 12 employees who travel throughout Louisiana, Texas, Arkansas, and Mississippi. Currently, the sales staff use their own cars to make sales calls, and CBI pays them 31 cents per mile for travel expenses. Each staff member currently travels approximately 2,000 miles a month, with cars getting 20 miles per gallon.

Gasoline prices, at the moment, are over $2.50 a gallon. With gasoline and car maintenance costs higher than ever, the current rate of 31 cents per mile means that CBI's sales employees are losing money. Something must be done to solve this problem. Toby has met with his staff, and they have decided to write a proposal to Andre Boussaint, CBI's CEO.

What must Toby and his staff include in the proposal—beyond the obvious proposal components (a title page, table of contents, abstract, introduction, and so forth)—to persuade the CEO to accept the suggestions? Suggest ways in which the problem can be solved.

WEB WORKSHOP

1. By typing "RFP," "proposal," "online proposal," or "online RFP" in an Internet search engine, you can find tips for writing proposals and requests for proposals (RFPs), software products offered to automatically generate e-proposals and winning RFPs, articles on how to write proposals, samples of RFPs and proposals, and online RFP and proposal forms.

To perform a more limited search, type in phrases like "automotive service RFP," "computer maintenance RFP," "desktop publishing RFP," "web design RFP," and many more topics. You will find examples of both proposals and RFPs from businesses, school systems, city governments, and various industries.

To enhance your understanding of business and industry's focus on proposal writing, search the Web for information on RFPs and proposals. Using the criteria in this chapter and your knowledge of effective technical communication techniques, analyze your findings.

- How do the online proposals or RFPs compare to those discussed in this textbook, in terms of content, tone, layout, and so on?
- What information provided in the textbook is missing in the online discussions?
- What are some of the industries that are requesting proposals, and what types of products or services are they interested in?
  a. Report your findings, either in an oral presentation or in writing (e-mail message, memo, letter, or report).
  b. Imagine that you are requesting a proposal for a product or service. Create your own online form to meet this need.
  c. Rewrite any of the proposals that you think can be improved, using the criteria in this chapter as your guide.
  d. Respond to an online RFP by writing a proposal. To complete this assignment, go online to research any information you need for your content.

QUIZ QUESTIONS

1. Explain the purpose of an internal proposal.
2. Explain the purpose of an external proposal.
3. What content do you include in the introduction?
4. What is an abstract in a proposal?
5. What do you include in a purpose statement?
6. What are the typical components of the discussion section?
7. What do you include in the recommendation?
8. Explain why you include a glossary.
9. What type of visuals can you include in a proposal?
10. In what ways can you organize the different sections of your proposal?

CHAPTER 18

....

Oral Communication

COMMUNICATION *at work*

In the TechStop scenario, a customer, Carolyn Jensen, complains about malfunctioning equipment. Her complaints are not handled successfully by the sales help, so Shuan Wang, the vice president of customer service, has decided to make an oral presentation to staff to address this issue.

TechStop, with 12 locations throughout the state, sells DVDs, VCRs, TVs, audio components, computers, and computer peripherals. Shuan Wang is the vice president

of customer service. Lately, Shuan has been receiving complaints from customers about poor service.

For example, one customer, Carolyn Jensen, owns a large electronics company. She has done business with Tech-Stop for six years. During that time, Carolyn's company has purchased over 1,000 pieces of equipment, including computers, printers, paper items, cell phones, pagers, and radios for a fleet of trucks.

Carolyn called a local TechStop to complain about problems her company was having with over a dozen printers. Her company's text looks fine on screen but hard copies print differently than they look. The company cannot print three different sized envelopes (as had been promised by the salesperson when Carolyn purchased the printers). Graphics are not printing in

Objectives

When you complete this chapter, you will be able to

1. Communicate effectively with customers, vendors, and coworkers.
2. Use the telephone to communicate successfully.
3. Use voice mail effectively.
4. Deliver effective informal oral presentations on the job.
5. Participate in teleconferences.
6. Communicate effectively in a videoconference or webconference.
7. Deliver formal oral presentations.
8. Use a variety of visual aids to enhance your oral presentations, including techniques for effective PowerPoint presentations.
9. Use the writing process—prewriting, writing, and rewriting—for informal and formal oral presentations.

---

color, though the color ink cartridges are full. Finally, printers are stopping their print jobs before all pages of a document have been printed. She purchased the printers 12 months and two weeks ago. The printer's in-store warranty was for 12 months, guaranteeing full replacement of parts and labor coverage.

Carolyn realizes that the warranty expired two weeks ago. However, she contends that several facts should negate this deadline. The deadline expired over a Christmas weekend, a severe snowstorm left many homes and businesses without power for several days, and other, more pressing business activities required her attention.

When she informed the sales help of her situation and her company's long-standing patronage of TechStop, the salesperson said, "Sorry, lady. The warranty's just no good anymore. Hey, we've all got problems. Anyway, there's no way my boss will even listen to you with this complaint. He told me to leave him alone when he's busy."

Unfortunately, Shuan has heard of other such complaints regarding rude and unresponsive sales personnel. Shuan fears that TechStop's staff is developing an overall corporate disregard for customer satisfaction.

To address this issue for the entire sales staff at TechStop's numerous locations throughout the state, Shuan plans to give a formal oral presentation on the following:

- Sales etiquette and customer interaction
- Store policy regarding customer satisfaction
- The impact on poor customer relations
- The consequences of failing to handle customers correctly

Shuan only hopes that his actions aren't too late.

Check Online Resources

www.prenhall.com/gerson
For more information about oral communication, visit our companion Web site.

Check out our quarterly newsletters TechCom at E-Notes www.prenhall.com/gerson for dot.com updates, new case studies, insights from business professionals, grammar exercises, and facts about technical communication.

THE IMPORTANCE OF ORAL COMMUNICATION

Many people, even the seemingly most confident, are afraid to speak in front of others. A recent Monster.com poll asked, "What is your biggest career-related phobia?" Here are the results.

Percentage	Phobia
42%	Giving a speech or presentation
32%	Confronting a coworker or boss
15%	Networking
11%	Writing a report or proposal

(*Monster Career Advice Newsletter*, 2006)

This chapter offers techniques to make your oral communication experience rewarding rather than frightening.

You may have to communicate orally with your peers, your subordinates, your supervisors, and the public. Oral communication is an important component of your business success because you will be required to speak formally and informally on an everyday basis.

EVERYDAY ORAL COMMUNICATION

"Hi. My name is Bill. How may I help you?" Think about how often you have spoken to someone today or this week at your job. You constantly speak to customers, vendors, and coworkers face to face, on the telephone, or by leaving messages on voice mail.

- If you work an 800 hotline, your primary job responsibility is oral communication.
- When you return the dozens of calls you receive or leave voice-mail messages, each instance reveals your communication abilities.
- As an employee, you must achieve rapport with your coworkers. Much of your communication to them will be verbal. What you say impacts your working relationships.

Every time you communicate orally, you reflect something about yourself and your company. The goal of effective oral communication is to ensure that your verbal skills make a good impression and communicate your messages effectively.

DOT-COM UPDATES

For more information about using voice mail effectively, check out the following link.

- "Top 10 Tips for Using Voice Mail Effectively" http://www.sideroad. com/Business_Etiquette/ voicemail-greeting.html

Telephone and Voice Mail

You speak on the telephone dozens of times each week. When speaking on a telephone, make sure that you do not waste either your time or your listener's time. See page 565 for 10 tips for telephone and voice-mail etiquette.

Informal Oral Presentations

As a team member, manager, supervisor, employee, or job applicant, you often will speak to a coworker, a group of colleagues, or a hiring committee. You will need to communicate orally in an informal setting for several reasons.

- Your boss needs your help preparing a presentation. You conduct research, interview appropriate sources, and prepare reports. When you have concluded your research, you might be asked to share your findings with your boss in a brief, informal oral presentation.

- Your company is planning corporate changes (staff layoffs, mergers, relocations, or increases in personnel). As a supervisor, you want to provide your input in an oral briefing to a corporate decision maker.

- At a departmental meeting, you are asked to report orally on the work you and your subordinates have completed and to explain future activities.

- In a team meeting, you participate in oral discussions regarding agenda items.

- Your company is involved in a project with coworkers, contractors, and customers from distant sites. To communicate with these individuals, you participate in a teleconference, orally sharing your ideas.

- You are applying for a job. Your interview, though not a formal, rehearsed presentation, requires that you speak effectively before a hiring committee.

- You are working from a remote location but need to participate in company-sponsored training. You cannot physically attend the workshop, but through a webconference (or webinar), you and other employees from around the country can be trained simultaneously on your computers.

These informal oral communication instances could be accomplished best through video-conferences, teleconferences, and webconferences (webinars).

Video and Teleconferences.   If you are communicating with a group and you want to hear what everyone else is saying simultaneously, video or teleconferences are an answer. Consider using a video or teleconference when three or more people at separate locations need to talk.

In a video or teleconference, you want all participants to feel as if they are in the same room facing each other. With expensive technology in place, such as cameras, audio components, coder/decoders, display monitors, and user interfaces, you want to avoid wasting time and money with poor communication.

Ten Tips for Video and Teleconferences

1. Inform participants of the conference date, time, time zone, and expected duration.

2. Make sure participants have printed materials before the teleconference.

3. Ensure that equipment has good audio quality.

4. Choose your room location carefully for quiet and privacy.

5. Consider arrangements for hearing impaired participants. You might need a text telephone (TTY) system or simultaneous transcription in a chat room.

6. Introduce all participants.

7. Direct questions and comments to specific individuals.

8. Do not talk too loudly, too softly, or too rapidly.

9. Turn off cell phones and pagers.

10. Limit side conversations.

Webconferencing.   Due to rising costs of airline tickets, time-consuming travel, and the need to complete projects or communicate information quickly, many companies and organizations are using webconferencing. Webconferencing, sometimes called webinars, e-seminars, webcasts and teleclasses, or virtual meetings, include web-based seminars, lectures, presentations, or workshops transmitted over the Internet (Coyner 2007; Murray 2004).

Ten Tips for Webconferencing

1. Limit webconferences to 60 to 90 minutes.

2. Limit a webconference's focus to three or four important ideas.

3. Start fast by limiting introductory comments.

4. Keep it simple. Instead of using too many Internet tools, limit yourself to simple and important features like polling and messaging.

5. Plan ahead. Make sure that all webconference participants have the correct Internet hardware and software requirements; know the correct date, time, agenda, and Web login information for your webconference; and have the correct Web URL or password.

6. Before the webconference, test your equipment, hypertext links, and PowerPoint slide controls.

7. Use both presenter and participant views. One way to ensure that all links and slides work is by setting up two computer stations. Have a computer set on the presenter's view and another computer logged in as a guest. This will allow you to view accurately what the audience sees and how long displays take to load.

8. Involve the audience interactively through questions and/or text messaging.

9. Personalize the presentation. Introduce yourself, other individuals involved in the presentation, and audience members.

10. Archive the presentation.

Formal Presentations

You might need to make a formal presentation for the following reasons.

- Your company asks you to visit a civic club meeting and to provide an oral presentation to maintain good corporate or community relations.
- Your company asks you to represent it at a city council meeting. You will give an oral presentation explaining your company's desired course of action or justifying activities already performed.
- Your company asks you to represent it at a local, regional, national, or international conference by giving a speech.
- A customer has requested a proposal. In addition to writing this long report, you and several coworkers also need to make an oral presentation promoting your service or product to the potential customer.

Types of Formal Oral Presentations.   Three types of formal oral presentations are as follows:

1. **The Memorized Speech**—The least effective type of oral presentation is the memorized speech. This is a well-prepared speech which has been committed to memory. Although such preparation might make you feel less anxious, too often these speeches sound mechanical and impersonal. They are often stiff and formal, and allow no speaker–audience interaction.

2. **The Manuscript Speech**—In a manuscript speech, you read from a carefully prepared manuscript. The entire speech is written on paper. This may lessen speaker anxiety and help you to present information accurately, but such a speech can seem monotonous, wooden, and boring to the listeners.

3. **The Extemporaneous Speech**—Extemporaneous speeches are probably the best and most widely used method of oral communication. You carefully prepare your oral presentation by conducting necessary research, and then you create a detailed outline. However, you avoid writing out the complete presentation. When you make your presentation, you rely on notes or PowerPoint slides with the major and minor headings for reference. This type of presentation helps you avoid seeming dull and mechanical, allows you to interact with the audience, and still ensures that you correctly present complex information. Figure 18.1 shows the oral presentation process.

FIGURE 18.1   The Oral Presentation Process (adapted from "Communications: Oral Presentations")

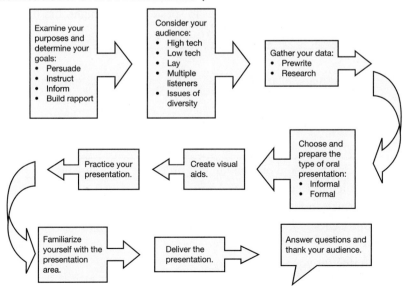

PARTS OF A FORMAL ORAL PRESENTATION

A formal oral presentation consists of an introduction, a discussion (or body for development), and a conclusion.

Introduction

The introduction should welcome your audience, clarify your intent by providing a verbal "road map," and capture your audience's attention and interest. This is the point in the presentation where you are drawing in your listener, hoping to create enthusiasm and a positive impression.

To create a positive impression, set the table. Address your audience politely by saying "Good morning" or "Good afternoon." Tell the audience your name and the names of others who might be speaking. Welcome the audience and thank them for inviting you to speak.

- Table Setters for Goodwill

example	"Good evening. Thank you for allowing us to speak to you tonight. I'm (name and title). You'll also hear from (name and title)."

Next, provide a "road map," clarifying what points you'll discuss and laying out the order of the topics.

- Road Map (thesis statement)

example	"What we are going to talk about, and in this order, are the following key points: • the issues that led us to consider road improvements • challenges to this construction task • optional approaches • costs • and a timeframe for your consideration."

You can use a variety of openings to capture your audience's attention, such as the following:

- Word Pictures—Anecdotes (specific in *time, place, person,* and *action*), Questions, Quotes, and Data (facts and figures)

Word Pictures and Data to Arouse Your Audience's Interest example

"From November 2006 to February 2007, our police department received over 75 reports of problems regarding Elm Street. These ranged from injuries related to hill jumping, cars sliding into the street's ditches, increased rush hour traffic, and tight turn lanes. One stretch of the road, from grid 39 to grid 47, is too narrow for snow removal crews to clean effectively. And these issues promise to get worse with residential growth anticipated to increase by 39%. As Sgt. Smith of the police department stated, 'Elm Street is a disaster waiting to happen.'"

- A Question or a Series of Questions—Asking questions involves the audience immediately. A training facilitator could begin a workshop as follows:

Questions to Involve Your Audience example

"How many reports do you write each week or month? How often do you receive and send e-mail messages to customers and colleagues? How much time do you spend on the telephone? Face it; technical communication is a larger part of your engineering job than you ever imagined."

These three questions are both personalized and pertinent. Through the use of the pronoun *you*, the facilitator speaks directly to each individual in the audience. By focusing on the listeners' job-related activities, the questions directly lead into the topic of conversation—technical communication.

- A Quotation from a Famous Person—The training facilitator in the previous example could have begun his speech with a quote from Warren Buffet, a famous businessperson.

A Quote to Arouse Audience Interest example

"How important is effective technical communication? Just listen to what Warren Buffet has to say on the topic:

'For more than forty years, I've studied the documents that public companies file. Too often, I've been unable to decipher just what is being said or, worse yet, had to conclude that nothing was being said . . . Perhaps the most common problem . . . is that a well-intentioned and informed writer simply fails to get the message across to an intelligent, interested reader. In that case, stilted jargon and complex constructions are usually the villain . . . When writing Berkshire Hathaway annual reports, I pretend that I'm talking to my sisters. I have no trouble picturing them: Though highly intelligent, they are not experts in accounting or finance. They will understand plain English, but jargon may puzzle them. My goal is simply to give them the information I would wish them to supply me if our positions were reversed. To succeed, I don't need to be Shakespeare; I must, though, have a sincere desire to inform.'

That's what I want to impart to you today: good writing is communication that is easy to understand. If simple language is good enough for Mr. Buffet, then that should be your goal."(Buffet 1-2)

Discussion (or Body)

After you have aroused your listeners' attention and clarified your goals, you have to prove your assertions. In the *discussion* section of your formal oral presentation, provide details to support your thesis statement. You can develop your content in a variety of ways, including the following:

Research

See Chapter 5 for more discussion of research.

- **Quotes, Testimony, Anecdotes**—You can find this type of information through primary and secondary research. For example, primary research, such as a survey or questionnaire, will help you substantiate content. By using people's direct comments, you help your audience relate to the content. You also can find quotes, testimony, or anecdotes in secondary research, such as periodicals, newspapers, books, or online. This type of information validates your comments.
- **Data**—Facts and figures, again found through research or interviews, develop and support your content. Stating exactly how often your company's computer system has been attacked by viruses will support the need for improved firewalls. A statistical analysis of the increase in insurance premiums will clarify the need for a new insurance carrier. Showing the exponential increase in the cost of gasoline over the last 10 years will show why your company should consider purchasing hybrids.

To help your audience follow your oral communication, present these details using any of the following modes of organization.

Methods of Organization

See Chapter 3 for more discussion of organization techniques.

Comparison/Contrast—In your presentation, you could compare different makes of office equipment, employees you are considering for promotion, different locations for a new office site, vendors to supply and maintain your computers, different employee benefit providers, and so forth. Comparison/contrast is a great way to make value judgments and provide your audience options.

Problem/Solution—You might develop your formal oral presentation by using a problem-to-solution analysis. For example, you might need to explain to your audience why your division needs to downsize. Your division has faced problems with unhappy customers, increased insurance premiums, decreased revenues, and several early retirements of top producers. In your speech, you can then suggest ways to solve these problems ("We need to downsize to lower outgo and ultimately increase morale"; "Let's create a 24-hour, 1-800 hotline to answer customer concerns"; "We should compare and contrast new employee benefits packages to find creative ways to lower our insurance costs").

Persuasive Communication

See Chapter 10 for more discussion of persuasive communication.

Argument/Persuasion—Almost every oral presentation has an element of argument/persuasion to it—as does all good written communication. You will usually be persuading your audience to do something based on the information you share with them in the presentation.

Importance—Prioritizing the information you present from least to most important (or most important to least) will help your listeners follow your reasoning more easily. To ensure the audience understands that you are prioritizing, provide verbal cues. These include simple words like *first, next, more important*, and *most important*. Do not assume that these cues are remedial or obvious. Remember, sometimes it is hard to follow a speaker's train of thought. Good speakers realize this and give the audience verbal signposts, reminding the listeners exactly where they are in the oral presentation and where the speaker is leading them.

Chronology—A chronological oral presentation can outline for your audience the order of the actions they need to follow. For example, you might need to prepare a

TABLE 18.1 The Purpose of Transitional Words

Purpose	Examples
Cause and effect	because, since, thus, therefore, for this reason, due to this, as a result of, consequently, in order to
Example	for instance, for example, another
Interpreting jargon	that is, in other words, more commonly called
Sequencing ideas	first, second, next, last, following, finally, above, below
Adding a point	furthermore, next, in addition, besides, not only . . . but also, similarly, likewise
Restating	in other words, that is, again, to clarify
Contrasting	but, instead, yet, however, on the other hand, nevertheless, in contrast, on the contrary, whereas, still
For emphasis	in fact, more importantly, clearly
Summarizing	to summarize, therefore, in summary, to sum up, consequently

yearly evaluation of all sales activities. Provide your audience with target deadlines and with the specific steps they must follow in their reports each quarter.

Maintaining Coherence—To maintain coherence, guide your audience through your speech as follows:

- **Use clear topic sentences**—Let your listeners know when you are beginning a new, key point: "Next, let's talk about the importance of conciseness in your technical writing."

- **Restate your topic often**—Constant restating of the topic is required because listeners have difficulty retaining spoken ideas. A reader can refer to a previously discussed point by turning back a page or two. Listeners do not have this option.

   Furthermore, a listener is easily distracted from a speech by noises, room temperature, uncomfortable chairs, or movement inside and outside the meeting site. Restating your topic helps your reader maintain focus. "Repeat major points. Reshow visuals, repeat points and ideas several times during your presentation. Put them in your summary, too" (O'Brien 2003).

- **Use transitional words and phrases**—This helps your listeners follow your speech. Transitional words and phrases, like those shown in Table 18.1, aid reader comprehension, emphasize key points, and highlight your speech's organization.

Conclusion

Conclude your speech by restating the main points, by recommending a future course of action, or by asking for questions or comments. A polite speaker leaves time for a few follow-up questions from the audience. Gauge your time well, however. You do not want to bore people with a lengthy discussion after a lengthy speech. You also do not want to cut short an important question-and-answer (Q and A) session. If you have given a controversial speech that you know will trouble some members of the audience, you owe them a chance to express their concerns.

VISUAL AIDS

Most speakers find that visual aids enhance their oral communication. "Visuals have the greatest, longest lasting impact—show as much or more visually as what you say. Use pictures; use color. Use diagrams and models" (O'Brien 2003).

Although PowerPoint slide shows, graphs, tables, flip charts, and overhead transparencies are powerful means of communication, you must be the judge of whether visual aids will enhance your presentation. Avoid using them if you think they will distract from your presentation or if you lack confidence in your ability to create them and integrate them effectively. However, with practice, you probably will find that visuals add immeasurably to the success of most presentations.

Table 18.2 lists the advantages, disadvantages, and helpful hints for using visual aids. For all types of visual aids, practice using them before you actually make your presentation. When you practice your speech, incorporate the visual aids you plan to share with the audience.

TABLE 18.2 Visual Aids—Advantages and Disadvantages

Type	Advantages	Disadvantages	Helpful Hints
Chalkboards	Are inexpensive. Help audiences take notes. Allow you to emphasize a point. Allow audiences to focus on a statement. Help you be spontaneous. Break up monotonous speeches.	Make a mess. Can be noisy. Make you turn your back to the audience. Can be hard to see from a distance. Can be hard to read if your handwriting is poor.	Clean the board well. Have extra chalk. Stand to the side as you write. Print in large letters. Write slowly. Avoid talking with your back to the audience. Don't erase too soon.
Chalkless Whiteboards	Same as above.	Are expensive. Require unique, erasable pens. Can stain clothing. Some pens can be hard to erase if left on the board too long. Pens that run low on ink create light, unreadable impressions.	Use blue, black, or red ink. Cap pens to avoid drying out. Use pens made especially for these boards. Erase soon after use.
Flip Charts	Can be prepared in advance. Are neat and clean. Can be reused. Are inexpensive. Are portable. Help you avoid a nonstop presentation. Allow for spontaneity. Help audiences take notes. Allow you to emphasize key points. Encourage audience participation. Allow easy reference by turning back to prior pages. Allow highlighting with different colors.	Are limited by small size. Require an easel. Require neat handwriting. Won't work well with large groups. You can run out of paper. Markers can run out of ink.	Have two pads. Have numerous markers. Use different colors for effect. Print in large letters. Turn pages when through with an idea so audience will not be distracted. Don't write on the back of pages where print bleeds through.

Overhead Transparencies	Are inexpensive. Can be used with lights on. Can be prepared in advance. Can be reused. Can be used for large audiences. Allow you to return to a prior point. Allow you to face audiences.	Require an overhead projector. Can be hard to focus. Require an electrical outlet and cords. Can become scratched and smudged. Can be too small for viewing. Bulbs burn out.	Use larger print. Protect the transparencies with separating sheets of paper. Frame transparencies for better handling. Turn off the overhead to avoid distractions. Focus the overhead before beginning your speech. Keep spare bulbs. Don't write on transparencies. Face the audience.
Slides (Slide Shows)	Are portable. Slides are easy to protect. Can be used for large groups. Can be prepared in advance. Are entertaining and colorful. Allow for later reference.	Require dark rooms. Hurt speaker–audience interaction (eye contact). Can be expensive. Can be challenging to create. Require screen, machinery, and electrical outlets. Slides can get out of order. Dark room makes taking notes challenging. Machinery can malfunction.	Use a pointer. Use a remote control for freedom of movement. Check slides to make sure none are out of order. Check working condition of the equipment.
Videotapes	Allow instant replay. Can be freeze-framed for emphasis. Can be economically duplicated. Can be rented or leased inexpensively. Are entertaining.	Require costly equipment (monitor and recorder). Are bulky and difficult to move. Can malfunction. Require dark rooms. Require compatible equipment. Deny easy note taking. Deny speaker–audience interaction.	Practice operating the equipment. Avoid long tapes.
Films	Are easy to use. Are entertaining. Offer many choices. Can be used for large groups.	Require dark rooms. Require equipment and outlets. Make note taking difficult. Deny speaker–audience interaction. Can malfunction and become dated.	Use up-to-date films. Avoid long films. Provide discussion time. Use to supplement the speech, not replace it. Practice with the equipment.
PowerPoint Presentations	Are entertaining. Offer flexibility, allowing you to move from topics with a mouse click. Can be customized and updated. Can be used for large groups. Allow for speaker–audience interaction. Can be supplemented with handouts easily generated by PPt. Can be prepared in advance. Can be reused. Allow you to return to a prior slide. Can include animation and hyperlinks.	Require computers, screens, and outlets. Work better with dark rooms. Computers can malfunction. Can be too small for viewing. Can distance the speaker from the audience.	Practice with the equipment. Bring spare computer cables. Be prepared with a backup plan if the system crashes. Have the correct computer equipment (cables, monitors, screens, etc.). Make backup transparencies. Practice your presentation.

POWERPOINT PRESENTATIONS

One of the most powerful oral communication tools is visual—Microsoft PowerPoint (PPt). Whether you are giving an informal or formal oral presentation, your communication will be enhanced by PowerPoint slides.

Today, you will attend very few meetings where the speakers do not use PowerPoint slides. PowerPoint slides are used frequently because they are simple to use, economical, and transportable. Even if you have never created a slide show before, you can use the templates in the software and the autolayouts to develop your own slide show easily. An added benefit of PowerPoint slides is that you can print them and create handouts for audience members.

FAQs

Q: Why is the use of PowerPoint so important in the workplace?
A: Widely used by businesspeople, educators, and trainers, PowerPoint is among the most prevalent forms of presentation technology. "According to Microsoft, 30 million PowerPoint presentations take place every day: 1.25 million every hour" (Mahin 2004). Employees in education, business, industry, technology, and government use PowerPoint not only for oral presentations but also as hard-copy text.

Q: Does everyone like PowerPoint? Aren't there any negative attitudes toward this technology?
A: Not everyone likes PowerPoint. Opposition to the use of this technology, however, usually stems from the following problems.

- Dull PowerPoint slides, lacking in variety and interest
- The use of Microsoft's standardized templates
- "Death by Bullet Point," caused by an excessive dependence on bullets

However, these challenges can be overcome easily through techniques discussed in this chapter.

Benefits of PowerPoint

When you become familiar with Microsoft PowerPoint, you will be able to achieve the following benefits.

1. Choose from many different presentation layouts and designs.

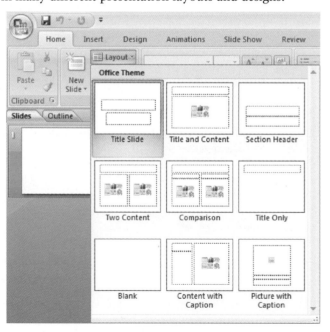

2. Create your own designs and layouts, changing colors and color schemes from preselected designs.

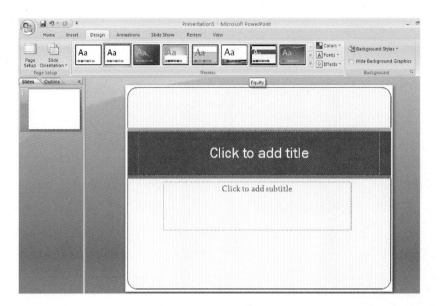

3. Add, delete, or rearrange slides as needed. By left-clicking on any slide, you can copy, paste, or delete it. By left-clicking between any of the slides, you can add a new blank slide for additional information.

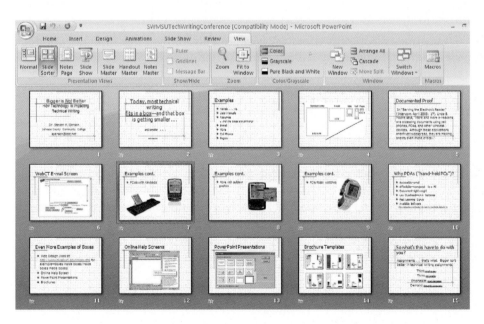

4. Insert art from the Web, add images, or create your own drawings.

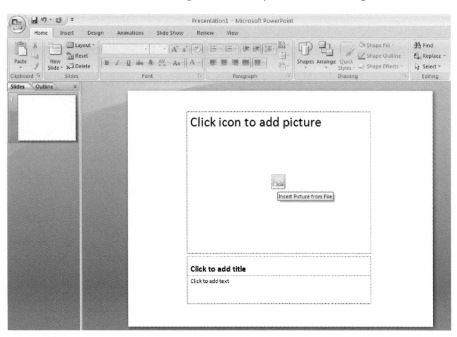

5. Add hyperlinks either to slides within your PowerPoint presentation or to external Web links.

Tips for Using PowerPoint

To make it easy for you to add PowerPoint slides to your presentations, consider the following hints.

1. **Create optimal contrast**—Use dark backgrounds for light text or light backgrounds for dark text. Avoid using red or green text (individuals who are color blind cannot see these colors). You should use color for emphasis only.

2. **Choose an easy-to-read font size and style**—Use common fonts, such as Times New Roman, Courier, or Arial. Arial is considered to be the best to use because sans serif fonts (those without feet) show up best in PowerPoint. Use no more than three font sizes per slide. Use at least a 24-point font size for text and 36-point font for headings.

3. **Limit the text to six or seven lines per slide and six or seven words per line**—Think 6 × 6. Two or more short, simple lines of text are better than one slide with many words. Also, use no more than 40 characters per line (a character is any letter, punctuation mark, or space). You can accomplish these goals by creating a screen for each major point discussed in your oral presentation.

4. **Use headings for readability**—To create a hierarchy of headings, use larger fonts for a first-level heading and smaller fonts for second-level headings. Each slide should have at least one heading to help the audience follow your thoughts.

5. **Use emphasis techniques**—To call attention to a word, phrase, or idea, use color (sparingly), boldface, all caps, or arrows. Use a layout that includes white space. Include figures, graphs, pictures from the Web, or other line drawings.

6. **End with an obvious concluding screen**—Often, if speakers do not have a final screen that *obviously* ends the presentation, the speakers will click to a blank screen and say, "Oh, I didn't realize I was through," or "Oh, I guess that's it." In contrast, an obvious ending screen will let you as the speaker end graciously—and without surprise.

7. **Prepare handouts**—Give every audience member a handout, and leave room on the handouts for note taking.

8. **Avoid reading your screens to your audience**—Remember they can read and will become quickly bored if you read slides to them. Speakers lose their dynamism when they resort to reading slides rather than speaking to the audience.

9. **Elaborate on each screen**—PowerPoint should not replace you as the speaker. In contrast, PowerPoint should add visual appeal, while you elaborate on the details. Give examples to explain fully the points in your presentation.

10. **Leave enough time for questions and comments**—Instead of rushing through each slide, leave sufficient time for the audience to consider what they have seen and heard. Both during and after the PowerPoint presentation, give the audience an opportunity for input.

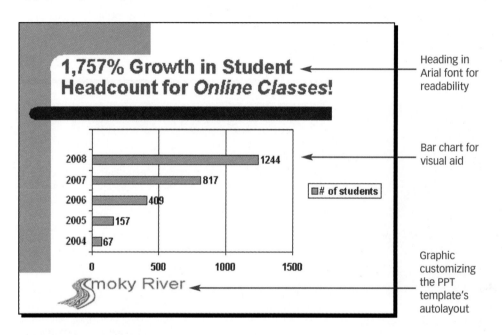

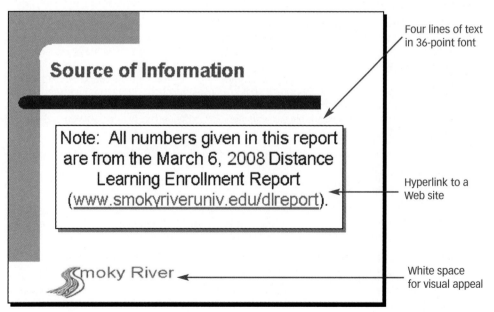

PowerPoint design template with autolayout and customized color →

Student Comments

"I could never have completed my degree program without online classes. They were fun and educational."

"DL gave me the chance to get college credit while working full time."

Smoky River

Photograph to add interest

How Is PowerPoint Used in Business?

Mark Uhlig is Chairperson and CEO of Uhlig Communications, Inc. This advanced-technology publishing company specializes in the management and creation of marketing and business documents for a diverse clientele, ranging from apartment and condominium communities to individual businesses.

Mark has to speak formally and informally every day to employees, customers, investors, and strategic partners. In many instances, Mark says, "Oral communication is superior to written correspondence. It's quicker, it requires less preparation, and it's more persuasive since you get immediate feedback from your audience and, thus, can react and adjust your message. The challenge," notes Mark, "is to speak with the same deliberation and structure as you would when writing."

That's where PowerPoint (PPt) comes in. According to Mr. Uhlig, PPt is the perfect hybrid, fusing the benefits of written and oral communication. With PPt, "you can talk from an outline or an agenda. You can prepare talking points to help you stay on message. Thus, you have the power of oral persuasion enhanced by the structure of written communication." In addition, PPt slides "allow the audience to use the printed slides as a record of the conversation, ensuring that their recollections are accurate." Finally, people happily will read PowerPoint slides "because they are short and can be accessed at a glance," Mark says.

Because PPt has become "the standard currency" for oral presentations, Mark uses PPt slides when speaking to his employees about their 401Ks, to investors about Uhlig Communications' financial performance, and to customers about the benefits of his products.

Mark sketched the following table to clarify how PPt is the perfect hybrid between oral and written communication.

Oral Communication		PowerPoint	Written Communication	
+	—	Easy	+	—
Fast	Hard for audience	Structured	Allows for	Takes time and
Easy	to recall key ideas	Combines	unlimited detail	effort to read
Persuasive—		oral persuasion	Allows for	Disallows give
allowing for	Heard in real time	and written record	random access	and take
give and	Not good for	Printed notes		
take	complex facts/figures	help recall		

PPt is just one of the many tools Mark employs in his publishing company to meet the needs of customers and other stakeholders. Clear communication is his goal, and PPt addresses this need.

PowerPoint Slides Checklist

_____ 1. Does the presentation include headings for each slide?

_____ 2. Have you used an appropriate font size for readability?

_____ 3. Did you choose an appropriate font type for readability?

_____ 4. Did you limit yourself to no more than three different font sizes per screen?

_____ 5. Has color been used effectively for readability and emphasis, including font color and slide background?

_____ 6. Did you use special effects effectively versus overusing them?

_____ 7. Have you limited text on each screen (remembering the 6 × 6 rule)?

_____ 8. Did you size your graphics correctly for readability, avoiding ones that are too small or too complex?

_____ 9. Have you used highlighting techniques (arrows, color, white space) to emphasize key points?

_____10. Have you edited for spelling and grammatical errors?

THE WRITING PROCESS AT WORK

As with writing memos, letters, e-mail messages, and reports, approach your oral communication project as a step-by-step process. Doing so will allow you to express yourself with confidence. Follow the writing process to organize your presentation effectively.

The Writing Process

Prewriting	Writing	Rewriting

• Decide whether your presentation will be formal or informal.
 • Decide whether you are writing to inform, instruct, persuade, or build rapport.
 • Determine whether your audience is high tech, low tech, or lay.
 • Gather information through primary and secondary research.
	• Prepare outlines or note cards.
	 • Organize your content using modes such as problem/solution, cause/effect, comparison, argument/persuasion, analysis, or chronology.
	 • Use visual aids to emphasize key points.
		• Consider all aspects of style, delivery, appearance, and body language and gestures.
		 • Practice in front of peers or colleagues for their input.
		 • Revise your content by
		   • adding details
		   • deleting wordiness
		   • simplifying words
		   • enhancing the tone
		   • reformatting your text
		   • proofreading and correcting errors

Prewriting

Prewriting gets you started with your presentation. Similar to prewriting for written communication, when you prewrite for a speech, you should do the following:

Consider the Purpose. Determine why you are making an oral presentation. Ask yourself the following questions.

- Are you selling a product or service to a client?
- Do you want to inform your audience of the features in your newly created software?

- Are you persuading your audience to increase corporate spending to enhance a benefits package?
- Are you giving a speech for one of your college classes?
- Has your boss asked for your help in preparing a presentation? After you research the content, will you have to present your information orally?
- Are you a supervisor justifying workforce cuts to your division?
- Did a customer request information on solutions to a problem?
- Are you representing your company at a conference by giving a speech?
- Are you running for an office on campus and giving a speech about your candidacy?
- At the division meeting, are you reporting orally on the work you and your team have completed and the future activities you plan for the project?

Determining the purpose of a presentation will ensure that you choose the appropriate content.

Inform—For example, when your speech is to *inform*, you want to update your listeners. Such a speech could be about new tax laws affecting listeners' pay, new management hirings or promotions, or budget constraints or cutbacks. Speeches that inform do not necessarily require any action on the part of your audience. Your listeners cannot alter tax laws, change hiring or promotion practices, or prevent cutbacks. The informative speech keeps your audience up to date.

Persuade—On the other hand, some speeches *persuade*. You will speak to motivate listeners. For instance, you might give an oral presentation about the need to hold more regular and constructive meetings. You might tell your audience that teamwork will enhance productivity. Maybe you are giving an oral presentation about the value of quality controls to enhance product development. In each instance, you want your audience to leave the speech ready and inspired to act on your suggestions.

Instruct—You might speak to *instruct*. In an instruction, you will teach an audience how to follow procedures. For instance, you could speak about new sales techniques, ways to handle customer complaints, implementation of software, manufacturing procedures, or how to prepare for on-campus interviews. When you instruct, your goal is both to inform and persuade. You will inform your audience how to follow steps in the procedure. In addition, you will motivate them, explaining why the procedure is important.

Build Trust—Finally, you might give an oral presentation to *build trust*. Let's say you are speaking at an annual meeting. Your goal not only might be to inform the audience of your company's status, but also to instill the audience with a sense of confidence about the company's practices. You could explain that the company is acting with the audience's best interests in mind. Similarly, in a departmental meeting, you might speak to build rapport. As a supervisor, you will want all employees to feel empowered and valued. Speaking to build trust will accomplish this goal.

Audience

See Chapter 4 for more discussion of audience.

Consider Your Audience.   When you plan your oral presentation, consider your audience. Ask yourself questions such as the following:

- Are you speaking to a high-tech, low-tech, or lay audience?
- Are you speaking up to supervisors?
- Are you speaking down to subordinates?
- Are you speaking laterally to peers?
- Are you speaking to the public?
- Are you addressing multiple audience types (supervisors, subordinates, and peers)?
- Is your audience friendly and receptive or hostile?

FIGURE 18.2 Oral Presentation Plan

Presentation Plan

Topic: _____

Objectives:

- What do you want your audience to believe or do as a result of your presentation?
- Are you trying to persuade, instruct, inform, build trust, or combinations of the above?

Development: What main points are you going to develop in your presentations?

1.

2.

3.

Organization: Will you organize your presentation using *analysis, comparison/contrast, chronology, importance,* or *problem/solution?*

Visuals: Which visual aids will you use?

- Are you speaking to a captive audience (one required to attend your presentation)?
- Is your audience diverse in terms of culture, gender, or age?
- Will you need translators for those with hearing impairments?

Considering your audience's level of knowledge and interest will help you prepare your presentation. You should consider whether your audience needs terms defined and what tone you should take. You cannot communicate effectively if your audience fails to understand you or if your tone offends or patronizes them. Plan how you will design your content and style to communicate most effectively with your audience.

A presentation plan, like the example shown in Figure 18.2, can help you accomplish these goals.

Gather Information. The best delivery by the most polished professional speaker will lack credibility if the speaker has little of value to communicate. As you plan your presentation, you must study and research your topic thoroughly before you package it.

You can rely on numerous sources when you research a topic for an oral presentation. For example, you could use any of the following sources.

- Interviews
- Questionnaires and surveys

Research

See Chapter 5 for more discussion of research techniques.

- Visits to job sites
- Conversations in meetings or on the telephone
- Company reports
- Internet research
- Library research including periodicals and books
- Market research

Using information from company reports or other sources such as the Internet, books, or periodicals requires that you read and document your research. Gathering information through interviews, questionnaires, surveys, or conversations, on the other hand, requires help from other people. To ensure that you get this assistance, consider doing the following:

- Ask politely for their assistance.
- Explain why you need the interview and information.
- Explain how you will use the information.
- Make a convenient appointment for the interview or to fill in the survey.
- Come prepared. Research the subject matter so you will be prepared to ask appropriate questions. Write your interview questions or the survey before you meet with the person. Take the necessary paper, forms, pencils, pens, laptop, electronic notepads, handheld PDAs, or recording devices you will need for the meeting.
- When the interview ends or the individuals complete the survey, thank them for their assistance.

Figure 18.3 shows a sample questionnaire used by a team. They were researching the feasibility of adding a child care center to their university campus for use by students and staff.

Writing the Presentation

After you obtain your information, your next step is to write a draft and consider visual aids for the presentation. The writing step in the communication process lets you use the research you gathered in the planning stage. When you organize your information, you will determine whether additional material is needed or if you can delete some of the material you have gathered.

Don't write out the complete text of the presentation. Too often when people have a complete text in front of them, they rely too heavily on the written words. They end up reading most of the paper to the audience rather than speaking more conversationally. Instead of writing out a complete copy of the presentation, use an outline or note cards to present your speech.

Outline.   You may want to write a more detailed outline focusing on your speech's major units of discussion and supporting information. A skeleton speech outline (Figure 18.4) provides you with a template for your presentation.

Note Cards.   If you decide that presenting your speech from the outline will not work for you, consider writing highlights on 3×5 note cards. Avoid writing complete sentences or filling in the cards from side to side. If you write complete sentences, you will be tempted to read the notes rather than speak to the audience. If you fill in the cards from side to side, you will have trouble finding key ideas. Write short notes (keywords or short phrases) that will aid your memory when you make your presentation.

Sample 3 × 5 Note Cards

Need for Parking Lot Expansion   ←——— Heading to maintain focus

- Safety

- Accessibility   ←——— Bulleted points for easy access

- Potential growth

Procedure for Parking Lot Expansion

- Stakeholders' meeting and vote

- City council approval

- Arrangement with contractors

FIGURE 18.3 Sample Research Questionnaire

Student and Staff Questionnaire for Proposed Child Care Center

1. Are you male or female?
2. Are you in a single- or double-income family?
3. Are you a student or staff?
4. How many children do you have?
5. What are the ages?
6. Would you be interested in having a child care center on campus?
7. How much would you be willing to pay per hour for child care at this center?
8. Do you think a child care center would increase enrollment at this university?
9. How many hours per week would you enroll your child/children?
10. What hours of operation should the child care center cover?
11. What credentials should the child care providers have?
12. What should be the number of children per classroom?

If student:
13. Are you enrolled full time or part time at the university?
14. Do you attend mornings, afternoons, evenings, weekends, or a combination of the above (please specify)?
15. Would you be willing to work in the child-care center part time?

If staff:
16. What hours do you work at the university?
17. What hours would you need to use for child care at the center?

Additional comments (if any):

Thank you for your assistance.

FIGURE 18.4 Skeleton Speech Outline

Skeleton Outline

Title:

Purpose:

    I. Introduction

        A. Attention getter:

        B. Focus statement:

    II. Body

        A. First main point:

            1. Documentation/subpoint:

                a. Documentation/subpoint:

                b. Documentation/subpoint:

            2. Documentation/subpoint:

            3. Documentation/subpoint:

        B. Second main point:

            1. Documentation/subpoint:

            2. Documentation/subpoint:

                a. Documentation/subpoint:

                b. Documentation/subpoint:

            3. Documentation/subpoint:

        C. Third main point:

            1. Documentation/subpoint:

            2. Documentation/subpoint:

            3. Documentation/subpoint:

                a. Documentation/subpoint:

                b. Documentation/subpoint:

    III. Conclusion

        A. Summary of main points:

        B. Recommended future course of action:

Rewriting the Presentation

In the rewriting step of the writing process, consider all aspects of style, delivery, appearance, and body language and gestures. Then, most importantly, practice. Even if you have excellent visual aids and well-organized content, if you fail to deliver effectively, your audience could miss your intended message.

Style.   As with good writing, effective oral communication demands clarity and conciseness. To achieve clarity, stick to the point. Your audience does not want to hear about your personal life or other irrelevant bits of information. You need to maintain focus on the topic. Concise oral presentations depend on the same skills evident in concise writing—word and sentence length. Trim your sentences of excess words (12 to 15 words per sentence is still the preferred length).

Remember to speak so that your audience can understand you and your level of vocabulary. You should speak to communicate rather than to impress your listeners.

Goals of Technical Communication

See Chapter 3 for more discussion of communication goals.

Delivery.   Effective oral communicators interact with and establish a dynamic relationship with their audiences. The most thorough research will be wasted if you are unable to create rapport and sustain your audience's interest. Although smaller audiences are usually easier to connect with, you can also establish a connection with much larger audiences through a variety of delivery techniques.

Eye Contact—Avoid keeping your eyes glued to your notes. You will find it easy to speak to one individual because you will naturally look him or her in the eye. The person will respond by looking back at you.

With a larger audience, whether the audience has 20 or 200-plus people, keeping eye contact is more difficult. Try looking into different people's eyes as you move through your presentation (or look slightly above their heads if that makes you more comfortable). Most of the audience has been in your position before and can sympathize.

Rate—Because your audience wants to listen and learn, you need to speak at a rate slow enough to achieve those two goals. Determine your normal rate of speech, and cut it in half. *Slow* is the best rate to follow in any oral presentation. You could speed up your delivery when you reach a section of less interesting facts. Slow down for the most important and most interesting parts. Match your rate of speaking to the content of your speech, just as actors vary their speech rate to reflect emotion and changes in content.

Enunciation—Speak each syllable of every word clearly and distinctly. Rarely will an audience ask you to repeat something even if they could not understand you the first time. It is up to you to avoid mumbling. Remember to speak more clearly than you might in a more conversational setting. Slowing your delivery will help you enunciate clearly.

Pitch—When you speak, your voice creates high and low sounds. That's *pitch*. In your presentation, capitalize on this fact. Vary your pitch by using even more high and low sounds than you do in your normal, day-to-day conversations. Modulate to stress certain keywords or major points in your oral presentations.

Pauses—One way to achieve a successful pace is to pause within the oral presentation. Pause to ask for and to answer questions, to allow ideas to sink in, and to use visual aids or give the audience handouts. These pauses will not lengthen your speech; they will only improve it.

A well-prepared speech will allow for pauses and will have budgeted time effectively. Know in advance if your speech is to be 5 minutes, 10 minutes, or an hour long. Then plan your speech according to time constraints, building pauses into your presentation. Practice the speech beforehand so you can determine when to pause and how often.

Emphasis   You will not be able to underline or boldface comments you make in oral presentations. However, just as in written communication, you will want to emphasize key ideas. Your body language, pitch, gestures, and enunciation will enable you to highlight words, phrases, or even entire sentences.

Interaction with Listeners   You might need your audience to be active participants at some point in your oral presentation, so you will want to encourage this response. Your attitude and the tone of your delivery are key elements contributing to an encouraging atmosphere.

Conflict Resolution   You might be confronted with a hostile listener who either disagrees with you or does not want to be in attendance. You need to be prepared to deal with such a person. If someone disagrees with you or takes issue with a comment you make, try these responses.

- "That is an interesting perspective."
- "Thanks for your input."
- "Let me think about that some more and get back to you."
- "I have got several more ideas to share. We could talk about that point later, during a break."

If you are confronted with a challenge,

- Put it off until later so it does not distract from your presentation.
- Let the situation diffuse.
- Give yourself some more time to think about it.
- Give the person time to cool off.

The important point to remember is to not allow a challenging person to take charge of your presentation. Be pleasant but firm and maintain control of the situation. You will be unable to please all of your listeners all of the time. However, you should not let one unhappy listener destroy the effect of your presentation for the rest of the audience.

Appearance.   When you speak to an audience, they see you as well as hear you. Therefore, avoid physical distractions. For example, avoid wearing clothes or jewelry that might distract the audience. You might be representing your company or trying to make a good impression for yourself when you speak, so dress appropriately.

Body Language and Gestures.   During an oral presentation, nonverbal communication can be as important as verbal communication. Your appearance and attitude are important. In addition, the way you present your speech through your movements and tone of voice will affect your listeners. If you are enthusiastic about your topic, your listeners will respond enthusiastically. If you are bored or ambivalent, your tone and mannerisms will reflect your attitude. If you are negative, your tone will communicate negativity to your audience.

To communicate effectively, be aware of your body language and your gestures.

- Avoid standing woodenly. Move around somewhat, scanning the room with your eyes, stopping occasionally to look at one person. Remember to look at all parts of the room as you make the oral presentation.
- If you are nervous and your hands shake, try holding on to a chair back, lectern, the top of the table, or a paper clip.
- Use hand motions to emphasize ideas and provide transitions. For instance, you could put one finger up for a first point, two fingers up for a second point, and so forth.
- Avoid folding your arms stiffly across your chest. This projects a negative, defensive attitude.

Post-speech Question and Answers. After your presentation, be prepared for a question-and-answer session. Politely invite your audience to participate by saying, "If you have any questions, I am happy to answer them." When an audience member asks a question, make sure everyone hears it. If not, you can repeat the question. If you fail to understand the question, ask the audience member to repeat it and clarify it.

If you have no answer, tell the audience. You could say that you will research the matter and get back with them. Faking an answer will only harm your credibility and detract from the overall effect of your presentation. Another valid option is to ask the audience what they think regarding the question or if they have any possible solutions or answers to the question. This is not only a good way to answer the question but also to encourage audience interaction.

Practice. Practice your speech including manipulation of your visual aids so you use them at appropriate times and places during the presentation. As you practice, you will grow more comfortable and less dependent on your note cards or outline. Use the Effective Oral Presentation Checklist to determine if you are sufficiently prepared for your oral presentation. You will find that the more you practice, the more comfortable you feel. Practicing will help you achieve the following:

- Decrease your fear
- Process your thoughts
- Become more comfortable with the topic
- Pronounce troublesome words
- Decide what to emphasize and how to emphasize it
- Enhance verbal and nonverbal cues
- Rearrange your content
- Add further details
- Know when and how to use your visual aids

Effective Oral Presentation Checklist

_____ 1. Does your speech have an introduction,
  - arousing the audience's attention?
  - clearly stating the topic of the presentation?

_____ 2. Does your speech have a body,
  - explaining what exactly you want to say?
  - developing your points thoroughly?

_____ 3. Does your speech have a conclusion, suggesting
  - what is next?
  - explaining when (due date) a follow-up should occur?
  - stating why that date is important?

_____ 4. Does your presentation provide visual aids to help you make and explain your points?

_____ 5. Do you modulate your pace and pitch?

_____ 6. Do you enunciate clearly so the audience will understand you?

_____ 7. Have you used body language effectively,
  - maintaining eye contact?
  - using hand gestures?
  - moving appropriately?
  - avoiding fidgeting with your hair or clothing?

_____ 8. Have you prepared for possible conflicts?

_____ 9. Do you speak slowly and remember to pause so the audience can think?

_____10. Have you practiced with any equipment you might use?

CHAPTER HIGHLIGHTS

1. Effective oral communication ensures that your message is conveyed successfully and that your verbal skills make a good impression.

2. You will communicate verbally on an everyday basis, informally and formally.

3. Some points to consider when you use the telephone are to script your telephone conversation, speak clearly and slowly, and avoid rambling.

4. To prepare for a teleconference, plan ahead, check equipment for audio quality, and choose a quiet and private location.

5. A webinar is a way to offer online training for employees at remote locations.

6. An effective oral presentation introduction might include an anecdote, question, quote, facts, or a "table setter."

7. In the discussion section of a formal oral presentation, organize your content according to comparison/contrast, problem/solution, argument/persuasion, importance, and chronology.

8. Conclude your formal speech by restating your main points, recommending a future course of action, or asking for questions and comments.

9. One of the most powerful oral communication tools is Microsoft PowerPoint.

10. Benefits of PowerPoint slides include different autolayouts and design templates, the ability to add and delete slides easily, hypertext links, and the incorporation of sound and graphics.

11. When you use PowerPoint slides, ensure contrast between the font color and background, choose an easy-to-read font size and type, use few words and lines per slide, use headings for navigation, and prepare print handouts for your audience.

12. During a PowerPoint presentation, avoid merely reading the slides, choosing instead to elaborate on each slide with additional details.

13. The writing process of prewriting, writing, and rewriting allows you to create effective oral communication.

14. When you prewrite to plan your presentation, consider purpose, recognize audience, and gather information.

15. To write your presentation, use a presentation plan, outline, note cards, and visual aids.

16. Many visual aids can enhance your oral presentation. Choose from chalkboards, chalkless whiteboards, flip charts, overhead projections, PowerPoint, videos, and film.

17. To perfect your oral presentation (the third part of the communication process), work on your delivery, including eye contact, rate of speech, enunciation, pitch, pauses and emphasis, interaction with the audience, and conflict resolution.

18. Appearance, body language, and gestures will impact your oral presentation.

19. Restate main points frequently in an oral presentation.

20. Be prepared for questions and discussion after you conclude your oral presentation.

APPLY YOUR KNOWLEDGE

CASE STUDIES

1. Read the **TechStop** scenario that begins this chapter. In this scenario, a customer is treated poorly by a salesperson. The vice president of customer service, Shuan Wang, has received reports of other problems with customer–sales staff interaction.

   Shuan needs to make a formal oral presentation to all sales personnel to improve customer service.

Assignment

Based on the scenario at the beginning of the chapter, outline the content for Shuan's oral presentation as follows:

   - Determine what kind of introduction should be used to arouse the audience's interest.
   - Provide a thesis statement.
   - Develop information to teach employees sales etiquette and customer interaction, store policy regarding customer satisfaction, the impact of poor customer relations, and the consequences of failing to handle customers correctly. Research these topics (either online, in a library, or through interviews) to find your content.
   - Organize the speech's content with appropriate transitions to aid coherence.
   - Determine which types of visual aids would work best for this presentation.

2. **Halfmoon** outdoor equipment is an online wholesaler of hiking, biking, and boating gear. Their staff is located in four northeastern states. Halfmoon's CEO, Montana Wildhack, wants his employees to excel in communication skills, oral and written. He has hired a consultant to train the employees. To accomplish this training, the consultant will provide a teleconference. Fifty employees will meet with the consultant in a training room, while another 150 Halfmoon staff members view the training session from their different work locations.

   During the presentation, the consultant highlights his comments by writing on a chalkless whiteboard, using green and yellow markers. He attempts humor by suggesting that Halfmoon's thermal stocking caps would be a perfect gift for one of the seminar's balding participants, Joe. In addition, the consultant has participants break into small groups to role play. They take mock customer orders, field complaints, and interact with vendors, using a prepared checklist of do's and don'ts.

   At one point, when a seminar participant calls in a question, the consultant says, "No, that would be wrong, wrong, wrong! Why would you ever respond to a customer like that? Common sense would dictate a different response." During a break, the consultant does not mute his clip-on microphone. To conclude the session, the consultant asks for questions and comments. When he cannot provide a good answer, he asks the participants to suggest options, and he says he will research the question and place an answer on his corporate Web site.

Assignment

Using the guidelines in the chapter, discuss the speaker's oral communication skills. Explain what went wrong and why. Explain what was effective.

INDIVIDUAL AND TEAM PROJECTS

Evaluating an Oral Presentation

Listen to a speech. You could do so on television; at a student union; at a church, synagogue, or mosque; at a civic event, city hall meeting, or community organization; at a company activity; or in your classroom. Answer the following questions.

1. What was the speaker's goal? Did the speaker try to persuade, instruct, inform, or build trust? Explain your answer by giving examples.

2. What type of introduction did the speaker use to arouse listener interest (anecdote, question, quote, facts, or startling comments)? Give examples to support your decision.

3. What visual aids were used in the presentation? Were these visual aids effective? Explain your answers.

4. How did the speaker develop the assertions? Did the speaker use analysis, comparison/contrast, argument/persuasion, problem/solution, or chronology? Give examples to prove your point.

5. Was the speaker's delivery effective? Use the following Presentation Delivery Rating Sheet to assess the speaker's performance by placing check marks in the appropriate columns.

Presentation Delivery Rating Sheet

Delivery Techniques	Good	Bad	Explanation
Eye Contact			
Rate of Speech			
Enunciation			
Pitch of Voice			
Use of Pauses			
Emphasis			
Interaction with Listeners			
Conflict Resolution			

Giving an Oral Presentation

1. Find an advertised job opening in your field of expertise or degree program. You could look either online, in the newspaper, at your school's career placement center, or at a company's human resource office. Perform a mock interview for this position. To do so, follow this procedure.

   • Designate one person as the applicant.
   • This person should prepare for the interview by making a list of potential questions (ones that the applicant believes he or she will be asked, as well as questions to ask the search committee).
   • Designate others in the class to represent the search committee.
   • The mock search committee should prepare a list of questions to ask the applicant.

2. Research a topic in your field of expertise or degree program. The topic could include a legal issue, a governmental regulation, a news item, an innovation in the field, or a published article in a professional journal or public magazine. Make an oral presentation about your findings.

Creating a PowerPoint Slide Presentation

For assignments 1 and 2 in "Giving an Oral Presentation," create PowerPoint slides to enhance your oral communication. To do so, follow the guidelines provided in this chapter.

Assessing PowerPoint Slides

After reviewing the following Microsoft PowerPoint slides, determine which are successful, which are unsuccessful, and explain your answers, based on the guidelines provided in this chapter.

A Library Perspective on Distance Learning

1. Library online research sites are rapidly evolving.
2. Instructors engaged in online course development should visit with library staff for updates on resources.
3. Many traditional library services are available to serve distance learning students, including course reserves, interlibrary loan, reference, reciprocal borrowing policies, document delivery, access to online databases and collections
4. Many new services, electronic books and journals, are available from libraries.
5. The Internet is not the online equivalent of an academic library.

A Technology Perspective on Online Education

<u>Students</u>—24-hour Call Centers answer student hardware/software needs

<u>Faculty</u>—The Tech Center helps faculty with course creations and tech resources

As **DISTANCE LEARNING** classes have risen in popularity, many ***diverse areas*** of interest across campuses have converged.

These include <u>counseling, library services, technology, faculty, administration, continuing education, etc.</u>

To ensure student success, *Distance Learning Coordinating Committees* need to draw from these multiple disciplines and collaborate on solutions to arising technology challenges.

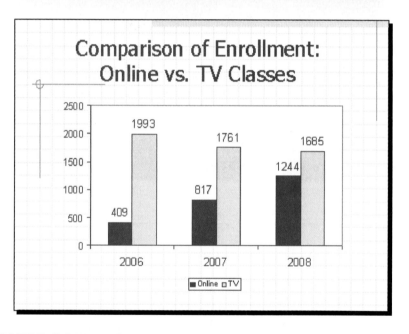

PROBLEM-SOLVING THINK PIECES

After reading the following scenarios, answer these questions.

- Would your oral presentation be everyday, informal, or formal?
- Would you use a videoconference, teleconference, or face-to-face meeting?
- Which visual aids would work best for your presentation?
- Is your oral presentation goal to persuade, instruct, inform, or build trust?

In addition, complete any or all of these assignments.

- Write a presentation plan.
- Write an outline for the presentation.
- Write a brief, introductory lead-in to arouse your listeners' interest.
- Create a questionnaire.

1. You work at FlashCom Electric. Your company has created a new interface for modems. Your boss has asked you to make a presentation to sales representatives from 20 potential vendors in the city. The oral presentation will explain to the vendors the benefits of your product, the sales breaks you will offer, and how the vendors can increase their sales.

2. After working for Friendly's, a major discount computer hardware and software retailer, for over a year, you have created a new organizational plan for their vast inventory. In an oral presentation, you plan to show the CEO and board of directors why your plan is cost effective and efficient.

3. You are the manager of an automotive parts supply company, Plugs, Lugs, & More. Your staff of 10 in-store employees lacks knowledge of the store's new merchandise, has not been meeting sales goals, and does not always treat customers with respect and care. It's time to address these concerns.

4. As CEO of your engineering/architectural firm (Levin, Lisk, and Lamb), you must downsize. Business is decreasing and costs are rising. To ensure third-quarter profitability, 10 percent of the staff must be laid off. That will amount to a layoff of over 500 employees. You now must make an oral presentation to your stockholders and the entire workforce (located in three states) at your company's annual meeting to report the situation.

5. Your Quality Circle Team is composed of account representatives, management, engineers, and technicians. You have been asked to research new ways to improve product delivery. You have concluded your research. Now, you need to share your suggestions with your 15 team members.

WEB WORKSHOP

PowerPoint has proven to be a useful tool in business presentations. However, PowerPoint must be used correctly, as discussed in the chapter, to be effective. Access an online search engine and type in phrases such as follows (include any additional ones you create):

- "using powerpoint in business"
- "effective powerpoint use"
- "tips for using powerpoint"
- "powerpoint + business presentations"
- "powerpoint + pros and cons"

Once you have found articles discussing this topic, summarize your findings and make an oral presentation (using PowerPoint slides).

QUIZ QUESTIONS

1. What are three ways you can achieve successful telephone conversations?
2. Why would you participate in a teleconference?
3. How can you create an effective teleconference or videoconference?
4. How many parts are in a formal oral presentation?
5. In what ways can you arouse audience attention in an oral presentation?
6. What does a thesis statement in your formal oral presentation achieve?
7. How can you organize the details in the discussion section of a formal oral presentation?
8. How can you achieve coherence in a formal oral presentation?
9. How can you conclude a formal oral presentation?
10. What are benefits of using Microsoft PowerPoint slide presentations?
11. How can you create an effective PowerPoint slide presentation?
12. What can you do during a PowerPoint presentation to make it more effective?
13. When prewriting a formal oral presentation, you should focus on which three considerations?
14. Instead of writing out an entire formal oral presentation, how can you prepare your speech?
15. In addition to PowerPoint, what are four other types of visual aids?

APPENDIX

....

Grammar, Punctuation, Mechanics, and Spelling

Correct organization and development of your memos, letters, or reports is important for the success of your technical communication. However, no one will be impressed with the quality of your work, or with you, if your writing is riddled with errors in sentence construction or punctuation. Your written correspondence is often your first contact with business associates. Many people mistakenly believe that only English teachers notice grammatical errors and wield red pens, but businesspeople as well take note of such errors and may see the writer as less competent.

We were working recently with a young executive who is employed by a branch of the federal government. This executive told us that whenever his supervisor found a spelling error in a subordinate's report, this report was paraded around the office. Everyone was shown the mistake and had a good laugh over it, and the report was then returned to the writer for correction. Our acquaintance assured us that all of this was in good-natured fun. However, he also said that employees quickly learned to edit and proofread their written communication to avoid such public displays of their errors. He went on to say that his dictionary was well thumbed and always on his desk.

Your writing at work may not be exposed to such scrutiny by coworkers. Instead, your writing may go directly to another firm, and those readers will see your mistakes. To avoid this problem, you must evaluate your writing for grammar, punctuation, and spelling errors. If you don't, your customers, bosses, and colleagues will.

GRAMMAR RULES

To understand the fundamentals of grammar, you must first understand the basic components of a sentence.

A correctly constructed sentence consists of a subject and a predicate (some sentences also include a phrase or phrases).

example	The *meeting*	began	at 4:00 A.M.
	subject	predicate	phrase

Subject: The *doer* of the action; the subject usually precedes the predicate
Predicate: The *action* in the sentence

He	ran	to the office to avoid being late.
doer	action	phrases

If the subject and the predicate (a) express a complete thought and (b) can stand alone, you have an *independent clause*.

The meeting began	at 4:00 P.M.
independent clause	phrase

A *phrase* is a group of related words that does not contain a subject and a predicate and cannot stand alone or be punctuated as a sentence. The following are examples of phrases.

at the house
in the box
on the job
during the interview

If a clause is dependent, it cannot stand alone.

Although he tried to hurry, he was late for the meeting.
dependent clause   independent clause

He was late for the meeting although he tried to hurry.
independent clause   dependent clause

NOTE: When a dependent clause begins a sentence, use a comma before the independent clause. However, when an independent clause begins a sentence, do not place a comma before the dependent clause.

Agreement between Pronoun and Antecedent (Referent)

A pronoun has to agree in gender and number with its antecedent.

Susan went on *her* vacation yesterday.
The *people* who quit said that *they* deserved raises.

Problems often arise when a singular indefinite pronoun is the antecedent. The following antecedents require singular pronouns: *anybody, each, everybody, everyone, somebody,* and *someone.*

incorrect

Anyone can pick up *their* applications at the job placement center.

correct

Anyone can pick up *his* or *her* applications at the job placement center.

Problems also arise when the antecedent is separated from the pronoun by numerous words.

incorrect

Even when the best *employee* is considered for a raise, *they* often do not receive it.

Even when the best *employee* is considered for a raise, *he* or *she* often does not receive it.

Agreement Between Subject and Verb

Writers sometimes create disagreement between subjects and verbs, especially if other words separate the subject from the verb. To ensure agreement, ignore the words that come between the subject and verb.

Her *boss* undoubtedly *think* that all the employees want promotions.

Her *boss* undoubtedly *thinks* that all the employees want promotions.

The *employees* who sell the most equipment *is* going to Hawaii for a week.

The *employees* who sell the most equipment *are* going to Hawaii for a week.

If a sentence contains two subjects (a compound subject) connected by *and,* use a plural verb.

Joe and Tiffany *was* both selected employee of the year.

Joe and Tiffany *were* both selected employee of the year.

The bench workers and their supervisor *is* going to work closely to complete this project.

The bench workers and their supervisor *are* going to work closely to complete this project.

Add a final *s* or *es* to create most plural subjects or singular verbs, as follows:

Plural Subjects	Singular Verbs
bosses hire	a boss hires
employees demand	an employee demands
experiments work	an experiment works
attitudes change	the attitude changes

If a sentence has two subjects connected by *either . . . or, neither . . . nor,* or *not only . . . but also,* the verb should agree with the closest subject. This also makes the sentence less awkward.

Either the salespeople or the warehouse worker deserv*es* raises.
Not only the warehouse worker but also the salespeople deserv*e* raises.
Neither the salespeople nor the warehouse worker deserv*es* raises.

Singular verbs are used after most indefinite pronouns such as the following:

another	everything
anybody	neither
anyone	nobody
anything	no one
each	nothing
either	somebody
everybody	someone
everyone	something

Anyone who works here *is* guaranteed maternity leave.
Everybody want*s* the company to declare a profit this quarter.

Singular verbs often follow collective nouns such as the following:

class	organization
corporation	platoon
department	staff
group	team

The *staff is* sending the boss a bouquet of roses.

Comma Splice

A *comma splice* occurs when two independent clauses are joined by a comma rather than separated by a period or semicolon.

incorrect

Sue was an excellent employee, she got a promotion.

Several remedies will correct this error.

1. Separate the two independent clauses with a semicolon.

correct

Sue was an excellent employee; she got a promotion.

2. Separate the two independent clauses with a period.

correct

Sue was an excellent employee. She got a promotion.

3. Separate the two independent clauses with a comma and a *coordinating conjunction (and, but, or, for, so, yet)*.

correct

Sue was an excellent employee, *so* she got a promotion.

4. Separate the two independent clauses with a semicolon (or a period), a conjunctive adverb, and a comma. *Conjunctive adverbs* include *also, additionally, consequently, furthermore, however, instead, moreover, nevertheless, therefore,* and *thus*.

correct

Sue was an excellent employee; *therefore,* she got a promotion.

or

Sue was an excellent employee. *Therefore,* she got a promotion.

5. Use a *subordinating conjunction* to make one of the independent clauses into a dependent clause. Subordinating conjunctions include *after, although, as, because, before, even though, if, once, since, so that, though, unless, until, when, where,* and *whether*.

correct

Because Sue was an excellent employee, she got a promotion.

Faulty or Vague Pronoun Reference

A pronoun must refer to a specific noun (its antecedent). Problems arise when (a) there is an excessive number of pronouns (causing vague pronoun reference) and (b) there is no specific noun as an antecedent. Notice that there seems to be an excessive number of pronouns in the following passage, and the antecedents are unclear.

example

> Although Bob had been hired over two years ago, *he* found that his boss did not approve *his* raise. In fact, *he* was also passed over for *his* promotion. The boss appears to have concluded that *he* had not exhibited zeal in *his* endeavors for their business. Instead of being a highly valued employee, *he* was not viewed with pleasure by those in authority. Perhaps it would be best if *he* considered *his* options and moved to some other company where *he* might be considered in a new light.

The excessive and vague use of *he* and *his* causes problems for readers. Do these words refer to Bob or to his boss? You are never completely sure. To avoid this problem, limit pronoun usage, as in the following revision.

example

> Although Bob had been hired over two years ago, he found that his boss, Joe, did not approve his raise. In fact, Bob was also passed over for promotion. Joe appears to have concluded that Bob had not exhibited zeal in his endeavors for their business. Instead of being a highly valued employee, Bob was not viewed with pleasure by those in authority. Perhaps it would be best if Bob considered his options and moved to some other company where he might be considered in a new light.

To make the preceding paragraph more precise, we have replaced vague pronouns (*he* and *his*) with exact names (*Bob* and *Joe*).

Fragments

A *fragment* occurs when a group of words is incorrectly used as an independent clause. Often the group of words begins with a capital letter and has end punctuation but is missing either a subject or a predicate.

incorrect

Working with computers.

(lacks a predicate and does not express a complete thought)

The group of words may have a subject and a predicate but be a dependent clause.

incorrect

Although he enjoyed working with computers.

(has a subject, *he*, and a predicate, *enjoyed*, but is a dependent clause because it is introduced by the subordinate conjunction *although*)

It is easy to remedy a fragment by doing one of the following:

- Add a subject.
- Add a predicate.
- Add both a subject and a predicate.
- Add an independent clause to a dependent clause.

correct

Joe found that working with computers used his training.

(subject, *Joe,* and predicate, *found*, have been added)

correct

Although he enjoyed working with computers, he could not find a job in a computer-related field.

(independent clause, *he could not find a job*, added to the dependent clause, *Although he enjoyed working with computers*)

Fused Sentence

A *fused sentence* occurs when two independent clauses are connected with no punctuation.

incorrect

The company performed well last quarter its stock rose several points.

There are several ways to correct this error.

1. Write two sentences separated by a period.

correct

The company performed well last quarter. Its stock rose several points.

2. Use a comma and a coordinating conjunction to separate the two independent clauses.

> **correct**

The company performed well last quarter, *so* its stock rose several points.

3. Use a subordinating conjunction to create a dependent clause.

> **correct**

Because the company performed well last quarter, its stock rose several points.

4. Use a semicolon to separate the two independent clauses.

> **correct**

The company performed well last quarter; its stock rose several points.

5. Separate the two independent clauses with a semicolon, a conjunctive adverb or a transitional word or phrase, and a comma.

> **correct**

The company performed well last quarter; *therefore*, its stock rose several points.

> **correct**

The company performed well last quarter; *for example*, its stock rose several points.

The following are transitional words and phrases, listed according to their use.

To Add

again	in addition
also	moreover
besides	next
first	second
furthermore	still

To Compare/Contrast

also	nevertheless
but	on the contrary
conversely	still
in contrast	

To Provide Examples

for example	of course
for instance	put another way
in fact	to illustrate

To Show Place

above	here
adjacent to	nearby
below	on the other side
elsewhere	there
further on	

To Reveal Time

afterward	second
first	shortly
meanwhile	subsequently
presently	thereafter

To Summarize

all in all	last
finally	on the whole
in conclusion	therefore
in summary	thus

Modification

A *modifier* is a word, phrase, or clause that explains or adds details about other words, phrases, or clauses.

Misplaced Modifiers.   A *misplaced modifier* is one that is not placed next to the word it modifies.

incorrect

He had a heart attack *almost* every time he was reviewed by his supervisor.

correct

He almost had a heart attack every time he was reviewed by his supervisor.

incorrect

The worker had to *frequently* miss work.

correct

The worker frequently had to miss work.

Dangling Modifiers.   A *dangling modifier* is a modifier that is not placed next to the word or phrase it modifies. To avoid confusing your readers, place modifiers next to the word(s) they refer to. Don't expect your readers to guess at your meaning.

incorrect

While working, tiredness overcame them.
(Who was working? Who was overcome by tiredness?)

correct

While working, the staff became tired.

incorrect

After soldering for two hours, the equipment was ready for shipping.
(Who had been soldering for two hours? Not the equipment!)

correct

After soldering for two hours, the technicians prepared the equipment for shipping.

Parallelism

All items in a list should be parallel in grammatical form. Avoid mixing phrases and sentences (independent clauses).

incorrect

We will discuss the following at the department meeting:

1. Entering mileage in logs (phrase)
2. All employees have to enroll in a training seminar. (sentence)
3. Purpose of quarterly reviews (phrase)
4. Some data processors will travel to job sites. (sentence)

correct

We will discuss the following at the department meeting:

1. Entering mileage in logs
2. Enrolling in training seminars
3. Reviewing employee performance quarterly
4. Traveling to job sites

} phrases

correct

At the department meeting, you will learn how to

1. Enter mileage in logs
2. Enroll in training seminars
3. Review employee performance quarterly
4. Travel to job sites

} phrases

PUNCTUATION

Apostrophe (')

Place an *apostrophe* before the final *s* in a singular word to indicate possession.

example   Jim's tool chest is next to the furnace.

Place the apostrophe after the final *s* if the word is plural.

example   The employees' reception will be held next week.

Don't use an apostrophe to make singular abbreviations plural.

The EXT's will be shipped today.

The EXTs will be shipped today.

Colon (:)

Use a *colon* after a salutation.

Dear Mr. Harken:

example

In addition, use a colon after an emphatic or cautionary word if explanations follow.

Note: Hand-tighten the nuts.
Caution: Wash thoroughly if any mixture touches your skin.

example

Finally, use a colon after an independent clause to precede a quotation, list, or example.

She said the following: "No comment."
These supplies for the experiment are on order: plastic hose, two batteries, and several chemicals.
The problem has two possible solutions: hire four more workers, or simply give everyone a raise.

example

NOTE: In the preceding examples, the colon follows an independent clause.

A common mistake is to place a colon after an incomplete sentence. Except for salutations and cautionary notes, whatever precedes a colon *must* be an independent clause.

The two keys to success are: earning money and spending wisely.

The two keys to success are earning money and spending wisely.
or
The two keys to success are as follows: earning money and spending wisely.
or
The two keys to success are as follows:
1. Earning money
2. Spending wisely

Comma (,)

Writers often get in trouble with *commas* when they employ one of two common "words of wisdom."

- When in doubt, leave it out.
- Use a comma when there is a pause.

Both rules are inexact. Writers use the first rule to justify the complete avoidance of commas; they use the second rule to sprinkle commas randomly throughout their writing. On the contrary, commas have several specific conventions that determine usage.

1. Place a comma before a coordinating conjunction *(and, but, or, for, so, yet)* linking two independent clauses.

> **example**
>
> You are the best person for the job, *so* I will hire you.
>
> We spent several hours discussing solutions to the problem, *but* we failed to decide on a course of action.

2. Use commas to set off introductory comments.

> **example**
>
> First, she soldered the components.
>
> In business, people often have to work long hours.
>
> To work well, you need to get along with your coworkers.
>
> If you want to test equipment, do so by 5:25 P.M.

3. Use commas to set off sentence interrupters.

> **example**
>
> The company, started by my father, did not survive the last recession.
>
> Mrs. Patel, the proprietor of the store, purchased a wide array of merchandise.

4. Set off parenthetical expressions with commas.

> **example**
>
> A worker, it seems, should be willing to try new techniques.
>
> The highway, by the way, needs repairs.

5. Use commas after each item in a series of three or more.

> **example**
>
> Prakash, Mirren, and Justin were chosen as employees of the year.
>
> We found the following problems: corrosion, excessive machinery breakdowns, and power failures.

6. Use commas to set off long numbers.

> **example**
>
> She earns $100,000 before taxes.

NOTE: Very large numbers are often written as words.

| Our business netted over $2 million in 2008. | **example** |

7. Use commas to set off the day and year when they are part of a sentence.

| The company hired her on September 7, 2008, to be its bookkeeper. | **example** |

NOTE: If the year is used as an adjective, do not follow it with a comma.

| The 2008 corporate report came out today. | **example** |

8. Use commas to set off the city from the state and the state from the rest of the sentence.

| The new warehouse in Austin, Texas, will promote increased revenues. | **example** |

NOTE: If you omit either the city or the state, you do not need commas.

| The new warehouse in Austin will promote increased revenues. | **example** |

Dash (—)

A *dash,* typed as two consecutive hyphens with no spaces before or after, is a versatile punctuation mark used in the following ways.

1. After a heading and before an explanation

| Forecasting—Joe and Joan will be in charge of researching fourth-quarter production quotas. | **example** |

2. To indicate an emphatic pause

| You will be fired—unless you obey company rules. | **example** |

3. To highlight a new idea

| Here's what we can do to improve production quality—provide on-the-job training, salary incentives, and quality controls. | **example** |

4. Before and after an explanatory or appositive series

| Three people—Sue, Luci, and Tom—are essential to the smooth functioning of our office. | **example** |

Ellipses (. . .)

Ellipses (three spaced periods) indicate omission of words within quoted materials.

| "Six years ago, prior to incorporating, the company had to pay extremely high federal taxes."
"Six years ago, . . . the company had to pay extremely high federal taxes." | **example** |

Exclamation Point (!)

Use an *exclamation point* after strong statements, commands, or interjections.

example	You must work harder! Do not use the machine! Danger!

Hyphen (-)

A *hyphen* is used in the following ways.

1. To indicate the division of a word at the end of a typed line. Remember, this division must occur between syllables.
2. To create a compound adjective

example	He is a well-known engineer. Until her death in 2008, she was a world-renowned chemist. Tom is a 24-hour-a-day student.

3. To join the numerator and denominator of fractions

example	Four-fifths of the company want to initiate profit sharing.

4. To write out two-word numbers

example	Twenty-six people attended the conference.

Parentheses ( )

Parentheses enclose abbreviations, numbers, words, or sentences for the following reasons.

1. To define a term or provide an abbreviation for later use

example	We belong to the Society for Technical Communication (STC).

2. To clarify preceding information in a sentence

example	The people in attendance (all regional sales managers) were proud of their accomplishments.

3. To number items in a series

example	The company should initiate (1) new personnel practices, (2) a probationary review board, and (3) biannual raises.

Period (.)

A *period* must end a declarative sentence (independent clause).

> I found the business trip rewarding.
>
> **example**

Periods are often used with abbreviations.

D.C.	Mrs.	A.M. or a.m.
e.g.	Ms.	P.M. or p.m.
Mr.		

It is incorrect to use periods with abbreviations for organizations and associations.

incorrect

S.T.C. (Society for Technical Communication)

correct

STC (Society for Technical Communication)

State abbreviations do not require periods if you use two capital letters. (Continue to use a period after a capital letter and a lowercase letter such as Tx., Ks., or Mo.)

incorrect

KS. (Kansas)
MO. (Missouri)
TX. (Texas)

correct

KS
MO
TX

Question Mark (?)

Use a *question mark* after direct questions.

> Do the lab results support your theory?
> Will you work at the main office or at the branch?
>
> **example**

Quotation Marks (" ")

Quotation marks are used in the following ways.

1. When citing direct quotations

> He said, "Your division sold the most compressors last year."
>
> **example**

NOTE: When you are citing a quotation within a quotation, use double quotation marks (" ") and single quotation marks (' ').

| example | Kim's supervisor, quoting the CEO, said the following to explain the new policy regarding raises: "'Only employees who deserve them will receive merit raises.'" |

2. To note the title of an article or a subdivision of a report

| example | The article "Robotics in Industry Today" was an excellent choice as the basis of your speech.
Section III, "Waste Water in District 9," is pertinent to our discussion. |

When using quotation marks, abide by the following punctuation conventions.

- Commas and periods always go *inside* quotation marks.

| example | She said, "Our percentages are fixed." |

- Colons and semicolons always go *outside* quotation marks.

| example | He said, "The supervisor hasn't decided yet"; however, he added that the decision would be made soon. |

- Exclamation points and question marks go inside the quotation marks if the quoted material is either exclamatory or a question. However, if the quoted material is not exclamatory or a question, then these punctuation marks go outside the quotation marks.

| example | John said, "Don't touch that liquid. It's boiling!" |

(Although the sentence isn't exclamatory, the quotation is. Thus, the exclamation point goes inside the quotation marks.)

| example | How could she say, "We haven't purchased the equipment yet"? |

(Although the quotation isn't a question, the sentence is. Thus, the question mark goes outside the quotation marks.)

Semicolon (;)

Semicolons are used in the following instances.

1. Between two independent clauses *not* joined by a coordinating conjunction

| example | The light source was unusual; it emanated from a crack in the plastic surrounding the cathode. |

2. To separate items in a series containing internal commas

| example | When the meeting was called to order, all members were present, including Susan Johnson, the president; Ruth Schneider, the vice president; Harold Holbert, the treasurer; and Linda Hamilton, the secretary. |

MECHANICS

Abbreviations

Never use an abbreviation that your reader will not understand. A key to clear technical communication is to write on a level appropriate to your reader. You may use the following familiar abbreviations without explanation: *Mrs., Dr., Mr., Ms.,* and *Jr.*

A common mistake is to abbreviate inappropriately. For example, some writers abbreviate *and* as follows:

I quit my job & planned to retire young.	**example**

This is too colloquial for professional technical communication. Spell out *and* when you write.

The majority of abbreviation errors occur when writers incorrectly abbreviate states and technical terms.

States. Use the U.S. Postal Service abbreviations in addresses.

Abbreviations for States

AL	Alabama	MT	Montana
AK	Alaska	NC	North Carolina
AZ	Arizona	ND	North Dakota
AR	Arkansas	NE	Nebraska
CA	California	NV	Nevada
CO	Colorado	NH	New Hampshire
CT	Connecticut	NJ	New Jersey
DE	Delaware	NM	New Mexico
FL	Florida	NY	New York
GA	Georgia	OH	Ohio
HI	Hawaii	OK	Oklahoma
IN	Indiana	OR	Oregon
IA	Iowa	PA	Pennsylvania
ID	Idaho	RI	Rhode Island
IL	Illinois	SC	South Carolina
KS	Kansas	SD	South Dakota
KY	Kentucky	TN	Tennessee
LA	Louisiana	TX	Texas
ME	Maine	UT	Utah
MD	Maryland	VT	Vermont
MA	Massachusetts	VA	Virginia
MI	Michigan	WA	Washington
MN	Minnesota	WV	West Virginia
MS	Mississippi	WI	Wisconsin
MO	Missouri	WY	Wyoming

Technical Terms. Units of measurement and scientific terms must be abbreviated accurately to ensure that they will be understood. Writers often use such abbreviations incorrectly. For example, "The unit measured 7.9 cent." is inaccurate. The correct abbreviation for centimeter is *cm*, not *cent*. Use the following abbreviation conventions.

Technical Abbreviations for Units of Measurement and Scientific Terms

absolute	abs	diameter	dia
alternating current	AC	direct current	DC
American wire gauge	AWG	dozen	doz (or dz)
ampere	amp	dram	dr
ampere-hour	amp-hr	electromagnetic force	emf
amplitude modulation	AM	electron volt	eV
angstrom unit	Å	elevation	el (or elev)
atmosphere	atm	equivalent	equiv
atomic weight	at wt	Fahrenheit	F
audio frequency	AF	farad	F
azimuth	az	faraday	f
barometer	bar.	feet, foot	ft
barrel, barrels	bbl	feet per second	ft/sec
billion electron volts	BeV	fluid ounce	fl oz
biochemical oxygen demand	BOD	foot board measure	fbm
board foot	bd ft	foot-candle	ft-c
Brinell hardness number	BHN	foot-pound	ft lb
British thermal unit	Btu	frequency modulation	FM
bushel	bu	gallon	gal
calorie	cal	gallons per day	GPD
candela	cd	gallons per minute	GPM
Celsius	C	grain	gr
center of gravity	cg	grams	g (or gm)
centimeter	cm	gravitational acceleration	g
circumference	cir	hectare	ha
cologarithm	colog	hectoliter	hl
continuous wave	CW	hectometer	hm
cosine	cos	henry	H
cotangent	cot	hertz	Hz
cubic centimeter	cc	high frequency	HF
cubic foot	cu ft (or ft^3)	horsepower	hp
cubic feet per second	cfs	horsepower-hours	hp-hr
cubic inch	cu in. (or in.^{3})	hour	hr
cubic meter	cu m (or m^3)	hundredweight	cwt
cubic yard	cu yd (or yd^3)	inch	in.
current (electric)	I	inch-pounds	in.-lb
cycles per second	CPS	infrared	IR
decibel	dB	inner diameter or inside dimensions	ID
decigram	dg	intermediate frequency	IF
deciliter	dl	international unit	IU
decimeter	dm	joule	J
degree	deg	Kelvin	K
dekagram	dkg	kilocalorie	kcal
dekaliter	dkl	kilocycle	kc
dekameter	dkm	kilocycles per second	kc/sec
dewpoint	DP	kilogram	kg

kilohertz	kHz	parts per billion	ppb
kilojoule	kJ	parts per million	ppm
kiloliter	kl	pascal	pas
kilometer	km	positive	pos or +
kilovolt	kV	pound	lb
kilovolt-amperes	kVa	pounds per square inch	psi
kilowatt-hours	kWH	pounds per square inch absolute	psia
lambert	L	pounds per square inch gauge	psig
latitude	lat	quart	qt
length	l	radio frequency	RF
linear	lin	radian	rad
linear foot	lin ft	radius	r
liter	l	resistance	r
logarithm	log.	revolution	rev
longitude	long.	revolutions per minute	rpm
low frequency	LF	second	s (or sec)
lumen	lm	secant	sec
lumen-hour	lm-hr	specific gravity	sp gr (or SG)
maximum	max	square foot	ft^2
megacycle	mc	square inch	$in.^2$
megahertz	MHz	square meter	m^2
megawatt	MW	square mile	mi^2
meter	m	tablespoon	tbs (or tbsp)
microampere	μamp	tangent	tan
microinch	μin.	teaspoon	tsp
microsecond	μsec	temperature	t
microwatt	μw	tensile strength	ts
miles per gallon	mpg	thousand	m
milliampere	mA	ton	t
millibar	mb	ultra high frequency	UHF
millifarad	mF	vacuum	vac
milligram	mg	very high frequency	VHF
milliliter	ml	volt-ampere	VA
millimeter	mm	volt	V
millivolt	mV	volts per meter	V/m
milliwatt	mW	volume	vol
minute	min	watt-hour	whr
nautical mile	NM	watt	W
negative	neg or −	wavelength	WL
number	no.	weight	wt
octance	oct	yards	y (or yd)
ounce	oz	years	y (or yr)
outside diameter	OD		

Capital Letters

Capitalize the following:

1. Proper nouns

example	people	cities	countries	companies	schools	buildings
	Susan	Houston	Italy	Bendix	Harvard	Oak Park Mall

2. People's titles (only when they precede the name)

example

Governor Sally Renfro
or
Sally Renfro, governor
Technical Supervisor Todd Blackman
or
Wes Schneider, the technical supervisor

3. Titles of books, magazines, plays, movies, television programs, and CDs (excluding the prepositions and all articles after the first article in the title)

example

Grey's Anatomy
The Colbert Report
The Catcher in the Rye
Walk the Line
American Idol

4. Names of organizations

example

Girl Scouts
Kansas City Regional Home Care Association
Kansas City Regional Council for Higher Education
Programs for Technical and Scientific Communication
American Civil Liberties Union
Students for a Democratic Society

5. Days of the week, months, and holidays

example

Monday
December
Thanksgiving

6. Races, religions, and nationalities

example

American Indian
Jewish
Polish

7. Events or eras in history

example

the Gulf War
the Vietnam War
World War II

8. North, South, East, and West (when used to indicate geographic locations)

| They moved from the North.
 People are moving to the Southwest. | **example** |

NOTE: Don't capitalize these words when giving directions.

| We were told to drive south three blocks and then to turn west. | **example** |

9. The first word of a sentence
Don't capitalize any of the following:

| Seasons—spring, fall, summer, winter
 Names of classes—sophomore, senior
 General groups—middle management, infielders, surgeons | **example** |

Numbers

Write out numbers one through nine. Use numerals for numbers 10 and above.

10	12
104	2,093
536	5,550,286

Although the preceding rules cover most situations, there are exceptions.

1. Use numerals for all percentages.

| 2 percent   18 percent   25 percent | **example** |

2. Use numerals for addresses.

| 12 Elm   935 W. Harding | **example** |

3. Use numerals for miles per hour.

| 5 mph   225 mph | **example** |

4. Use numerals for time.

| 3:15 A.M. | **example** |

5. Use numerals for dates.

| May 31, 2008 | **example** |

6. Use numerals for monetary values.

| $45 $.95 $2 million | **example** |

7. Use numerals for units of measurement.

14'   6 3/4"   16 mm 10 V

8. Do not use numerals to begin sentences.

incorrect

568 people were fired last August.

correct

Five hundred sixty-eight people were fired last August.

9. Do not mix numerals and words when writing numbers. When two or more numbers appear in a sentence and one of them is 10 or more, figures are used.

example

We attended 4 meetings over a 16-day period.

10. Use numerals and words in a compound number adjective to avoid confusion.

example

The worker needed six 2-inch nails.

SPELLING

The following is a list of commonly misspelled or misused words. You can avoid many common spelling errors if you familiarize yourself with these words. Remember to run spell check; also remember that spell check will not understand context. The incorrect word contextually could be spelled correctly.

accept, except	incite, insight
addition, edition	its, it's
access, excess	loose, lose
advise, advice	miner, minor
affect, effect	passed, past
all ready, already	patients, patience
assistants, assistance	personal, personnel
bare, bear	principal, principle
brake, break	quiet, quite
coarse, course	rite, right, write
cite, site, sight	stationery, stationary
council, counsel	their, there, they're
desert, dessert	to, too, two
disburse, disperse	whose, who's
fiscal, physical	your, you're
forth, fourth	

PROOFREADER'S MARKS

The following figure illustrates proofreader's marks and how to use them.

PROOFREADER'S MARKS			
Symbol	Meaning	Mark on Copy	Revision
e	delete	hire better peoples*e*	hire better people
(tr) or ∩	transpose	computer systme	computer system
⊃⊂	close space	we ne ed 40	we need 40
#	insert space	three#mistakes	three mistakes
¶	begin paragraph	¶The first year	The first year
RUN IN	no paragraph	financial. We earned twelve dollars.	financial. We earned twelve dollars.
⊏	move left	[Next year	Next year
⊐	move right	⌐Next year	Next year
∧	insert period	happy employee ⊙	happy employee.
∧	insert comma	For two years⌃	For two years,
∧	insert semicolon	We need you⌃ come to work.	We need you; come to work.
∧	insert colon	dogs⌃brown, black, and white.	dogs: brown, black, and white.
∧	insert hyphen	first⌃rate	first-rate
⌄	insert apostrophe	Marys car	Mary's car
⌄" / ⌄"	insert quotation marks	"Like a Rolling Stone	"Like a Rolling Stone"
(SP)	spell out	④chips	four chips
(cap) or ≡	capitalize	the meeting	The meeting
(lc)	lower case	the Boss	the boss
∧	insert	the ⌃left margin	the left margin

APPLY YOUR KNOWLEDGE

Spelling

In the following sentences, circle the correctly spelled words within the parentheses.

1. Each of the employees attended the meeting (accept except) the line supervisor, who was out of town for job-related travel.
2. The (advise advice) he gave will help us all do a better job.
3. Management must (affect effect) a change in employees' attitudes toward absenteeism.
4. Let me (site cite sight) this most recent case as an example.
5. (Its It's) too early to tell if our personnel changes will help create a better office environment.
6. If we (lose loose) another good employee to our competitor, our production capabilities will suffer.
7. I'm not (quite quiet) sure what she meant by that comment.
8. (Their There They're) budget has gotten too large to ensure a successful profit margin.
9. We had wanted to attend the conference (to too two), but our tight schedule prevented us from doing so.
10. (You're Your) best chance for landing this contract is to manufacture a better product.

In the letter on page below, correct the misspelled words.

March 5, 2008

Joanna Freeman
Personel Director
United Teletype
1111 E. Street
Kansas City, MO 68114

Dear Ms. Freeman:

Your advertizmemt in the Febuary 18, 2008, Kansas City Star is just the opening I have been looking for. I would like to submit my quallifications.

As you will note in the inclosed resum, I recieved an Enginneering degree from the Missouri Institute of Technology in 2005 and have worked in the electronic enginneering department of General Accounts for three years. I have worked a great deal in design electronics for microprocesors, controll systems, ect.

Because your company has invented many extrordinary design projects, working at your company would give me more chances to use my knowlege aquired in school and through my expirences. If you are interested in my quallifications, I would be happy to discuss them futher with you. I look forward to hearing from you.

Sincerly,

Bob Cottrell

Bob Cottrell

Fragments and Comma Splices

In the following sentences, correct the fragments and comma splices by inserting the appropriate punctuation or adding any necessary words.

1. She kept her appointment with the salesperson, however, the rest of her staff came late.
2. When the CEO presented his fiscal year projections, he tried to motivate his employees, many were not excited about the proposed cuts.
3. Even though the company's sales were up 25 percent.
4. The supervisor wanted the staff members to make suggestions for improving their work environment, the employees, however, felt that any grievances should be taken directly to their union representatives.
5. Which he decided was an excellent idea.
6. Because their machinery was prone to malfunctions and often caused hazards to the workers.
7. They needed the equipment to complete their job responsibilities, further delays would cause production slowdowns.
8. Their client who was a major distributor of high-tech machinery.
9. Robotics should help us maintain schedules, we'll need to avoid equipment malfunctions, though.
10. The company, careful not to make false promises, advertising their product in media releases.

In the letter on page below, correct the fragments and comma splices.

May 12, 2008

Maurene Pierce
Dean of Residence Life
Mann College
Mannsville, NY 10012

Subject: Report on Dormitory Damage Systems

Here is the report you authorized on April 5 for an analysis of the current dormitory damage system used in this college.

The purpose of the report was to determine the effectiveness of the system. And to offer any concrete recommendations for improvement. To do this, I analyzed in detail the damage cost figures for the past three years, I also did an extensive study of dormitory conditions. Although I had limited manpower. I gathered information on all seven dormitories, focusing specifically on the men's athletic dorm, located at 1201 Chester. In this dorm, bathroom facilities, carpeting, and air conditioning are most susceptible to damage. Along with closet doors.

Nonetheless, my immediate findings indicate that the system is functioning well, however, improvements in the physical characteristics of the dormitories, such as new carpeting and paint, would make the system even more efficient.

I have enjoyed conducting this study, I hope my findings help you make your final decision. Please contact me. If I can be of further assistance.

Rob Harken

Rob Harken

Punctuation

In the following sentences, circle the correct punctuation marks. If no punctuation is needed, draw a slash mark through both options.

1. John took an hour for lunch (, ;) but Joan stayed at her desk to eat so she could complete the project.

2. Sally wrote the specifications (, ;) Randy was responsible for adding any needed graphics.

3. Manufacturing maintained a 93.5 percent production rating in July (, ;) therefore, the department earned the Golden Circle Award at the quarterly meeting.

4. In their year-end requests to management (, ;) supervisors asked for new office equipment (, ;) and a 10 percent budget increase for staffing.

5. The following employees attended the training session on stress management (, :) Steve Janasz, purchasing agent (, ;) Jeremy Kreisler, personnel director (, ;) and Prakash Patel, staff supervisor.

6. Promotions were given to all sales personnel (, ;) secretaries, however, received only cost-of-living raises.

7. The technicians voted for better work benefits (, ;) as an incentive to improve morale.

8. Although the salespeople were happy with their salary increases (, ;) the technicians felt slighted.

9. First (, ;) let's remember that meeting schedules should be a priority (, ;) and not an afterthought.

10. The employee (, ;) who achieves the highest rating this month (, ;) will earn 10 bonus points (, ;) therefore (, ;) competition should be intense.

In the letter below, no punctuation has been added. Instead, there are blanks where punctuation might be inserted. First, decide whether any punctuation is needed (not every blank requires punctuation). Then, insert the correct punctuation—a comma, colon, period, semicolon, or question mark.

January 8_ 2008

Mr_ Ron Schaefer
1324 Homes
Carbondale_ IL_ 34198

Dear Mr_ Schaefer_

Yesterday_ my partners_ and I read about your invention in the Herald Tribune_ and we want to congratulate you on this new idea_ and ask you to work with us on a similar project_

We cannot wait to begin our project_ however_ before we can do so_ I would like you to answer the following questions_

- Has your invention been tested in salt water_
- What is the cost of replacement parts_
- What is your fee for consulting_

Once_ I receive your answers to these questions_ my partners and I will contact you regarding a schedule for operations_ We appreciate your design concept_ and know it will help our business tremendously_ We look forward to hearing from you_

Sincerely_

Elias Agamenyon

Elias Agamenyon

Agreement (Subject/Verb and Pronoun/Antecedent)

In the following sentences, circle the correct choice to achieve agreement between subject and verb or pronoun and antecedent.

1. The employees, though encouraged by the possibility of increased overtime, (was were) still dissatisfied with their current salaries.
2. The supervisor wants to manufacture better products, but (they he) doesn't know how to motivate the technicians to improve their work habits.
3. The staff (was were) happy when the new manager canceled the proposed meeting.
4. Anyone who wants (his or her their) vote recorded must attend the annual board meeting.
5. According to the printed work schedule, Susan and Tom (work works) today on the manufacturing line.
6. According to the printed work schedule, either Susan or Tom (work works) today on the manufacturing line.
7. Although Tamara is responsible for distributing all monthly activity reports, (she they) failed to mail them.
8. Every one of the engineers asked if (he or she they) could be assigned to the project.
9. Either the supervisor or the technicians (is are) at fault.
10. The CEO, known for her generosity to employees and their families, (has have) been nominated for the humanitarian award.

In the following memo, find and correct the errors in agreement.

MEMO

DATE: October 30, 2008
TO: Tammy West
FROM: Susan Lisk
SUBJECT: REPORT ON AIR HANDLING UNIT

There has been several incidents involving the unit which has resulted in water damage to the computer systems located below the air handler.

The occurrences yesterday was caused when a valve was closed creating condensation to be forced through a humidifier element into the supply air duct. Water then leaked from the duct into the room below causing substantial damage to four disc-drive units.

To prevent recurrence of this type of damage, the following actions has been initiated by maintenance supervision:

- Each supervisor must ensure that their subordinates remove condensation valves to avoid unauthorized operation.
- Everyone must be made aware that they are responsible for closing condensation valves.
- The supply air duct, modified to carry away harmful sediments, are to be drained monthly.

Maintenance supervision recommend that air handlers not be installed above critical equipment. This will avoid the possibility of coil failure and water damage.

Capitalization

In the following memo, nothing has been capitalized. Capitalize those words requiring capitalization.

---

date: december 5, 2008
to: jordan cottrell
from: richard davis
subject: self-contained breathing apparatus (scba) and negative pressure respirator evaluation and fit-testing report

the evaluation and fit-testing have been accomplished. the attached list identifies the following:

- supervisors and electronic technicians who have used the scba successfully.
- the negative pressure respirators used for testing in an isoamyl acetate atmosphere.

fit-testing of waste management personnel will be accomplished annually, according to president chuck carlson. new supervisors and technicians will be fit-tested when hired.

all apex corporation personnel located in the new york district (12304 parkview lane) must submit a request form when requesting a respirator or scba for use. any waste management personnel in the north and south facilities not identified on the attached list will be fit-tested when use of scba is required. if you have any questions, contact chuck carlson or me (richard davis, district manager) at ext. 4036.

GRAMMAR QUIZ

The following sentences contain errors in spelling, punctuation, verb and pronoun agreement, sentence structure (fragments and run-ons), and modification. Circle the letter corresponding to the section of the sentence containing the error.

example	When you recieve the salary increase, your family will celebrate the occasion.
	Ⓐ B C

1. Each department manager should tell <u>his</u> subordinates to <u>advise</u> of any negative
                                          A   B
<u>occurrences</u> regarding <u>in-house</u> training.
   C   D

2. The <u>lawyers'</u> new offices were similar to <u>their</u> former ones _____ the offices
      A   B   C
were on a <u>quiet</u> street.
         D

3. New York City is <u>divided</u> into five <u>boroughs</u> ; Manhattan, the Bronx, Queens,
                      A   B   C

   Brooklyn, <u>and</u> Staten Island.
            D

4. Fashion consultants explain that <u>clothes</u> create a strong impression _____
                                A   B

   so they advise executives _____ to <u>choose</u> wardrobes carefully.
                           C   D

5. Everyone should make sure that <u>they are</u> well represented in union meetings ;
                              A   B

   otherwise , management could become <u>too</u> powerful.
          C   D

6. The employment agency , <u>too</u> busy to return the telephone calls from prospective
                        A B

   clients , <u>are</u> harming business opportunities.
         C D

7. The supervisory staff <u>are</u> making decisions based on scheduling , but all <u>employees</u>
                         A   B   C

   want to <u>ensure</u> quality control.
             D

8. Because the price of cars has risen dramatically _____ most people keep
                                               A

   <u>their</u> cars longer _____ to save <u>capital</u> expenditures.
      B   C   D

9. The <u>department's</u> manager heard that a merger was possible _____ but he
       A   B

   decided <u>to</u> keep the news <u>quiet</u>.
           C   D

10. The manager of the department believed that her <u>employee's</u> were excellent , but
                                      A   B

   she decided that no raises could be given _____ because the price of stocks
                                     C

   <u>was</u> falling.
    D

11. The detailed report from the audit _____ of the department gave good
                              A

   <u>advise</u> about how to restructure , so we are <u>all ready</u> to do so.
      B   C   D

12. When my boss and <u>I</u> looked at the books , we found these problems : lost invoices,
                 A   B   C

   unpaid bills , and late payments.
            D

13. The most dedicated staff members <u>are</u> <u>accepted</u> for the on-site training sessions
                              A   B

   because : they are responsive to criticism, represent the <u>company's</u> future, and
           C   D

   strive for improvements.

14. Despite her assurances to the contrary , there is still three unanswered questions :
                                            A   B   C
who will make the payments, when will these payments occur, and why is there a
                                                                               D
delay?

15. The reputation of many companies often depend on one employee who represents
                              A   B   C   D
that company.

16. Either my monthly activity report or my year-end report are due today , but my
                                            A   B   C
computer is broken ; therefore, I need to use yours.
                   D

17. Today's American manpower , according to many foreign governments, suffers
                     A   B   C   D
from lack of discipline.

18. Rates are increasing next year because : fuel, maintenance, and insurance are all
                                           A   B   C
higher than last year.
       D

19. Many colleges have long-standing football rivalries , one of the most famous ones
                        A   B
is between KU and KSU (two universities in Kansas).
C   D

20. Everyone who wants to enroll in the business school should do so before June if
    A   B
they can to ensure getting the best classes.
C   D

21. Because John wanted high visibility for his two businesses , he paid top dollar
                                                    A   B
_____ and spent long hours looking for appropriate cites.
     C   D

22. Many people apply for jobs at Apex , however , only a few are accepted.
                                      A   B   C   D

23. Because most cars break down occasionally , all drivers should know how to
                                 A   B
change a flat tire , and how to signal for assistance.
                  C   D

24. Arriving on time, working diligently, and closing the office securely— everyone
                             A   B
needs to know that they are responsible for these job duties.
C   D

25. Mark McGwire not only hit the ball further than other players, but also he hit
                                          A   B
more dingers than other players who hit fewer homers.
                                C   D

REFERENCES

PREFACE

Burton, Susan. "'Technical Communicator,' Your Time Has Come." *Intercom* (April 2007): 3.

CHAPTER 1

"Business Identity." *Crane's*. February 13, 2004. www.crane.com/business/businessidentity/.

Costas, Bob. "Importance of Writing." *The National Commission on Writing*. 2007. www.nationalcommission.org.

Fisher, Lori, and Lindsay Bennion. "Organizational Implications of the Future Development of Technical Communication: Fostering Communities of Practice in the Workplace." *Technical Communication* 52 (August 2005): 277–288.

Hughes, Michael A. "Managers: Move from Silos to Channels." *Intercom* (March 2003): 9–11.

"Individual's and Teams' Roles and Responsibilities." GOAL/QPC. February 24, 2003. www.goalqpc.com/index.htm.

Mader, Stewart. "Using Wiki in Education." *The Science of Spectroscopy*. December 14, 2005. www.scienceofspectroscopy.info/edit/index.php?title=Using_wiki_in_education.

McNeill, Angie E. "Self-Directed Work Teams." *Intercom* (September/October 2000): 15–17.

Miller, Carolyn R., et al. "Communication in the Workplace." *Center for Communication in Science, Technology, and Management*. October 1996. www.chass.ncsu.edu/ccstm/pubs/no2/index.html.

Nesbitt, Pamela, and Elizabeth Bagley-Woodward. "Practical Tips for Working with Global Teams." *Intercom* (June 2006): 25–30.

"Planning Job Choices." NaceWeb. January 15, 2004. www.naceweb.org/press/display.asp?year=2004&prid=184.

"Top Ten Qualities/Skills Employers Want." *Job Outlook 2006 Student Version*. National Association of Colleges and Employers. 2005: 5. http://career.clemson.edu/pdf_docs/NACE_JO6.pdf.

"TWiki—Enterprise Wiki & Collaboration Platform." *TWiki*. March 4, 2007. http://twiki.org.

"We Are Smarter Than Me." December 18, 2006. www.wearesmarter.org.

"What Is Wikipedia." *Wikipedia*. January 9, 2007. http://en.wikipedia.org/wiki/Wikipedia:Introduction.

"Wiki." *Wikipedia*. January 9, 2007. www.wikipedia.org.

"Writing: A Ticket to Work . . . Or a Ticket Out, A Survey of Business Leaders." *National Commission on Writing*. 2007. www.writingcommission.org.

"Writing Skills Necessary for Employment, Says Big Businesses." *National Commission on Writing*. 2005. www.writingcommission.org/pr/writing_for_employ.html.

CHAPTER 2

Albers, Michael J. "Single Sourcing and the Technical Communication Career Path." *Technical Communication* 50 (August 2003): 336–343.

Campbell, Kim S., et al. "Leader-Member Relations as a Function of Rapport Management." *Journal of Business Communication* 40 (2003): 170–194.

Cargile-Cook, Kelli. "Usability Testing in the Technical Communication Classroom." *Intercom* (September/October 1999): 36–37.

Carter, Locke. "The Implications of Single Sourcing for Writers and Writing." *Technical Communication* 50 (August 2003): 317–320.

Dorazio, Pat. Interview. June 2000.

Rizzo, Tony. "Dropped Decimal Point Becomes a Costly Point of Contention." *The Kansas City Star*. February 19, 2005: A1, A6.

"Writing: A Powerful Message from State Government." *Report of the National Commission on Writing*. College Board. 2005.

CHAPTER 3

Adams, Rae, et al. "Ethics and the Internet." *Proceedings: 42nd Annual Technical Communication Conference*. April 23–26, 1995: 328.

Anderson-Hancock, Shirley A. "Society Unveils New Ethical Guidelines." *Intercom* 42 (July/August 1995): 6–7.

Barker, Thomas, et al. "Coming into the Workplace: What Every Technical Communicator Should Know—Besides Writing." *Proceedings: 42nd Annual Technical Communication Conference*. April 23–26, 1995: 38–39.

Bowman, George, and Arthur E. Walzer. "Ethics and Technical Communication." *Proceedings: 34th Annual Technical Communication Conference*. May 10–13, 1987: MPD-93.

Bremer, Otto A., et al. "Ethics and Values in Management Thought." *Business Environment and Business Ethics: The Social, Moral, and Political Dimensions of Management*. Edited by Karen Paul. Cambridge, MA: Ballinger, 1987: 61–86.

"Circular 66, Copyright Registration for Online Works." *U.S. Copyright Office*. July 2006. www.copyright. gov/circs/circ66.html.

"Code for Communicators." Society for Technical Communication. March 10, 2004. www.stermc.org/resources/resource_code.htm.

"College Degree Nearly Doubles Annual Earnings, Census Bureau Reports." *U.S. Census Bureau News*. March 28, 2005. March 5, 2007. www.census.gov/Press-release/www/releases/archives/education.

Dillon, Sam. "Literacy Falls for Graduates from College, Testing Finds." *The New York Times*. December 16, 2005. www.nytimes.com.

Gerson, Steven M., and Sharon J. Gerson. "A Survey of Technical Writing Practitioners and Professors: Are We on the Same Page?" *Proceedings: 42nd Annual Technical Communication Conference*. April 23–26, 1995: 44–47.

Girill, T. R. "Technical Communications and Ethics." *Technical Communication* 34 (August 1987): 178–179.

Guy, Mary E. *Ethical Decision Making in Everyday Work Situations*. New York: Quorum Books, 1990.

Harkness, Holly E. "Sarbanes-Oxley & New Opportunities." *Intercom* (January 2004): 16–18.

Hartman, Diane B., and Karen S. Nantz. "Send the Right Messages About E-Mail." *Training & Development* (May 1995): 60–65.

Johnson, Dana R. "Copyright Issues on the Internet." *Intercom* (June 1999): 17.

LeVie, Donald S. "Internet Technology and Intellectual Property." *Intercom* (January 2000): 20–23.

Mollison, Andrew. "U.S. Loses Its Lead in High School and College Grads." *Cox Newspapers*. November 6, 2001: 1–3.

Moore, Linda E. "Serving the Electronic Reader." *Intercom* (April 2003): 17.

Perlin, Neil. "Technical Communication: The Next Wave." *Intercom* (January 2001): 4–8.

"STC Ethical Principles for Technical Communicators." *Society for Technical Communication*. September 2007. www.stc.org/pdf_files/EthicalPrinciples.pdf.

"Summary of Sarbanes-Oxley Act of 2002." American Institute of Certified Public Accountants. January 8, 2004. www.aicpa.org/index.htm.

Turner, John R. "Ethics Online: Looking Toward the Future." *Proceedings: 42nd Annual Technical Communication Conference*. April 23–26, 1995: 59–62.

Wilson, Catherine Mason. "Product Liability and User Manuals." *Proceedings: 34th International Technical Communication Conference*. May 10, 1987: WE-68–71.

"Writing: A Ticket to Work . . . Or a Ticket Out, A Survey of Business Leaders." *National Commission on Writing*. 2007. www.writingcommission.org.

CHAPTER 4

Cardarella, Toni. "Business." *The Kansas City Star*. September 30, 2003. www.kansascity.com/mld/kansascity/business/6877765.htm?lc.

"Civilian Labor Force by Age, Sex, Race, and Hispanic Origin, 1994, 2004, and Projected 2014." *The United States Department of Labor*. December 7, 2005. March 6, 2007. www.bls.gov/news.release/ecopro.t07.htm.

Courtis, John K., and Salleh Hassan. "Reading Ease of Bilingual Annual Reports." *Journal of Business Communication* 39 (October 2002): 394–413.

Flint, Patricia, et al. "Helping Technical Communicators Help Translators." *Technical Communication* 46 (May 1999): 238–248.

Gardner, Amanda. "Language a Widening Barrier to Health Care." *Health on the Net Foundation*. January 9, 2007. March 6, 2007. www.hon.ch/News/HSN/533896.html.

Grimes, Diane Susan, and Orlando C. Richard. "Could Communication Form Impact Organizations' Experience with Diversity?" *Journal of Business Communication* 40 (January 2003): 7–27.

Horton, William. "The Almost Universal Language: Graphics for International Documents." *Technical Communication* 40 (November 1993): 682–693.

Hussey, Tim, and Mark Homnack. "Foreign Language Software Localization." *Proceedings: 37th International Technical Communication Conference*. May 20–23, 1990: RT-44–47.

Jordan, Katrina. "Diversity Training: What Works, What Doesn't, and Why?" *Civil Right Journal* (Fall 1999): 1.

King, Janice M. "The Challenge of Communicating in the Global Marketplace." *Intercom* 42.5 (May/June 1995): 1, 28.

McInnes, R. "Workforce Diversity: Changing the Way You Do Business." *Diversity World*. February 12, 2003. www.diversityworld.com/workforce_diversity.htm.

Melgoza, Cesar M. "Geoscape Findings Confirm Hispanic Market Growth." *Hispanic MPR.com.* March 6, 2007. www.hispanicmpr.com/2006/02/14/.2006.

"Multilingual Features in Windows XP Professional." August 24, 2001. www.microsoft.com/windowsxp/pro/techinfo/planning/multilingual/default.asp.

Nethery, Kent. "Let's Talk Business." February 16, 2003. www.cuspomona.edu/~cljones/powerpoints/chap02/sld001.htm.

"Occupational Employment in Private Industry by Race/Ethnic Group/Sex and by Industry, United States 2005." *The U.S. Equal Employment Opportunity Commission.* May 9, 2005. March 6, 2007. www.eeoc.gov/stats/jobpat/2003/national.html.

Rains, Nancy E. "Prepare Your Documents for Better Translation." *Intercom* 41.5 (December 1994): 12.

Sanchez, Mary. "KC Hospitals Seek to Overcome Language Barriers." *The Kansas City Star.* January 7, 2003: A1, A4.

Scott, Julie S. "When English Isn't English." *Intercom* (May 2000): 20–21.

Skettino, Michele. "America in the 21st Century: A Marketer's Guide to the Nation." *Interep Research Division.* August 24, 2004: 13–15. http://site.interep.com/apps/Research/resweb.nsf/bf25ab0f47ba5dd785256499006b15a4/dd35caf6283fb50585256a4c0058a200/$FILE/A21C001.pdf.

St. Amant, Kirk R. "Communication in International Virtual Offices." *Intercom* (April 2003): 27–28.

St. Amant, Kirk R. "Designing Web Sites for International Audiences." *Intercom* (May 2003): 15–18.

St. Amant, Kirk R. "International Integers and Intercultural Expectations." *Intercom* (May 1999): 25–28.

Swenson, Lynne V. "How to Make (American) English Documents Easy to Translate." *Proceedings: 34th International Technical Communication Conference.* May 10, 1987: WE-193–195.

"Swiss Fight Encroachment of English." *The Kansas City Star.* December 7, 2002: A16.

Walmer, Daphne. "One Company's Efforts to Improve Translation and Localization." *Technical Communication* 46 (May 1999): 230–237.

Weiss, Edmund H. "Twenty-five Tactics to 'Internationalize' Your English." *Intercom* (May 1998): 11–15.

"What Is the 'Business Case' For Diversity?" *Society for Human Resource Management.* February 12, 2003. www.shrm.org/diversity/businesscase.asp.

CHAPTER 5

The American Psychological Association. March 6, 2007. www.apa.org.

The Modern Language Association: MLA. March 6, 2007. www.mla.org.

CHAPTER 6

Broughton, Philip Delves. "Boss's Angry Email Sends Shares Plunging." *London Daily Telegraph.* April 6, 2001.

Cobbs, Chris. "r u using 2 much IM style in ur writing at school?" *The Kansas City Star.* November 6, 2002: F5.

Gerson, Steven M., et al. "Core Competencies." Survey. Prentice Hall Publishing Company. 2004.

Hoffman, Jeff. "Instant Messaging in the Workplace." *Intercom* (February 2004): 16–17.

Miller, Carolyn, et al. "Communication in the Workplace." *Center for Communication in Science, Technology, and Management.* October 1996.

Munter, Mary, et al. "Business E-mail: Guidelines for Users." *Business Communication Quarterly* 66 (March 2003): 29.

Shinder, Deb. "Instant Messaging: Does It Have a Place in Business Networks?" *WindowSecurity.com.* June 8, 2005. www.windowsecurity.com/articles/Instant-Messaging-Business-Networks.html.

"The Transition to General Management." Harvard Business School. 1998. November 12, 2002. www.hbs.edu/gm/index.html.

"Writing: A Ticket to Work . . . Or a Ticket Out, A Survey of Business Leaders." *National Commission on Writing.* 2007. www.writingcommission.org.

CHAPTER 7

Bloch, Janel M. "Online Job Searching: Clicking Your Way to Employment." *Intercom* (September/October 2003): 11–14.

Dikel, Margaret F. "The Online Job Application: Preparing Your Resume for the Internet." September 1999. www.dbm.com/jobguide/eresume.html.

Dixson, Kirsten. "Crafting an E-mail Resume." *BusinessWeek Online.* November 13, 2001. www.businessweek.com.

Drakeley, Caroline A. "Viral Networking: Tactics in Today's Job Market." *Intercom* (September/October 2003): 5–7.

Hartman, Peter J. "You Got the Interview. Now Get the Job!" *Intercom* (September/October 2003): 23–25.

Isaacs, Kim. "Resume Critique Checklist." *MonsterTRAK Career Advice Archives.* September 5, 2006. www.monster.com.

Kallick, Rob. "Research Pays Off During Interview." *The Kansas City Star.* March 23, 2003: D1.

Kendall, Pat. "Electronic Resumes." December 9, 2003. www.reslady.com.

LeVie, Donald S., Jr. "Resumes: You Can't Escape." *Intercom* (April 2000): 8–11.

Mulvaney, Mary Kay. "The Information Interview: Bridging College and Beyond." *Business Communication Quarterly* 66 (September 2003): 66–72.

Ralston, Steven M., et al. "Helping Interviewees Tell Their Stories." *Business Communication Quarterly* 66 (September 2003): 8–22.

Robart, Kay. "Submitting Resumes via E-Mail." *Intercom* (July/August 1998): 12–14.

Stafford, Diane. "Show Up Armed with Answers." *The Kansas City Star*. August 3, 2003: L1.

Stern, Linda. "New Rules of the Hunt." *Newsweek* (February 17, 2003): 67.

"Top Ten Qualities/Skills Employers Want." *Job Outlook 2005 Student Version*. National Association of Colleges and Employers. 2005: 1–13. www.jobweb.com/joboutlook/2005outlook/JO5Student.pdf.

Vogt, Peter. "Your Resume's Look Is as Important as Its Content." *MonsterTRAK Career Advice Archives*. March 6, 2007. www.monster.com.

"Writing: A Powerful Message from State Government." *Report of the National Commission on Writing*. College Board. 2005.

CHAPTER 8

Schriver, Karen A. "Quality in Document Design." *Technical Communication* (Second Quarter 1993): 250–251.

CHAPTER 9

Horton, William. "The Almost Universal Language: Graphics for International Documents." *Technical Communication* (November 1993): 682–693.

Reynolds, Michael, and Liz Marchetta. "Color for Technical Documents." *Intercom* (April 1998): 5–7.

CHAPTER 10

"CDC and CPSC Warn of Winter Home Heating Hazards." *Centers for Disease Control and Prevention*. March 6, 2007. www.cdc.gov/nceh/pressroom/2006/COwarning.htm.

Dorazio, Pat. Interview. June 2000.

CHAPTER 13

Berst, Jesse. "Seven Deadly Web Site Sins." *ZDNet*. January 30, 1998. December 15, 2005. www.zdnet.com/anchordesk/story/story_1716.html.

Blumberg, Max. "Blogging Statistics." *Max Blumberg Positioning Game*. January 3, 2005. http://maxblumberg.typepad.com/dailymusings/2005/01/usa_blog_usage.html.

Bruner, Rick. "Blogging Statistics." *Idiotprogrammer*. April 23, 2004. June 6, 2005. www.imaginaryplanet.net/weblogs/idiotprogrammer/?p=83397841.

Cross, Jay. "Blogging for Business." *Learning Circuits*. August 2003. May 23, 2005. www.learningcircuits.org/2003/aug2003/cross.htm.

Dorazio, Pat. Interview. June 2000.

Eddings, Earl, Kim Buckley, Sharon Coleman Bock, and Nathaniel Williams. Software Documentation Specialists at PDA. Interview. January 12, 1998.

Foremski, Tom. "IBM Is Preparing to Launch a Massive Corporate Wide Blogging Initiative." *Silicon Valley Watcher*. May 13, 2005. May 23, 2005. www.siliconvalleywatcher.com/mt/archives/2005/05/can_blogging_bo.php.

Gard, Lauren. "The Business of Blogging." *BusinessWeek Online*. December 13, 2004. June 6, 2005. www.businesswekk.com/magazine/content/04_50/b3912115_mz016.htm.

"Gates Backs Blogs for Businesses." *BBC News*. May 21, 2004. May 23, 2005. http://news.bbc.co.uk/2/hi/technology/3734981.stm.

Goldenbaum, Don, and George Calvert. Applied Communications Group. Interview. January 10, 1998.

Hemmi, Jane A. "Differentiating Online Help from Printed Documentation." *Intercom* (July/August 2002): 10–12.

"Human-Computer Interaction." Rensselaer Polytechnic Institute: Academic Programs. April 21, 1998. www.llc.rpi.edu/acad/hcicert.htm.

Kharif, Olga. "Blogging for Business." *BusinessWeek Online*. August 9, 2004. May 23, 2005. www.businessweek.com/technology/content/aug2004/tc2004089_3601_tco24.htm.

Li, Charlene. "Blogging: Bubble or Big Deal?" *Forrester*. November 5, 2004. May 23, 2005. www.forrester.com/Research/Print/Document/0,7211,35000,00.html.

McAlpine, Rachel. "Passing the Ten-Second Test." *Wise-Women.com*. April 22, 2002. www.wise-women.org/features/tenseconds/index.shtml.

McGowan, Kevin S. "The Leap Online." *Intercom* (September/October 2000): 22–25.

Moore, Linda E. "Serving the Electronic Reader." *Intercom* (April 2003): 16–17.

Pratt, Jean A. "Where Is the Instruction in Online Help Systems?" *Technical Communication* (February 1998): 33–37.

Ray, Ramon. "Blogging for Business." *Inc.com*. September 2004. May 23, 2005. www.inc.com/partners/sbc/articles/20040929-blogging.html.

Spyridakis, Jan H. "Guidelines for Authoring Comprehensible Web Pages and Evaluating Their Success." *Technical Communication* 47 (August 2000): 359–382.

Timpone, Donna. "Help! Six Fixes to Improve the Usability of Your Online Help." *Society for Technical Communication: 44th Annual Conference, 1997 Proceedings*. May 1997: 306.

Wagner, Carol A. "Using HCI Skills to Create Online Message Help." *Society for Technical Communication: 44th Annual Conference, 1997 Proceedings*. May 1997: 35.

Williams, Thomas R. "Guidelines for Designing and Evaluating the Display of Information on the Web." *Technical Communication* 47 (August 2000): 383–396.

Wuorio, Jeff. "Blogging for Business: 7 Tips for Getting Started." *Microsoft*. May 23, 2005. www.microsoft.com/smallbusiness/issues/marketing/online_marketing/blogging_for_business.

Wuorio, Jeff. "5 Ways Blogging Can Help Your Business." *Microsoft*. May 23, 2005. www.microsoft.com/smallbusiness/issues/marketing/online_marketing/5_ways_blog.htm.

Yeo, Sarah C. "Designing Web Pages That Bring Them Back." *Intercom* (March 1996): 12–14.

CHAPTER 15

"Robert's Rules of Order Revised." *Constitution Society*. September 15, 2004. www.constitution.org/rror/rror—00.htm.

CHAPTER 18

Buffet, Warren. "Preface." In *A Plain English Handbook: How to Create Clear SEC Disclosure Documents*. Office of Investor Education and Assistance. Washington, D.C.: U.S. Securities and Exchange Commission. August 1998: 1–2.

"Communications: Oral Presentations." 2004. www.surrey.ac.uk/Skills/pack.pres.html.

Coyner, Dale. "Webconferencing 101." March 7, 2007. www.isquare.com/webconf.cfm.

Mahin, Linda. "PowerPoint Pedagogy." *Business Communication Quarterly* 67 (June 2004): 219–222.

Monster Career Advice Newsletter. August 29, 2006.

Murray, Krysta. "Web Conferencing Tips From a Pro." July 1, 2004. March 7, 2007. www.successmtgs.com/successmtgs/magazine/search_display.jsp?vnu_content_id=1000542704.

O'Brien, Tim. "In Front of a Group." *The Kansas City Star*. July 15, 2003: C5.

INDEX